A LABORATORY GUIDE TO

THIRTEENTH EDITION
Human Physiology
Concepts and Clinical Applications

Stuart Ira Fox

PIERCE COLLEGE

A LABORATORY GUIDE TO HUMAN PHYSIOLOGY: CONCEPTS AND CLINICAL APPLICATIONS, THIRTEENTH EDITION

Published by McGraw-Hill, a business unit of The McGraw-Hill Companies, Inc., 1221 Avenue of the Americas, New York, NY 10020. Copyright © 2013 by The McGraw-Hill Companies, Inc. All rights reserved. Printed in the United States of America. Previous editions © 2011, 2009, and 2008. No part of this publication may be reproduced or distributed in any form or by any means, or stored in a database or retrieval system, without the prior written consent of The McGraw-Hill Companies, Inc., including, but not limited to, in any network or other electronic storage or transmission, or broadcast for distance learning.

Some ancillaries, including electronic and print components, may not be available to customers outside the United States.

This book is printed on acid-free paper.

6 7 8 9 LMN 21 20 19

ISBN 978-0-07-742732-0
MHID 0-07-742732-7

Senior Vice President, Products & Markets: *Kurt L. Strand*
Vice President, General Manager, Products & Markets: *Marty Lange*
Vice President, Content Production & Technology Services: *Kimberly Meriwether David*
Managing Director: *Micheal S. Hackett*
Director: *James F. Connely*
Brand Manager: *Marija Magner*
Director of Development: *Rose Koos*
Senior Development Editor: *Fran Simon*
Marketing Manager: *Chris Loewenberg*
Project Manager: *Sherry L. Kane*
Senior Buyer: *Sandy Ludovissy*
Designer: *Colleen P. Havens*
Interior Designer: *Amanda Kavanagh, ARK Design Studio*
Cover Illustration: *© 2012 William B. Westwood, all rights reserved*
Senior Content Licensing Specialist: *John C. Leland*
Photo Research: *Mary Reeg*
Compositor: *Electronic Publishing Services Inc., NYC*
Typeface: *9.5/12 Utopia*
Printer: *LSC Communications*

All credits appearing on page or at the end of the book are considered to be an extension of the copyright page.

Library of Congress Cataloging-in-Publication Data

Cataloging-in-Publication Data has been requested from the Library of Congress.

The Internet addresses listed in the text were accurate at the time of publication. The inclusion of a website does not indicate an endorsement by the authors or McGraw-Hill, and McGraw-Hill does not guarantee the accuracy of the information presented at these sites.

www.mhhe.com

Contents

Preface v

1 Introduction: Structure and Physiological Control Systems 1

Exercise 1.1 Microscopic Examination of Cells 2
A. The Inverted Image 2
B. The Metric System: Estimating the Size of Microscopic Objects 3
C. Microscopic Examination of Cheek Cells 5
D. Cell Structure and Cell Division 6

Exercise 1.2 Microscopic Examination of Tissues and Organs 13
A. Epithelial Tissue 13
B. Connective Tissues 16
C. Muscle Tissue 19
D. Nervous Tissue 19
E. An Organ: The Skin 20

Exercise 1.3 Homeostasis and Negative Feedback 25
A. Negative Feedback Control of Water-Bath Temperature 26
B. Resting Pulse Rate: Negative Feedback Control and Normal Range 27

2 Cell Function and Biochemical Measurements 33

Exercise 2.1 Measurements of Plasma Glucose, Cholesterol, and Protein 34
A. Carbohydrates: Measurement of Plasma Glucose Concentration 37
B. Lipids: Measurement of Plasma Cholesterol Concentration 38
C. Proteins: Measurement of Plasma Protein Concentration 39

Exercise 2.2 Thin-Layer Chromatography of Amino Acids 49

Exercise 2.3 Electrophoresis of Serum Proteins 53

Exercise 2.4 Measurements of Enzyme Activity 61
A. Catalase in Liver 63
B. Measurement of Alkaline Phosphatase in Serum 63
C. Measurement of Lactate Dehydrogenase in Serum 63

Exercise 2.5 Genetic Control of Metabolism 69
A. Phenylketonuria 70
B. Alkaptonuria 71
C. Cystinuria 71

Exercise 2.6 Diffusion, Osmosis, and Tonicity 75
A. Solubility of Compounds in Polar and Nonpolar Solvents 76
B. Osmosis Across an Artificial Semipermeable Membrane 77
C. Concentration and Tonicity 79

3 The Nervous System and Sensory Physiology 85

Exercise 3.1 Recording the Nerve Action Potential 86
Exercise 3.2 Electroencephalogram (EEG) 93
Exercise 3.3 Reflex Arc 99
A. Tests for Spinal Nerve Stretch Reflexes 99
B. A Cutaneous Reflex: The Plantar Reflex and Babinski's Sign 101

Exercise 3.4 Cutaneous Receptors and Referred Pain 105
A. Mapping the Temperature and Touch Receptors of the Skin 105
B. The Two-Point Threshold in Touch Perception 107
C. Adaptation of Temperature Receptors 108
D. Referred Pain 108

Exercise 3.5 Eyes and Vision 115
A. Refraction: Test for Visual Acuity and Astigmatism 115
B. Accommodation 118
C. Extrinsic Muscles of the Eye and Nystagmus 120
D. Pupillary Reflex 120
E. Examination of the Eye with an Ophthalmoscope 121
F. The Blind Spot 122
G. The Afterimage 123
H. Color Vision and Color Blindness 123

Exercise 3.6 Ears: Cochlea and Hearing 129
A. Conduction of Sound Waves Through Bone: Rinne's and Weber's Tests 129
B. Binaural Localization of Sound 132

Exercise 3.7 Ears: Vestibular Apparatus—Balance and Equilibrium 135

Exercise 3.8 Taste Perception 139

4 The Endocrine System 145

Exercise 4.1 Histology of the Endocrine Glands 148
A. Ovary 148
B. Testis 150
C. Pancreatic Islets (of Langerhans) 150
D. Adrenal Gland 151
E. Thyroid 152
F. Pituitary Gland 152

Exercise 4.2 Thin-Layer Chromatography of Steroid Hormones 157
A. Steroid Hormones 157
B. Thin-Layer Chromatography 160

Exercise 4.3 Insulin Shock 167

5 SKELETAL MUSCLES 173

Exercise 5.1 Neural Control of Muscle Contraction 175
A. Frog Muscle Preparation and Stimulation 177
B. Stimulation of a Motor Nerve 180

Exercise 5.2 Summation, Tetanus, and Fatigue 185
A. Twitch, Summation, Tetanus, and Fatigue in the Gastrocnemius Muscle of the Frog 187
B. Twitch, Summation, and Tetanus in Human Muscle 187

Exercise 5.3 Electromyogram (EMG) 193
A. Electromyogram Recording 195
B. Biofeedback and the Electromyograph 196

6 BLOOD: GAS TRANSPORT, IMMUNITY, AND CLOTTING FUNCTIONS 203

Exercise 6.1 Red Blood Cells and Oxygen Transport 205
A. Hematocrit 206
B. Red Blood Cell Count 207
C. Hemoglobin Concentration 207
D. Calculation of Mean Corpuscular Volume (MCV) and Mean Corpuscular Hemoglobin Concentration (MCHC) 207

Exercise 6.2 White Blood Cell Count, Differential Count, and Immunity 213
A. Total White Blood Cell Count 215
B. Differential White Blood Cell Count 216

Exercise 6.3 Blood Types 223
A. The Rh Factor 223
B. The ABO Antigen System 225

Exercise 6.4 Blood Clotting System 229
A. Test for Prothrombin Time 231
B. Test for Activated Partial Thromboplastin Time (APTT) 232

7 THE CARDIOVASCULAR SYSTEM 235

Exercise 7.1 Effects of Drugs on the Frog Heart 237
A. Effect of Caffeine on the Heart 240
B. Effect of Nicotine on the Heart 240
C. Effect of Epinephrine on the Heart 241
D. Effect of Atropine on the Heart 241
E. Effect of Pilocarpine on the Heart 241
F. Effect of Digitalis on the Heart 242
G. Effect of Calcium Ions on the Heart 242
H. Effect of Potassium Ions on the Heart 242

Exercise 7.2 Electrocardiogram (ECG) 247
Exercise 7.3 Effects of Exercise on the Electrocardiogram 259
Exercise 7.4 Mean Electrical Axis of the Ventricles 265
Exercise 7.5 Heart Sounds 273
A. Auscultation of Heart Sounds with the Stethoscope 273
B. Correlation of the Phonocardiogram with the Electrocardiogram 275

Exercise 7.6 Measurements of Blood Pressure 281
Exercise 7.7 Cardiovascular System and Physical Fitness 287

8 RESPIRATION AND METABOLISM 293

Exercise 8.1 Measurements of Pulmonary Function 295
A. Measurement of Simple Lung Volumes and Capacities 298
B. Measurement of Forced Expiratory Volume 305

Exercise 8.2 Effect of Exercise on the Respiratory System 313
Exercise 8.3 Oxyhemoglobin Saturation 321
Exercise 8.4 Respiration and Acid-Base Balance 327
A. Ability of Buffers to Stabilize the pH of Solutions 328
B. Effect of Exercise on the Rate of CO_2 Production 330
C. Role of Carbon Dioxide in the Regulation of Ventilation 330

9 RENAL FUNCTION AND HOMEOSTASIS 335

Exercise 9.1 Renal Regulation of Fluid and Electrolyte Balance 337
Exercise 9.2 Renal Plasma Clearance of Urea 345
Exercise 9.3 Clinical Examination of Urine 351
A. Test for Proteinuria 352
B. Test for Glycosuria 352
C. Test for Ketonuria 352
D. Test for Hemoglobinuria 353
E. Test for Bilirubinuria 353
F. Microscopic Examination of Urine Sediment 353

10 DIGESTION AND NUTRITION 361

Exercise 10.1 Histology of the Gastrointestinal Tract, Liver, and Pancreas 363
A. Esophagus and Stomach 364
B. Small Intestine and Large Intestine 365
C. Liver 368
D. Pancreas 369

Exercise 10.2 Digestion of Carbohydrate, Protein, and Fat 373
A. Digestion of Carbohydrate (Starch) by Salivary Amylase 374
B. Digestion of Protein (Egg Albumin) by Pepsin 375
C. Digestion of Triglycerides by Pancreatic Juice and Bile 378

Exercise 10.3 Nutrient Assessment, BMR, and Body Composition 385
A. Body Composition Analysis 386
B. Energy Intake and the Three-Day Dietary Record 387
C. Energy Output: Estimates of the BMR and Activity 388

11 REPRODUCTIVE SYSTEM 397

Exercise 11.1 Ovarian Cycle as Studied Using a Vaginal Smear of the Rat 399
Exercise 11.2 Human Chorionic Gonadotropin and the Pregnancy Test 405
Exercise 11.3 Patterns of Heredity 411
A. Sickle-Cell Anemia 412
B. Inheritance of PTC Taste 413
C. Sex-Linked Traits: Inheritance of Color Blindness 413

Appendix 1 Basic Chemistry A-1
Appendix 2* Sources of Equipment and Solutions
Appendix 3* Multimedia Correlations to the Laboratory Exercises
Credits C-1
Index I-1

---

*Appendixes 2 & 3 can be found online at www.mhhe.com/fox13

Preface

The thirteenth edition, like the previous editions, is a stand-alone human physiology manual that can be used in conjunction with any human physiology textbook. It includes a wide variety of exercises that support most areas covered in a human physiology course, allowing instructors the flexibility to choose those exercises best suited to meet their particular instructional goals. Background information needed to understand the principles and significance of each exercise is presented in a concise manner, so that little or no support is needed from the lecture text.

However, lecture and laboratory segments of a human physiology course are most effectively wedded when they cover topics in a similar manner and sequence. Thus, this laboratory guide is best used in conjunction with the textbook *Human Physiology*, 13/e, by Stuart Ira Fox (McGraw-Hill, © 2013).

The laboratory experiences provided by this guide allow students to become familiar—in an intimate way that cannot be achieved by lecture and text alone—with many fundamental concepts of physiology. In addition to providing hands-on experience in applying physiological concepts, the laboratory sessions allow students to interact with the subject matter; with other students; and with the instructor in a personal, less formal way. Active participation is required to carry out the exercise procedures, collect data, and to complete the laboratory report.

The questions in the laboratory reports, like those at the end of each chapter in the textbook *Human Physiology*, by Stuart Ira Fox, are organized into four levels. These are (1) *Test Your Knowledge*, (2) *Test Your Understanding*, (3) *Test Your Analytical Ability*, and where appropriate, (4) *Test Your Quantitative Ability*. This organization promotes higher-order learning and understanding in the laboratory and helps to better integrate the laboratory with information learned in the lecture portion of the physiology course. In addition, the thirteenth edition has a new element in the laboratory report entitled *Clinical Investigation Questions*. These test student understanding of the new element, *Clinical Investigation* (discussed shortly), that now begins most exercises.

Clinically oriented laboratory exercises that heighten student interest and demonstrate the health applications of physiology have been a hallmark of previous editions and continue to be featured in this latest edition. Change is required, however, because vendors change and available laboratory equipment and supplies change. This thirteenth edition accommodates such changes and makes new advances in improving the ability of students to benefit from the physiology laboratory experience.

The Thirteenth Edition

Integration of the Laboratory Guide with the Textbook

This laboratory guide contains all of the information students need to understand and perform the laboratory exercises. It is thus a self-contained, stand-alone laboratory guide. This benefits students because they don't have to bring the larger and heavier textbook to the laboratory section and sift through the textbook to find the information particularly relevant to the laboratory exercise.

However, students benefit when the laboratory is well integrated with the lecture portion of the physiology course. To facilitate the interaction between lecture and laboratory, this guide uses three devices to allow students to cross-reference the material in the laboratory to the information in the lecture textbook, *Human Physiology*, 13/e, by Stuart Ira Fox:

1. **Textbook Correlations** boxes are found at the beginning of each laboratory exercise. These provide specific page numbers in the textbook that correspond to the laboratory exercise. Students don't need this information to answer the questions in their laboratory report, but will benefit from greater depth and wider perspective when their textbook is used in conjunction with the laboratory guide.

2. **Figure cross-references** between the laboratory guide and the textbook are updated in the thirteenth edition. Whenever a figure in this laboratory guide has a full-color counterpart in the textbook, the specific number of the full-color text figure is provided in the caption of the laboratory manual figure. This allows students to better integrate the laboratory exercise with the concepts discussed in the textbook and lecture portion of their course. Figures in the textbook *Human Physiology*, 13/e, by Stuart Ira Fox, together with a correlation list of these to figures in this laboratory guide, are available to instructors on the text website at www.mhhe.com/fox13.

3. **Clinical Investigations**. *Human Physiology,* 13e, by Stuart Ira Fox, and *Connect* website have *Clinical Investigations* that begin every chapter, so that students can use their understanding of physiological concepts to solve medical mysteries pertinent to the material in the chapters. This provides for more active learning—and the more actively students use the information and concepts the better they can learn them. This principle is now carried forward in the thirteenth edition of this *Laboratory Guide.* A Clinical Investigation is now provided at the beginning of almost every exercise, and questions relating to it are included in the *Laboratory Report* at the end of the exercise. In addition to enhancing learning, this feature highlights the usefulness of the laboratory concepts to the students' intended health careers.

Changes in the Laboratory Exercises

The thirteenth edition retains the procedures using the Biopac, Intelitool, iWorx, and PowerLab systems where appropriate. These are systems for performing computerized data acquisition and analysis that can be adapted for use with this laboratory guide. Also retained from the previous editions are the multimedia correlations for the exercises. These multimedia correlations include *MediaPhys,* a supplementary CD that supports concepts presented in lecture and laboratory, and *Physiology Interactive Lab Simulations (Ph.I.L.S.)* references. Ph.I.L.S. contains simulated lab exercises that can be performed in addition to (or instead of) the noted exercises in this laboratory guide. In this thirteenth edition, these multimedia correlations (where available) now appear with the textbook correlations at the beginning of the exercises rather than in footnotes.

As previously mentioned, an important new feature of this thirteenth edition of the *Laboratory Guide* is a *Clinical Investigation* at the beginning of most exercises. These clinical investigations utilize information contained in the exercises so that students can actively learn the new concepts as they solve medical mysteries. This heightens interest and provides connections between the information learned in the laboratory, lecture, textbook, *Connect* website for the text, and medical applications relevant to health.

Learning Outcomes are provided at the beginning of each exercise, and learning levels similar to the organization of the textbook are presented in the Review Activities at the end of each exercise. However, a new element has been added to the Review Activities sections of this thirteenth edition of the *Laboratory Guide.* This element—*Clinical Investigation Questions*—tests students' understanding of the clinical investigation as it relates to the physiological concepts presented in the laboratory exercise.

Some of the specific changes in the exercises include the following:

- **Exercise 1.1**   Revised figure 1.3.
- **Exercise 1.3**   Clinical Investigation added at the beginning and Clinical Investigation Questions added at end of exercise; set point of body temperature added; uterine contractions added as example of positive feedback; normal ranges of pulse rate and blood glucose added.
- **Exercise 2.1**   Clinical Investigation added at beginning and Clinical Investigation Questions added at end of exercise; revised figure 2.3.
- **Exercise 2.3**   Clinical Investigation added at beginning and Clinical Investigation Questions added at end of exercise.
- **Exercise 2.4**   Clinical Investigation added at beginning and Clinical Investigation Questions added at end of exercise.
- **Exercise 2.5**   Clinical Investigation added at beginning and Clinical Investigation Questions added at end of exercise.
- **Exercise 2.6**   Clinical Investigation added at beginning and Clinical Investigation Questions added at end of exercise; new clinical applications box added on edema; revised information on intravenous fluids; new quantitative ability question.
- **Exercise 3.1**   Clinical Investigation added at beginning and Clinical Investigation Questions added at end of exercise; updated figure 3.1; revised description of neuron; new Clinical Application box on local anesthetics.
- **Exercise 3.3**   Clinical Investigation added at beginning and Clinical Investigation Questions added at end of exercise.
- **Exercise 3.4**   Clinical Investigation added at beginning and Clinical Investigation Questions added at end of exercise; table 3.1 updated; figures 3.15 and 3.16 revised; new table 3.2 added on distance between touch receptors; added discussion of referred pain of gallstone.
- **Exercise 3.5**   Clinical Investigation added at beginning and Clinical Investigation Questions added at end of exercise; new clinical applications box on LASIK; updated discussion of glaucoma; updated discussion of color vision.
- **Exercise 3.6**   Clinical Investigation added at beginning and Clinical Investigation Questions added at end of exercise; updated and expanded discussions of hearing transduction, cochlear implants, and sound localization; added question on sound localization.
- **Exercise 3.7**   Clinical Investigation added at beginning and Clinical Investigation Questions added at end of exercise; new clinical applications box on Ménière's disease.
- **Exercise 3.8**   Updated discussion of physiology of taste.
- **Exercise 4.1**   Clinical Investigation added at beginning and Clinical Investigation Questions added at end of exercise; updated table 4.1; added discussion of the regulation of adrenal gland and expanded discussion of thyroid hormones.

- **Exercise 4.2** Clinical Investigation added at beginning and Clinical Investigation Questions added at end of exercise; updated and expanded discussion of DHT.
- **Exercise 4.3** Clinical Investigation added at beginning and Clinical Investigation Questions added at end of exercise; updated discussion of type 2 diabetes mellitus.
- **Exercise 5.1** Clinical Investigation added at the beginning and Clinical Investigation Questions added at end of exercise; revised figure 5.1; new Clinical Application box on botulinum and Botox.
- **Exercise 5.2** Clinical Investigation added at the beginning and Clinical Investigation Questions added at end of exercise; updated discussion of muscle fatigue.
- **Exercise 5.3** Clinical Investigation added at the beginning and Clinical Investigation Questions added at end of exercise; updated discussion of results of damage to upper motoneurons.
- **Exercise 6.1** Clinical Investigation added at the beginning and Clinical Investigation Questions added at end of exercise; Unopette procedures for rbc count and hemoglobin concentration moved to Web; new discussion of iron homeostasis; new question on blood doping.
- **Exercise 6.2** Clinical Investigation added at the beginning and Clinical Investigation Questions added at end of exercise; updated discussion of neutrophil functions.
- **Exercise 6.3** Clinical Investigation added at the beginning and Clinical Investigation Questions added at end of exercise; artificial blood kits added to Materials section.
- **Exercise 6.4** Clinical Investigation added at the beginning and Clinical Investigation Questions added at end of exercise; updated discussions of the role of vitamin D in blood clotting and the action of coumarin drugs.
- **Exercise 7.1** Clinical Investigation added at the beginning and Clinical Investigation Questions added at end of exercise; new figure 7.6b showing Biopac recording of frog heart contractions.
- **Exercise 7.2** Clinical Investigation added at the beginning and Clinical Investigation Questions added at end of exercise; revised figure 7.7; updated discussion of atrial fibrillation; expanded discussion of circus rhythms.
- **Exercise 7.3** Clinical Investigation added at the beginning and Clinical Investigation Questions added at end of exercise; expanded discussion of cardiac output during exercise; revised figure 7.14.
- **Exercise 7.4** Clinical Investigation added at the beginning and Clinical Investigation Questions added at end of exercise; addition of Biopac and PowerLab to Materials section.
- **Exercise 7.5** Clinical Investigation added at the beginning and Clinical Investigation Questions added at end of exercise; new discussion of mitral valve prolapse; revised figures 7.20 and 7.21.
- **Exercise 7.6** Clinical Investigation added at the beginning and Clinical Investigation Questions added at end of exercise; definition of hypertension added.
- **Exercise 7.7** Clinical Investigation added at the beginning and Clinical Investigation Questions added at end of exercise; new discussion of cardiovascular regulation at start of exercise; added discussion of ejection fraction.
- **Exercise 8.1** Clinical Investigation added at the beginning and Clinical Investigation Questions added at end of exercise; new figure 8.1 showing airways and alveoli; updated discussion of COPD; new figure 8.15 showing Biopac recordings of FEV.
- **Exercise 8.2** Clinical Investigation added at the beginning and Clinical Investigation Questions added at end of exercise; updated discussion of the central control of breathing.
- **Exercise 8.3** Clinical Investigation added at the beginning and Clinical Investigation Questions added at end of exercise.
- **Exercise 8.4** Clinical Investigation added at the beginning and Clinical Investigation Questions added at end of exercise.
- **Exercise 9.1** Clinical Investigation added at the beginning and Clinical Investigation Questions added at end of exercise; new Clinical Application box on diabetes insipidus; new information on hypokalemia; new question added for Test Your Analytical Abilities.
- **Exercise 9.2** Clinical Investigation added at the beginning and Clinical Investigation Questions added at end of exercise; updated and expanded information on BUN test.
- **Exercise 9.3** Clinical Investigation added at the beginning and Clinical Investigation Questions added at end of exercise; addition of dog and cat urine to Materials section.
- **Exercise 10.1** Clinical Investigation added at the beginning and Clinical Investigation Questions added at end of exercise; new discussion of Paneth cell function; updated and expanded discussion of inflammatory bowel disease.
- **Exercise 10.2** Clinical Investigation added at the beginning and Clinical Investigation Questions added at end of exercise; expanded discussion of peptic ulcers; new discussion of GERD; updated and expanded discussion of gallstones.
- **Exercise 10.3** Clinical Investigation added at the beginning and Clinical Investigation Questions added at end of exercise; updated discussions of essential fatty acids and omega-3 fatty acids; updated website for tracking calories.
- **Exercise 11.1** Clinical Investigation added at the beginning and Clinical Investigation Questions added at end of exercise.

- **Exercise 11.2** Clinical Investigation added at the beginning and Clinical Investigation Questions added at end of exercise.
- **Exercise 11.3** Clinical Investigation added at the beginning and Clinical Investigation Questions added at end of exercise.

New and Revised Figures in the Thirteenth Edition

The new thirteenth edition of this *Laboratory Guide* contains many figures that correspond to figures in the textbook, *Human Physiology,* 13/e, by Stuart I. Fox. These figures help students better understand the physiological basis for the laboratory exercise, thereby improving the self-contained nature of the *Laboratory Guide* and increasing the degree of cross-referencing between lecture and laboratory.

Many of these figures have been modified for the new editions of the textbook and the *Laboratory Guide* by the inclusion of numbered steps, new renditions, and other changes. The correspondence of figures between the *Human Physiology* textbook and the *Laboratory Guide,* as well as the correlations to topics in the textbook that are presented at the beginning of each exercise, help students better integrate their learning experiences in laboratory and lecture.

Safety

Special effort has been made to address concerns about the safe use and disposal of body fluids. For example, normal and abnormal artificial serum can be used as a substitute for blood in section 2 (plasma chemistry); artificial saliva is suggested in exercise 10.2 (digestion); and in section 9 (renal function) both normal and abnormal artificial urine is now available. In the interest of safety, a substitute for the use of benzene (previously required in two exercises) is provided.

The international symbol for caution is used throughout the laboratory guide to alert the reader when special attention is necessary while preparing for or performing a laboratory exercise. For reference, laboratory safety guidelines appear on the inside front cover.

Technology

Computer-assisted and computer-guided instruction in human physiology laboratories has greatly increased in recent years. Computer programs provide a number of benefits: some experiments that require animal sacrifice can be simulated; data can be analyzed against a data bank and displayed in an appealing and informative manner; class data records can be analyzed; and costs can be reduced by eliminating the use of some of the most expensive equipment.

This edition continues to reference programs offered by Biopac, Intelitool, and iWorx. In addition, correlations to McGraw-Hill's *MediaPhys* and to the *Physiology Interactive Lab Simulations* (Ph.I.L.S.) have been incorporated into the thirteenth edition.

Organization of the Laboratory Guide

The exercises in this guide are organized in this manner:

1. Each exercise begins with a list of **materials** needed to perform the exercise, so that it is easier to set up the laboratory. This section is identified by a materials icon.

2. Following the materials section is an overview paragraph describing the **concept** behind the laboratory exercise.

3. Following the concept paragraph is a list of **learning outcomes.** This helps students guide their learning while performing the exercise, and helps instructors determine and document student mastery of the laboratory material.

4. A box providing **textbook correlations** is placed near the beginning of each exercise. This section can be used to help integrate the lecture textbook (*Human Physiology,* 13/e, by Stuart Ira Fox) with the laboratory material. In the new thirteenth edition, this box now includes **multimedia correlations** for relevant sections of *MediaPhys* and *Physiology Interactive Lab Simulations* (Ph.I.L.S.) by McGraw-Hill.

5. A **Clinical Investigation** presents a medical mystery at the beginning of most exercises. This mystery can be solved using the physiological concepts and information provided in the exercise, allowing students to make connections between their laboratory experience and its health applications. This heightens interest and fosters active learning of the information in the exercise. Questions relating to the Clinical Investigation are asked in the last section of the laboratory report.

6. A brief **introduction** to the exercise presents the essential information for understanding the physiological significance of the exercise. This concisely written section eliminates the need to consult the lecture text.

7. Boxed information, set off as screened insets, provides the **clinical applications** of different aspects of the laboratory exercise. This approach was pioneered by this laboratory manual and the current edition continues that tradition.

8. The **procedure** is stated in the form of easy-to-follow steps. These directions are set off from the textual material through the use of a distinctive typeface,

making it easier for students to locate them as they perform the exercise.

9. **Normal values** boxes are placed following the procedures in cases where students obtain clinically applicable measurements. These boxes emphasize the relationship between clinical measurements for diagnosis and physiological regulation of homeostasis.

10. A **laboratory report** follows each exercise. Students enter data here when appropriate and answer questions. The questions in the laboratory report begin with the most simple form (objective questions) in most exercises and progress to essay questions. The essay questions are designed to stimulate conceptual learning and to maximize the educational opportunity provided by the laboratory experience. A section entitled *Clinical Investigation Questions* follows, testing students' comprehension of the physiological concepts in the Clinical Investigation and their medical applications.

SUPPLEMENTAL MATERIALS

A comprehensive selection of supplemental materials is available for use in conjunction with *Human Physiology,* 13/e, and *A Laboratory Guide to Human Physiology: Concepts and Clinical Applications,* 13/e, by Stuart Ira Fox. Instructors can obtain teaching aids by calling the McGraw-Hill Customer Service Department at 1-800-338-3987, visiting our catalog at www.mhhe.com, or contacting their local McGraw-Hill sales representative. Students can order supplemental materials by contacting their campus bookstore.

INSTRUCTOR'S MANUAL FOR THE LABORATORY GUIDE

Accessed via the Instructor Center of the *Human Physiology* textbook website at www.mhhe.com/fox13, this helpful manual includes suggestions for coordinating lab exercises with the textbook, introductions to each exercise, materials lists, approximate completion times, and solutions to the laboratory reports for each exercise. A listing of laboratory supply houses and instructions for mixing commonly used solutions are also provided.

Instructors using this laboratory manual independently of *Human Physiology,* 13/e, by Stuart Ira Fox, can access the Instructor's Manual on McGraw-Hill's Lab Central at www.mhhe.com/labcentral.

McGraw-Hill ConnectPlus® Anatomy & Physiology is a web-based assignment and assessment platform that gives students the means to better connect with their coursework, with their instructors, and with the important concepts that they will need to know for success now and in the future. With Connect® Anatomy & Physiology, instructors can deliver assignments, quizzes, and tests easily online. Students can practice important skills at their own pace and on their own schedule. With ConnectPlus Anatomy & Physiology, students also get 24/7 online access to an eBook—an online edition of the text—to aid them in successfully completing their work, wherever and whenever they choose. (www.mhhe.com/fox13).

NEW! **ConnectPlus®** with **LearnSmart™** makes teaching easier and learning smarter.

PHYSIOLOGY INTERACTIVE LAB SIMULATIONS (PH.I.L.S.) 4.0

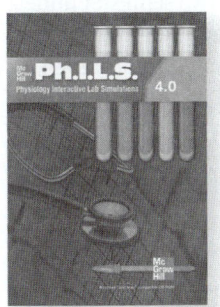

NEW! Ph.I.L.S. 4.0 has been updated! Users have requested and we are providing *five new exercises* (Respiratory Quotient, Weight & Contraction, Insulin and Glucose Tolerance, Blood Typing, and Anti-Diuretic Hormone). Ph.I.L.S. 4.0 is the perfect way to reinforce key physiology concepts with powerful lab experiments. Created by Dr. Phil Stephens at Villanova University, this program offers *42 laboratory simulations* that may be used to supplement or substitute for wet labs. All 42 labs are self-contained experiments—no lengthy instruction manual required. Users can adjust variables, view outcomes, make predictions, draw conclusions, and print lab reports. This easy-to-use software offers the flexibility to change the parameters of the lab experiment. There is no limit!

MEDIAPHYS 3.0

This physiology tutorial offers detailed explanations, high quality illustrations, and amazing animations to provide a thorough introduction to the world of physiology. MediaPhys is filled with interactive activities and quizzes to help reinforce physiology concepts that are often difficult to understand.

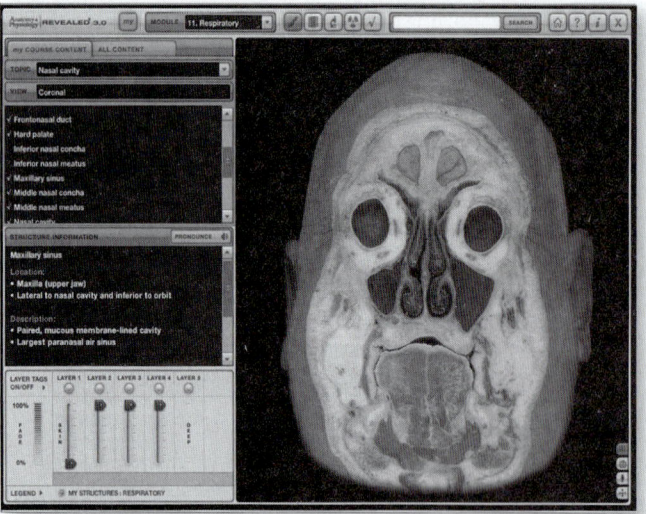

ANATOMY & PHYSIOLOGY REVEALED® STUDENT TUTORIAL

Anatomy & Physiology Revealed is the ultimate online interactive cadaver dissection experience. Now fully customizable to fit any course or lab, this state-of-the-art program uses cadaver photos combined with a layering technique that allows the student to peel away layers of the human body to reveal structures beneath the surface. *Anatomy & Physiology Revealed* also offers animations, histologic and radiologic imaging, audio pronunciations, and comprehensive quizzing. It can be used as part of any one or two semester undergraduate Anatomy & Physiology or Human Anatomy course; *Anatomy & Physiology Revealed* is available stand-alone, or can be combined with any McGraw-Hill product.

ACKNOWLEDGMENTS

A number of users of this laboratory guide provided thoughtful reviews that greatly aided in the revision. Their names are listed here. Also, the thirteenth edition was greatly benefited by input from my colleague Dr. Laurence G. Thouin, Jr. His numerous suggestions helped to make the thirteenth edition more accurate and student-friendly.

The shaping of the thirteenth edition was also aided by suggestions from other colleagues and students. I greatly appreciate the support of the editors at McGraw-Hill. Their contributions helped make this the best edition yet of *A Laboratory Guide to Human Physiology: Concepts and Clinical Applications*.

REVIEWERS

Miguel Angeles
Merritt College; University of California–Berkeley

K. M. Cianci
Fresno Pacific University

Julie Collins
Eastern Oklahoma State College

Alyssa Kiesow
Northern State University

Susan Landon-Arnold
Northern State University

Ann M. Mills
Minnesota West Community and Technical College

Section 1

Introduction: Structure and Physiological Control Systems

The cell is the basic unit of structure and function in the body. Each cell is surrounded by a *plasma* (or *cell*) *membrane* and contains specialized structures called *organelles* within the cell fluid, or *cytoplasm*. The structure and functions of a cell are largely determined by genetic information contained within the membrane-bound *nucleus*. This genetic information is coded by the specific chemical structure of *deoxyribonucleic acid (DNA)* molecules, the major component of *chromosomes*. Through genetic control of *ribonucleic acid (RNA)* and the synthesis of proteins (such as enzymes described in section 2), DNA within the cell nucleus directs the functions of the cell and, ultimately, those of the entire body.

Cells with similar specializations are grouped together to form **tissues,** and tissues are grouped together to form larger units of structure and function known as **organs.** Organs located in different parts of the body but that cooperate in the service of a common function are called **organ systems** (e.g., the cardiovascular system).

The complex activities of cells, tissues, organs, and systems are coordinated by a wide variety of regulatory mechanisms that act to maintain **homeostasis**—a state of dynamic constancy in the internal environment. **Physiology** is largely the study of the control mechanisms that participate in maintaining homeostasis.

Exercise 1.1	Microscopic Examination of Cells
Exercise 1.2	Microscopic Examination of Tissues and Organs
Exercise 1.3	Homeostasis and Negative Feedback

EXERCISE 1.1

Microscopic Examination of Cells

MATERIALS

1. Compound microscopes
2. Prepared microscope slides, including whitefish blastula (early embryo); clean slides; and coverslips (Note: Slides with dots, lines, or the letter *e* can be prepared with dry transfer patterns used in artwork.)
3. Lens paper
4. Methylene blue stain
5. Cotton-tipped applicator sticks or toothpicks

The microscope and the metric system are important tools in the study of cells. Cells contain numerous organelles with specific functions and are capable of reproducing themselves by mitosis. However, there is also a special type of cell division called meiosis used in the gonads to produce sperm or ova.

LEARNING OUTCOMES

You should be able to:

1. Identify the major parts of a microscope and demonstrate proper technique in the care and handling of this instrument.
2. Define and interconvert units of measure in the metric system and estimate the size of microscopic objects.
3. Describe the general structure of a cell and the specific functions of the principal organelles.
4. Describe the processes of mitosis and meiosis and explain their significance.

Textbook/Multimedia Correlations

Before performing this exercise, you should study the introductory material presented here. Further information relating to this exercise can be found in these pages of *Human Physiology*, thirteenth edition, by Stuart I. Fox:

- *Cytoplasm and Its Organelles.* Chapter 3, p. 57.
- *DNA Synthesis and Cell Division.* Chapter 3, p. 73.

The **microscope** is the most basic and widely used instrument in the life science laboratory. The average binocular microscope for student use, as shown in figure 1.1, includes these parts:

1. eyepieces, each with an ocular lens (usually 10× magnification, and may have a pointer)
2. a stage platform with manual or mechanical stage controls
3. a substage condenser lens and iris diaphragm, each with controls
4. coarse focus and fine focus adjustment controls
5. objective lenses on a revolving nosepiece (usually include: a scanning lens, 4×; a low-power lens, 10×; a high-power lens, 45×; and often an even more powerful oil-immersion lens, 100×).

CARE AND CLEANING

The microscope is an expensive, delicate instrument. To maintain it in good condition, always take these precautions:

1. Carry the microscope by holding on to the arm and base.
2. Use the *coarse focus* knob *only* with low power; after focusing on low power with the coarse focus and *fine focus* knob, you can change to higher magnification objective lenses and focus using only the fine focus knob.
3. Clean the ocular and objective lenses with lens paper moistened with distilled water before and after use. (Use alcohol only if oil has been used with an oil-immersion, 100× lens.)
4. Always leave the lowest power objective lens (usually 4× or 10×) facing the stage before putting the microscope away.

A. THE INVERTED IMAGE

Obtain a slide with the letter *e* mounted on it. Place the slide on the microscope stage, and rotate the nosepiece until the 10× objective clicks into the vertical position. Using the coarse adjustment, carefully lower the objective lens until it almost touches the slide. Now, looking through the ocular lens, slowly raise the objective lens until the letter *e* comes into focus.

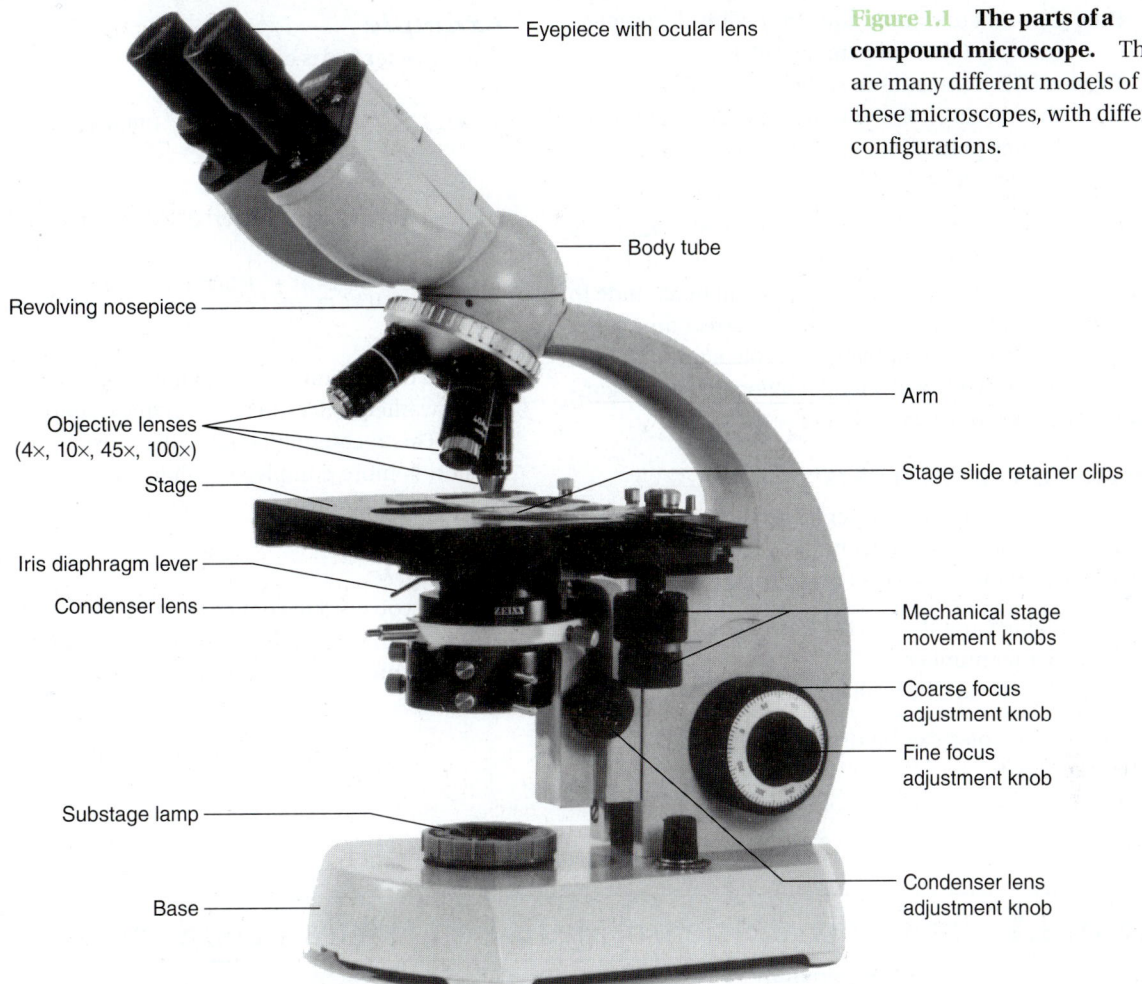

Figure 1.1 **The parts of a compound microscope.** There are many different models of these microscopes, with different configurations.

Procedure

1. If the visual field is dark, increase the light by adjusting the lever that opens (and closes) the iris diaphragm. If there is still not enough light, move the substage condenser lens closer to the slide by rotating its control knob. Bring the image into sharp focus using the fine focus control. Now, draw the letter *e* as it appears in the microscope.

2. While looking through the ocular lens, rotate the mechanical stage controls so that the mechanical stage moves to the *right.* In which direction does the *e* move?

3. While looking through the ocular lens, rotate the mechanical stage controls so that the mechanical stage moves *toward* you. In which direction does the *e* move?

B. The Metric System: Estimating the Size of Microscopic Objects

It is important in microscopy, as in other fields of science, that units of measure are standardized and easy to use. The **metric system** (from the Greek word *metrikos*, meaning "measure") first developed in late eighteenth-century France, is the most commonly used measurement system in scientific literature. The modern definitions of the units used in the metric system are those adopted by the General Conference on Weights and Measures, which in 1960 established the International System of Units, also known (in French) as Système International d'Unités, and abbreviated SI (in all languages). The definitions for the metric units of *length, mass, volume,* and *temperature* are as listed here:

meter (m)—unit of length equal to 1,650,763.73 wavelengths in a vacuum of the orange-red line of the spectrum of krypton-86

gram (g)—unit of mass based on the mass of 1 cubic centimeter (cm^3) of water at the temperature (4° C) of its maximum density

3

liter (L)—unit of volume equal to 1 cubic decimeter (dm^3) or 0.001 cubic meter (m^3)

Celsius (C)—temperature scale in which 0° is the freezing point of water and 100° is the boiling point of water; this is equivalent to the centigrade scale

Conversions between different orders of magnitude in the metric system are based on powers of ten (table 1.1). Therefore, you can convert from one order of magnitude to another by moving the decimal point the correct number of places to the right (for multiplying by whole numbers) or to the left (for multiplying by decimal fractions). Sample conversions are illustrated in table 1.2.

DIMENSIONAL ANALYSIS

If you are unsure about the proper factor for making a metric conversion, you can use a technique called *dimensional analysis*. This technique is based on two principles:

1. Multiplying a number by 1 does not change the value of that number.
2. A number divided by itself is equal to 1.

These principles can be used to change the units of any measurement.

Example

Since 1 meter (m) is equivalent to 1,000 millimeters (mm),

$$\frac{1 \text{ m}}{1,000 \text{ mm}} = 1 \quad \text{and} \quad \frac{1,000 \text{ mm}}{1 \text{ m}} = 1$$

Suppose you want to convert 0.032 meter to millimeters:

$$0.032 \text{ m} \times \frac{1,000 \text{ mm}}{1 \text{ m}} = 32.0 \text{ mm}$$

In dimensional analysis the problem is set up so that the unwanted units (meter, *m* in this example) cancel each other. This technique is particularly useful when the conversion is more complex or when some of the conversion factors are unknown.

Example

Suppose you want to convert 0.1 milliliter (mL) to microliter (μL) units. If you remember that 1 mL = 1,000 μL, you can set up the problem as shown here:

$$0.1 \text{ mL} \times \frac{1,000 \text{ μL}}{1 \text{ mL}} = 100 \text{ μL}$$

Table 1.1 International System of Metric Units, Prefixes, and Symbols

Multiplication Factor	Prefix	Symbol	Term
$1,000,000 = 10^6$	Mega	M	One million
$1,000 = 10^3$	Kilo	k	One thousand
$100 = 10^2$	Hecto	h	One hundred
$10 = 10^1$	Deka	da	Ten
$1 = 10^0$			
$0.1 = 10^{-1}$	Deci	d	One-tenth
$0.01 = 10^{-2}$	Centi	c	One-hundredth
$0.001 = 10^{-3}$	Milli	m	One-thousandth
$0.000001 = 10^{-6}$	Micro	μ	One-millionth
$0.000000001 = 10^{-9}$	Nano	n	One-billionth
$0.000000000001 = 10^{-12}$	Pico	p	One-trillionth
$0.000000000000001 = 10^{-15}$	Femto	f	One-quadrillionth

Table 1.2 Sample Metric Conversions

To Convert From	To	Factor	Move Decimal Point
Meter (Liter, gram)	Milli-	× 1,000 (10^3)	3 places to right
Meter (Liter, gram)	Micro-	× 1,000,000 (10^6)	6 places to right
Milli-	Meter (Liter, gram)	÷ 1,000 (10^{-3})	3 places to left
Micro-	Meter (Liter, gram)	÷ 1,000,000 (10^{-6})	6 places to left
Milli-	Micro-	× 1,000 (10^3)	3 places to right
Micro-	Milli-	÷ 1,000 (10^{-3})	3 places to left

If you remember that a milliliter is one-thousandth of a liter and that a microliter is one-millionth of a liter, you can set up the problem in this way:

$$0.1 \text{ mL} \times \frac{1.0 \text{ L}}{1,000 \text{ mL}} \times \frac{1,000,000 \text{ μL}}{1.0 \text{ L}} = 100 \text{ μL}$$

VISUAL FIELD AND THE ESTIMATION OF MICROSCOPIC SIZE

If the magnification power of your ocular lens is 10× and you use the 10× objective lens, the total magnification of the visual field will be 100×. At this magnification, the diameter of the visual field is approximately 1,600 *micrometers* (μm), or *microns*.

You can estimate the size of an object in the visual field by comparing it with the total diameter (line *AB*) of the visual field. Using this diagram:

How long is line *AC* in micrometers (μm)? _____
How long is line *AD* in micrometers (μm)? _____
How long is line *AE* in micrometers (μm)? _____

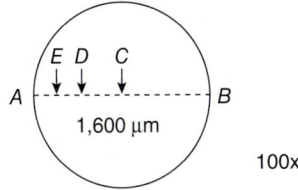

The diameter of the field of vision using the 45× objective lens (total magnification 450×) is approximately 356 micrometers. Using the diagram previously shown and applying the same technique, answer these questions assuming use of a 45× objective lens:

How long is line *AC* in micrometers (μm)? _____
How long is line *AD* in nanometers (nm)? _____

Since the diameter of the visual field is inversely proportional to the magnification, when the 4× scanning objective lens is used (total magnification 40×), the visual field is 4,000 μm; when the 100× oil-immersion lens is used (total magnification 1,000×), the visual field is 160 μm.

Procedure

From your instructor, obtain a slide that contains a pattern of small dots and a pattern of thin lines.

1. Using the 10× objective lens:
   (a) estimate the diameter of one dot: _____ m
   (b) estimate the distance between the *nearest* edges of two adjacent dots: _____ m
2. Using the 45× objective lens:
   (a) estimate the width of one line: _____ m
   (b) estimate the distance between the *nearest* edges of two adjacent lines: _____ m

C. MICROSCOPIC EXAMINATION OF CHEEK CELLS

The surfaces of the body are covered and lined with *epithelial membranes* (one of the primary tissues described in exercise 1.2). In membranes that are several cell layers thick, such as the membrane lining of the cheeks, cells are continuously lost from the surface and replaced through cell division in deeper layers. In contrast to cells in the outer layer of the epidermis of the skin, which die before they are lost, the cells in the outer layer of epithelial tissue in the cheeks are still alive. You can therefore easily collect and observe living human cells by rubbing the inside of the cheeks.

Most living cells are difficult to observe under the microscope unless they are stained. In this exercise, the stain *methylene blue* will be used. Methylene blue is positively charged and combines with negative charges in the chromosomes to stain the nucleus blue. The cytoplasm contains a lower concentration of negatively charged organic molecules and so appears almost clear.

Procedure

1. Rub the inside of one cheek with the cotton tip of an applicator stick (or a toothpick).
2. Press the cotton tip of the applicator stick (or the end of the toothpick) against a clean glass slide. Maintaining pressure, rotate the cotton tip against the slide and then push the cheek smear across the slide about ½ inch.
3. Observe the *unstained* cells under 100× and 450× total magnification.
4. Remove the slide from the microscope. Holding it over a sink or special receptacle, place a drop of methylene blue stain on the smear.
5. Place a coverslip over the stained smear and observe the stained cheek cells at 100× and 450× total magnification.
6. Using the procedure described in the previous section, estimate the size of the average cheek cell using both 100× and 450× total magnification. 100× _____ μm; 450× _____ μm. Are they the same? _____
7. If you have an oil-immersion lens, you can observe the cheek cells under a total magnification of 1,000×. Follow these steps:
   (a) Focus the cells using the 45× objective lens, as before.
   (b) Move that objective lens out of the way, but do not yet place a different lens over the slide.
   (c) Place a drop of immersion oil on the spot of the slide where the light shines through.
   (d) Turn the 100× objective lens until it clicks into position; the lens should be immersed in the oil drop but not touching the slide.

5

(e) Wait a moment for movement of the oil to subside, and then adjust the focusing using only the fine focus knob. It may also help to adjust the light intensity.

D. Cell Structure and Cell Division

Cells vary greatly in size and shape. The largest cell, an *ovum* (egg cell), can barely be seen with the unaided eye; other cells can be observed only through a microscope. Each cell has an outer *plasma membrane* (or cell membrane) and generally one *nucleus,* surrounded by a fluid matrix, the *cytoplasm.* Within the nucleus and the cytoplasm are a variety of subcellular structures, called **organelles** (fig. 1.2). The structures and principal functions of important organelles and other cellular components are listed in table 1.3.

The process of cell division is called **mitosis** (fig. 1.3*a*). This process allows new cells to be formed to replace those that are dying and also permits body growth. Mitosis consists of a continuous sequence of four stages (table 1.4 and fig. 1.3*a*) in which both the nucleus and cytoplasm of a cell split to form two identical *daughter cells.* During mitotic cell division, the chromosomes (duplicated earlier) separate, and one of the duplicate sets of chromosomes goes to each daughter cell. The two daughter cells therefore have the same number of chromosomes as the parent cell.

The 46 chromosomes present in most human cells represent 23 *pairs* of chromosomes; one set of 23 was inherited from the mother and the other set of 23 from the father. A cell with 46 chromosomes is said to be *diploid,* or *2n.*

In the process of *gamete* (sperm and ova) production in the *gonads* (testes and ovaries), specialized germinal cells undergo a type of division called **meiosis** (fig. 1.3*b*). Meiosis comprises two cell divisions of the germinal cells (table 1.5). After the DNA duplicates, the first meiotic cell division separates homologous chromosomes (maternal and paternal) into two daughter cells, and the second meiotic cell division separates the duplicated chromosomes into two more daughter cells. The daughter gamete cells formed by the two divisions of meiosis thus have only one set of 23 chromosomes; they are said to be *haploid,* or *1n.* In this way the original diploid number of 46 chromosomes can be restored when the sperm and egg unite in the process of fertilization.

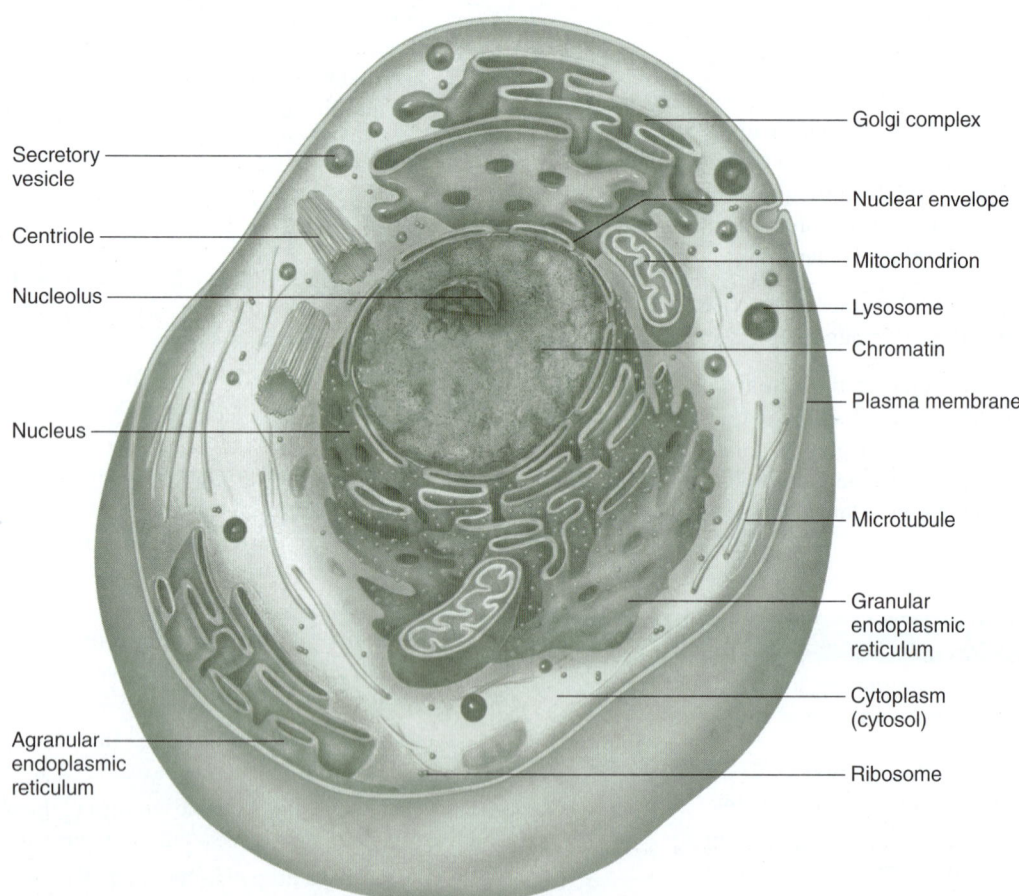

Figure 1.2 **A generalized human cell showing the principal organelles.** Most cells of the body are highly specialized, so they may have structures that differ from those shown. AP|R

(For a full-color version of this figure, see fig. 3.1 in *Human Physiology*, thirteenth edition, by Stuart I. Fox.)

Table 1.3 Cellular Components: Structure and Function

Component	Structure	Function
Plasma (cell) membrane	Membrane composed of double layer of phospholipids in which proteins are embedded	Gives form to cell and controls passage of materials into and out of cell
Cytoplasm	Fluid, jellylike substance between the cell membrane and the nucleus in which organelles are suspended	Serves as matrix substance in which chemical reactions occur
Endoplasmic reticulum	System of interconnected membrane-forming canals and tubules	Agranular (smooth) endoplasmic reticulum metabolizes nonpolar compounds and stores Ca^{2+} in striated muscle cells, granular (rough) endoplasmic reticulum assists in protein synthesis
Ribosomes	Granular particles composed of protein and RNA	Synthesize proteins
Golgi complex	Cluster of flattened membranous sacs	Synthesizes carbohydrates and packages molecules for secretion, secretes lipids and glycoproteins
Mitochondria	Membranous sacs with folded inner partitions	Release energy from food molecules and transform energy into usable ATP
Lysosomes	Membranous sacs	Digest foreign molecules and worn and damaged organelles
Peroxisomes	Spherical membranous vesicles	Contain enzymes that detoxify harmful molecules and break down hydrogen peroxide
Centrosome	Nonmembranous mass of two rodlike centrioles	Helps organize spindle fibers and distribute chromosomes during mitosis
Vacuoles	Membranous sacs	Store and release various substances within the cytoplasm
Microfilaments and microtubules	Thin, hollow tubes	Support cytoplasm and transport materials within the cytoplasm
Cilia and flagella	Minute cytoplasmic projections that extend from the cell surface	Move particles along cell surface or move the cell
Nuclear envelope	Double-layered membrane that surrounds the nucleus, composed of protein and lipid molecules	Supports nucleus and controls passage of materials between nucleus and cytoplasm
Nucleolus	Dense, nonmembranous mass composed of protein and RNA molecules	Produces ribosomal RNA for ribosomes
Chromatin	Fibrous strands composed of protein and DNA	Contains genetic code that determines which proteins (including enzymes) will be manufactured by the cell

Table 1.4 Major Events in Mitosis

Stage	Major Events
Prophase	Chromosomes form from the chromatin material, centrioles migrate to opposite sides of the nucleus, the nucleolus and nuclear membrane disappear, and spindles appear and become associated with centrioles and centromeres.
Metaphase	Duplicated chromosomes align themselves on the equatorial plane of the cell between the centrioles, and spindle fibers become attached to duplicate parts of chromosomes.
Anaphase	Duplicated chromosomes separate, and spindles shorten and pull individual chromosomes toward the centrioles.
Telophase	Chromosomes elongate and form chromatin threads, nucleoli and nuclear membranes appear for each chromosome mass, and spindles disappear.

Table 1.5 Stages of Meiosis

Stage	Events
First Meiotic Division	
Prophase I	Chromosomes appear double-stranded.
	Each strand, called a chromatid, contains duplicate DNA joined together by a structure known as a centromere.
	Homologous chromosomes pair up side by side.
Metaphase I	Homologous chromosome pairs line up at equator.
	Splindle apparatus is complete.
Anaphase I	Homologous chromosomes separate; the two members of a homologous pair move to opposite poles.
Telophase I	Cytoplasm divides to produce two haploid cells.
Second Meiotic Division	
Prophase II	Chromosomes appear, each containing two chromatids.
Metaphase II	Chromosomes line up single file along equator as spindle formation is completed.
Anaphase II	Centromeres split and chromatids move to opposite poles.
Telophase II	Cytoplasm divides to produce two haploid cells from each of the haploid cells formed at telophase I.

(a) Mitosis

(a) **Interphase**
- The chromosomes are in an extended form and seen as chromatin in the electron microscope.
- The nucleolus is visible.

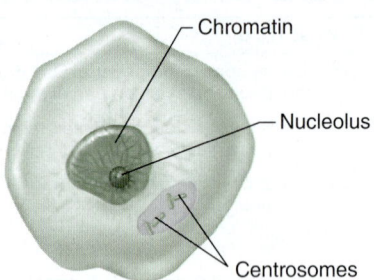

(b) **Prophase**
- The chromosomes are seen to consist of two chromatids joined by a centromere.
- The centrioles move apart toward opposite poles of the cell.
- Spindle fibers are produced and extend from each centrosome.
- The nuclear membrane starts to disappear.
- The nucleolus is no longer visible.

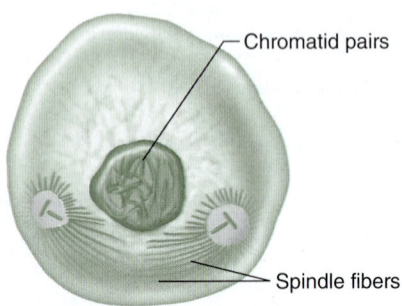

(c) **Metaphase**
- The chromosomes are lined up at the equator of the cell.
- The spindle fibers from each centriole are attached to the centromeres of the chromosomes.
- The nuclear membrane has disappeared.

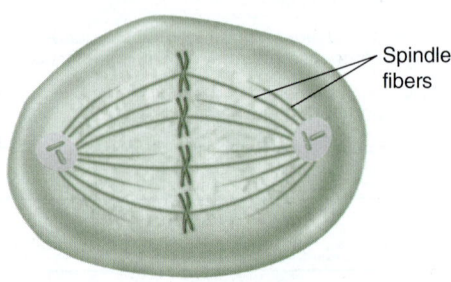

(d) **Anaphase**
- The centromere split, and the sister chromatids separate as each is pulled to an opposite pole.

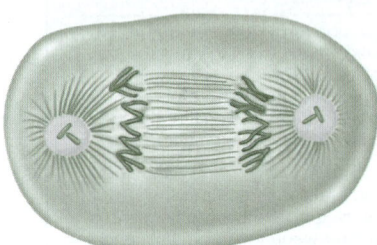

(e) **Telophase**
- The chromosomes become longer, thinner, and less distinct.
- New nuclear membranes form.
- The nucleolus reappears.
- Cell division is nearly complete.

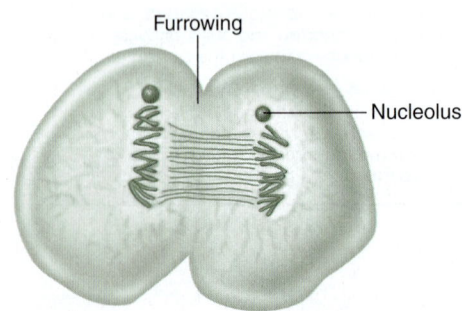

Figure 1.3 Cell division. (a) The stages of mitosis.

(For a full-color version of this figure, in *Human Physiology*, thirteenth edition, by Stuart I. Fox, see figs. 3.27 and 3.30; for figures of spermatogenesis and oogenesis, see figs. 20.15 and 20.30.)

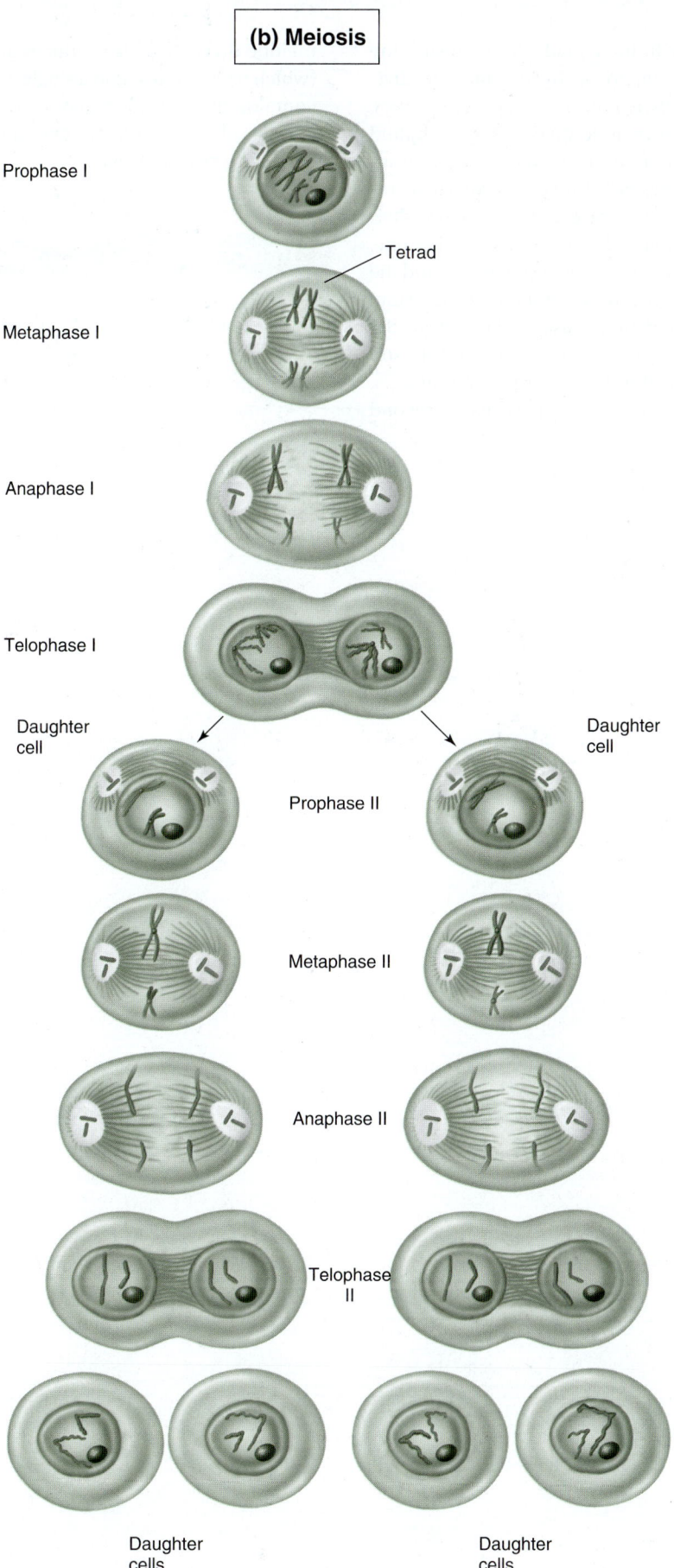

Figure 1.3 *Continued* (b) The stages of meiosis. Mitosis is the type of cell division that occurs when organs grow or cells within organs need to be replaced. Meiosis is the type of cell division that only occurs in the gonads for the production of gametes (sperm and ova).

Meiosis occurs only in the gonads. In the testes, the formation of haploid spermatozoa by the meiotic division of diploid parent cells is known as *spermatogenesis*. As expected from the two meiotic divisions, one diploid parent cell gives rise to four spermatozoa. In the ovaries, meiosis produces haploid egg cells (ova or oocytes) in a process known as *oogenesis*. In this case, the first meiotic division results in a large haploid cell and a much smaller one, known as a *polar body*. The polar body degenerates, and the remaining haploid cell begins the second meiotic division. However, its division stops at metaphase II. At this stage, the cell (called a *secondary oocyte*) may be ovulated. If it isn't fertilized, it degenerates without ever completing meiosis. If the secondary oocyte is fertilized, it completes the second meiotic division. This produces another small polar body (which degenerates) and a single, large haploid cell that also contains the sperm's haploid set of chromosomes. These two sets of chromosomes combine in the newly diploid fertilized egg cell, or *zygote*.

Procedure

1. Study figure 1.2. Cover the labels with a blank sheet of paper, and try to write them in (watch spelling!).
2. Examine a slide of a whitefish blastula (or similar early embryo), and observe the different stages of mitosis as shown in figure 1.3.

Laboratory Report 1.1

Name _____
Date _____
Section _____

REVIEW ACTIVITIES FOR EXERCISE 1.1

Test Your Knowledge

1. Give the *total* magnification when you use the
   (a) low-power objective lens _____
   (b) high-dry power objective lens _____
   (c) oil-immersion objective lens _____

2. Give the metric units for the
   (a) weight of 1 cubic centimeter of water at its maximum density _____
   (b) temperature at which water freezes _____
   (c) unit of volume equal to 0.001 cubic meter _____

3. Identify the principal organelle or cell component described here:
   (a) helps organize spindle fibers during cell division (mitosis) _____
   (b) the major site of energy production in the cell _____
   (c) a system of membranous tubules in the cytoplasm; often involved with protein synthesis _____
   (d) the location of genetic information _____
   (e) the vesicle that contains digestive enzymes _____
   (f) the site of protein synthesis _____

4. Match these events of mitosis with the correct name of the stage:
   ____ 1. the nuclear membrane disappears; spindles appear
   ____ 2. chromosomes line up along the equator of the cell
   ____ 3. duplicated chromosomes separate and are pulled toward the centrioles
   ____ 4. chromosomes elongate into chromatin threads; nuclear membranes and nucleoli reappear

   (a) metaphase
   (b) telophase
   (c) anaphase
   (d) prophase

Test Your Understanding

5. Compare and contrast *mitosis* and *meiosis* in terms of where and when they occur and their products. What are the ways that mitosis and meiosis are used in the body?

Test Your Analytical Ability

6. In metaphase I of meiosis, the homologous chromosomes line up side by side along the equator so that (a) crossing-over (exchange of DNA regions) can occur between the homologous pairs and (b) the homologous chromosomes can be pulled to opposite poles during anaphase I. In mitosis, by contrast, homologous chromosomes line up single-file along the equator. What benefits are derived from these two different ways that homologous chromosomes are positioned at metaphase in meiosis and mitosis?

7. Why do you think it is that scientists prefer to use the metric system over the English system of measurements? What problems might result if a country uses both systems of measurement?

Test Your Quantitative Ability

8. Match these equivalent measurements:
    ____ 1. 100 mL   (a) 100 µL
    ____ 2. 0.10 mL   (b) 0.00001 L
    ____ 3. 0.0001 mL   (c) 1.0 dL
    ____ 4. 0.01 mL   (d) 100 nL

EXERCISE 1.2

Microscopic Examination of Tissues and Organs

MATERIALS

1. Compound microscopes
2. Lens paper
3. Prepared microscope slides of tissues

The body is composed of only four primary tissues, and each is specialized for specific functions. Most organs of the body are composed of all four primary tissues, which cooperate in determining the overall structure and function of the organ.

LEARNING OUTCOMES

You should be able to:

1. Define the terms *tissue* and *organ*.
2. List the distinguishing characteristics of the four primary tissues.
3. Identify and describe the subcategories of the primary tissues.
4. In general terms, correlate the structures of the primary tissues with their function.

Textbook/Multimedia Correlations

Before performing this exercise, you should study the introductory material presented here. Further information relating to this exercise can be found in these pages of *Human Physiology*, thirteenth edition, by Stuart I. Fox:

- *The Primary Tissues.* Chapter 1, p. 10.
- *Organs and Systems.* Chapter 1, p. 19.

The trillions of cells that compose the human body have many basic features in common, but they differ considerably in size, structure, and function. Furthermore, cells neither function as isolated units nor are they haphazardly arranged in the body. An aggregation of cells that are similar in structure and that work together to perform a specialized activity is referred to as a tissue. Groups of tissues integrated to perform one or more common functions constitute organs. Tissues are categorized into four principal types, or **primary tissues:** (1) *epithelial,* (2) *connective,* (3) *muscle,* and (4) *nervous.*

A. Epithelial Tissue

Epithelial tissue, or *epithelium,* functions to protect, secrete, or absorb. Epithelial membranes cover the outer surface of the body (epidermis of the skin) and the outer surfaces of internal organs; they also line the body cavities and the *lumina* (the inner hollow portions) of ducts, vessels, and tubes. All *glands* are derived from epithelial tissue. Epithelial tissues share these characteristics:

1. The cells are closely joined together and have little intercellular substance (matrix) between them.
2. There is an exposed surface facing either the external environment or the *lumen* (cavity) of an organ.
3. A *basement membrane* (composed of glycoproteins) is present to anchor the epithelium to underlying connective tissue.

Epithelial tissues composed of a single layer of cells are called *simple;* those that contain more than one layer are known as *stratified.* Epithelial tissues may be further classified by the shape of their surface cells: *squamous* (if the cells are flat), *cuboidal,* or *columnar.* Using these criteria, one can identify these types of epithelia:

1. **Simple squamous epithelium** (fig. 1.4, *top*). Adapted for diffusion, absorption, filtration, and secretion—present in such places as the lining of air sacs, or *alveoli,* within the lungs (where gas exchange occurs); parts of the kidney (where blood is filtered); and the lining, or *endothelium,* of blood vessels (where exchange between blood and tissues occurs).
2. **Stratified squamous epithelium** (fig. 1.4, *middle*). Found in areas that receive a lot of wear and tear. The outer cells are sloughed off and replaced by new cells, produced by mitosis in the deepest layers. Stratified squamous epithelium is found in the mouth; esophagus; nasal cavity; and in the openings into the ears, anus, and vagina. A special *keratinized,* or *cornified,* layer of dead surface cells is found in the stratified squamous epithelium of the skin (the epidermis).
3. **Simple cuboidal epithelium** (fig. 1.4, *bottom*). Found lining such structures as small tubules of the kidneys and the ducts of the salivary glands or of the pancreas.
4. **Simple columnar epithelium** (fig. 1.5, *top*). Simple epithelium of tall columnar cells found lining the lumen of the gastrointestinal tract, where it is

13

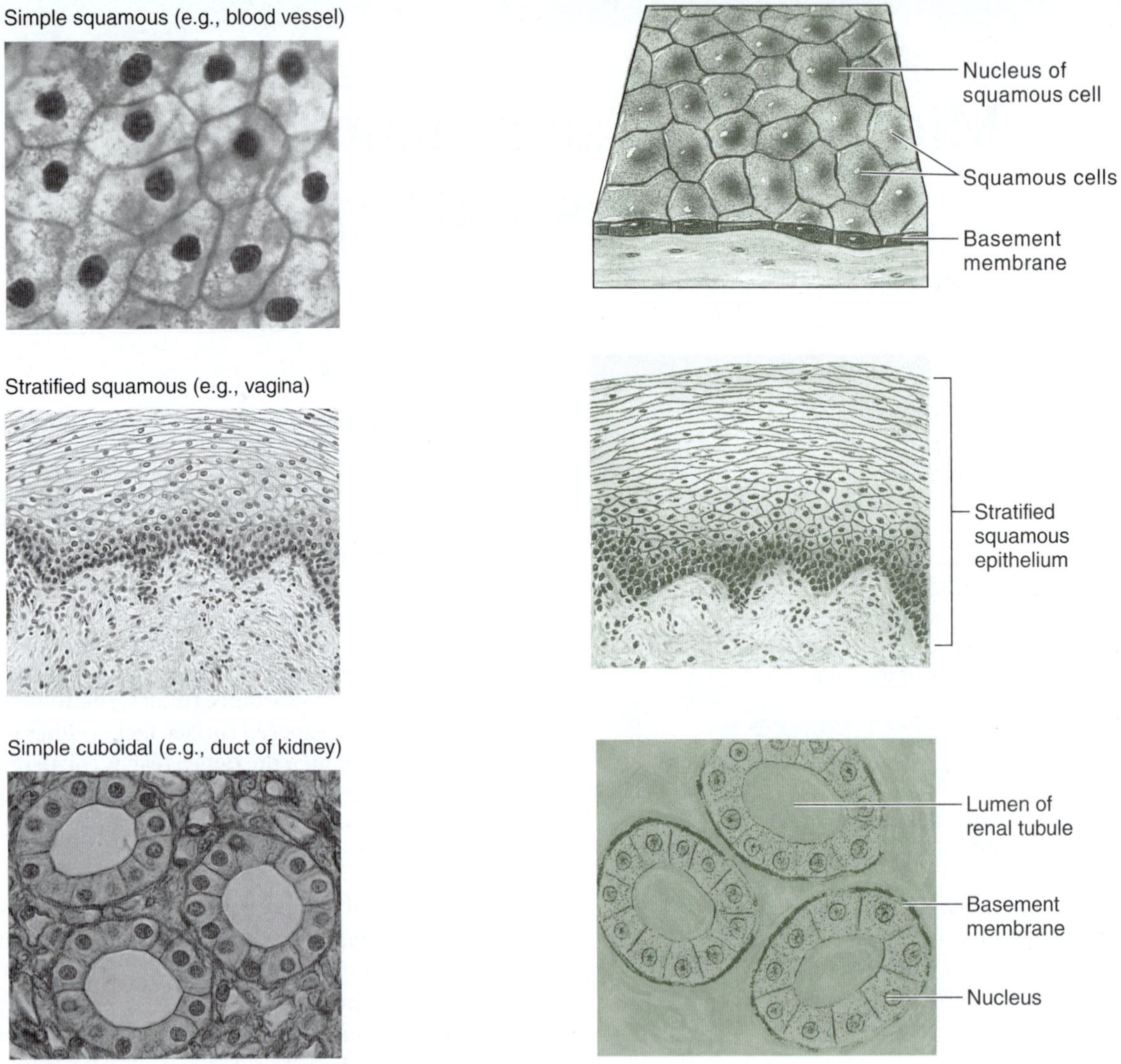

Figure 1.4 Squamous and cuboidal epithelial membranes.
The structures shown in each photomicrograph are depicted in the accompanying diagrams.

specialized to absorb the products of digestion. It also contains mucus-secreting *goblet cells*.

5. **Simple ciliated columnar epithelium** (fig. 1.5, *upper middle*). Columnar cells that support hair-like *cilia* on the exposed surface. These cilia produce wave-like movements along the luminal surface of female uterine tubes and the ductus deferens (vas deferens) of the male.

6. **Pseudostratified ciliated columnar epithelium** (fig. 1.5, *lower middle*). Simple but appears stratified because the nuclei are at different levels. Also characterized by hair-like cilia, this epithelium is found lining the respiratory passages of the trachea and bronchial tubes.

7. **Transitional epithelium** (fig. 1.5, *bottom*). Found only in the urinary bladder, ureters, and a small portion of the urethra; uniquely stratified to permit periodic distension (stretching).

Epithelial membranes cover all body surfaces. Therefore, in order for a molecule or ion to move from the external environment into the blood (and from there to the body organs), it must first pass through an epithelial membrane. The transport of digestion products across the intestinal epithelium into the blood is called *absorption*. The transport of molecules and ions out of the urinary filtrate (originally derived from the blood) back into the blood is called *reabsorption*.

Transport between cells is limited by the **junctional complexes** that connect adjacent epithelial cells. Junctional complexes consist of three structures: (1) *tight junctions*, where the plasma membranes of the two cells are physically joined together and proteins penetrate the membranes of the two cells to bridge their cytoskeletons; (2) *adherens junctions*, where the plasma membranes of the two cells come very close together and are "glued" by interactions between proteins that span each membrane and connect the cytoskeleton of each cell; and (3) *desmosomes*, where the plasma membranes of the two cells are "buttoned together" by interactions between particular desmosomal proteins (fig. 1.6).

The extent to which the junctional complexes surround each epithelial cell will determine how much

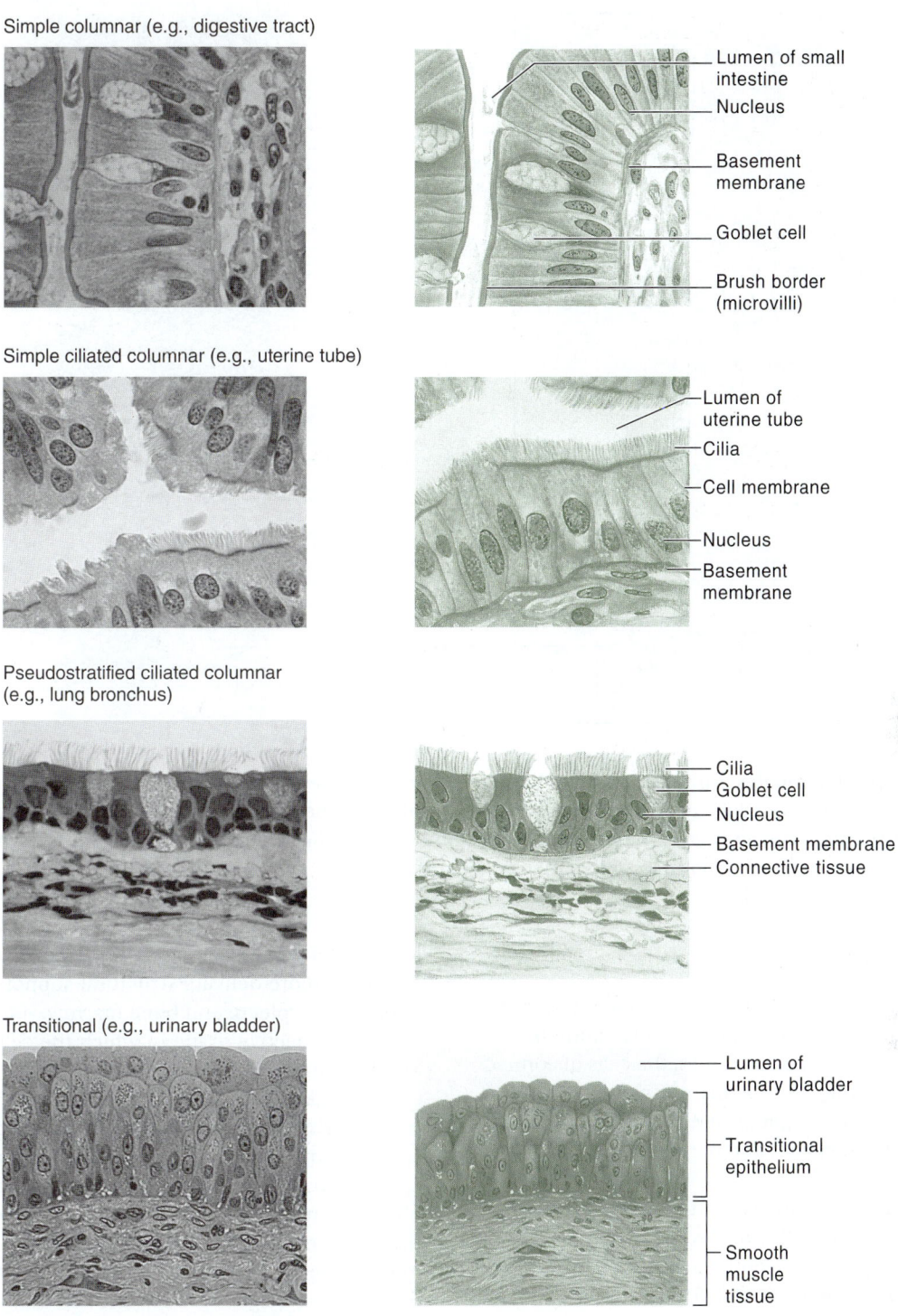

Figure I.5 Columnar and transitional epithelial membranes. The structures shown in each photomicrograph are depicted in the accompanying diagrams.

transport will be possible between them—a process termed *paracellular transport*. Some molecules and ions can also be transported in a highly selective manner across the epithelial cell's plasma membrane. This is termed *transcellular transport* because the molecules and ions transported in this way cross the cytoplasm of the epithelial cells. Membrane transport processes are described in exercise 2.6.

Procedure

1. Observe slides of the mesentery, esophagus, skin, pancreas, vas deferens or uterine tube, trachea, and urinary bladder.
2. Identify the type of epithelium in each of the slides.

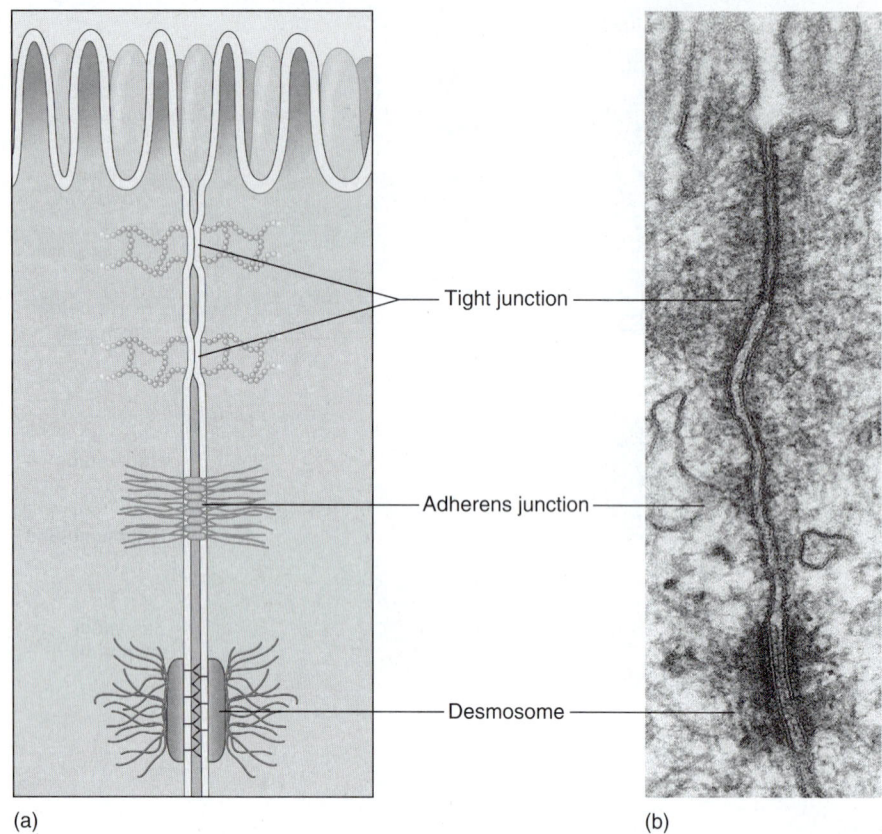

Figure 1.6 **Junctional complexes provide a barrier between adjacent epithelial cells.** Proteins penetrate the plasma membranes of two cells and are joined to the cytoskeleton of each cell. Junctional complexes consist of three components: tight junctions, adherens junctions, and desmosomes. Epithelial membranes differ, however, in the number and arrangement of these components, illustrated in (a) and shown in an electron micrograph in (b).

(For a full-color version of this figure, see fig. 6.22 in *Human Physiology*, thirteenth edition, by Stuart I. Fox.)

B. Connective Tissues

Connective tissue is characterized by abundant amounts of extracellular material, or *matrix*. Unlike epithelial tissue, composed of tightly packed cells, the cells of connective tissue (which may be of many types) are spread out. The large extracellular spaces in connective tissue provide room for blood vessels and nerves to enter and leave organs.

There are five major types of connective tissues: (1) *mesenchyme,* an undifferentiated tissue found primarily during embryonic development; (2) *connective tissue proper;* (3) *cartilage;* (4) *bone;* and (5) *blood.*

Connective tissue proper (fig. 1.7) refers to a broad category of tissues with a somewhat loose, flexible matrix. This tissue may be *loose (areolar),* which serves as a general binding and packaging material in such areas as the skin and the fascia of muscle, or *dense,* as is found in tendons and ligaments. The degree of denseness relates to the relative proportion of protein fibers to the semi-fluid *ground substance* in the matrix. These protein fibers may be made of *collagen,* which gives tensile strength to tendons and ligaments; they may be made of *elastin (elastic fibers),* which are prominent in large arteries and the lower respiratory system; or they may be *reticular fibers* providing more delicate structural support to the lymph nodes, liver, spleen, and bone marrow. *Adipose tissue* is a type of connective tissue in which the cells *(adipocytes)* are specialized to store fat.

Cartilage consists of cells *(chondrocytes)* and a semi-solid matrix that imparts strength and elasticity to the tissue. The three types of cartilage are shown in figure 1.8. *Hyaline cartilage* has a clear matrix that stains a uniform blue. The most abundant form of cartilage, hyaline cartilage is found on the articular surfaces of bones (commonly called "gristle"), in the trachea, bronchi, nose, and the costal cartilages between the ventral ends of the first ten ribs and the sternum. *Fibrocartilage* matrix is reinforced with collagen fibers to resist compression. It is found in the symphysis pubis, where the two pelvic bones articulate, and between the vertebrae, where it forms intervertebral discs. *Elastic cartilage* contains abundant elastic fibers for flexibility. It is found in the external ear, portions of the larynx, and in the auditory canal (eustachian tube).

Bone (fig. 1.9*a*) contains mature cells called *osteocytes,* surrounded by an extremely hard matrix impregnated with calcium phosphate. Arranged in concentric layers, the osteocytes surround a *central canal,* containing nerves and blood vessels, and obtain nourishment via small channels in the matrix called *canaliculi.*

Blood (fig. 1.9b) is considered a unique type of connective tissue because its extracellular matrix is fluid *(plasma)* that suspends and transports blood cells *(erythrocytes, leukocytes, and thrombocytes)* within blood vessels. The composition of blood will be described in more detail in later exercises.

Procedure

1. Observe slides of skin, mesentery, a tendon, the spleen, cartilage, and bone.
2. Identify the types of connective tissue in each slide.

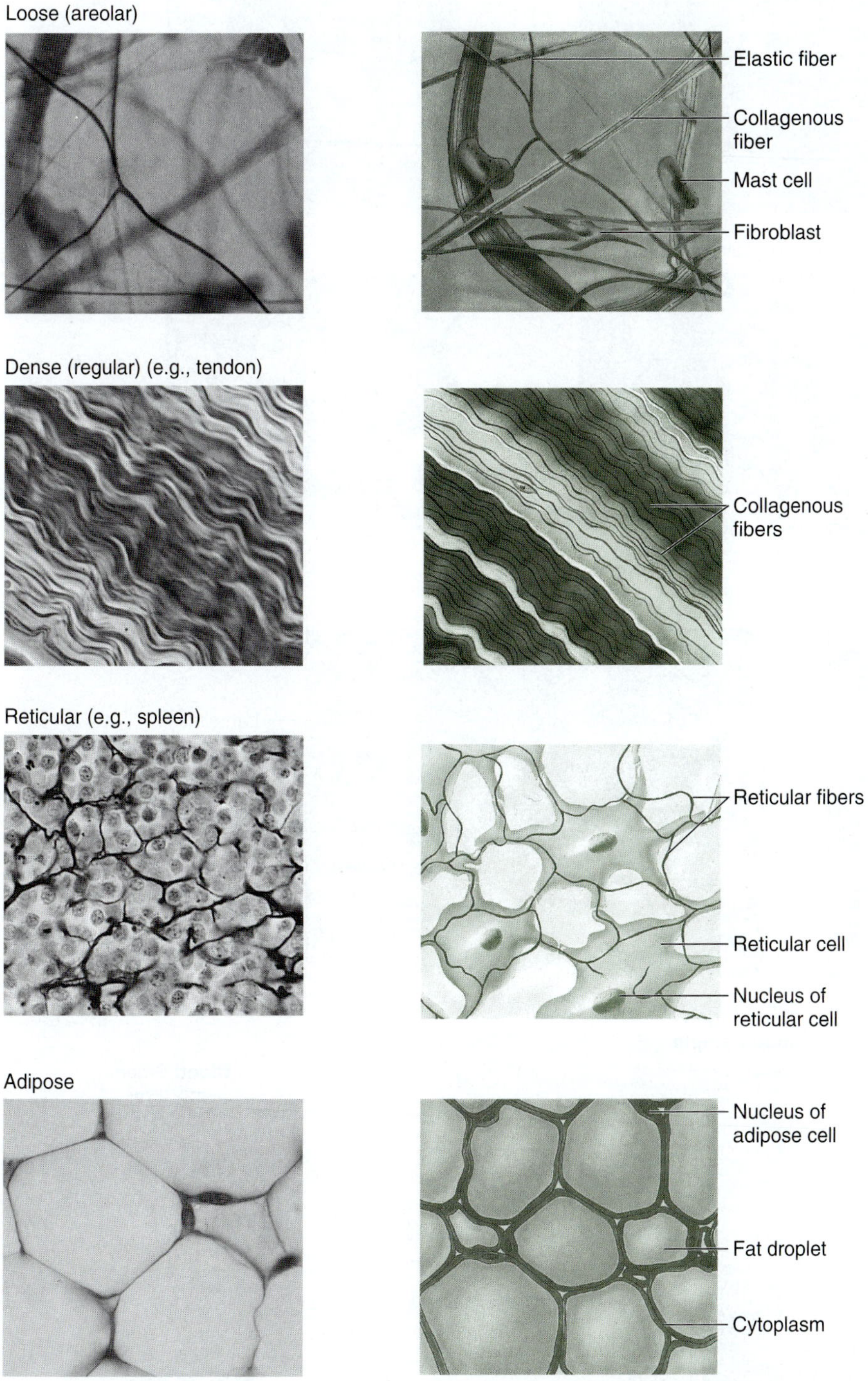

Loose (areolar)

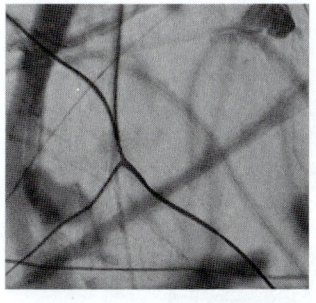

Dense (regular) (e.g., tendon)

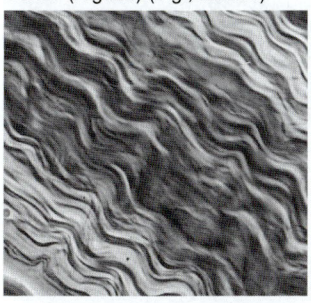

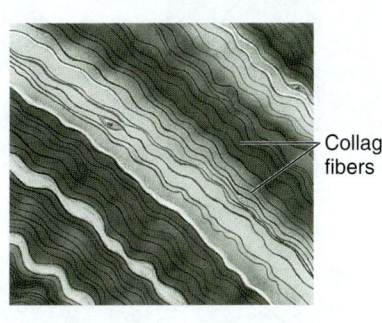

Reticular (e.g., spleen)

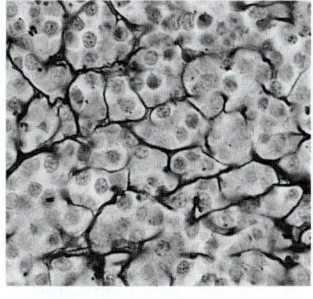

Adipose

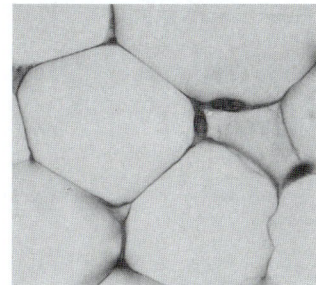

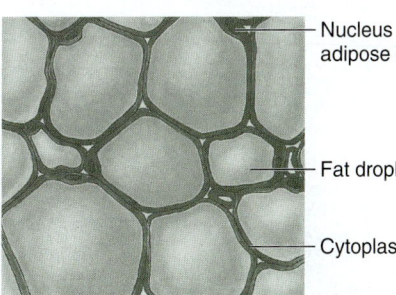

Figure 1.7 Connective tissue proper. The structures shown in each photomicrograph are depicted in the accompanying diagrams.

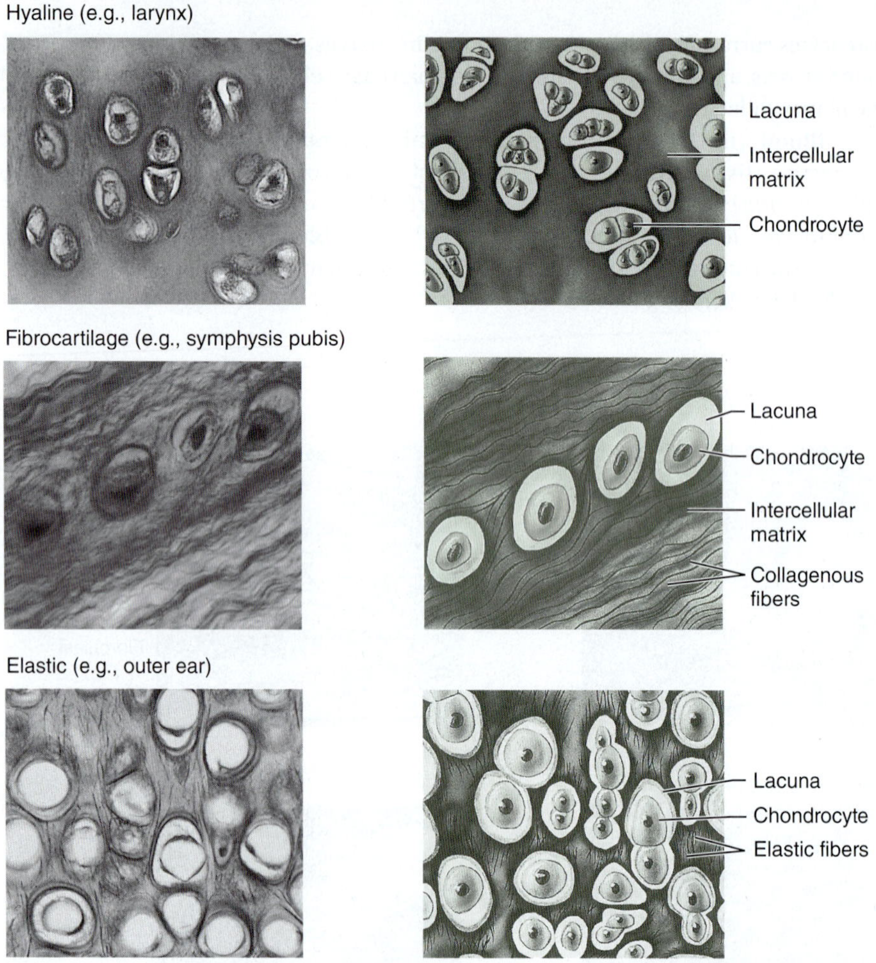

Figure 1.8 **Different forms of cartilage.** The structures shown in each photomicrograph are depicted in the accompanying diagrams.

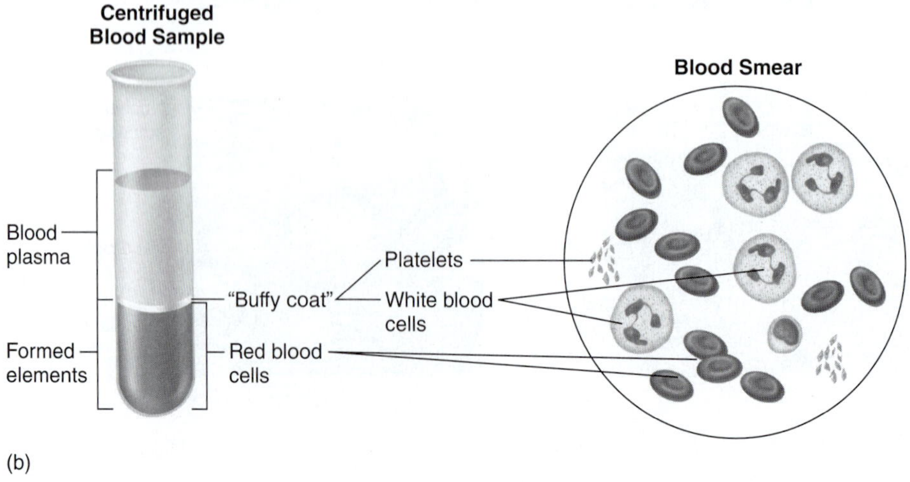

Hyaline (e.g., larynx)

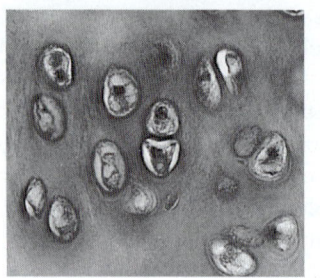

Fibrocartilage (e.g., symphysis pubis)

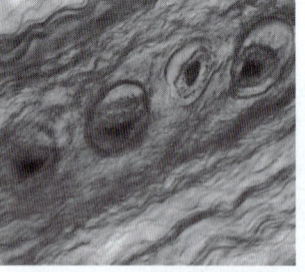

Elastic (e.g., outer ear)

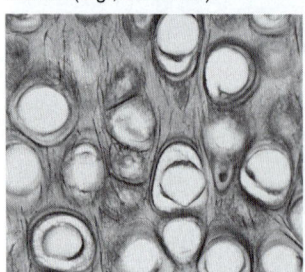

Bone (osseous)

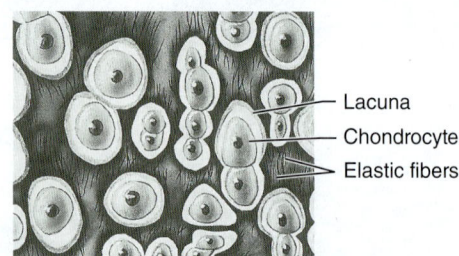

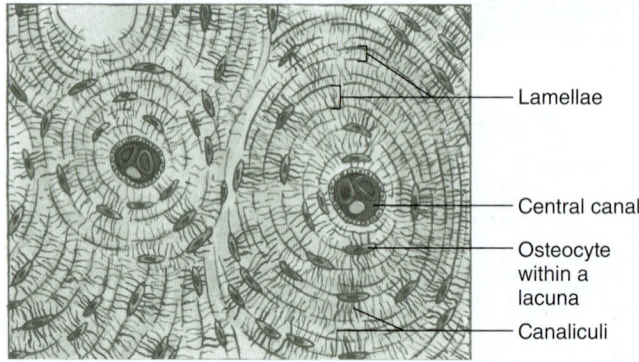

(a)

Figure 1.9 **Bone and blood.** (a) The structure of bone is shown in a photomicrograph and illustration. (b) The constituents of a centrifuged sample of blood are illustrated. When centrifuged, the solid elements—consisting of red blood cells, white blood cells, and platelets—are packed at the bottom of the tube, separating them from the fluid plasma. AP|R

(For a full-color version of part [b] of this figure, see fig. 13.1 in *Human Physiology*, thirteenth edition, by Stuart I. Fox.)

(b)

C. Muscle Tissue

Muscles are responsible for heat production; body posture and support; and for a wide variety of movements, including locomotion. Muscle tissues, which are contractile, are composed of muscle cells, or *fibers,* elongated in the direction of contraction. The three types of muscle tissues—*smooth, cardiac,* and *skeletal*—are shown in figure 1.10.

Smooth muscle tissue is found in the digestive tract, blood vessels, respiratory passages, and the walls of the urinary and reproductive ducts. Smooth muscle fibers are long and spindle shaped, with a single nucleus near the center. **Cardiac muscle** tissue, found in the heart, is characterized by striated fibers that are branched and interconnected by *intercalated discs.* These interconnections allow electrical impulses to pass from one *myocardial (heart muscle) cell* to the next. **Skeletal muscle** tissue attaches to the skeleton and is responsible for voluntary movements. Skeletal muscle fibers are long and thin and contain numerous nuclei. Skeletal muscle is under voluntary control, whereas cardiac and smooth muscles are classified as involuntary. This distinction relates to the type of nerves involved (innervation) and not to the characteristics of the muscles. Both skeletal muscle and cardiac muscle cells are categorized as *striated muscle* because they contain cross striations.

Procedure

1. Observe prepared slides of smooth, cardiac, and skeletal muscles.
2. Identify the major distinguishing features of each type of muscle.

D. Nervous Tissue

Nervous tissue, which forms the nervous system, consists of two major categories of cells. The nerve cell, or **neuron** (fig. 1.11), is the functional unit of the nervous system. The typical neuron has a *cell body* with a nucleus; projections called *dendrites* branching from the cell body; and a single cytoplasmic extension called an *axon,* or *nerve fiber.* Dendrites and axons are best distinguished functionally:

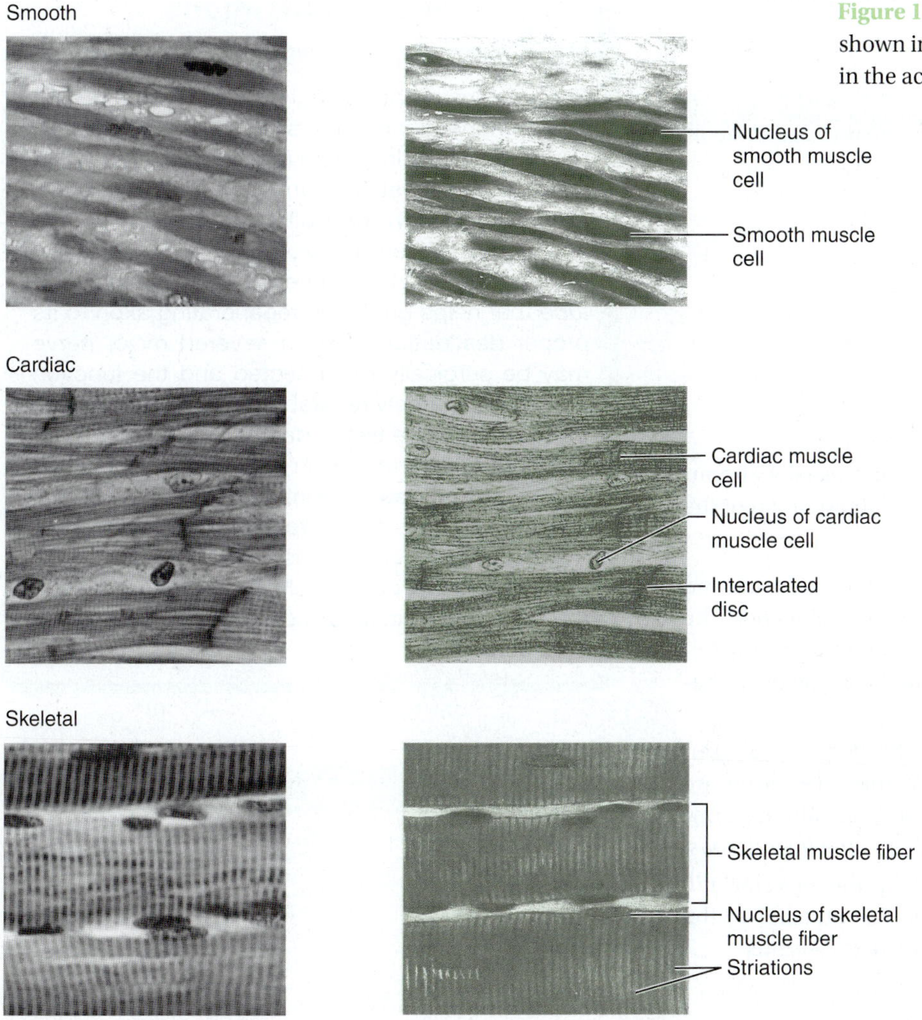

Figure 1.10 **Muscle tissue.** The structures shown in each photomicrograph are depicted in the accompanying diagrams.

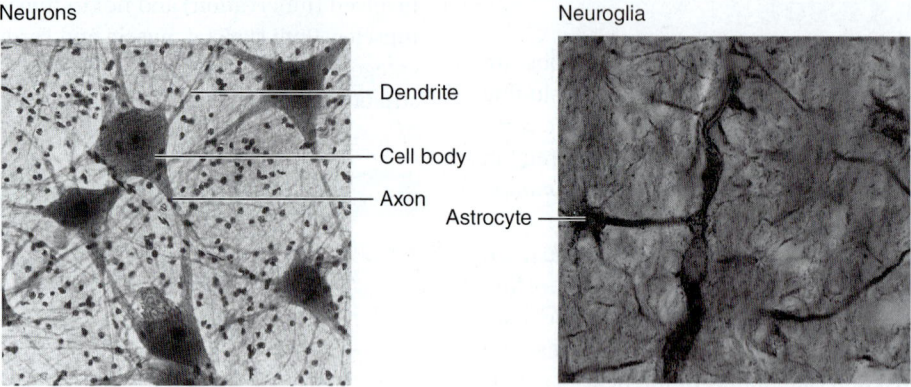

Figure 1.11 **Nervous tissue.** Photomicrographs of representative neurons and neuroglia in the CNS.

dendrites conduct impulses to the cell body, whereas the axon conducts impulses away from the cell body. The neuron is generally capable of receiving, producing, and conducting electrical impulses. Most neurons release specialized chemicals from the axon endings. A second category of cell found in the nervous system is a **neuroglial cell.** Various types of neuroglia support the neurons structurally and functionally.

Procedure

1. Observe prepared slides of the spinal cord and the brain.
2. Identify the parts of a neuron.
3. Distinguish neurons from neuroglial cells.
4. Without referring to the caption, identify the various tissue types in the photomicrographs in figure 1.11.

E. An Organ: The Skin

Organs contain more than one type—and usually all four types—of primary tissue. The **skin,** the largest organ of the body, provides an excellent example.

- Epithelial tissue is represented by the *epidermis* and the *hair follicles* (fig. 1.12). Like all glands, the oily *sebaceous glands* associated with hair follicles and the *sweat glands* are a type of epithelial tissue.
- Connective tissue is seen in the *dermis*. Collagen fibers that form dense connective tissue are located in the dermis, whereas adipose connective tissue is embedded in the *hypodermis*.
- Muscle tissue is represented by the *arrector pili muscle,* a smooth muscle that attaches to the hair follicle and the matrix of the dermis.

Nervous tissue is featured within skin by the sensory and motor nerves and by *Meissner's corpuscle* (the oval structure in the dermis near the start of the sensory nerve, fig. 1.12), a sensory structure sensitive to textures and vibrations.

 Clinical Applications

Some axons of the central nervous system (CNS) and peripheral nervous system (PNS) are surrounded by myelin sheath (are *myelinated*); others lack a myelin sheath (are *unmyelinated*). Neuroglial cells called **Schwann cells** form myelin sheaths in the PNS. When an axon in a peripheral neuron is cut, the Schwann cells form a *regeneration tube* that helps guide the regenerating axon to its proper destination. Even a severed major nerve may be surgically reconnected and the function of the nerve largely reestablished, if the surgery is performed before tissue death. Neuroglial cells of the CNS that form myelin sheaths are known as **oligodendrocytes.** In contrast to Schwann cells, oligodendrocytes do not form regeneration tubes. For this and other reasons that are incompletely understood, cut or severely damaged neurons of the brain and spinal cord usually result in permanent damage.

Procedure

1. Observe a prepared slide of the skin or scalp.
2. Identify the structures of the skin, and try to find all four types of primary tissue.

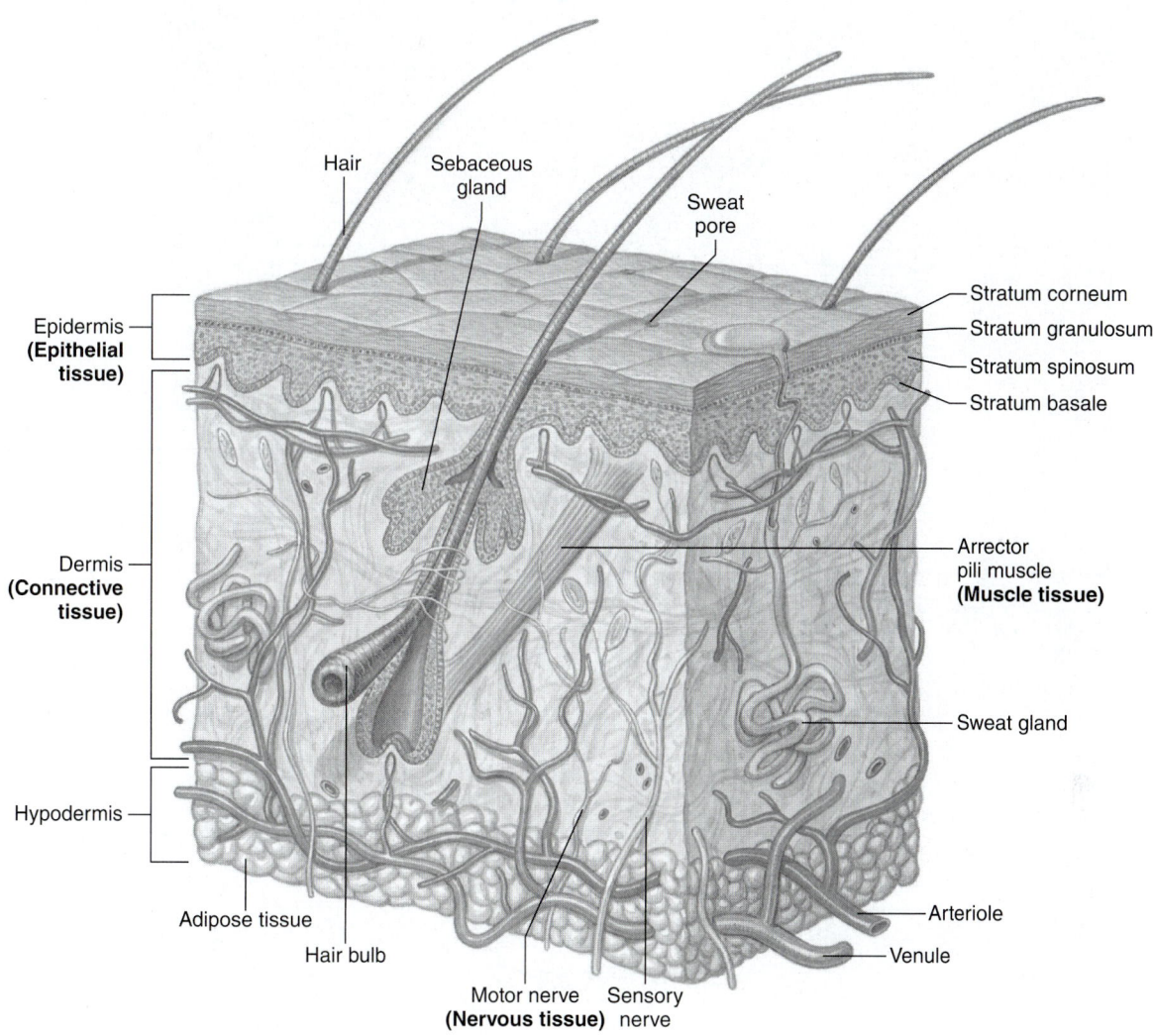

Figure 1.12 **A diagram of the skin.** The skin is an organ that contains all four types of primary tissues. AP|R

(For a full-color version of this figure, see fig. 1.22 in *Human Physiology,* thirteenth edition, by Stuart I. Fox.)

Laboratory Report 1.2

Name _____
Date _____
Section _____

REVIEW ACTIVITIES FOR EXERCISE 1.2

Test Your Knowledge

1. Define *tissue*. _____
2. Define *organ*. _____
3. Describe and give examples of these epithelial membranes:
   (a) simple squamous

   (b) stratified squamous

   (c) simple columnar

   (d) pseudostratified

4. What is the common characteristic of connective tissues? _____
   _____

5. Fill in the blanks.
   (a) Two examples of dense connective tissues are _____ and _____.
   (b) The connective tissue of the dermis is classified as _____.
   (c) Hyaline cartilage is found in the _____.
   (d) Fibrocartilage is found in the _____.

6. Skeletal and cardiac muscles are categorized as _____ muscles.
7. What type of muscle fiber is found in the walls of blood vessels? _____

Test Your Understanding

8. Compare and contrast the structure and function of
   (a) the epithelium of the skin and the epithelium of the intestine

23

(b) cardiac muscle and skeletal muscle

9. Identify the distinguishing characteristic of connective tissues. Give examples of three connective tissues, and describe how they fit into the connective tissue category.

Test Your Analytical Ability

10. Would you expect the muscle fibers of the tongue to be striated or smooth? What about the muscle of the diaphragm? Explain your answer.

11. Blood vessels and nerves travel in connective tissues, not between the cells of epithelial membranes. Why? Would you expect to see strands of connective tissue within the pancreas and liver? Explain your answer.

EXERCISE 1.3

Homeostasis and Negative Feedback

MATERIALS

1. Watch or clock with a second hand
2. Hot plate; beaker; thermometer; crushed ice; constant-temperature water bath

The regulatory mechanisms of the body help to maintain a state of dynamic constancy of the internal environment known as homeostasis. Most systems of the body maintain homeostasis by operating negative feedback mechanisms that control effectors (muscles and glands).

LEARNING OUTCOMES

You should be able to:

1. Define the term *homeostasis*.
2. Explain how the negative feedback control of effectors helps to maintain homeostasis.
3. Explain why the internal environment is in a state of dynamic, rather than static, constancy.
4. Define the terms *set point* and *sensitivity*.
5. Explain how a normal range of values for temperature or heart rate is obtained, and discuss the significance of these values.

Textbook/Multimedia Correlations

Before performing this exercise, you should study the introductory material presented here. Further information relating to this exercise can be found in these pages of *Human Physiology*, thirteenth edition, by Stuart I. Fox:

- *Negative Feedback Loops.* Chapter 1, p. 6.
- *Feedback Control of Hormone Secretion.* Chapter 1, p. 9.
- Multimedia correlations are provided in *MediaPhys 3.0:* Topics 1.3–1.9

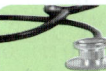

Clinical Investigation

A patient was examined by a nurse and found to have a temperature of 100.9°C and a resting pulse of 100 beats per minute. A laboratory test measured a fasting blood glucose concentration of 85 mg/dL.

- Describe where these patient values fall relative to the normal range and what they may indicate.
- Explain, using the terms *sensor, integrating center,* and *effector,* how these clinical measurements relate to homeostasis, health, and disease.

Although the structure of the body is functional, the study of body function involves much more than a study of body structure. The extent to which each organ performs the functions endowed by its genetic programming is determined by regulatory mechanisms that coordinate body functions in the service of the entire organism. The primary prerequisite for a healthy organism is the maintenance of **homeostasis,** the dynamic constancy of the internal environment.

For internal constancy to be maintained, changes in the body must stimulate **sensors** that can send information to an **integrating center.** This allows the integrating center to detect changes from a **set point.** The set point is analogous to the temperature set on a house thermostat. In a similar manner, there is a set point for body temperature, blood glucose concentration, the tension on a tendon, and so on. The integrating center is often a particular region of the brain or spinal cord, but it can also be a group of cells in an endocrine gland. A number of different sensors may send information to a particular integrating center, which can then integrate this information and direct the responses of **effectors**—generally, muscles or glands. The integrating center may cause increases or decreases in effector action to counter the deviations from the set point and defend homeostasis. Homeostasis is thus maintained as a state of *dynamic,* rather than absolute, constancy (fig. 1.13). Homeostasis of body temperature, for example, is maintained about a set point average of 37°C by physiological mechanisms (sweating, shivering, and others) that operate in a negative feedback manner (discussed next).

25

Since a disturbance in homeostasis initiates events that lead to changes in the opposite direction, the cause-and-effect sequence is described as a **negative feedback mechanism** (or a *negative feedback loop*). A constant-temperature water bath, for example, uses negative feedback mechanisms to maintain the temperature at which the bath is set (the **set point**). Deviations from the set point are detected by a thermostat (temperature sensor), which turns on a heating unit (the effector) when the temperature drops below the set point and turns off the unit when the temperature rises above the set point (fig. 1.13).

By means of the negative feedback control of the heating unit, the water-bath temperature is not allowed to rise or fall too far from the set point. Keep in mind, however, that the temperature of the water is at the set point only briefly. The set point is only the *average value* within a *range* (from the highest to the lowest value) of temperatures. The *sensitivity* of this negative feedback mechanism is measured by the temperature deviation from the set point required to activate the compensatory (negative feedback) response (turning the heater on or off).

As you might expect, **positive feedback** has the opposite effect to negative feedback. Whereas negative feedback combats changes, positive feedback acts to increase or accelerate changes. An avalanche is an example of positive feedback. Clearly, homeostasis requires negative rather than positive feedback mechanisms. There are some examples of positive feedback mechanisms in the body, however. For example, injury to a blood vessel causes an avalanche-like sequence of changes that ends in the formation of a blood clot. The clot itself, however, stops the continued leakage of blood and so maintains homeostasis. There is also a positive feedback mechanism that culminates in the expulsion of a mature egg cell from the ovary in the process of ovulation, and positive feedback mechanisms that promote increasingly strong contractions of the uterus during labor.

A. Negative Feedback Control of Water-bath Temperature

Modern constant-temperature water-baths are so sensitive that tiny deviations from the set point are quickly corrected by very efficient electronic mechanisms. This makes it difficult to observe the negative feedback corrections. However, you can observe the way that the sensor and effector (heating element) work to defend constancy of temperature by adding cold water or crushed ice to the bath. This is analogous to the way that negative feedback mechanisms in your body maintain homeostasis of body temperature when you are exposed to a cold ambient temperature.

A variation on this theme of temperature homeostasis can be performed using a hot plate, a beaker of water, and a thermometer. In this procedure, you are the sensor and the integrating center because you adjust the effector (hot plate) to maintain a constant temperature of the water. You can also observe the effectiveness of **antagonistic effectors** by using some crushed ice (as well as decreased heat settings of the hot plate) to lower the water temperature. This is analogous to the antagonistic actions of sweating (which lowers body temperature) and shivering (which raises it).

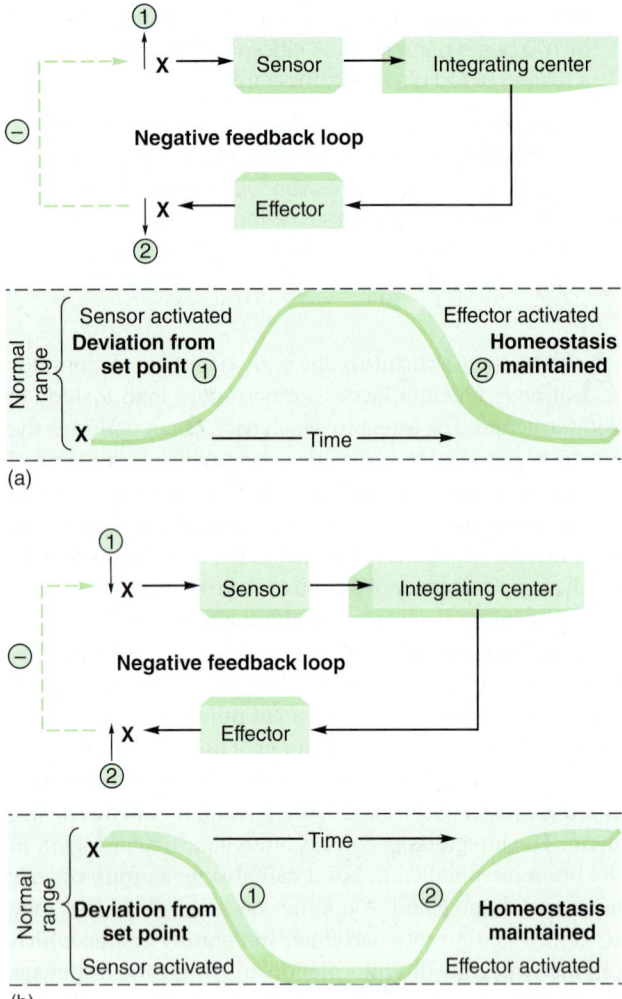

Figure 1.13 Homeostasis is maintained by negative feedback loops. (a) A rise in some factor (1) in the internal environment is corrected (2) by mechanisms that bring it back down. (b) A fall in some factor (1) in the internal environment is corrected (2) by mechanisms that help to raise it. In both cases, the negative feedback loops compensate for the changes to help maintain a state of dynamic constancy.

(For a full-color version of this figure, see figs. 1.1 and 1.2 in *Human Physiology*, thirteenth edition, by Stuart I. Fox.)

Procedure

1. The temperature of the constant-temperature water-bath is set by the instructor at a certain temperature, somewhere between 40° C and 60° C. A red indicator light goes on when the heating element is activated and off when the heating element is inactivated.

2. Observe the temperature of the constant-temperature water-bath over the course of 5 minutes, recording its temperature once each minute in your laboratory report. Use these measurements to determine the set point of the water-bath (this can be taken as the average of your measurements). Record this value in the laboratory report.

3. Add a fairly large amount of cold water or crushed ice to the constant-temperature water-bath. Record the temperature of the water-bath once a minute for 5 minutes after adding the cold water or crushed ice. Continue observing until the water temperature has returned to the previously established set point. In your laboratory report, record how long this takes.

4. Using a hot plate, a beaker of tap water, and a thermometer, attempt to heat the water to 37° C and then maintain this temperature by either decreasing or increasing the heat setting on the hot plate. After the 37° C temperature was first attained, record the temperature readings of the water every 5 minutes for 30 minutes.

5. Now, use some crushed ice (as well as a decreased setting on the hot plate) to lower the temperature once the water temperature exceeds 37° C. Record the temperature readings of the water every 5 minutes for another 30 minutes.

B. Resting Pulse Rate: Negative Feedback Control and Normal Range

Homeostasis—the dynamic constancy of the internal environment—is maintained by negative feedback mechanisms far more complex than those involved in maintaining a constant-temperature water bath. In most cases, several effectors, many with antagonistic actions, are involved in maintaining homeostasis. It is as if the temperature of a water bath were determined by the antagonistic actions of both a heater and a cooling system.

However, this simple example gives a wrong impression: the effectors in the body are generally increased or decreased in activity, *not* just turned on or off. Because of this, negative feedback control in the body works far more efficiently than a thermostat. For example, the blood pressure is greatly affected by the cardiac rate (and thus the pulse rate), which in turn is largely determined by the levels of activity of two antagonistic nerves. One of these (a *sympathetic nerve,* described in section 7) stimulates an increase in cardiac rate. A different nerve (a *parasympathetic nerve*) produces inhibitory effects that slow the cardiac rate.

We have negative feedback loops that help to maintain homeostasis of arterial blood pressure, in part by adjusting the heart rate, or pulse rate. If everything else is equal, the blood pressure is lowered by a decreased heart rate and is raised by an increased heart rate. This is accomplished by regulating the activity of the sympathetic and parasympathetic divisions of the autonomic nervous system. Thus, a fall in blood pressure—produced daily as we go from a lying to a standing position—is compensated by a faster heart rate (fig. 1.14). As a consequence of this negative feedback loop, our heart rate varies as we go through our day (within a normal range of 60–90 beats per minute), speeding up and slowing down, so that we can maintain homeostasis of blood pressure and keep it within normal limits.

Therefore, the resting pulse rate (measured in *beats per minute*) is not absolutely constant but instead varies about a set point value. This exercise will demonstrate that your pulse rate is in a state of dynamic constancy (implying negative feedback controls). From the data you can determine your pulse-rate set point as the average value of the measurements.

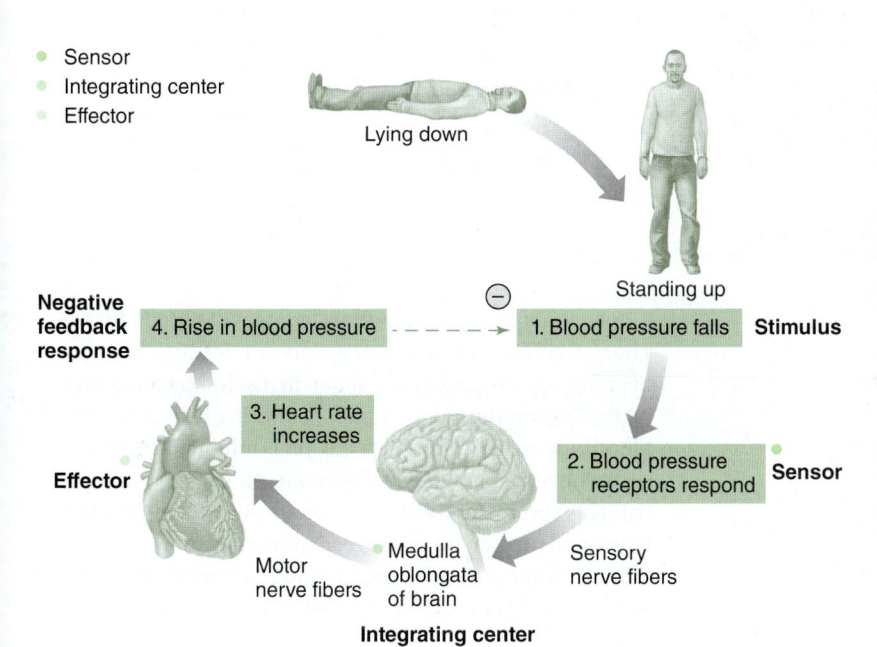

Figure 1.14 **Negative feedback control of blood pressure.** Blood pressure influences the activity of sensory neurons from the blood pressure receptors (sensors); a rise in pressure increases the firing rate of nerve impulses, and a fall in pressure decreases the firing rate. When a person stands up from a lying-down position, the blood pressure momentarily falls. The resulting decreased firing rate of nerve impulses in sensory neurons affects the medulla oblongata of the brain (the integrating center). This causes the motor nerves to the heart (effector) to increase the heart rate, helping to raise the blood pressure.

(For a full-color version of this figure, see fig. 1.6 in *Human Physiology,* thirteenth edition, by Stuart I. Fox.)

- Sensor
- Integrating center
- Effector

27

Procedure

1. Gently press your index and middle fingers (not your thumb) against the radial artery in your wrist until you feel a pulse. Alternatively, the carotid pulse in the neck may be used for these measurements.
2. The pulse rate is usually expressed as beats per minute. However, only the number of beats per 15-second interval (quarter minute) need be measured; multiplying this by four gives the number of beats per minute. Record the number of beats per 15-second interval in the data table provided in the laboratory report.
3. Pause 15 seconds, and then count your pulse during the next 15-second interval. Repeat this procedure over a 5-minute period. Recording your count once every half minute for 5 minutes, a total of 10 measurements (expressed as beats per minute) will be obtained.
4. Using the grid provided in the laboratory report, graph your results by placing a dot at the point corresponding to the pulse rate for each measurement, and then connect the dots.

> **Normal Values** Students often ask, How do my measurements compare with those of others? and Are my measurements normal? The normal resting pulse is 60–90 beats per minute.

Normal values are those that healthy people have. Since healthy people differ to some degree in their particular values, what is considered normal is usually expressed as a range of values that encompasses the measurements of most healthy people. An estimate of the **normal range** is a statistical determination subject to statistical errors and also subject to questions about what is meant by the term *healthy*.

Healthy, in this context, means the absence of cardiovascular disease. Included in the healthy category, however, are endurance-trained athletes, who usually have lower than average cardiac rates, and relatively inactive people, who have higher than average cardiac rates. For this reason, determinations of normal ranges can vary, depending on the relative proportion of each group in the sample tested. A given class of students may therefore have an average value and a range of values that differ somewhat from those of the general population.

Procedure

1. Each student in the class determines his or her average cardiac rate (pulse rate) from the previous data either by taking an arithmetic average or by observing the average value of the fluctuations in the previously constructed graph. Record your average in the laboratory report.
2. Record the number of students in the class with average pulse rates in each of the rate categories shown in the laboratory report. Also, calculate the percentage of students in the class within each category and record this percentage in the laboratory report.
3. Divide the class into two groups: those who exercise on a regular basis (at least three times a week) and those who do not. Determine the average pulse rate and range of values for each of these groups. Enter this information in the given spaces in the laboratory report.

Clinical Applications

The concept of homeostasis is central to medical diagnostic procedures. Through the measurement of body temperature, blood pressure, concentrations of specific substances in the blood, and many other variables, the clinical examiner samples the internal environment. If a particular measurement deviates significantly from the range of normal values—that is, if that individual is not able to maintain homeostasis—the cause of the illness may be traced and proper treatment determined to bring the measurement back within the normal range.

For example, measurements of fasting blood glucose levels are commonly performed to detect **diabetes mellitus,** a disorder in the secretion or action of the hormone *insulin*. The negative feedback loops that maintain homeostasis involve clusters of cells (islets) in the pancreas that secrete the hormone insulin and another hormone, glucagon (fig. 1.15). These hormones normally maintain a fasting blood glucose concentration in the range of 70 to 99 mg/dL (milligrams per deciliter).

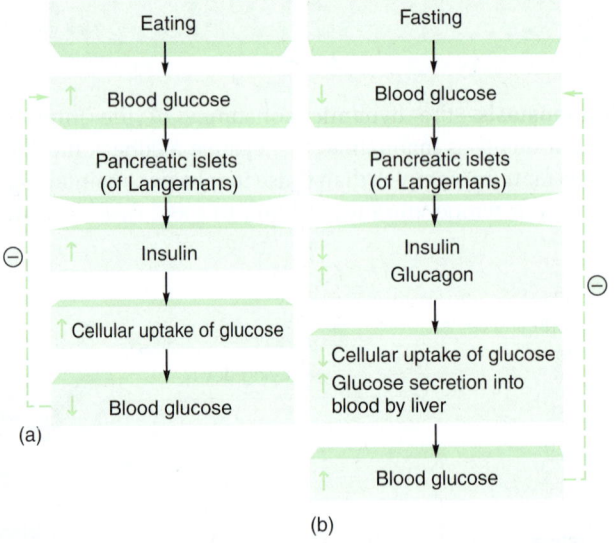

(a) Eating → Blood glucose ↑ → Pancreatic islets (of Langerhans) → Insulin ↑ → ↑Cellular uptake of glucose → ↓ Blood glucose

(b) Fasting → Blood glucose ↓ → Pancreatic islets (of Langerhans) → Insulin ↓ / Glucagon ↑ → ↓Cellular uptake of glucose / ↑Glucose secretion into blood by liver → ↑ Blood glucose

Figure 1.15 Negative feedback control of blood glucose. The rise in blood glucose that occurs after eating carbohydrates is corrected by the action of insulin, secreted in increasing amounts (a) at that time. During fasting, when blood glucose falls, insulin secretion is inhibited and the secretion of an antagonistic hormone, glucagon, is increased (b). This stimulates the liver to secrete glucose into the blood, helping to prevent blood glucose from continuing to fall. In this way, blood glucose concentrations are maintained within a homeostatic range following eating and during fasting.

(For a full-color version of this figure, see fig. 1.7 in *Human Physiology,* thirteenth edition, by Stuart I. Fox.)

Laboratory Report 1.3

Name _____

Date _____

Section _____

DATA FROM EXERCISE 1.3

A. Negative Feedback Control of Water-Bath Temperature

1. Measurements of constant-temperature water-bath temperature (once each minute for 5 minutes):
   _____; _____; _____; _____; _____.

2. Calculate the **average** temperature of the constant-temperature water-bath: _____. (*Note:* Do this by adding all of your measurements and dividing by 5, the total number of measurements you made.) This is taken as the set point of the water-bath.

3. What is the **range** of values in the 5 measurements? _____ to _____. (*Note:* This is the lowest to the highest of your measurements.)

4. What was the temperature of the constant-temperature water-bath after adding the cold water or crushed ice? _____

5. How long did it take for the temperature to return to the set point? _____

6. Record the temperature reading of the water in the beaker on the hot plate after first attaining 37° C (one measurement every 5 minutes for 30 minutes, adjusting only the hot plate settings): _____; _____; _____; _____; _____; _____.

    (a) Calculate the average of these measurements: _____.

    (b) Indicate the range of these measurements: _____.

7. Record the temperature reading of the water in the beaker on the hot plate (one measurement every 5 minutes for 30 minutes) when you used crushed ice as well as lower hot plate settings to lower the water temperature:
   _____; _____; _____; _____; _____.

    (a) Calculate the average of these measurements: _____.

    (b) Indicate the range of these measurements: _____ to _____.

8. Which negative feedback mechanism was more effective at maintaining constant temperature—the electronic mechanism of the constant-temperature water-bath (measurements in no. 1) or the human manual methods (measurements in nos. 6 and 7)?
   _____

9. Which was more effective at maintaining a constant temperature—adjustments of one effector only (the hot plate) or adjustments of antagonistic effectors (hot plate and crushed ice)?
   _____

B. Resting Pulse Rate: Negative Feedback Control and Normal Range

Measurement	1	2	3	4	5	6	7	8	9	10
Beats per 15 seconds										
Beats per minute										

Pulse rate (beats per minute) vs. Measurements (1–10); y-axis 50 to 100.

29

1. Calculate your **average** pulse rate: _____ pulses/minute
   (*Note:* Do this by adding all of your measurements and dividing by 10, the total number of measurements you made.)
2. What is the **range** of values in the 10 measurements? _____ pulses/minute to _____ pulses/minute
   (*Note:* This is the lowest to the highest of your measurements.)
3. What is the **sensitivity** of your negative feedback mechanism maintaining homeostasis of your pulse rate?
   ± _____ pulses/minute
   (*Note:* To obtain this, first take the difference between your average pulse rate and each of your individual measurements by subtracting the higher from the lower value. Then, add all of the differences together. Finally, divide this sum by 10, the number of measurements.)
4. Pulse rate values of the class:

Pulse Rate (beats per minute)	Number of Students	Percentage of the Total
Over 100 bpm		
90–100		
80–89		
70-79		
60-69		
50-59		
Under 50 bpm		

5. Data for the exercise and nonexercise groups:

	Exercise Group	Nonexercise Group
Individual pulse rate averages		
Range of pulse rates		
Average of pulse rates		

REVIEW ACTIVITIES FOR EXERCISE 1.3

Test Your Knowledge

1. Define *homeostasis*. _____
   _____
2. Define *set point*. _____
   _____

Test Your Understanding

3. Draw a flow diagram, illustrating cause and effect with arrows, to show how constant temperature is maintained in a water bath. (*Note:* Flow diagrams are pictorial displays of processes that occur in sequence, using arrows to indicate the direction of cause-and-effect sequences.) Use the terms *sensor, integrating center,* and *effector* in your answer.

4. Suppose that a constant-temperature water-bath contained two antagonistic effectors: a heater and a cooler. Draw a flow diagram to show how this dual system could operate to maintain a constant temperature about some set point.

5. Distinguish positive feedback from negative feedback. Does positive feedback occur in the body? If so, give two examples.

Test Your Analytical Ability

6. Explain why your graph of pulse-rate measurements suggests the presence of negative feedback control mechanisms.

7. Sympathetic nerves to the heart increase the rate of beat, while parasympathetic nerves decrease the rate of beat. Draw a negative feedback loop showing how sympathetic and parasympathetic nerves are affected in someone experiencing a fall in blood pressure (the initial stimulus). (*Note:* The sensor detects the fall in blood pressure.)

8. Why would there be different published values for the normal range of a particular measurement? Why do these values have to be updated periodically?

Clinical Investigation Questions

9. How do the patient's measurements relate to the normal ranges for these measurements? What do the patient's measurements suggest?

10. What type of physiological mechanism is responsible for homeostasis of body temperature, pulse rate, and fasting blood glucose? Explain.

Section 2

Cell Function and Biochemical Measurements

Physiological control systems maintain homeostasis of the internal chemical environment to which the organ systems are exposed. The concentrations of *glucose, protein,* and *cholesterol* in plasma (the fluid portion of the blood), for example, are maintained within certain limits despite the variety in dietary food selections and variations in our eating schedules. This regulation is necessary for health. If plasma glucose levels fall too low, for example, the brain may "starve" and a coma may result. A drop in plasma protein, as another example, may disturb the normal distribution of fluid between the blood and tissues. An abnormal rise in these values, or other abnormal changes in the chemical composition of plasma, can endanger a person's health in various ways.

Abnormal changes in the internal chemical environment, which can contribute to disease processes, are usually the result of disorders in cell function. For example, because most plasma proteins are produced by liver cells, diseases of the liver can result in the lowering of plasma protein concentrations. Similarly, abnormal lowering of plasma glucose levels may result from excess secretion of the hormone insulin by certain cells of the pancreas. Thus, homeostasis of the internal chemical environment depends on proper cell function.

All of the molecules found in the body's internal environment, aside from those few obtained directly from food, are produced within the cells. Some molecules remain within the cells; others are secreted into the tissue fluids and blood. Almost all of these molecules are produced by chemical reactions catalyzed by special proteins known as **enzymes.**

All enzymes in the body are produced within tissue cells according to information contained in the **DNA** (genes). In this way, the overall metabolism of carbohydrates, lipids, proteins, and other molecules in the cell is regulated largely by genes. Defects in these genes can result in the production of defective enzymes, which result in impaired metabolism. Thus, the study of organ system physiology is intertwined with the study of cell function and biochemistry, as well as with the study of genetics.

Proper cell function also depends upon the integrity of the plasma (cell) membrane. Composed primarily of two semifluid phospholipid layers, cell membranes can regulate the passive transport of molecules moving from higher to lower concentration by diffusion. Special membrane proteins can serve as channels for the passage of larger or more polar molecules, whereas other membrane proteins serve as carriers that require metabolic energy to "pump" molecules across the membrane "uphill" from lower to higher concentrations (a process called active transport).

Exercise 2.1	Measurements of Plasma Glucose, Cholesterol, and Protein
Exercise 2.2	Thin-Layer Chromatography of Amino Acids
Exercise 2.3	Electrophoresis of Serum Proteins
Exercise 2.4	Measurements of Enzyme Activity
Exercise 2.5	Genetic Control of Metabolism
Exercise 2.6	Diffusion, Osmosis, and Tonicity

EXERCISE 2.1

Measurements of Plasma Glucose, Cholesterol, and Protein

MATERIALS

1. Pyrex (or Kimax) test tubes, mechanical pipettors for 40 μL, 50 μL, 100 μL, and 5.0 mL volumes; and corresponding pipettes (0.10 mL and 5.0 mL total volume—see fig. 2.1)
2. Constant-temperature water-bath, set at 37° C
3. Colorimeter and cuvettes
4. Glucose kit ("Glucose LiquiColor Test," Stanbio Laboratory, Inc.)
5. Cholesterol kit ("Cholesterol LiquiColor Test," Stanbio Laboratory, Inc.)
6. Total Protein Standard (10 g/dL from Stanbio Laboratory, Inc.); the following concentrations: 2, 4, 6, 8 g/dL can be prepared by dilution
7. Biuret reagent. To a 1.0-L volumetric flask, add 45 g of sodium potassium tartrate and 15 g of $CuSO_4 \cdot 5\ H_2O$. Fill 2/3 full with 0.2N NaOH and shake to dissolve. Add 5 g of potassium iodide and fill to 1.0 L volume with 0.2N NaOH
8. Serum (Artificial "Normal" and "Abnormal Control" sera can be purchased from Stanbio Laboratory, Inc.)

The concentrations of glucose, cholesterol, and protein in plasma (or serum) can be measured using colorimetric techniques in the laboratory. Abnormal concentrations of these molecules are associated with specific disease states.

LEARNING OUTCOMES

You should be able to:

1. Describe how Beer's law can be used to determine the concentration of molecules in solution.
2. Use the formula method and graphic method to determine the concentration of molecules in plasma (serum) samples.
3. Explain the physiological roles of glucose, protein, and cholesterol in the blood.
4. Explain why abnormal measurements of plasma glucose, cholesterol, and protein are clinically significant.

Textbook/Multimedia Correlations

Before performing this exercise, you should study the introductory material presented here. Further information relating to this exercise can be found in these pages of *Human Physiology,* thirteenth edition, by Stuart I. Fox:

- *Carbohydrates and Lipids.* Chapter 2, p. 33.
- *Proteins.* Chapter 2, p. 41.
- *Exchange of Fluid Between Capillaries and Tissues.* Chapter 14, p. 457.

Clinical Investigation

A patient had a fasting plasma glucose concentration of 250 mg/dl, a fasting plasma cholesterol concentration of 300 mg/dl, and a plasma protein concentration of 9.5 g/dl.

- Determine if this plasma glucose level is normal, and identify the likely cause of abnormal levels.
- Determine if the plasma cholesterol level is normal, and describe the danger that abnormal levels present.
- Determine if the plasma protein concentration is normal, and if not, identify what might account for this measurement.

Organic molecules found in the body contain the atoms carbon (C), hydrogen (H), and oxygen (O) in various ratios, and some of these molecules also contain the atoms nitrogen (N), phosphorus (P), and sulfur (S). Many organic molecules are very large. They consist of smaller repeating subunits chemically bonded to each other. The term *monomer* refers to the individual subunits; the term *polymer* refers to the long chain formed from these repeating subunits.

When two monomers are bonded together, a molecule of water (HOH) is released. This reaction is called **condensation,** or **dehydration synthesis.**

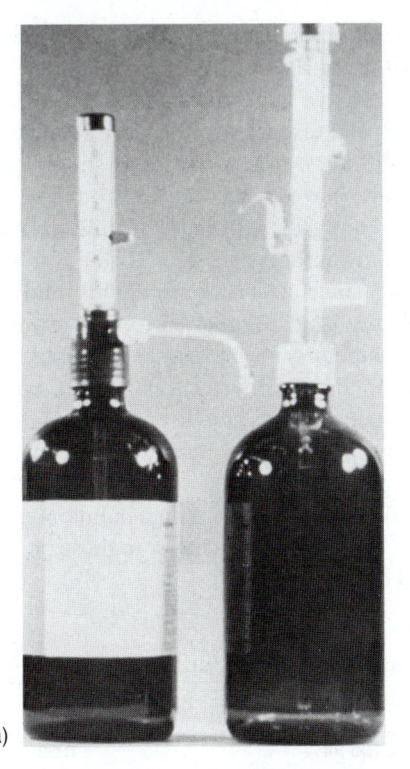

Figure 2.1 Automatic devices for dispensing fluids.
(a) Device to dispense milliliters (such as 5.0 mL) of reagent.
(b) An automatic microliter pipettor (Eppendorf) for dispensing 100 μL (0.10 mL) of solution, or similar volumes.

$$A—OH + HO—B \rightarrow A—B + HOH$$

The new molecule (**A—B**) formed from the two monomers (**A** and **B**) is called a *dimer*. This dimer may participate in a condensation reaction with a third monomer to form a *trimer*. The stepwise addition of new monomers to the growing chain by condensation reactions will result in the elongation of the chain and the formation of the full polymer. Examples of monomers and polymers are given in table 2.1.

When the chemical bond between monomers is broken, a molecule of water is consumed. This **hydrolysis reaction** is the reverse of a condensation reaction.

$$A—B + HOH \rightarrow A—OH + B—OH$$

Ingested foods are usually polymers—mainly proteins, carbohydrates, and triglycerides. In the stomach and small intestine, these polymers are hydrolyzed (in the process of *digestion*) into their respective monomers: amino acids, monosaccharides, fatty acids, and glycerol. These monomers are then moved across the wall of the small intestine into the blood of the capillaries (a process called *absorption*). The vascular system transports them primarily to the liver and then to all the other organs of the body.

Once inside the cells of the body, the monomers can be either hydrolyzed into smaller molecules, by a process that yields energy for the cell, or condensed to form new, larger polymers in the cytoplasm. Some of these new polymers are released into the blood (e.g., hormones and the plasma proteins), whereas others remain inside the cell and contribute to its structure and function. In turn, some of the new polymers of the cell can eventually be hydrolyzed to form new monomers, which may be used by the cell or released into the blood for use by other cells in the body.

In a healthy person, the concentrations of the different classes of monomers and polymers in the blood plasma are held remarkably constant and vary only within narrow limits. When the concentration of one of these molecules in the blood deviates from the normal range, specific compensatory mechanisms are activated that bring the concentration back to normal (negative feedback). Homeostasis is thus maintained.

When the concentration of any of the monomers or polymers in the blood remains consistently above or below normal, the health of the person may be threatened. Abnormal concentrations of different molecules in the blood are characteristic of different diseases and aid in their diagnosis. The disease *diabetes mellitus*, for example, is characterized by a high blood glucose concentration. Therefore, accurate measurement of the concentrations of different molecules in the blood is extremely important in physiology and clinical laboratories.

THE COLORIMETER

The colorimeter is a device used in physiology and clinical laboratories to measure the concentration of a substance in a solution. This is accomplished by the application of **Beer's law**, which states that the concentration of a substance in a solution is directly proportional to the amount of light absorbed *(Absorbance, A)* by the solution and inversely proportional to the logarithm of the amount of light transmitted *(Percent Transmittance, %T)* by the solution.

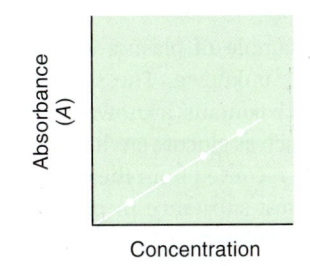

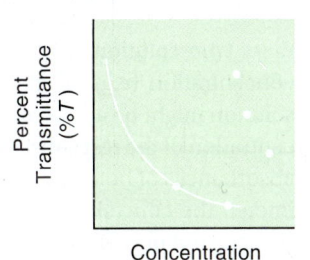

Table 2.1 Examples of Monomers and Polymers

Monomer	Examples	Polymer	Examples
Monosaccharides	Glucose, fructose	Polysaccharides	Starch, glycogen
Amino acids	Glycine, phenylalanine	Proteins	Hemoglobin, albumin
Fatty acids and glycerol		Triglycerides	Fats, oils
Ribonucleotides and deoxyribonucleotides		Nucleic acids	DNA and RNA

Beer's law will apply only if the incident light (the light entering the solution) is monochromatic—light composed of a single wavelength. White light is a mixture of many different wavelengths between 380 and 750 nanometers (nm), or millimicrons (mμ). The rods and cones within the eyes respond to the light waves, and the brain interprets these different wavelengths as different colors.

Violet	380–435 nanometers (nm)
Blue	436–480 nanometers (nm)
Green	481–580 nanometers (nm)
Yellow	581–595 nanometers (nm)
Orange	596–610 nanometers (nm)
Red	611–750 nanometers (nm)

By means of a prism or diffraction grating, the colorimeter can separate white light into its component wavelengths. The operator of this device can select incident light of any wavelength by turning the appropriate dial to that wavelength. This light enters a specific tube, the *cuvette*, which contains the test solution. A given fraction of the incident light is absorbed by the solution and the remainder of the light passes through the cuvette. The transmitted light generates an electric current by means of a photoelectric cell, and the amount of this current is registered on a galvanometer scale.

The colorimeter scale indicates the percent transmittance (%). Since the amount of light that goes into the solution and the amount of light that leaves the solution are known, a ratio of the two indicates the light absorbance (A) of that solution. The colorimeter also includes an absorbance scale. In the following exercises, the absorbance scale will be used rather than the percent transmittance scale because absorbance and concentration are directly proportional to each other. This relationship can be described in a simple formula, where 1 and 2 represent different solutions:

$$\frac{\text{Concentration}_1}{\text{Absorbance}_1} = \frac{\text{Concentration}_2}{\text{Absorbance}_2}$$

One solution might be a sample of plasma whose concentration (e.g., of glucose) is unknown. The second solution might be a *standard*, which contains a known concentration of the test substance (such as glucose). When the absorbances of both solutions are recorded from the colorimeter, the concentration of the test substance in plasma (i.e., the unknown) can easily be calculated:

$$C_x = C_{std} \frac{A_x}{A_{std}}$$

where

x = unknown plasma
std = standard solution
A = absorbance value
C = concentration

Suppose there are four standards. Standard 1 has a concentration of 30 mg per 100 mL (or mg per deciliter, dL). Standards 2, 3, and 4 have concentrations of 50 mg/dL, 60 mg/dL, and 70 mg/dL, respectively. Because standard 3 has twice the concentration of standard 1, it should (according to Beer's law) have twice the absorbance. The second standard (at 50 mg/dL), similarly, should have an absorbance value midway between that of the first and the fourth standard, since its concentration is midway between 30 and 70 mg/dL. Experimental errors, however, make this unlikely. Therefore, it is necessary to average the answers obtained for the unknown concentration when different standards are used. This can be done either arithmetically by applying the previous formula or by means of a graph.

A graph plotting the four standard data points, including a straight line of "best fit" drawn closest to these points, is called a **standard curve.**

Standard	Concentration (mg/dL)	Absorbance
1	30 mg/100 mL	0.25
2	50 mg/100 mL	0.38
3	60 mg/100 mL	0.41
4	70 mg/100 mL	0.57
Unknown	??	0.35

Now suppose that a solution of unknown concentration has an absorbance of 0.35. The standard curve graph can be used to determine its concentration.

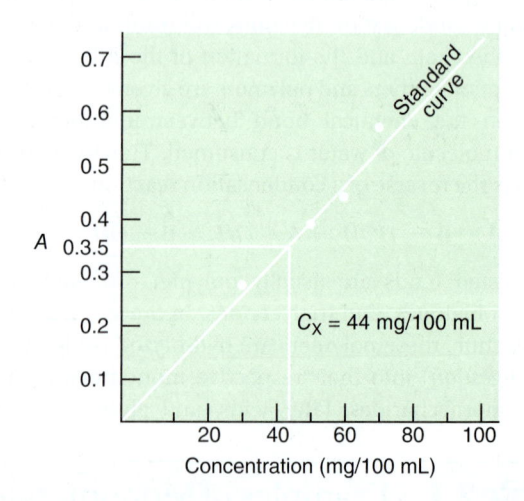

C_x = 44 mg/100 mL

Standardizing the Colorimeter

This procedure is intended specifically for the Spectronic 20 (Bausch & Lomb) colorimeter (fig. 2.2a). Although the general procedure is similar for all colorimeters, specific details may vary between different models.

Procedure

1. Set the wavelength dial so that the correct wavelength in nanometers (provided in each exercise) is lined up with the indicator in the window adjacent to this dial.
2. When there is no cuvette in the cuvette holder, the light source is blocked. The pointer should read zero transmittance or infinite absorbance at the left end of the scale. Turn knob to align the pointer with the *left* end of the scale.
3. Place the cuvette, which contains all the reagents *except* the test solution (e.g., glucose), into the cuvette holder. This tube is called the **blank** because it has a concentration of test substance equal to zero. It should therefore have an absorbance of zero (or a transmittance of 100%) read at the right end of the scale. Turn knob to align the pointer with the *right* end of the scale.
4. Repeat steps 2 and 3 to confirm settings.
5. Place the other cuvettes, which contain the standard solutions and the unknown, in the cuvette holder. Close the hatch and read the absorbance value of each solution.

Note: *Before placing each cuvette in the chamber, wipe it with a lint-free, soft paper towel. If the cuvette has a white indicator line, place the cuvette so that this line is even with the line at the front of the cuvette holder.*

A. Carbohydrates: Measurement of Plasma Glucose Concentration

The monomers of the carbohydrates are the **monosaccharides,** or simple sugars. The general formula for these molecules is $C_nH_{2n}O_n$, where n can be any number. *Glucose,* for example, has the formula $C_6H_{12}O_6$. The monosaccharide *fructose* has the same formula but differs from glucose in the arrangement of the atoms (glucose and fructose are *isomers*).

Two monosaccharides can join together by means of a dehydration synthesis (condensation) reaction to form a **disaccharide.** *Sucrose* (common table sugar), for example, is a disaccharide of glucose and fructose, whereas *maltose* is a disaccharide of two glucose subunits.

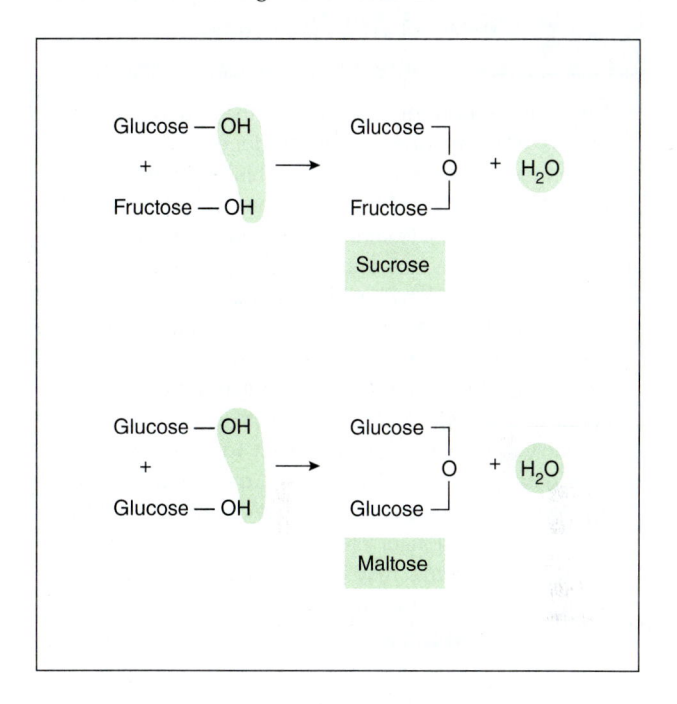

Glucose — OH + Fructose — OH → Glucose—O—Fructose (Sucrose) + H_2O

Glucose — OH + Glucose — OH → Glucose—O—Glucose (Maltose) + H_2O

The continued addition of glucose subunits to maltose will result in the production of a long, branched chain of repeating glucose subunits, forming the **polysaccharide** *glycogen* (or animal starch). This polysaccharide is formed inside muscle and liver cells and serves as an efficient storage form of glucose. When the blood glucose level drops below normal, the liver cells can hydrolyze stored glycogen and release glucose into the blood. Conversely, when the blood sugar level rises above normal, the liver cells can take glucose from the blood

(a)

(b)

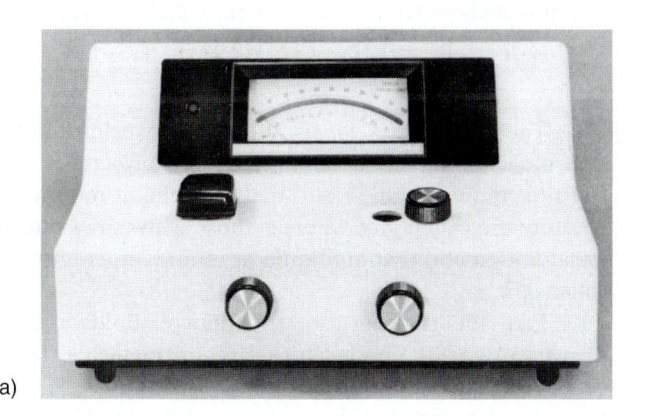

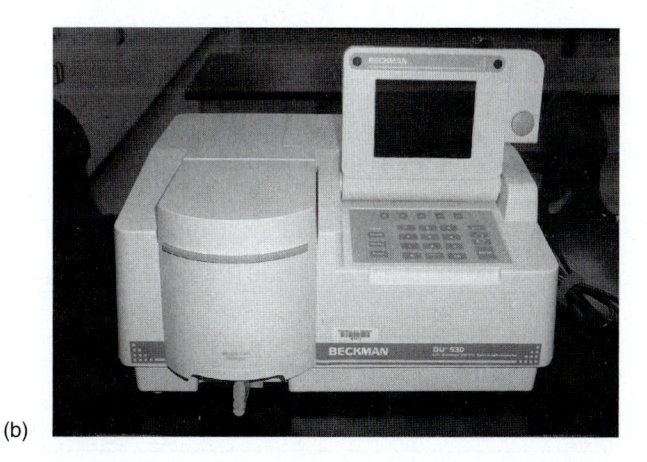

Figure 2.2 Devices used to measure the concentrations of solutions by their light absorbances. (a) A colorimeter, used to measure light absorbance in the visible spectrum only. (b) A spectrophotometer that can measure the absorbance of light in the ultraviolet as well as the visible region.

37

and store it as glycogen for later use. In this way, the equilibrium between blood glucose and liver glycogen helps to maintain constancy *(homeostasis)* of the blood sugar level. This process is regulated by hormones that include epinephrine (adrenaline), insulin, hydrocortisone, and glucagon.

Clinical Applications

The most important regulator of the blood glucose level is the hormone *insulin,* produced by the islets of Langerhans in the pancreas. This hormone stimulates the transport of blood glucose into the cells of the body; hence, it lowers the blood glucose level (see fig. 1.15). An elevated fasting blood glucose concentration *(hyperglycemia),* resulting in the "spilling over" of glucose into the urine, is the hallmark of **diabetes mellitus.** This disease may be due to the lack of insulin, so that the affected person requires insulin injections; this is *type 1 diabetes.* More commonly, it is primarily the result of low sensitivity to insulin, or insulin resistance, in *type 2 diabetes.* Low blood glucose, *hypoglycemia,* is clinically rare, but can result from the excessive secretion (or injection) of insulin. In addition, slight hypoglycemia may be associated with arthritis, renal disease, and the late stages of pregnancy.

Normal Values The normal fasting range of glucose in the plasma is 70–100 mg per 100 mL (or 70–100 mg/dL).

Procedure

Measurement of Plasma Glucose Concentration

1. Obtain three test tubes, and label them *U* (unknown), *S* (standard), *B* (blank).
2. Using a mechanical pipettor (do not pipette by mouth), pipette 5.0 mL of the glucose reagent into *each* tube.
3. Use a microliter pipettor to add 40 µL (0.04 mL) of the following solutions into each of the indicated test tubes to avoid contamination. Use different pipette tips for adding each solution.

Tube	Serum	Standard	Water	Reagent
Unknown (*U*)	40 µL	—	—	5.0 mL
Standard (*S*) (100 mg/dL)	—	40 µL	—	5.0 mL
Blank (*B*)	—	—	40 µL	5.0 mL

Note: All tubes must contain equal volumes (5.04 mL) of solution: 1 deciliter (dL) = 100 milliliters (mL).

4. Gently tap each tube to mix the contents and allow the tubes to stand at room temperature for 10 minutes.
5. Set the monochromator (wavelength) dial at 500 nm and standardize the colorimeter, using solution *B* as the blank.
6. Record the absorbance values of solutions *U* and *S* in the chart in the laboratory report.
7. Using Beer's law formula, calculate the concentration of glucose in the unknown plasma sample and enter the value in the laboratory report.
8. Using graph paper that follows this exercise, draw a graph of absorbance versus glucose concentration (mg/dL). Plot the standard *(S)* absorbance value and draw a standard curve. Then determine the unknown glucose concentration from the graph.

B. Lipids: Measurement of Plasma Cholesterol Concentration

The lipids are an extremely diverse family of molecules that share the common property of being soluble (dissolvable) in organic solvents such as benzene, ether, chloroform, and carbon tetrachloride, but *are not soluble in water or plasma.* The lipids found in blood can be classified as **free fatty acids (FFA),** also known as nonesterified fatty acids (NEFA); **triglycerides** (or neutral fats); **phospholipids;** and **steroids.** Carbon, hydrogen, and oxygen form the basic structure of lipids, but these atoms are not present in the same predictable ratio as they are in carbohydrates.

Fatty acids are long chains, ranging from sixteen to twenty-four carbons in length. When adjacent carbons are linked by single bonds, the fatty acid is said to be *saturated;* when adjacent carbons are linked by double bonds, the fatty acid is said to be *unsaturated.*

Saturated: $-CH_2-CH_2-CH_2-CH_2-CH_2-$
Unsaturated: $=CH-CH_2-CH=CH-CH_2-$

The triglycerides consist of three fatty acids bonded to a molecule of the alcohol *glycerol.* Triglycerides with few sites of unsaturation and that are solid at room temperature are called *fats,* whereas those with many sites of unsaturation and that are liquid at room temperature are called *oils.*

Like the triglycerides, the phospholipids consist of two fatty acids bonded to a glycerol molecule. However, as their name implies, each phospholipid also contains the element phosphorus (in the form of phosphate, PO_4^{3-}) bonded to the third position on the glycerol molecule. Phospholipids are important components of cell membranes.

The steroids are characterized by a structure consisting of four rings. One of the most important steroids in the body is cholesterol.

Cholesterol

Clinical Applications

There is evidence that high blood cholesterol, together with other risk factors such as hypertension and cigarette smoking, contributes to **atherosclerosis.** In atherosclerosis, deposits of cholesterol and other lipids, calcium salts, and smooth muscle cells build up in the walls of arteries and reduce blood flow. These deposits—called *atheromas*—also serve as sites for the production of *thrombi* (blood clots), which further occlude blood flow. The reduction in blood flow through the artery may result in heart disease or cerebrovascular accident (stroke). Plasma cholesterol attached to LDL carrier proteins (the so-called "bad cholesterol") enters arteries and promotes atherosclerosis, whereas blood cholesterol attached to HDL carriers (the "good cholesterol") travels from arteries to the liver. Thus, the risk of atherosclerosis is increased by high LDL cholesterol and reduced by high HDL cholesterol. A lowering of LDL and a rise in HDL cholesterol can be promoted by regular exercise and a healthful diet—one low in saturated fat, but that contains adequate amounts of omega-3 fatty acids (provided by such foods as oily fish and walnuts).

Normal Values Normal values for plasma cholesterol are 130–250 mg/dL.

Procedure

Measurement of Plasma Cholesterol Concentration

1. Obtain three test tubes, and label them *U* (unknown), *S* (standard), *B* (blank).
2. Using a mechanical pipettor, pipette 5.0 mL of the cholesterol reagent into each tube.
3. Use a microliter pipettor to add 50 μL (0.05 mL) of the following solutions into each of the indicated test tubes. Use different pipette tips for adding each solution to avoid contamination.

Tube	Serum	Standard	Water	Reagent
Unknown *(U)*	50 μL	—	—	5.0 mL
Standard *(S)* (200 mg/dL)	—	50 μL	—	5.0 mL
Blank *(B)*	—	—	50 μL	5.0 mL

Note: *All tubes must contain equal volumes (5.05 mL) of solution: 1 deciliter (dL) = 100 milliliters (mL).*

4. Gently tap each tube to mix the contents and allow the tubes to stand at room temperature for 10 minutes.
5. Transfer the solutions to three cuvettes. Standardize the spectrophotometer at 500 nm, using solution *B* as the blank.
6. Record the absorbance values of solutions *U* and *S* in the laboratory report.
7. Using Beer's law formula, calculate the concentration of cholesterol in the unknown plasma sample. Enter this value in the laboratory report.
8. Using graph paper that follows this exercise, draw a graph of absorbance versus cholesterol concentration (mg/dL). Plot the standard *(S)* absorbance value and draw a standard curve. Then determine the unknown cholesterol concentration from the graph.

$$C_{plasma} = \frac{A_{plasma}}{A_{standard}} \times C_{standard}$$

C. PROTEINS: MEASUREMENT OF PLASMA PROTEIN CONCENTRATION

Proteins are long chains of amino acids bonded to one another by condensation reactions. Each amino acid has an amino ($-NH_2$) end and a carboxyl ($-COOH$) end, as shown by the general formula:

When amino acids bond (through a **peptide bond**) to form a protein, one end of the protein will have a free amino group and the other end will have a free carboxyl group.

There are more than twenty-two different amino acids in nature, each differing from the others with respect to the combination of atoms in the *R group,* sometimes known as the *functional group.* The amino acid **glycine,** for example, has a hydrogen atom (H) in the *R* position, whereas the amino acid **alanine** has a methyl group (CH_3) in the *R* position.

Proteins in the plasma serve a variety of functions. Some proteins may be active as enzymes, hormones, or carrier molecules (transporting lipids, iron, or steroid hormones in the blood), while others have an immune function (antibodies). The **plasma proteins** are classified according to their behavior during biochemical separation procedures. These classes include the *albumins,* the *alpha* and *beta globulins* (synthesized mainly in the liver from amino acids absorbed by the intestine), and *gamma globulins* (antibodies produced by the lymphoid tissue). The albumins, produced by the liver, constitute 70% of the plasma proteins.

Clinical Applications

An abnormally low concentration of total blood protein *(hypoproteinemia)* may be due to an inadequate production of protein by the liver caused by liver disease (such as cirrhosis or hepatitis), the loss of protein in the urine *(albuminuria)* caused by kidney diseases, or the loss of plasma proteins through the intestine. Hypoproteinemia decreases the colloid osmotic pressure of the blood and may lead to accumulation of excess fluid in the tissue spaces, a condition called **edema.**

An abnormally high concentration of total plasma protein *(hyperproteinemia)* may be due to dehydration or to an increased production of the plasma proteins. An increased production of gamma globulins (antibodies), for example, is characteristic of pneumonia and many other infections, and of parasitic diseases such as malaria.

In addition to the separate functions of the different plasma proteins, the total concentration of proteins in the plasma is physiologically important. The plasma proteins exert an osmotic pressure (described in exercise 2.6), the **colloid osmotic (or oncotic) pressure,** which pulls fluid from the tissue spaces into the capillary blood. This force compensates for the continuous filtration of fluid from the capillaries into the tissue spaces produced by the *hydrostatic pressure* of the blood (fig. 2.3).

Normal Values The normal fasting total protein level is 6.0–8.4 g/dL.

Tube Number	Serum	Standard	Water	Reagent
1 Blank (B)	—	—	50 µL	5.0 mL
2 Standard (S) (2.0 g/dL)	—	50 µL	—	5.0 mL
3 (4.0 g/dL)	—	50 µL	—	5.0 mL
4 (6.0 g/dL)	—	50 µL	—	5.0 mL
5 (8.0 g/dL)	—	50 µL	—	5.0 mL
6 (10.0 g/dL)	—	50 µL	—	5.0 mL
7 Unknown (U)	50 µL	—	—	5.0 mL

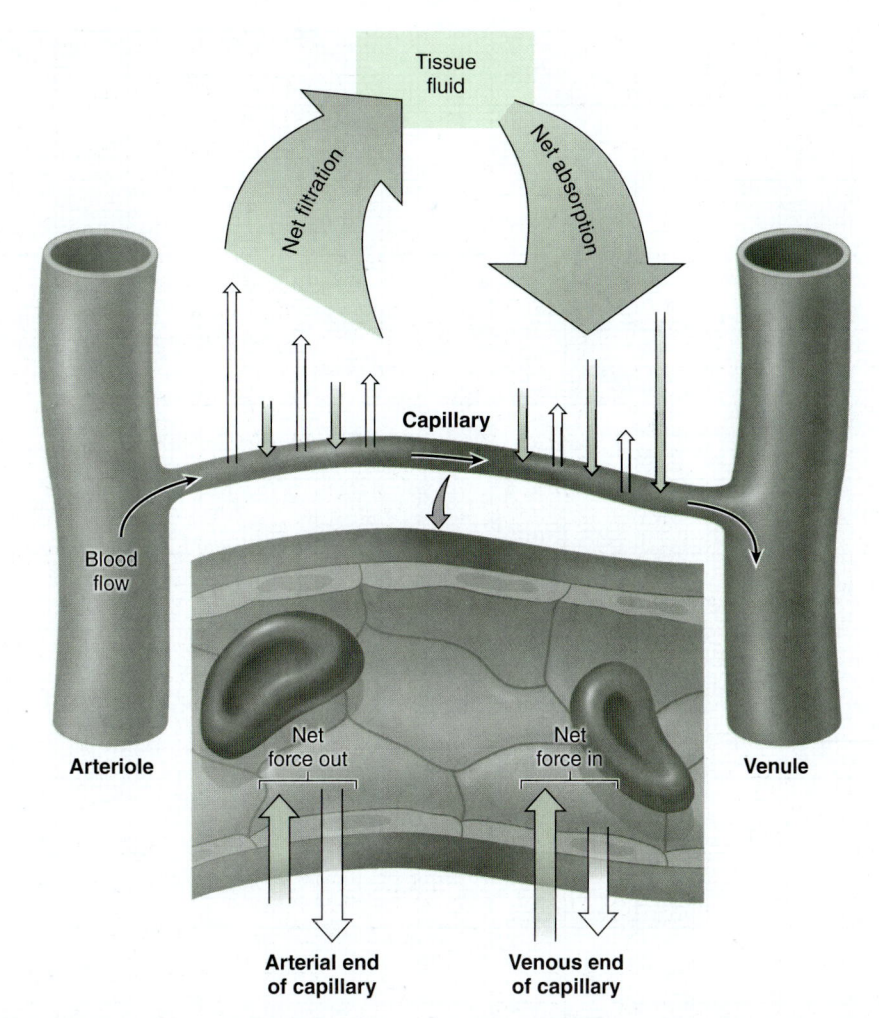

Figure 2.3 **Circulation of fluid between the blood plasma in a capillary and the tissues.** Arrows pointing away from the capillary indicate the force exerted by the blood pressure, whereas arrows pointing toward the capillary indicate the force exerted by the colloid osmotic pressure of the plasma proteins. The arrow labeled "blood flow" indicates the direction of flow along the capillary from arteriole to venule. APǀR

(For a full-color version of this figure, see fig. 14.9 in *Human Physiology*, thirteenth edition, by Stuart I. Fox.)

Procedure

Measurement of Total Plasma Protein Concentration

1. Obtain seven test tubes, and label them 1–7.
2. Using a mechanical pipettor (do not pipette by mouth), pipette 5.0 mL of biuret reagent into each tube.
3. Use a microliter pipettor to add 50 μL (0.05 mL) of the following solutions into each of the indicated test tubes. Use different pipette tips for adding each solution to avoid contamination.

Note: *All tubes must contain equal volumes (5.05 mL) of solution. (The expression g/dL is equivalent to g per 100 mL or g%.)*

4. Gently tap each tube to mix the contents, and allow the tubes to stand at room temperature for at least 5 minutes.
5. Transfer the solutions to seven cuvettes. Standardize the spectrophotometer at 550 nm, using solution *B* as the blank.
6. Record the absorbance values of the unknown *(U)* and the five standard solutions (*S* tubes 2–6) in the laboratory report.
7. Using Beer's law formula, calculate the concentration of total protein in the unknown plasma sample. Enter this value in the laboratory report.
8. Using graph paper that follows this exercise, draw a graph of absorbance versus total protein concentration (g/dL). Plot the standard (*S* tubes 2–6) absorbance values and draw a standard curve. Then determine the unknown total protein concentration from the graph.

41

Laboratory Report 2.1

Name _____

Date _____

Section _____

DATA FROM EXERCISE 2.1

A. Carbohydrates: Measurement of Plasma Glucose Concentration

1. Record the absorbance values of solutions *U* and *S*.
   Absorbance of solution *U*: _____
   Absorbance of solution *S*: _____

2. Enter the unknown serum glucose concentration derived from the equation for Beer's law: _____ mg/dL. How does this value for the unknown compare with that derived from the graph? Explain.

3. Is the glucose concentration of the serum sample normal or abnormal? Explain.

B. Lipids: Measurement of Plasma Cholesterol Concentration

1. Record the absorbance values of solutions *U* (unknown) and *S* (standard).
   Absorbance of solution *U*: _____
   Absorbance of solution *S*: _____

2. Enter the unknown serum cholesterol concentration derived from the equation for Beer's law: _____ mg/dL. How does this value for the unknown compare with that derived from the graph? Explain.

3. Is the cholesterol concentration of the serum sample normal or abnormal? Explain.

C. Proteins: Measurement of Plasma Protein Concentration

1. Record your absorbance values in this data table.

Tube Number	Protein Concentration (g/dL)	Absorbance
1	0 (blank)	0
2	2.0	
3	4.0	
4	6.0	
5	8.0	
6	10.0	
7	**Unknown** (serum sample)	

2. Use the graph paper on page 42 or 43 to plot a standard curve.
3. Derive an estimated protein concentration for the unknown serum sample.

    Unknown protein concentration from the curve: _____ g/dL.

4. Use the equation for Beer's law to determine the protein concentration of the unknown serum.

    Unknown protein concentration from Beer's equation: _____ g/dL.

5. How does this value for the unknown compare with that derived from the graph? Explain.

6. Is the total protein concentration of the unknown serum sample normal or abnormal? Explain: What clinical condition(s) might have caused an abnormal result?

REVIEW ACTIVITIES FOR EXERCISE 2.1

Test Your Knowledge

1. The concentration of a solution is _____ proportional to its absorbance.
2. The relationship in question number 1 is described by _____ law.
3. Hyperglycemia is characteristic of the disease _____.
4. Hyperglycemia may be caused by a deficiency of the hormone _____.
5. Cholesterol belongs to the general category of molecules known as _____ and to the specific category of molecules known as _____.
6. High blood cholesterol, along with other risk factors, is a contributing factor in the disease _____.
7. Most of the plasma proteins are produced by the _____ (organ).
8. Low plasma protein concentration is described clinically as _____ and can produce a physical condition called _____.

9. The colloid osmotic pressure of the blood is related to the plasma concentration of _____.

10. "All fats are lipids, but not all lipids are fats." Explain.

Test Your Understanding

11. Describe the functions of the plasma proteins. Where are these proteins produced?

12. What does the blank tube contain, and what is its function in a colorimetric assay?

Test Your Analytical Ability

13. Do any of the proteins in your plasma come from food proteins? Does the starch (glycogen) in your liver come from food starch? Explain your answers.

14. Why do you draw a linear (straight line) graph of absorbance versus concentration even though your experimental values deviated slightly from a straight line? Why must your line intersect the origin of the graph (zero concentration equals zero absorbance)? Explain.

Test Your Quantitative Ability

15. Suppose a standard cholesterol solution with a concentration of 150 mg/dL has an absorbance of 0.32, and the unknown plasma sample has an absorbance of 0.46. Calculate the cholesterol concentration in the plasma.

16. Draw a standard curve, given the absorbances of the following protein standards:

    2.0 g/dl Standard 0.06
    4.0 g/dl Standard 0.13
    6.0 g/dl Standard 0.17
    8.0 g/dl Standard 0.26
    10.0 g/dl Standard 0.31

    If the unknown plasma sample has an absorbance of 0.22, what is its concentration?

Clinical Investigation Questions

17. Is the fasting plasma sample of this patient normal? If not, what is the likely reason for the abnormal level?

18. Is the fasting plasma triglyceride level normal? If not, what danger does this level present? Might this patient have *metabolic syndrome* with both diabetes and high triglycerides (*Human Physiology,* 13th ed., chapter 19, section 19.2)?

19. What does this patient's plasma protein level indicate? Given that a person with uncorrected diabetes urinates excessively, how might that relate to the measured plasma protein concentration?

EXERCISE 2.2

Thin-Layer Chromatography of Amino Acids

MATERIALS

1. Silica gel plates (F-254 rapid, adhered to plastic or glass); capillary tubes; chromatography (or hair) dryers; rulers
2. Alternatively, silica gel plates with plastic backing and fluorescence indicator that are only 2.5 cm by 7.5 cm in size can be obtained from Carolina (#HT-68-9460).

Note: *Chromatography paper can be used as an alternative.*

3. Developing chambers or oven set at about 60° C
4. Amino acid solutions: arginine, cysteine, aspartic acid, phenylalanine—1.0 mg/mL of each dissolved in 0.1N HCl:isopropyl alcohol (9:1); "unknown" solution of amino acids, containing two of these amino acids in the same solution
5. Developing solvent: 20 mL of 17% NH_4OH (dilute concentrated NH_4OH with an equal amount of water), 40 mL of ethyl acetate, and 40 mL of methanol per developing chamber
6. Ninhydrin spray

Amino acids, the subunits of protein structure and function, constitute approximately twenty chemically unique molecules. The distinctive properties of each amino acid provide the basis for its separation from the others and its identification. This information can be clinically useful in the diagnosis of genetic diseases that involve amino acid metabolism.

LEARNING OUTCOMES

You should be able to:

1. Using the general formula for amino acids, explain how one amino acid differs from another.
2. Explain how thin-layer chromatography can separate different amino acids present together in a single solution.
3. Explain what the R_f value signifies. Calculate the R_f values for different spots and use this information to identify unknown amino acids.

Textbook/Multimedia Correlations

Before performing this exercise, you should study the introductory material presented here. Further information relating to this exercise can be found in these pages of *Human Physiology*, thirteenth edition, by Stuart I. Fox:

- *Proteins.* Chapter 2, p. 41.
- *Inborn Errors of Metabolism.* Chapter 4, p. 95.

In this exercise, you will attempt to identify two unknown amino acids present in the same solution. To accomplish this task, you must (1) *separate* the two amino acids, and (2) *identify* these amino acids by comparing their behavior with that of known amino acids.

Because each amino acid has a chemically different R group, each will dissolve in a given solvent to a different degree. These differences will be used to separate and identify the amino acids on a **thin-layer plate.**

The thin-layer plate consists of a thin layer of porous material (in this procedure, silica gel) coated on one side of a plastic, glass, or aluminum plate. The solutions of amino acids are applied to different spots on the plate (a procedure called *spotting*) and allowed to dry. The plate is then placed on edge in a solvent bath with the spots above the solvent.

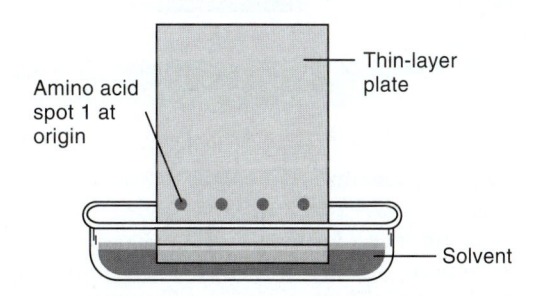

As the solvent creeps up the plate by capillary action, it will wash the amino acids off their original spots (or *origins*) and carry them upward toward the other end of the plate. Because each amino acid has a different solubility, size, and shape, the ability of the solvent to dissolve and carry a particular amino acid through the pores of the silica gel will vary with the properties of the amino acid. The process is halted shortly before the solvent front reaches the top of the plate, so that some amino acids will have migrated farther from the origin than others.

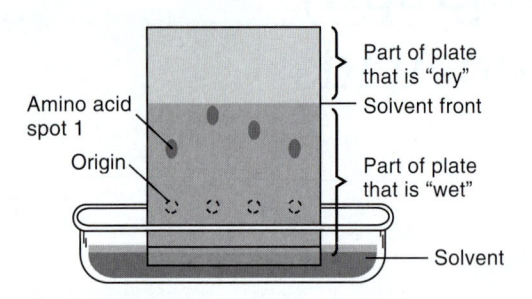

If this chromatography were repeated using the same amino acids and the same solvent, the final pattern *(chromatogram)* would be the same as that obtained previously. In other words, the distance that a given amino acid migrates in a given solvent, relative to the *solvent front*, can be used as an identifying characteristic of that amino acid. A numerical value (the R_f value) can be assigned to this characteristic by calculating the distance the amino acid traveled relative to that traveled by the solvent front, as:

$$R_f = \frac{\text{distance from origin to spot}}{\text{distance from origin to solvent front}}$$

An unknown amino acid can be identified by comparing its R_f value in a given solvent with the R_f values of known amino acids in the same solvent.

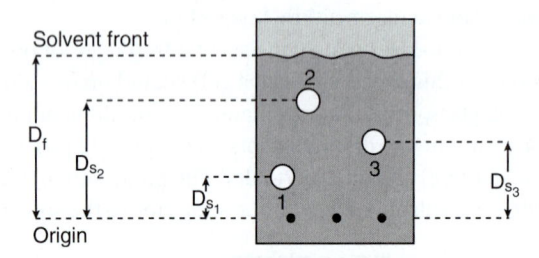

Procedure

1. Fill a capillary tube with amino acid solution 1 (arginine). Gently touch it to the first spot on the plate 1½ inches from the bottom and ½ inch from the left-hand edge. Dry the spot with a hair dryer. Repeat the spotting and drying procedure until you have made *five* applications of the *same* amino acid to the *same* spot.

Note: *(1) Be careful to apply the amino acid solution to exactly the same spot each time (use a spotting guide, or make X's lightly with pencil before spotting); (2) dry the spot thoroughly between applications; and (3) spot gently so that you do not gouge out the silica gel.*

Clinical Applications

An abnormally high concentration of certain amino acids or their metabolites in the blood frequently results in deterioration of the central nervous system and mental retardation. These abnormal concentrations are usually due to defective enzymes involved in the degradation of the amino acids.

In the disease **phenylketonuria (PKU)**, for example, the enzyme that converts the amino acid phenylalanine to tyrosine is defective, leaving phenylalanine and its other metabolites to accumulate in the body. This condition, which affects one baby in every 10,000 to 20,000, results in severe mental retardation. Other diseases of similar etiology (cause) include *homocystinuria*, *alkaptonuria*, and *maple syrup disease* (the name refers to a characteristic odor of the urine). The defective enzymes are synthesized by defective genes; hence, the diseases of amino acid metabolism that have this etiology are referred to as **inborn errors of metabolism** (see exercise 2.5).

2. Repeat the spotting procedure for amino acid 2 (cysteine), using a new capillary tube and applying the spot about ½ inch to the right of the first amino acid.
3. Repeat the spotting procedure for amino acids 3 (aspartic acid) and 4 (phenylalanine), applying them about ½ inch to the right of the preceding spot.
4. Repeat the spotting procedure with the solution containing two unknown amino acids, applying it about ½ inch to the right of amino acid 4.
5. Carefully place the thin-layer plate in a chromatography developing chamber previously filled with solvent. Cover the chamber and allow the solvent to migrate up the plate for 1 hour.
6. Remove the plate from the developing chamber, dry it, and then spray it with ninhydrin in a well-ventilated area.

Note: *Amino acids are colorless, so it is necessary to react them with ninhydrin, a reagent that combines with the amino acids to produce a blue-colored complex.*

7. Heat the plate in an oven set at approximately 60° C for *10–15* minutes.
8. Remove the plate and measure the distance, in centimeters, from the origin to the solvent front and from the origin to the center of each amino acid spot. Record these values and calculate R_f values. Enter your data in the laboratory report.

Laboratory Report 2.2

Name _____

Date _____

Section _____

DATA FROM EXERCISE 2.2

1. Record your data in this table and calculate the R_f value for each spot.

Amino Acid	D_s	D_f	R_1
Arginine			
Cysteine			
Aspartic acid			
Phenylalanine			
Unknown 1			
Unknown 2			

2. The unknown solution contained the two amino acids _____ and _____.

REVIEW ACTIVITIES FOR EXERCISE 2.2

Test Your Knowledge

1. Which part of an amino acid distinguishes it and grants it chemical specificity? _____

2. Inherited defects in the ability to convert one amino acid into another are in a class of disorders called _____

3. Define the term R_f. _____

Test Your Understanding

4. Why do different amino acids have different R_f values?

5. Suppose an amino acid is 8 cm from the origin and the solvent front is 12 cm from the origin. What is the R_f value for this amino acid?

6. What is the maximum R_f value that a spot can have? Explain.

Test Your Analytical Ability

7. Suppose amino acid "A" has a higher R_f value than amino acid "B" in a solvent system "1" and the order is reversed in solvent system "2." Further, suppose that solvent system "1" has a higher ratio of methanol to ethyl acetate (and is thus more polar) than solvent system "2." What can you conclude about the structure of amino acid "A" compared to the structure of amino acid "B"?

8. Describe how amino acids differ from each other, and how they can be grouped into chemical subcategories. What are the benefits of this diversity of chemical structure? What might happen if a mutation caused one amino acid to be substituted for a different one in a particular enzyme protein? Explain.

Test Your Quantitative Ability

9. Suppose on a thin-layer chromatography plate the solvent front migrates to a distance of 12 cm from the origin, while amino acids (a) and (b) migrate 4 cm and 9 cm, respectively. Calculate the R_f values of each amino acid.

EXERCISE 2.3

Electrophoresis of Serum Proteins

MATERIALS

1. Sepra Tek System (Gelman), with forceps to handle the cellulose acetate strips; Ponceau S stain (Gelman) in 5% acetic acid (v/v)
2. Alternatively, vertical protein electrophoresis gel boxes can be used for SDS (sodium dodecyl sulfate)-polyacrylamide gel electrophoresis. These employ premade gels in sealed combs (cassettes) that accommodate ten samples of 30 µL or less. After electrophoresis, gels are stained with one-step Coomassie stain kits. Available suppliers are Invitrogen Corporation and Bio-Rad Laboratories, Inc. In this case, microliter pipettes (able to dispense samples of 30 µL or less) are also needed.
3. Sterile lancets, 70% ethanol, disposal receptacle for all blood-contaminated objects (Alternatively, test tubes containing previously prepared serum samples may be used.)
4. Unheparinized capillary tubes, clay sealant, and microhematocrit centrifuge

Plasma contains classes of proteins that differ in structure and function. These classes can be separated from one another and identified by electrophoresis.

LEARNING OUTCOMES

You should be able to:

1. Explain what is meant by the term *amphoteric*, and describe how amphoteric molecules can be separated by electrophoresis.
2. Identify the different bands of serum proteins in an electrophoresis pattern.
3. Describe the origin and functions of the different classes of plasma proteins.

Textbook/Multimedia Correlations

Before performing this exercise, you should study the introductory material presented here. Further information relating to this exercise can be found in these pages of *Human Physiology,* thirteenth edition, by Stuart I. Fox:

- *Proteins.* Chapter 2, p. 41.
- *Plasma Proteins.* Chapter 13, p. 407.
- *Antibodies.* Chapter 15, p. 504.

Clinical Investigation

A patient's plasma had a slightly low level of albumin but an elevated concentration of gamma globulin.

- Identify plasma albumin and some of its functions.
- Describe gamma globulin and what might account for elevated levels.

The unique structure and physiological role of each type of protein is determined by the specific number, type, and sequence of its component amino acids. Proteins differ in size, range in shape from elliptical (globular) to fibrous, and contain different numbers of positive and negative charges.

Under acidic conditions, the amino group ($-NH_2$) of an amino acid tends to gain a H^+ and become positively charged ($-NH_3^+$) whereas under basic conditions, the carboxyl group ($-COOH$) loses a H^+ and becomes negatively charged ($-COO^-$). Because amino acids can have either polarity, depending on the pH, they are said to be **amphoteric.**

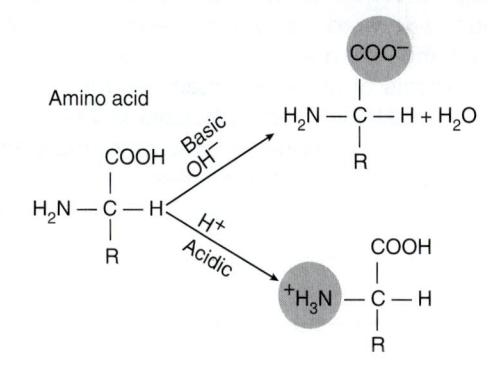

Amino acid

$H_2N-\underset{R}{\overset{|}{C}}-H$

Basic OH^- → $H_2N-\underset{R}{\overset{COO^-}{\underset{|}{C}}}-H + H_2O$

H^+ Acidic → $^+H_3N-\underset{R}{\overset{COOH}{\underset{|}{C}}}-H$

When an amino acid is electrically neutral, its amphoteric nature can be shown by the *zwitterion formula,* in which neutrality is indicated by a balance between positive and negative charges.

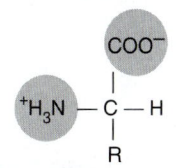

$^+H_3N-\underset{R}{\overset{COO^-}{\underset{|}{C}}}-H$

Amino acid (zwitterion formula)

In addition to the amino and carboxyl ends of a protein, the functional groups (*R* groups) of many amino acids have either an amino-containing functional group (such as lysine or arginine) or a carboxyl-containing functional group (such as aspartic acid or glutamic acid). Since each type of protein has a characteristic ratio of these two types of amino acids, each protein will have a characteristic net charge at a given pH.

At a pH of 8.8 (slightly basic), each of the variety of proteins found in plasma will have a different degree of net negative charge. When plasma proteins are placed in an electric field, each protein will migrate away from the negative pole (cathode) and toward the positive pole (anode) at different rates. The rates at which they travel will also be influenced by their size and shape. This technique, known as **electrophoresis,** can be used to separate and identify the different classes of plasma proteins.

There are two main types of plasma proteins: albumin and the globulins (see section C, exercise 2.1). The latter is composed of four primary subclasses: *alpha*-1 (α_1), *alpha*-2 (α_2), *beta* (β), and *gamma* (γ) *globulins* (fig. 2.4).

The fluid part of the blood as it circulates in the vessels is **plasma.** When blood clots, a soluble protein in the plasma *(fibrinogen)* is converted into an insoluble thread-like protein called *fibrin*. The strands of fibrin intertwine to form the meshwork of the blood clot. **Serum,** the fluid that remains after the clot has formed, does not contain fibrinogen and is incapable of further clotting.

Albumins are the most abundant of the serum proteins. They serve as carrier molecules for hormones, lipids, and bile pigment; they are also responsible for most of the colloid osmotic pressure exerted by the blood. The *alpha* and *beta globulins* serve a variety of functions and, like albumin and fibrinogen, are synthesized by the liver. The *gamma globulins* are **antibodies,** produced by white blood cells known as lymphocytes.

Clinical Applications

In many instances, the diagnosis of a disease can be aided by an analysis of the electrophoresis pattern made by migrating plasma proteins. Although direct observation is useful, more reliable information can be gained by a quantitative measurement of the proteins in each band. These measurements are made by a *densitometer*, a device that optically scans the electrophoresis pattern and graphically records the absorbance of different regions of the strip.

Diseases that can be accurately diagnosed through electrophoresis include acute inflammations (elevated alpha-2 proteins), viral hepatitis (change in gamma globulin and albumin), and cirrhosis of the liver (broad gamma globulin band). In addition, electrophoresis is valuable in the diagnosis of nephrotic syndrome, malignant tumors, and many other diseases.

Procedure

SDS-Polyacrylamide Gel Electrophoresis

1. Obtain a ten-sample cassette (comb) of precast polyacrylamide gel (fig. 2.5), and identify the sample wells. (**Caution:** polyacrylamide is a neurotoxin and a carcinogen.)
2. Using an automatic microliter pipette, deliver samples of plasma or serum to the wells. The instructor will specify the amount of each sample.
3. Place the gel in the vertical electrophoresis box (fig. 2.6) and attach the red and black electrodes to the power supply.
4. Set the power supply (fig. 2.7) to the voltage and for the length of time specified by the instructor.
5. At the end of electrophoresis, unplug the unit from the power supply and remove the gel. Use the one-step Coomassie stain kit to stain and visualize the protein bands in the polyacrylamide gel. Identify each band.

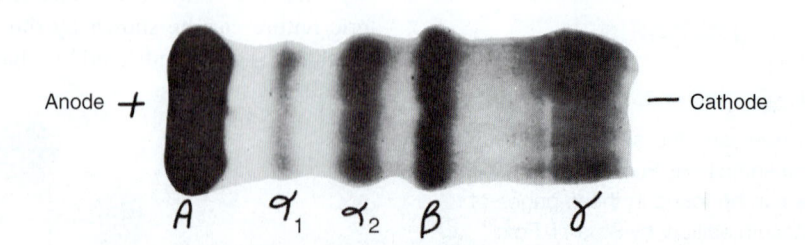

Figure 2.4 **Electrophoresis pattern of normal serum.** The bands include albumin (A) and the α_1, α_2, β, and γ globulins.

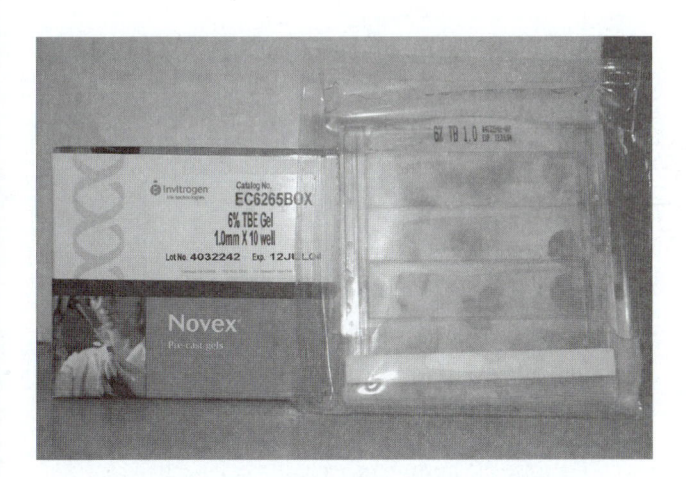

Figure 2.5 Cassettes of precast SDS-polyacrylamide gel for protein electrophoresis. These have ten wells for samples. They can be obtained from different suppliers, but the one in the photograph is from Invitrogen Corporation.

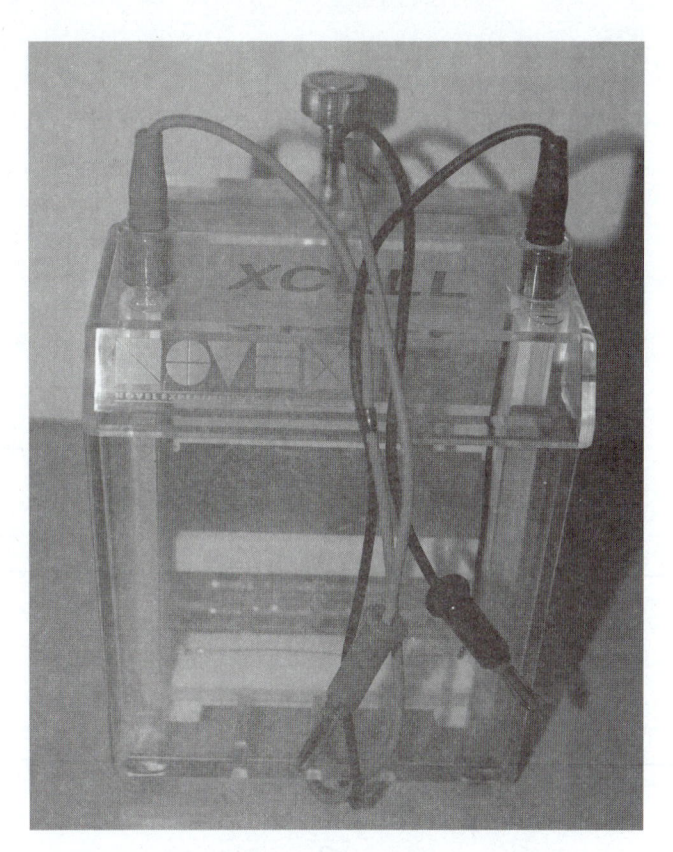

Figure 2.6 Vertical electrophoresis box for precast polyacrylamide gel cassettes. These can be obtained from different suppliers, but the one in the photograph is from Invitrogen Corporation.

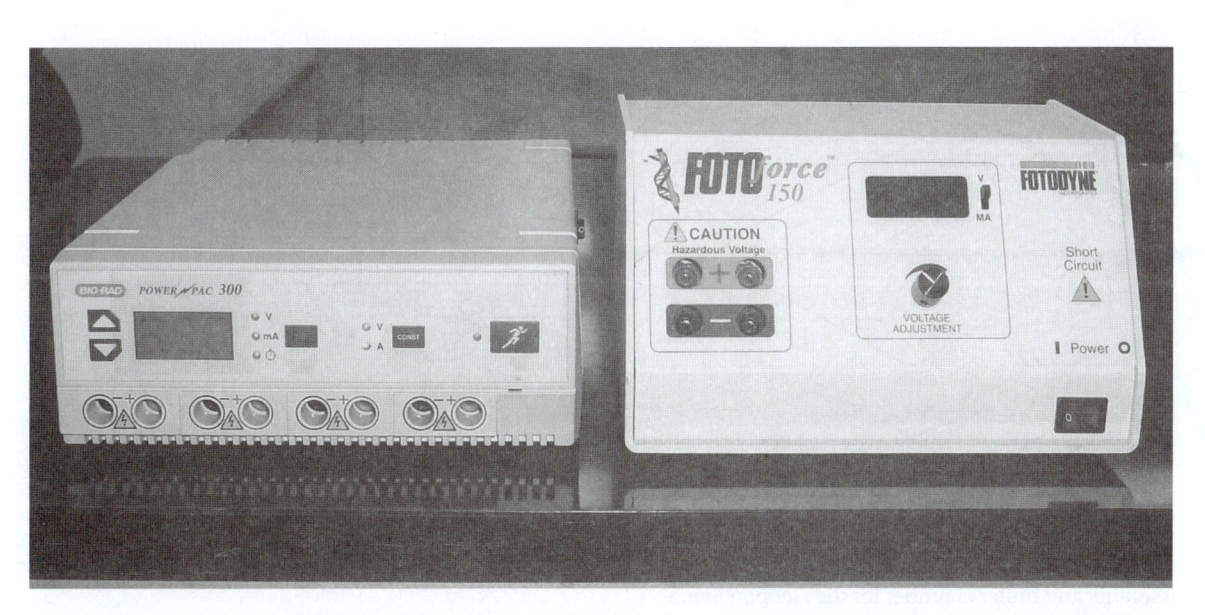

Figure 2.7 Electrophoresis power supplies. The models shown here are examples of equipment with digital displays.

Procedure

Cellulose Acetate Electrophoresis

1. Float a strip of cellulose acetate in the buffer for 1 minute; then immerse it in the buffer for *10 minutes*.
2. Using forceps, remove the cellulose acetate strip from the buffer, and blot it with filter paper premoistened with the buffer.
3. Raise the tension latch of the frame assembly, placing the movable support bridge in the vertical position (figs. 2.8 and 2.9*a*).*
4. Center the cellulose acetate strip on the support bridges, and fasten it with the membrane clamps. Tension the membrane by releasing the latch (fig. 2.9*b,c*).
5. Place the membrane frame assembly in the chamber, bringing the strip ends into contact with the buffer, and position the cover on the chamber.
6. Cleanse the tip of a finger with 70% ethanol, and puncture it with a sterile lancet.

*For Gelman Sepra Tek System.

Note: Extreme caution must be exercised when handling blood to guard against contracting infectious agents. Handle only your own blood, and discard all objects containing blood into the receptacles provided by the instructor.

7. Quickly fill an *unheparinized* capillary tube with blood, seal one end, and immediately centrifuge for 3 minutes. (Do *not* use heparinized capillary tubes—the anticoagulant heparin is a protein and will interfere with the test.)
8. Break the capillary tube at the junction of the packed red blood cells and the pale yellow serum. Place a drop of serum on the sample well of the applicator block by lightly touching it with the capillary tube. Similarly, a sample of serum from six different students can be placed in each of six wells (fig. 2.9*d*).
9. Fill a new capillary tube with the serum provided by the instructor, and place it on sample wells 7 and 8.
10. Position the applicator on the applicator block, and load it by depressing the button for 4 seconds (fig. 2.9*e*).
11. Position the loaded applicator on the chamber cover. Depress the button for 4 seconds to place the serum on the cellulose acetate strip (fig. 2.9*f*).
12. Connect the chamber to the power supply and electrophorese for *20 minutes* at 200 V (figs. 2.7 and 2.9*g,h*).

Note: *During this time, clean the applicator by placing it on filter paper moistened with the buffer. Rinse with tap water and distilled water.*

13. When the voltage is off, open the chamber and remove the membrane frame assembly. Raise the tension latch and strip clamps and remove the cellulose acetate strip with forceps.
14. Float the strip on Ponceau S stain for *1 minute;* then immerse it completely in the stain for *10 minutes*.
15. Remove the strip with forceps and rinse it in two successive baths of 5% acetic acid. Tape the cellulose acetate strip in the space provided (or draw a facsimile), and label the bands.

Note: *The albumin band (nearest the anode) is the darkest band; the gamma globulin band (nearest the cathode) is the widest band.*

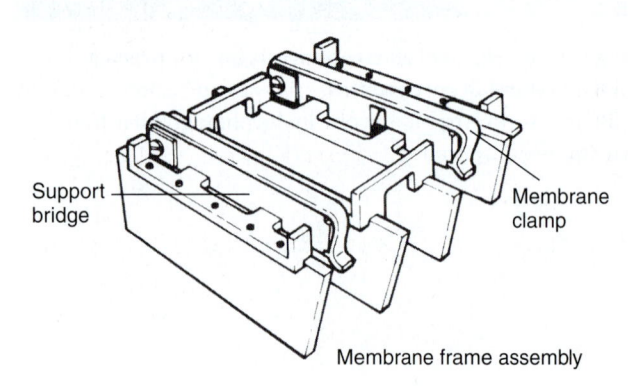

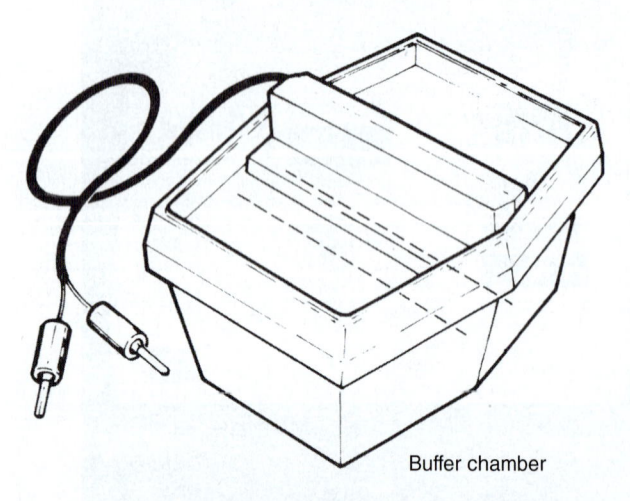

Figure 2.8 An electrophoresis system. A membrane frame assembly and buffer chamber for the Gelman Sepra Tek electrophoresis system.

(a)

(b)

(c)

(d)

(e)

(f)

(g)

(h)

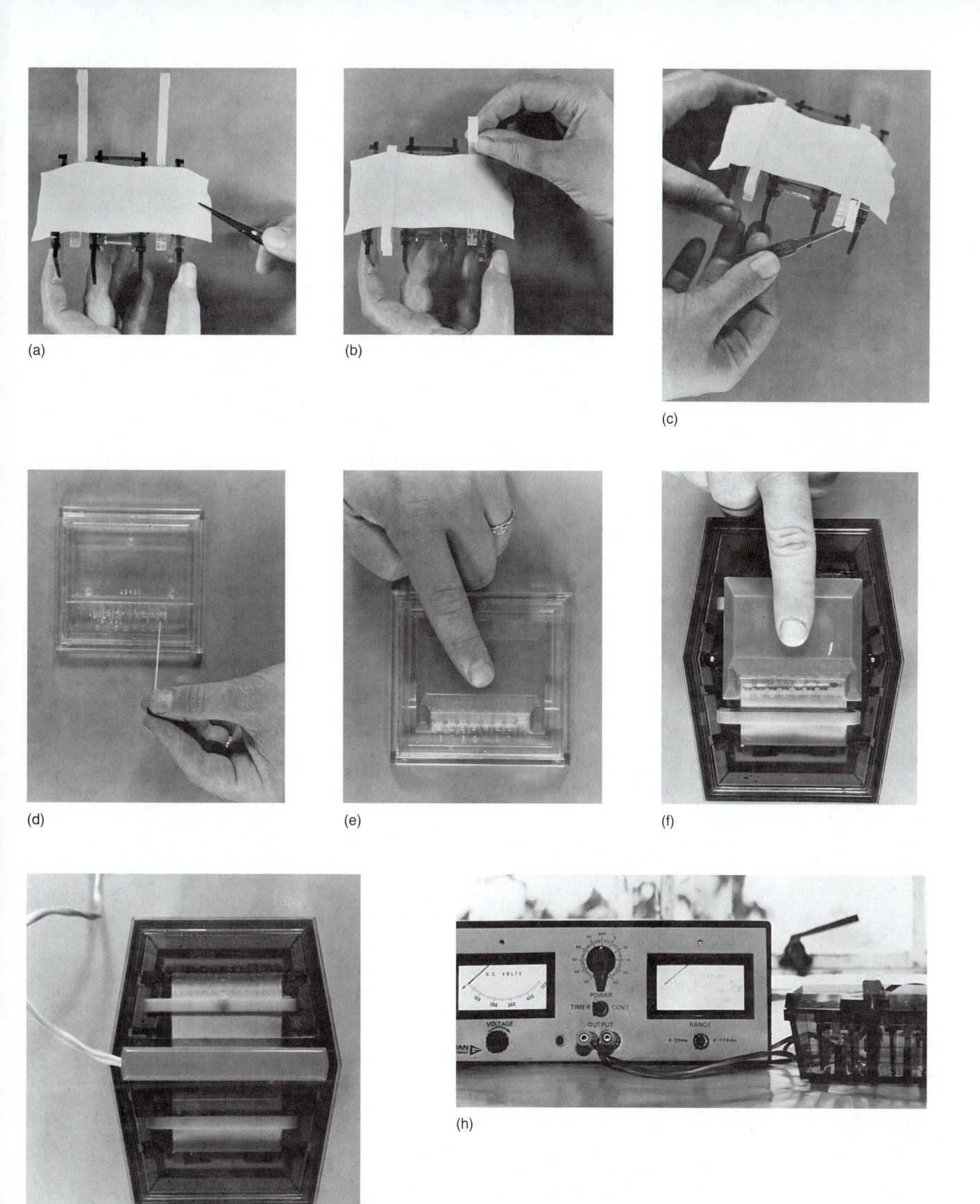

Figure 2.9 **Procedure for performing electrophoresis of serum proteins.** Steps a–h are described in the text.

57

Laboratory Report 2.3

Name _____
Date _____
Section _____

DATA FROM EXERCISE 2.3

1. In this space, tape your electrophoresis strip (or draw a facsimile).

2. Label the protein bands in your strip.

REVIEW ACTIVITIES FOR EXERCISE 2.3

Test Your Knowledge

1. Describe the property of an amphoteric molecule. _____
   _____

2. The amphoteric nature of an electrically neutral amino acid can best be shown by its _____ formula. Draw one here.

3. Name two amino acids that have
   (a) an extra amino group: _____ and
   _____.
   (b) an extra carboxyl group: _____
   and _____.

4. Proteins have a net _____ charge in acidic solutions and a net _____ charge in basic solutions.

5. The soluble protein in plasma that aids in the formation of blood clots is _____.

6. The insoluble protein that aids in the formation of blood clots is _____.

7. The most abundant class of plasma proteins is the _____.

8. Albumin and many other plasma proteins are made by the _____.

9. The class of plasma proteins containing antibodies is the _____.

Test Your Understanding

10. What is the difference between serum and plasma? Explain.

11. What factors contribute to the distance that a protein will migrate during electrophoresis?

Test Your Analytical Ability

12. Why is the albumin band the darkest of the bands? What does the width of the gamma globulin band reveal about the structure of antibodies?

13. How might liver disease affect the density of the bands? Which bands would be the most affected? What effects might these changes in plasma proteins have on the body?

Clinical Investigation Questions

14. What is plasma albumin, and what are some of its functions?

15. What is plasma gamma globulin, and what might an increase in the gamma globulin concentration indicate?

EXERCISE 2.4

Measurements of Enzyme Activity

MATERIALS

1. Beakers, rusty nails, chicken liver
2. Hydrogen peroxide
3. Test tubes, mechanical pipettors, automatic microliter pipettes
4. Constant-temperature water-bath set at 37° C
5. Cuvettes and spectrophotometer (colorimeter)
6. Serum or reconstituted normal and abnormal serum (available, for example, from Bio-Analytic Laboratories, Inc. or Stanbio Laboratory, Inc.).
7. Alkaline Phosphatase Test and Lactate Dehydrogenase Test (available, for example, from Biotron Diagnostics, Inc. or from Bio-Analytic Laboratories, Inc.)

The presence of a specific enzyme can be detected by the reaction it catalyzes, and the enzyme concentration can be measured by the amount of product it forms in a given time. Enzyme activity is affected by pH, temperature, and the availability of substrates and coenzymes.

LEARNING OUTCOMES

You should be able to:

1. Describe the lock-and-key model of enzyme activity, and use this model to explain enzyme specificity.
2. Describe the effects of pH and temperature on enzyme activity.
3. Describe how enzyme concentration is measured.

Textbook/Multimedia Correlations

Before performing this exercise, you should study the introductory material presented here. Further information relating to this exercise can be found in these pages of *Human Physiology*, thirteenth edition, by Stuart I. Fox:

- *Enzymes as Catalysts.* Chapter 4, p. 89.
- *Effects of Temperature and pH.* Chapter 4, p. 92.
- *Cofactors and Coenzymes.* Chapter 4, p. 93.

Clinical Investigation

A patient had normal plasma acid phosphatase activity, but had elevated levels of alkaline phosphatase and lactate dehydrogenase activities.

- Explain the activity of phosphatases, how acid and alkaline phosphatase differ, and what can cause abnormal levels of these enzymes.
- Describe the activity of lactate dehydrogenase and the possible causes of an abnormally elevated plasma level of this enzyme.

Enzymes are biological **catalysts;** that is, enzymes are substances that increase the rate of a chemical reaction without changing the nature of the reaction and without being altered by the reaction. The catalytic process occurs in two stages: (1) the reactants (hereafter referred to as the *substrates* of the enzyme) bond in a specific manner to the enzyme, forming an *enzyme-substrate complex,* and (2) the enzyme-substrate complex dissociates into the free, unaltered enzyme and *products*.

All enzymes are proteins (although not all proteins are enzymes). The polypeptide chain of each enzyme bends and folds in a unique way to produce a characteristic three-dimensional structure. The substrates interact with a specific part of this structure, the **active site,** which is complementary in shape to the substrate molecules. This is most easily visualized by the *lock-and-key model* of enzyme action (fig. 2.10). In some cases, the fit between an enzyme and its substrates may not be perfect at first. A perfect fit may be induced, however, as the substrate gradually slips into the active site in much the same way as the shape of a thin glove changes to fit a hand entering it. New bonds are more easily formed as substrates are brought close together in the active sites of the enzyme.

The shape of the active site is determined by the amino acid sequence of the protein; this shape is different for different enzymes. Enzymes are *relatively specific,* interacting only with specific substrates and catalyzing selective reactions.

The relative specificity of an enzyme can often be deduced from its name. Thus, the enzyme *lactate dehydrogenase* removes a hydrogen from lactic acid (the suffix *-ase* denotes

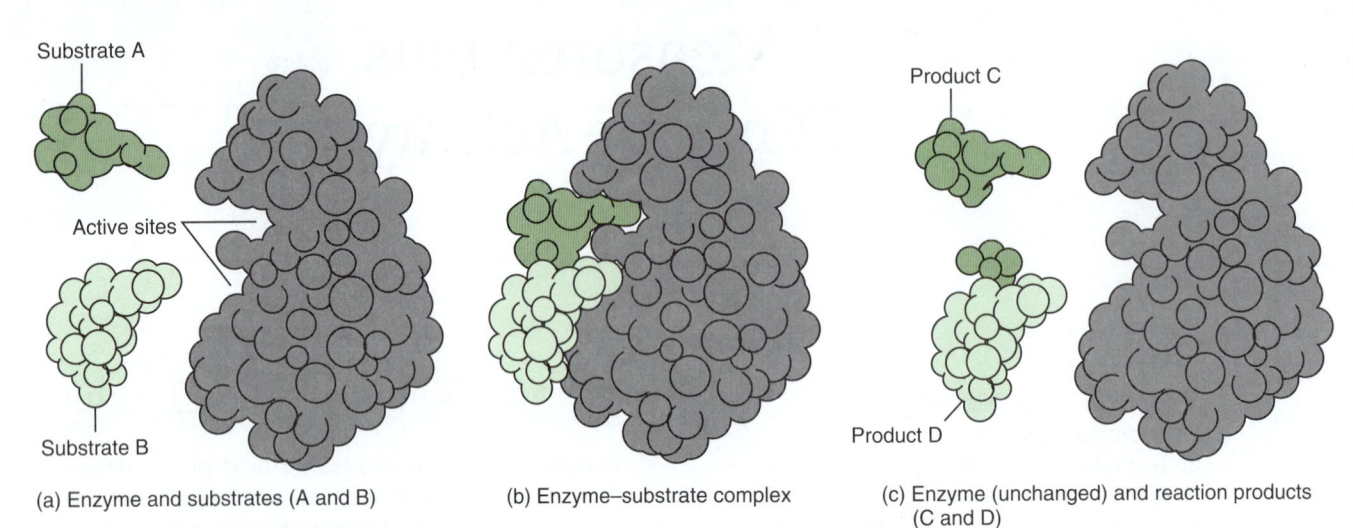

A + B → C + D
(Reactants) Enzyme (Products)

Substrate A

Active sites

Substrate B

Product C

Product D

(a) Enzyme and substrates (A and B) (b) Enzyme–substrate complex (c) Enzyme (unchanged) and reaction products (C and D)

Figure 2.10 **The lock-and-key model of enzyme activity.** (a) Substrates A and B fit into the active sites of the enzyme, forming (b) an enzyme-substrate complex. This complex then dissociates (c) to yield free enzyme and the products of the reaction. AP|R

(For a full-color version of this figure, see fig. 4.2 in *Human Physiology*, thirteenth edition, by Stuart I. Fox.)

an enzyme), whereas *phosphatase* enzymes hydrolyze the phosphate group from a wide variety of organic compounds. These guidelines do not apply to enzymes discovered before a systematic terminology was developed, such as the digestive enzymes *pepsin* and *trypsin* that hydrolyze peptide bonds.

Enzymatic activity is dependent on the delicate bending and folding of polypeptide chains, so changes in pH and temperature, which affect the three-dimensional structure of proteins, also affect enzymatic activity. Although most enzymes display maximum activity at pH 7 and at 40° C to 45° C, the **pH optimum** and **temperature optimum** of one enzyme may be significantly different from those of another enzyme. Enzymatic activity diminishes when the pH or temperature varies from the optimal value for that enzyme.

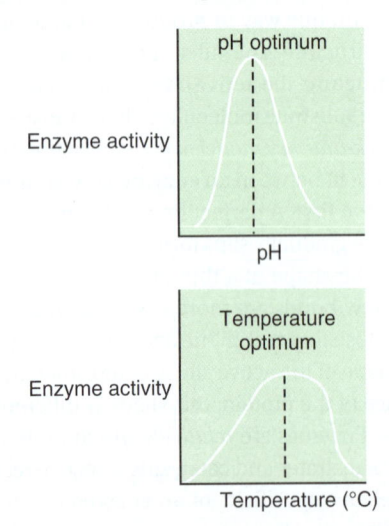

Enzyme activity vs pH (pH optimum)

Enzyme activity vs Temperature (°C) (Temperature optimum)

The activity of many enzymes is absolutely dependent on the presence of specific, smaller, nonprotein molecules called *cofactors* or *coenzymes*. Cofactors are specific inorganic ions (e.g., Mg^{2+}) required for enzyme activity, and coenzymes are organic compounds derived from the water-soluble vitamins that play a similar role.

To assay (measure) the enzyme activity in an unknown sample, the appropriate cofactors or coenzymes must be provided and the pH and temperature must be standardized. In addition, the concentration of substrate should be very large relative to the concentration of enzyme, so that the availability of substrate does not limit the rate of the reaction. Under these conditions, the amount of product doubles when the concentration of enzyme doubles or when the reaction time doubles. If the reaction time is held constant, the amount of product formed is *directly proportional* to the enzyme concentration.

The concentration of the enzyme in a sample is usually measured in **units of activity,** determined under conditions of specified pH, temperature, reaction time, and other controlled factors. A unit of enzyme activity is a measure of the reaction rate and may be determined by either the quantity of substrate consumed or the quantity of product formed in a given time interval. Specifically, one **international unit (IU)** is defined as *the quantity of enzyme required to convert one micromole of substrate per minute into products* under specified conditions of pH, temperature, and other controlled conditions. For example, the normal concentration of the enzyme alkaline phosphatase in serum ranges from 9 to 35 international units per liter (IU/L).

In these exercises, you will assay (measure) the activity of three enzymes found in serum. These enzymes are not normally active in serum but are enzymes released into the blood from ruptured cells of damaged tissues. These enzymatic assays are often valuable in the diagnosis of certain diseases.

A. Catalase in Liver

Catalase is an enzyme that converts hydrogen peroxide (H_2O_2) into water and oxygen gas. Found in many tissues, catalase is one of the most rapidly acting enzymes in the body.

$$2H_2O_2 \xrightarrow{\text{Catalase}} 2H_2O + O_2$$
$$\text{Substrate} \qquad\qquad \text{Products}$$

Procedure

1. Fill two small 100 mL beakers to the halfway mark with hydrogen peroxide.
2. Immerse a rusty nail or similar object in the solution in one beaker and then gently stir. Observe the effect of an inorganic catalyst and record your observations in the laboratory report.
3. Mince a fresh chicken liver with scissors, and add it to the hydrogen peroxide in the second beaker. Stir the solution and record your observations in the laboratory report.

B. Measurement of Alkaline Phosphatase in Serum

Plasma contains a number of enzymes released from damaged tissue cells. Assays of these enzymes can thus be useful in clinical diagnosis. In this exercise, you will determine the concentration of plasma **alkaline phosphatase** by measuring the increase in the absorbance of a solution caused by the accumulation of the product of the enzymatic reaction over time. The increase in absorbance of the solution with time is due to the formation of the yellow product, p-nitrophenol, which occurs when alkaline phosphatase hydrolyzes the phosphate from the substrate p-nitrophenylphosphate at a high pH.

Clinical Applications

There are two enzymes in serum that display phosphatase activity (remove phosphate from organic compounds). One of these has a pH optimum of 4.9 *(acid phosphatase)*; the other has a pH optimum of 9.8 *(alkaline phosphatase)*. Abnormally high levels of serum acid phosphatase activity may be noted in patients with cancer of the prostate, whereas elevated alkaline phosphatase activity is primarily associated with various liver and bone diseases.

Normal Values The normal range for alkaline phosphatase concentrations in plasma is 9–35 IU/L.

Procedure

1. Obtain three cuvettes. Label one *U* (for unknown serum), one *S* (for standard reference), and one *B* (for the water blank).
2. Pipette 0.5 mL of substrate reagent into each cuvette, and place in a 37° C water bath. Let the solutions sit in the water bath for approximately 5 minutes.
3. Pipette 50 μL (0.05 mL) of the unknown serum to the cuvette marked *U*. Wait 30 seconds, and pipette 50 μL of the enzyme standard (containing 25 IU/L) to the cuvette marked *S*. Wait 30 seconds, and pipette 50 μL of distilled water to the cuvette marked *B*. Be sure to return the cuvettes to the water bath immediately.
4. Ten minutes after having added the serum to cuvette *U*, add 2.5 mL of color developer to this cuvette, cap, and mix. Repeat this procedure 30 seconds later with cuvette *S*, and then with cuvette *B*. Each cuvette, in this way, will have incubated for exactly 10 minutes before you terminated the reaction by adding color developer.
5. Set the colorimeter at a wavelength of 590 nm. Standardize the colorimeter using the blank cuvette *(B)* and read the absorbance values of cuvettes *U* and *S*. Record these values in your laboratory report.
6. Calculate the concentration of alkaline phosphatase in the unknown serum sample (cuvette *U*) using the Beer's law formula, and enter your value in the laboratory report.

Example

Suppose the enzyme reference standard (25 IU/L) had an absorbance of 0.28, and the unknown serum had an absorbance of 0.34. The concentration of the unknown could be calculated as:

$$\frac{0.34}{0.28} \times 25 \text{ IU/L} = 30 \text{ IU/L}$$

C. Measurement of Lactate Dehydrogenase in Serum

An enzyme of great importance in physiology and medicine is **lactate dehydrogenase (LDH)**. In skeletal muscles during heavy exercise, this enzyme catalyzes the conversion of pyruvic acid to lactic acid. In the process of this reaction, the coenzyme $NADH + H^+$ is oxidized to NAD:

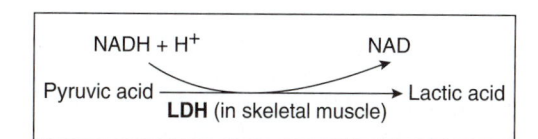

$$\text{Pyruvic acid} \xrightarrow[\text{LDH (in skeletal muscle)}]{NADH + H^+ \quad\to\quad NAD} \text{Lactic acid}$$

The same enzyme also catalyzes the reverse reaction, whereby lactic acid is converted into pyruvic acid. This reverse reaction occurs in normal heart tissue. It also occurs in the liver, which can use lactic acid for energy production (cell respiration) or in the formation of glucose. This reaction is accompanied by the reduction of NAD to NADH + H$^+$ as:

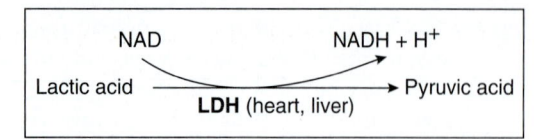

Lactic acid $\xrightarrow[\text{LDH (heart, liver)}]{\text{NAD} \quad\quad \text{NADH + H}^+}$ Pyruvic acid

Clinical Applications

Serum **lactate dehydrogenase (LDH)** concentrations are elevated following a myocardial infarction (MI or "heart attack"). Measurements of these levels, in conjunction with other tests, can help diagnose heart attacks. Serum LDH levels are also elevated in some renal (kidney) diseases, cirrhosis of the liver, and hepatitis.

Normal Values The normal LDH concentration for healthy adults is 100–225 IU/L.

Procedure

1. Using a mechanical pipettor (do not pipette by mouth), pipette 1.0 mL of working reagent (containing lactate and NAD) into a cuvette.
2. Place the cuvette in a 37° C water-bath for 5 minutes. This is a preincubation to bring the substrate to the proper temperature.
3. While waiting, set the spectrophotometer to a wavelength of 340 nm.
4. While waiting, fill a cuvette with water, and using this as the blank, set the spectrophotometer to read zero absorbance.
5. Pipette 0.05 ml (50 μL) of the serum sample into the cuvette that was prewarmed in the water-bath. Immediately return it to the water-bath.
6. Wait *30 seconds,* and then quickly remove the cuvette from the water-bath, insert it into the spectrophotometer, and take an absorbance reading.
7. Return the cuvette to the water-bath, and then, *1 minute* after your first reading, take a second absorbance reading. The second absorbance reading should be higher, because the conversion of NAD to NADH in the reaction results in an increase in absorbance.
8. Subtract the first absorbance reading from the second absorbance reading to obtain the change in absorbance per minute: _____.
9. To obtain the enzyme concentration in IU/L, multiply the change in absorbance by 3,376 (a conversion factor). Enter your measurement in the laboratory report.

Example

Suppose the first absorbance reading was 0.10 and the second absorbance reading was 0.15. The difference in absorbance is 0.05. Then,

$$0.05 \times 3{,}376 = 169 \text{ IU/L}$$

Laboratory Report 2.4

Name _____

Date _____

Section _____

DATA FROM EXERCISE 2.4

A. Catalase in Liver

1. Describe what occurred when the rusty nail was placed in the beaker of hydrogen peroxide. Write a simple equation that might describe this reaction.

2. Describe what occurred when the chicken liver was placed in the beaker of hydrogen peroxide. Explain what occurred, and write a simple equation that might describe this reaction.

B. Measurement of Alkaline Phosphatase in Serum

1. Record your absorbance values for cuvettes U and S in these spaces.
   Absorbance of U: _____
   Absorbance of S: _____
   Calculate the alkaline phosphatase concentration of the unknown serum: _____ IU/L.

2. Was the value you obtained in the normal range? What might an abnormally high value indicate?

65

C. Measurement of Lactate Dehydrogenase in Serum

1. Calculate the lactate dehydrogenase concentration of the unknown serum: _____ IU/L.

2. Was the value you obtained in the normal range? What might an abnormally high value indicate?

REVIEW ACTIVITIES FOR EXERCISE 2.4

Test Your Knowledge

1. Define an *international unit (IU)* of enzyme activity. _____
   _____
   _____

2. Define the *pH optimum* of an enzyme. _____
   _____
   _____

3. Define the *temperature optimum* of an enzyme. _____
   _____
   _____

4. What are cofactors and coenzymes? _____
   _____
   _____

5. Judging from its name, describe the activity of these enzymes:

   (a) phosphatase _____
   _____
   _____

   (b) glycogen synthetase _____
   _____
   _____

   (c) lactate dehydrogenase _____
   _____
   _____

   (d) DNA polymerase _____
   _____
   _____

Test Your Understanding

6. What must you add to a sample of plasma to measure the activity of a particular enzyme in plasma? How might you measure the activity of a different enzyme in the same tube of plasma?

7. What happens to the structure of an enzyme when the pH and temperature are changed from the optima for that enzyme?

8. Describe the reactions catalyzed by lactate dehydrogenase in the skeletal muscles, liver, and heart. What is the metabolic significance of these reactions? What happens to NAD and NADH in these reactions?

Test Your Analytical Ability

9. Why were you so careful to have each tube incubate for exactly 10 minutes? What might have occurred if you allowed the tubes to incubate overnight before finally stopping the reaction? Would this be an accurate test? Explain.

10. Pancreatic amylase is an enzyme normally secreted by the pancreas into the small intestine where it catalyzes the hydrolysis of starch. Should it be in the blood? If not, what might explain its presence in the blood? Would it have any activity as it circulates in the blood plasma? Explain.

Test Your Quantitative Ability

11. Suppose a reference standard of a particular enzyme contains 50 IU/L and has an absorbance of 0.15, while an unknown plasma sample has an absorbance of 0.21. Calculate the enzyme activity per liter in the plasma.

Clinical Investigation Questions

12. What is the activity of phosphatases, how do acid and alkaline phosphatase differ, and what are some possible causes of abnormal levels?

13. What is the activity of lactate dehydrogenase, and what might cause plasma levels to be abnormally elevated?

14. Given the abnormally elevated levels of both alkaline phosphatase and lactate dehydrogenase, is there one particular organ that is likely affected? Explain.

EXERCISE 2.5

Genetic Control of Metabolism

MATERIALS

1. Test tubes, Pasteur pipettes (droppers), urine collection cups
2. Phenistix (Ames Laboratories), silver nitrate (3 g per 100 mL), 10% ammonium hydroxide, 40% sodium hydroxide, lead acetate (saturated)

A different gene codes for the production of each enzyme, so a defective gene can result in a specific metabolic disorder. These inborn errors of metabolism can be detected by tests for specific enzyme products.

LEARNING OUTCOMES

You should be able to:

1. Define the terms *phenotype, genotype, transcription,* and *translation*.
2. Describe how genes regulate metabolic pathways.
3. Describe how inborn errors of metabolism are produced.
4. Explain the etiology (cause) of phenylketonuria (PKU).

Textbook/Multimedia Correlations

Before performing this exercise, you should study the introductory material presented here. Further information relating to this exercise can also be found in these pages of *Human Physiology,* thirteenth edition, by Stuart I. Fox:

- *Cell Nucleus and Gene Expression.* Chapter 3, p. 62.
- *Metabolic Pathways.* Chapter 4, p. 95.

Clinical Investigation

An albino student performing these exercises tested negative for phenylketonuria (PKU) and alkaptonuria, but tested positive for cystine in the urine.

- Explain the cause of the student's albinism, the reason the test for phenylketonuria was negative, and what significance a positive PKU would have.
- Explain the possible medical significance of a positive test for cystine in the urine.

The physical appearance of an individual (**phenotype**) is largely determined by the individual's genetic endowment (**genotype**). The control of the phenotype by the genotype is achieved by means of the genetic regulation of cellular metabolism.

All of the chemical reactions of cellular metabolism are catalyzed by specific enzymes, and the information for the synthesis of each specific enzyme is coded by a specific gene. That is, one gene makes one enzyme, capable of catalyzing one type of reaction (changing *A* to *B*, for example).

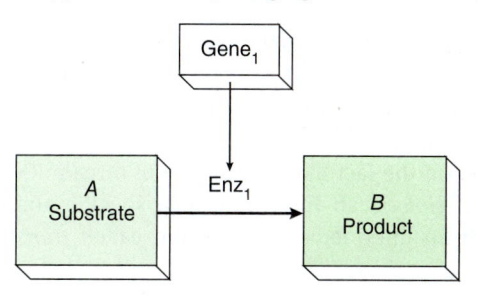

The product of this reaction may become the substrate of a second enzyme, made by a second gene, which converts *B* into a new product, *C*. A third enzyme, made by a third gene, may then convert *C* into *D*. Thus, a **metabolic pathway** is formed where the product of one enzyme becomes the substrate of the next. In a metabolic pathway, some initial substrate, *A*, is converted into a final product (such as *D*) through a number of *intermediates* (*B* and *C*, for example).

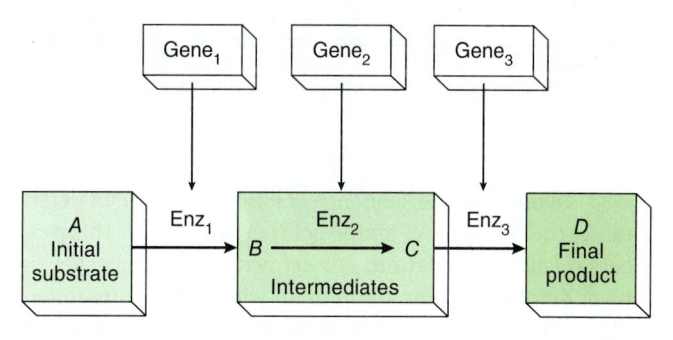

Genes contain the information for the synthesis of all proteins, not only those with enzymatic activity. The genetic code is based on the sequence of **DNA** components known as **nucleotide bases** *(adenine, guanine, cytosine, thymine)*. Thus, the sequence of these bases is different in different genes.

69

The sequence of bases in the DNA that composes one gene is used as a template for the synthesis of one **RNA** molecule. The RNA molecule consists of a linear sequence of nucleotide bases (uracil, cytosine, guanine, and adenine), precisely complementary to the sequence of bases in the region of the DNA (gene) on which it was made. This complementarity is ensured because only a specific base on the RNA can bind to a specific base on the DNA.

A part of the genetic code is thus transcribed by the synthesis of a specific RNA molecule—a process called **transcription.** This RNA contains a part of the genetic message and is called *messenger RNA (mRNA)*.

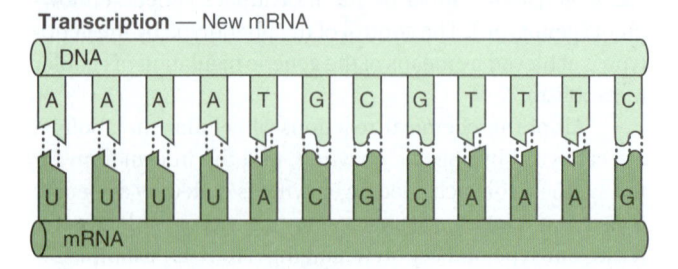

Transcription — New mRNA

The messenger RNA, in association with ribosomes, forms the template for **protein synthesis,** where the sequence of amino acids in the protein is specified by the sequence of bases in the mRNA. The genetic code is based on the fact that a sequence of three mRNA bases (a *base triplet*) can bind only to a specific amino acid through an intermediate compound called *transfer RNA (tRNA)*.

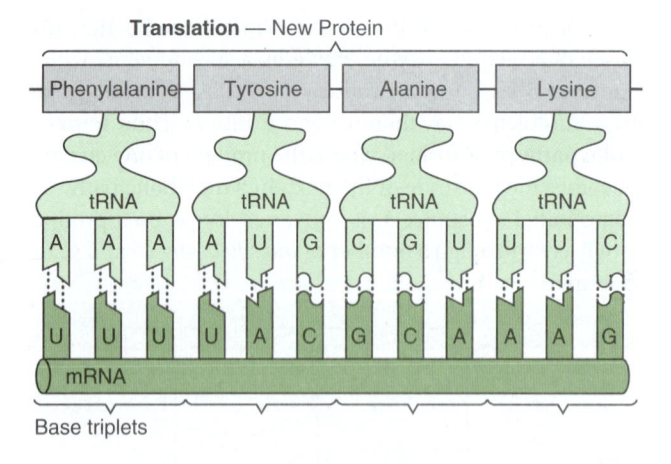

Translation — New Protein

In this way, the sequence of bases in the mRNA determines the sequence of amino acids in the protein. This process is called **translation.** The sequence of amino acids, in turn, determines how the protein will fold (i.e., its three-dimensional structure). The three-dimensional structure of a protein directly determines its function.

INBORN ERRORS OF METABOLISM

When a gene is missing or defective, the enzyme that it makes will also be missing or defective. This will result in a hereditary metabolic disorder in which there will be a decrease in the intermediates formed *after* the step normally catalyzed by that enzyme. The intermediates formed *before* this step will accumulate in the blood and body tissues and will be excreted in the urine.

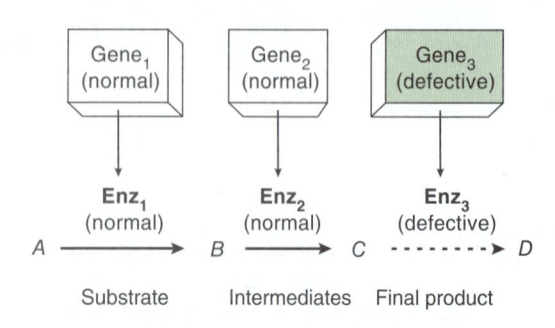

Thus, intermediate D decreases, while intermediates A, B, and C increase.

There are a number of genetic defects associated with the metabolism of the amino acids *phenylalanine* and *tyrosine*. The phenotypic (expressed) effects of these **inborn errors of metabolism** depend on which enzymes are defective and therefore on which intermediate products either are absent or accumulate abnormally in the body tissues.

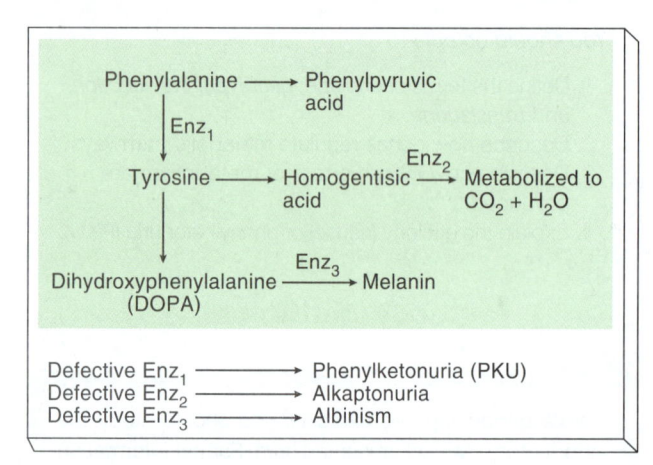

Probably the best known phenotypic effect of this group of hereditary disorders is the lack of *melanin* pigment characteristic of **albinism. Phenylketonuria (PKU)** is the most clinically serious disorder of this group, since the accumulation of phenylpyruvic acid can affect the developing central nervous system and produce mental retardation. The odious effects of PKU can be avoided only by placing the affected child on a special low phenylalanine diet.

A. PHENYLKETONURIA

A person with PKU excretes large amounts of phenylpyruvic acid because of the inability to convert the amino acid phenylalanine into the amino acid tyrosine.

Procedure

1. Dip the test end of a Phenistix strip into a sample of urine.

   Note: *Use care when handling all body fluids, including urine. Clean spills and dispose of urine containers properly, as described by your instructor. Alternatively, synthetic urine can be used.*

2. Compare the color of the strip with the color chart provided.

Note: *Federal law now mandates the Guthrie test using a spot of blood from newborn babies for diagnostic screening for PKU. Therefore, this procedural kit may not be readily available.*

B. ALKAPTONURIA

A person with **alkaptonuria** excretes large amounts of homogentisic acid, which reacts with silver to form a black precipitate. This condition does not have immediate adverse effects on health, but may cause the development of a characteristic type of joint degeneration later in life.

Procedure

1. Add 10 drops of urine (or synthetic urine) to a test tube containing 5 drops of 3% silver nitrate ($AgNO_3$).
2. Add 5 drops of 10% ammonium hydroxide (NH_4OH) to the tube and mix.
3. A positive test for homogentisic acid is indicated by the presence of a black precipitate.

C. CYSTINURIA

The renal tubules can normally reabsorb all of the amino acids filtered in the glomerulus. People with the recessive trait known as **cystinuria,** however, have an impaired ability to reabsorb the amino acid *cystine* and the related amino acids lysine, arginine, and ornithine.

Cystine is the least soluble amino acid and may precipitate in the urinary tract to form stones. This condition accounts for 1% of the cases of renal stones in the United States. (About 10% are uric acid stones, and the remainder are formed from calcium salts, primarily calcium oxalate.)

Procedure

1. Add 1.0 mL of 40% NaOH to a test tube containing 5.0 mL of urine (or synthetic urine). Allow the tube to cool.
2. Add 3.0 mL of lead acetate. A brown-to-black precipitate indicates the presence of cystine in the urine.

Clinical Applications

Defective genes generally cannot be replaced with "good" genes, although a few such treatments have now been reported. At present, therefore, treatment of those with genetic disorders depends upon early diagnosis and preventive measures. Newborn babies are routinely tested for PKU; and those with the enzyme defect can be placed on low-phenylalanine diets. Unfortunately, many other inborn errors of metabolism cannot be treated by dietary restrictions. The only way to prevent some of these diseases is through genetic counseling for prospective parents who are carriers of the disease.

When there is a defective enzyme in a metabolic pathway, the molecule that is the substrate of that enzyme accumulates in particular tissues of the body. Such inborn errors of metabolism are not restricted to pathways of amino acid metabolism. In *Tay-Sachs disease,* for example, there is a defect in the enzyme (hexosaminidase A) that breaks down a complex type of lipid known as a ganglioside, resulting in lipid accumulation in the brain and retina. This disease, which occurs primarily in certain families of European Jewish origin (specifically, Ashkenazic Jews), is invariably fatal. Inborn errors also occur in carbohydrate metabolism. In *glycogen storage disease,* for example, the enzyme that breaks down glycogen may be defective, resulting in the excessive accumulation of glycogen, which causes liver damage.

Laboratory Report 2.5

Name _____
Date _____
Section _____

REVIEW ACTIVITIES FOR EXERCISE 2.5

Test Your Knowledge

1. Define these terms:

    (a) *genetic transcription* _____

    (b) *genetic translation* _____

2. In one sentence, distinguish between phenotype and genotype. _____

Test Your Understanding

3. Using letters *A, B, C, D,* and *E* for molecules, illustrate how these would interrelate in an unbranched metabolic pathway. Then, illustrate a metabolic pathway in which *C* branches to form *D* and *E*. Describe how your two pathways would be affected by an inborn error of metabolism between *C* and *D*.

4. Explain the etiology of phenylketonuria (PKU).

Test Your Analytical Ability

5. Propose reasons why it would be difficult to correct an inborn error of metabolism. Explain why a simple pill or even an injection probably wouldn't be effective.

6. Speculate as to how the information obtained from the sequencing of the human genome might be used to someday eliminate inborn errors of metabolism.

Clinical Investigation Questions

7. What caused the student's albinism, and how does that relate to phenylketonuria? Do you think the student was surprised to have a negative PKU test? Explain.

8. What is cystinuria, and what is its medical significance?

EXERCISE 2.6

Diffusion, Osmosis, and Tonicity

MATERIALS

1. Test tubes, thistle tubes, dialysis tubing
2. Beakers, ring stands, burette clamp
3. Lancets and alcohol swabs; or animal (dog, cat, or rat) blood from veterinary or other appropriate source. Biohazard receptacle for all blood contaminated items.
4. Microscopes, slides, coverslips, and transfer pipettes
5. Make two dilutions of molasses, 200 mL molasses (a 12-oz. jar) diluted with water to 1 L (20% solution); and 250 mL molasses diluted with water to make 1 L (25% solution)
6. Sodium chloride solutions in dropper bottles at these concentrations: 0.20 g per 100 mL; 0.45 g per 100 mL; 0.85 g per 100 mL; 3.5 g per 100 mL; and 10 g per 100 mL
7. Toluene, potassium permanganate crystals, vegetable oil, laboratory detergent

Osmosis is the net diffusion of water (solvent) through a membrane that separates two solutions. Osmosis occurs passively when the two solutions have different total concentrations of solutes to which the membrane is relatively impermeable. If there is no osmosis when a membrane separates two solutions, those solutions are said to be isotonic to each other.

LEARNING OUTCOMES

You should be able to:

1. Distinguish between *solute, solvent,* and *solution.*
2. Define *passive transport, diffusion,* and *active transport.*
3. Define *osmosis, osmotic pressure,* and *osmolality.*
4. Define *isotonic, hypotonic,* and *hypertonic.*
5. Calculate the osmolality of solutions when the concentration of solute (in g/L) and the molecular weight of a solute are known.
6. Describe how red blood cells (RBCs) are affected when they are placed in isotonic, hypotonic, and hypertonic solutions.

Textbook/Multimedia Correlations

Before performing this exercise, you should study the introductory material presented here. Further information relating to this exercise can be found in these pages of *Human Physiology,* thirteenth edition, by Stuart I. Fox:

- *Osmosis.* Chapter 6, p. 136.
- *Regulation of Blood Osmolality.* Chapter 6, p. 141.
- Multimedia correlations are provided in *MediaPhys 3.0:* Topics 3.9–3.24.

Clinical Investigation

A child with an intestinal disorder that caused a low plasma albumin concentration developed edema (see exercise 2.1, fig. 2.3). She was given an isotonic intravenous solution containing albumin and normal saline.

- Explain the osmotic importance of plasma albumin and why edema occurred.
- Identify the significance of normal saline and the meaning of the terms *osmotic pressure* and *tonicity.*

If you were to drop a pinch of sugar (the *solute*) into a beaker of water (the *solvent*), the resulting *solution* would, after a time, have a uniform sweetness. The uniform sweetness would result from the constant state of motion of all of the solute and solvent molecules in the solution, producing a net movement of solute molecules from regions of higher concentration to regions of lower concentration. This net movement of molecules is known as **diffusion.**

The rate of diffusion is proportional to the concentration differences that exist in the solution. The diffusion rate will steadily decrease as the solute becomes evenly distributed in the solvent, and the net movement of molecules by diffusion (but not their random motion) will cease entirely when the concentration differences are abolished (when the solution is uniformly concentrated).

A molecule may move into or out of a cell by diffusion if (1) a difference in the concentration of that molecule *(concentration gradient)* exists between the intracellular and extracellular compartments and (2) the plasma (cell) membrane will allow the passage of that molecule.

The movement of a molecule across the plasma (cell) membrane by diffusion is called **passive transport.** The term *passive* is used because the cell need not expend metabolic energy in the process. By contrast, cells often must move molecules across the plasma membrane from lower to higher concentrations; that is, cells must "fight" diffusion in the attempt to maintain a concentration difference across the membrane. However, to move molecules "uphill" against their concentration gradients, the cells must expend metabolic energy. This process is called **active transport.** Sodium, for example, is maintained at a higher concentration outside the cell than inside the cell, whereas potassium is maintained at a higher concentration inside the cell than outside the cell (fig. 2.11).

The *permeability* of a membrane refers to the ease with which substances can pass through (permeate) it. A membrane completely permeable to all molecules is not a barrier to diffusion, whereas a membrane completely impermeable to all molecules essentially divides the solutions into two noncommunicating compartments. A living cell must selectively interact with its environment, taking in raw materials and excreting waste products, so the cell is surrounded by a membrane that is **semi-** (or **selectively**) **permeable.** A semipermeable membrane is completely permeable to some molecules, slightly permeable to others, and completely impermeable to still others.

The **plasma (cell) membrane** is composed primarily of two semifluid phospholipid layers with proteins. Some proteins are partially submerged; others span the complete thickness of the membrane. In this way, the membrane is not continuous but behaves as if tiny protein channels were serving as waterways for diffusion, allowing the passage of ions and smaller molecules while excluding the passage of molecules larger than the channels. (fig. 2.12).

A. Solubility of Compounds in Polar and Nonpolar Solvents

Most of the molecules that a cell encounters are water soluble (easily dissolved in water). Such molecules have charged groups and are said to be *polar*. The lipids of the cell membrane are *nonpolar* and serve as a barrier to the passage of polar molecules across the membrane. Small polar molecules may pass through the pores in the lipid barrier, but large polar molecules, such as proteins and polysaccharides, are restricted by the pore size.

Many organic solvents (benzene, or toluene, for example) are nonpolar; that is, they are soluble in lipids but not in water. Such nonpolar molecules are not limited in their passage by the membrane pores and can rapidly diffuse into the cells by passing through the lipid layers of the membrane.

Procedure

1. Pour about 2.0 mL of water and 2.0 mL of toluene into a test tube.
2. Shake the tube, and record your observations in the laboratory report.
3. Using forceps, drop 2–3 crystals of potassium permanganate (KMnO$_4$) into the tube. Shake the tube, and record your observations in the laboratory report.
4. Add about 1.0 mL of yellow vegetable oil to the tube. Shake the tube, and record your observations in the laboratory report.
5. Add a pinch of laboratory detergent to the tube. Shake the tube, and record your observations in the laboratory report.

Note: *One end of the detergent molecule is polar (charged); the other end is nonpolar. The detergent can thus act as a bridge between the two phases.*

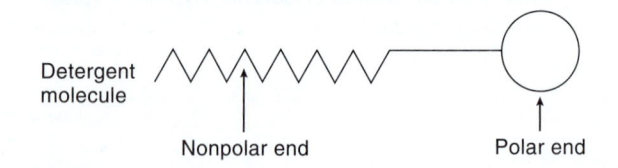

Intracellular fluid concentrations	Ion	Extracellular fluid concentrations
12 mM	Na$^+$	145 mM
150 mM	K$^+$	5 mM
9 mM	Cl$^-$	125 mM
0.0001 mM	Ca^{2+}	2.5 mM

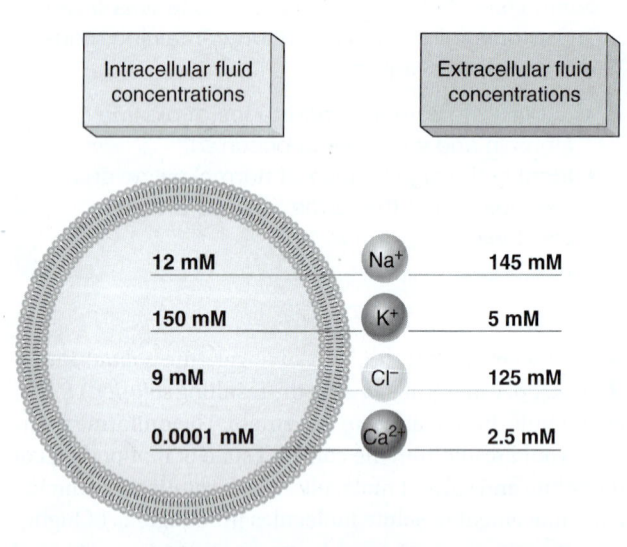

Figure 2.11 Concentrations of ions in the intracellular and extracellular fluids. This distribution of ions, and the different permeabilities of the plasma membrane to these ions, affects the membrane potential and other physiological processes. (Molarity, symbolized by M, is a unit of concentration based on the atomic weight.)

(For a full-color version of this figure, see fig. 6.26 in *Human Physiology*, thirteenth edition, by Stuart I. Fox.)

Detergent molecule — Nonpolar end / Polar end

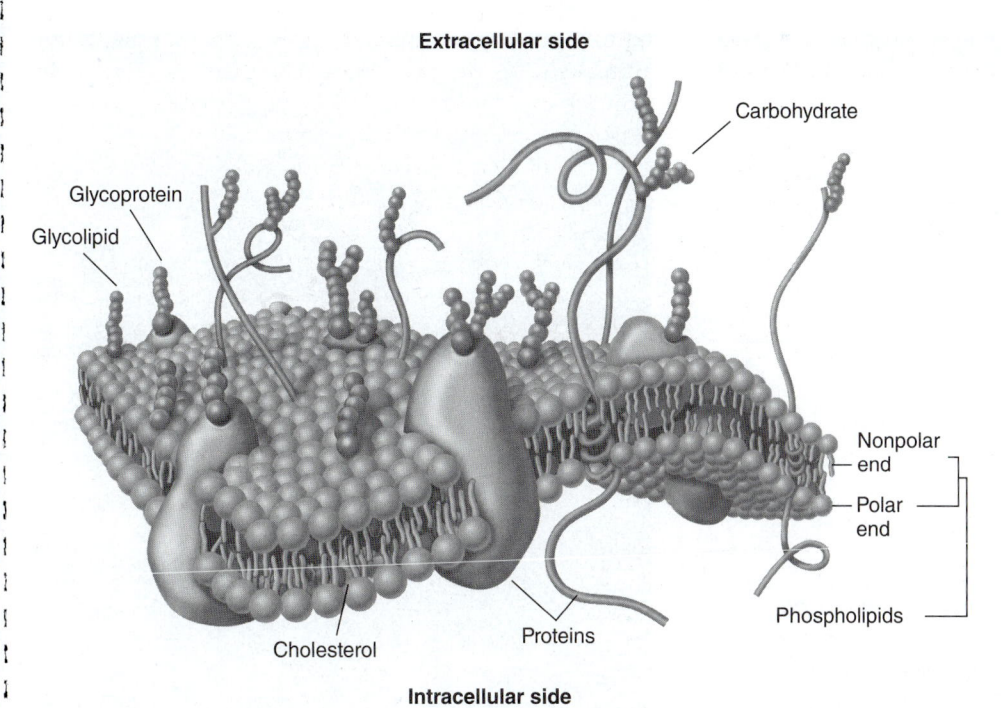

Figure 2.12 **Structure of the plasma (cell) membrane.** The plasma membrane consists of a double layer of phospholipids, with the phosphate-containing polar ends (spheres) oriented outward and the nonpolar portions of the molecules (wavy lines) oriented toward the center. Proteins are interposed between the phospholipids, and carbohydrates are often bound to the external surface of these proteins. AP|R

(For a full-color version of this figure, see fig. 3.2 in *Human Physiology*, thirteenth edition, by Stuart I. Fox.)

B. Osmosis Across an Artificial Semipermeable Membrane

Imagine a solution divided into two compartments by a membrane. If the membrane is completely permeable to solute and solvent molecules, these molecules will be able to diffuse across it so that the solute/solvent ratio (concentration) will be the same on both sides of the membrane. Suppose, however, that the membrane is permeable to the solvent but not to the solute. If the solvent is water, the water will diffuse from the region where the solute/solvent ratio is *lower* (relatively more water) to the region where the solute/solvent ratio is *higher* (relatively less water), until the solute/solvent ratio (concentration) is the same on both sides of the membrane. The net diffusion of water across a membrane is called **osmosis** (fig. 2.13).

In osmosis, water diffuses into the more concentrated (greater solute/solvent ratio) solution from the less concentrated solution. The more highly concentrated solution is said to have a greater **osmotic pressure** than the less concentrated solution. The osmotic pressure is a measure of the ability of a solution to "pull in" water from another solution separated from it by a semipermeable membrane. Keep in mind, however, that the "pulling" is a metaphor; because osmosis is the simple diffusion of water through a membrane, water moves into the more concentrated solution as a result of the higher-to-lower *water* concentration gradient. Since the osmotic pressure of a solution is proportional to its solute concentration, the osmotic pressure of distilled water is zero.

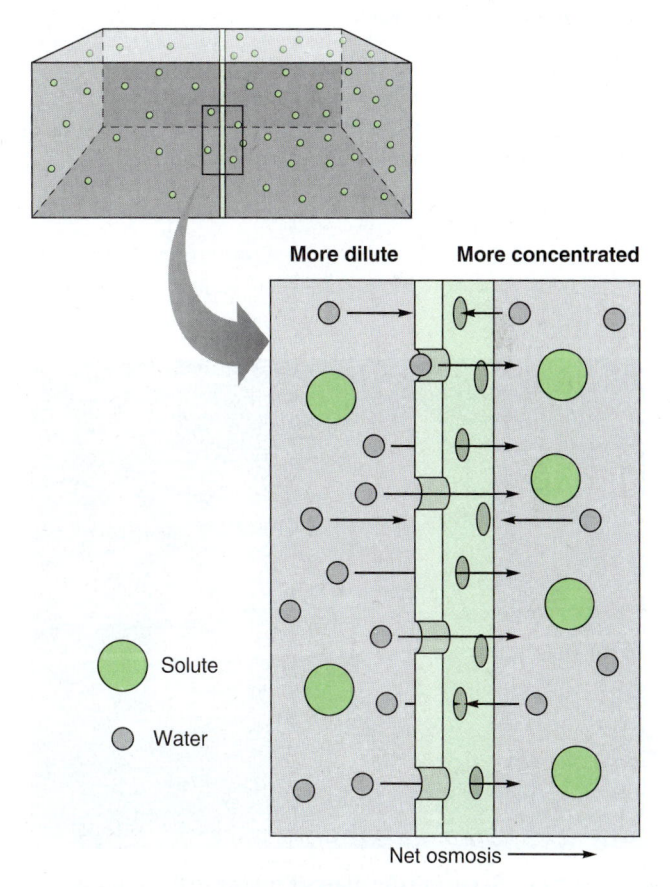

Figure 2.13 **A model of osmosis.** The solution on the left is more dilute than the one on the right, causing water to diffuse from left to right by osmosis across the semipermeable membrane. AP|R

(For a full-color version of this figure, see fig. 6.7 in *Human Physiology*, thirteenth edition, by Stuart I. Fox.)

Procedure

1. Cut a 2½-inch piece of dialysis tubing. Soak this piece in tap water until the layers separate. (Speed this process by rotating the tubing between two fingers.) Slide one blade of the scissors inside the tube and cut lengthwise, producing a single rectangular sheet of dialysis membrane.

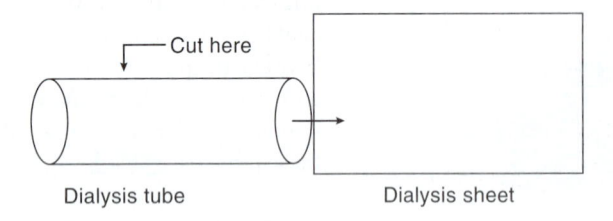

Dialysis tube — Cut here — Dialysis sheet

Note: *Dialysis tubing is a plastic porous material used to separate molecules on the basis of their size. It is an artificial semipermeable membrane. Molecules larger than the pore size remain inside the tubing, whereas smaller molecules (including water) can move through the membrane by diffusion. This technique of physical separation is called* **dialysis.**

2. Place the rectangular piece of dialysis tubing tightly over the mouth of the thistle tube so that no air is trapped between the dialysis tubing and the molasses solution. Keeping the dialysis tubing taut, secure it to the thistle tube with several wrappings of a rubber band (fig. 2.14).
3. Divide the class into two molasses groups (20% and 25%). With the thistle tube secured in a vertical position in the mouth of a large beaker, thread a narrow-diameter plastic tube longer than the thistle tube down into it. Then, slowly pour the molasses into the other end (fig. 2.15).
4. Pour water into the beaker until the level is just below the dialysis tubing over the mouth of the thistle tube. Then, gently lower the thistle tube so that the dialysis membrane is in the water (fig. 2.16). Do not place the membrane all the way to the bottom of the beaker.
5. Mark the meniscus of the molasses solution on the thistle tube with a grease pencil. Every 15 minutes for a 1-hour period, record the change in the level of this meniscus (in centimeters) in the laboratory report.

Figure 2.15 Pouring molasses into the thistle tube. The narrow plastic tube is optional, helping to prevent air bubbles. The thistle tube should be filled only until the meniscus of the molasses is about one-third up the narrow portion. Once the lower part of the thistle tube is immersed in water, the meniscus of the molasses on the thin segment of the thistle tube should be marked.

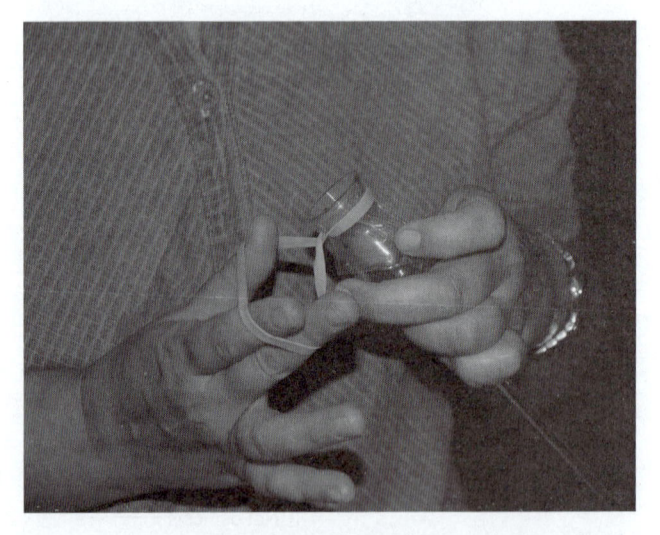

Figure 2.14 Securing the dialysis tubing to the mouth of the thistle tube. The dialysis membrane is pulled tight. A broad rubber band is best for this purpose.

Thistle tube

Wax mark on tube

Dialysis membrane

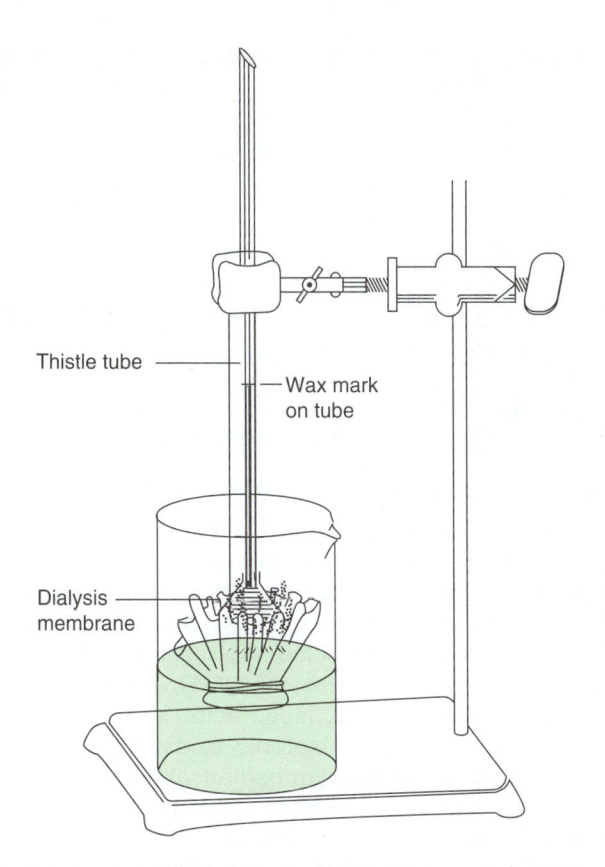

Figure 2.16 A thistle tube setup for the osmosis exercise.

C. Concentration and Tonicity

Osmosis, the net diffusion of water across a membrane, requires that the membrane be completely permeable to water but only partially permeable or completely impermeable to solute molecules. In this case, the nonpenetrating solutes are said to be **osmotically active**. Osmosis is then caused by a difference in the concentration of osmotically active solutes, and the osmotic pressure of a solution is proportional to the concentration of osmotically active solutes. Solutes that are as freely permeable as water are not osmotically active.

The simplest way to express the concentration of a solution is the weight of the solute, in grams or milligrams, per 100 mL of solution. For example, 150 mg in 100 mL may be expressed as 150 mg per 100 mL—or as 150 mg%, or 150 mg/dL.

It is frequently more useful to express concentration in terms of *molarity* or *molality*. These measurements take into account the different molecular weights of the solutes; for example, a one-molar (1 M) or a one-molal (1 m) solution of sodium chloride (NaCl) would require you to weigh out a different amount of solute than you would for a 1 M or a 1 m solution of glucose ($C_6H_{12}O_6$). The common unit of weight in a 1 M NaCl solution and a 1 M glucose solution is the **mole**. The significance of the mole value is that *solutions with equal molarities have equal numbers of molecules.* Although they weigh different amounts, a mole of NaCl contains the same number of molecules (Avogadro's number—6.02×10^{23}) as a mole of glucose.

One mole is equal to the molecular weight of the solute in grams. The molecular weight is obtained by adding the atomic weights of each element in the molecule.

	Sodium chloride (NaCl)	*Glucose* ($C_6H_{12}O_6$)
Atomic weights:	Na = 23.0	C_6 12 × 6 = 72
	Cl = 35.5	H_{12} 1 × 12 = 12
		O_6 16 × 6 = 96
Molecular weights:	58.5	180

A one-molar (1 M) solution contains 1 mole of solute in 1 L of solution. Thus, 1 M NaCl contains 58.5 g of NaCl per liter, whereas 1 M glucose contains 180 g of glucose per liter. A one-molal (1 m) solution of NaCl contains 1 mole of NaCl (58.5 g) dissolved in 1,000 g of solvent. If water is the solvent, 1,000 g equals 1,000 mL (at maximum density, 4° C). A 1 M solution has a final volume of 1,000 mL, whereas a 1 m solution has a final volume that slightly exceeds 1,000 mL.

Decreasing the value of both the solute and the solvent by the same proportion does not change the concentration of the solution. A 1 M glucose solution can be made by dissolving 90 g of glucose in a final volume of 500 mL or by dissolving 180 mg of glucose in 1 mL of solvent.

If the concentration of osmotically active solute is the *same* on both sides of a membrane (if the osmotic pressures are equal), osmosis will not occur. These two solutions are said to be **isotonic** to each other (*iso* means "same"). If the concentration of a third solution is *less* than that of the first two solutions, it is said to be **hypotonic** to the first two solutions (*hypo* means "below"). Water will diffuse from the third solution into the first two solutions if these solutions are separated by a semipermeable membrane. If the concentration of a fourth solution is *greater* than that of the first two solutions, it is said to be **hypertonic** (*hyper* means

Clinical Applications

Water returns from the tissue fluid to the blood capillaries because the protein concentration of blood plasma is higher than the protein concentration of tissue fluid. In contrast to other plasma solutes, plasma proteins cannot freely pass from the capillaries into the tissue fluid. Therefore, plasma proteins are *osmotically active* and cause osmotic return of tissue fluid to the capillaries (exercise 2.1; see fig. 2.3). If a person has an abnormally low concentration of plasma proteins, excessive accumulation of fluid in the tissues—a condition called **edema**—will result.

"above") to the first two solutions. Water will diffuse out of the first two solutions and into the fourth if these solutions are separated by a semipermeable membrane.

In all cases, water diffuses from the solution of lower osmotic pressure (lower solute concentration) to the solution of greater osmotic pressure (greater solute concentration). The osmotic pressure of a solution is proportional to the number of solute molecules in solution. A one-molal solution of glucose, for example, has 1 mole of glucose molecules in solution. This solution would have the same osmotic pressure as a 1 *m* solution of sucrose or a 1 *m* solution of urea. These solutions are said to have the same **osmolality** (1.0 osmole/kg water, or 1.0 Osm). A solution containing 1 mole of glucose plus 1 mole of urea would have an osmolality of 2 (2.0 Osm). The osmolality of a solution is determined by *the sum of all the moles of solute in a solution.*

Some molecules *dissociate* (come apart) when they are dissolved in solution. Common table salt (NaCl), for example, completely dissociates in solution to Na$^+$ and Cl$^-$ ions. Thus, a one-molal solution of NaCl has a total osmolality of 2 (one mole of Na$^+$ plus one mole of Cl$^-$), written 2.0 Osm. This one-molal solution of NaCl is isotonic to a two-molal solution of glucose, a molecule that does not dissociate when dissolved in solution.

It is frequently convenient to express concentration in terms of *milliosmolality (mOsm)*. A 0.1 *m* solution of NaCl, for example, has an osmolal concentration of 200 mOsm, whereas a 0.1 *m* solution of glucose has an osmolal concentration of 100 mOsm.

TONICITY OF SALINE SOLUTIONS USING RED BLOOD CELLS AS OSMOMETERS

The **red blood cell (RBC)** has the same osmolality and the same osmotic pressure as plasma. When a red blood cell is placed in a hypotonic solution, it will expand or perhaps even burst (a process called *hemolysis*) as a result of the influx of water, extruding its hemoglobin into the solution. When placed in a hypertonic solution, a red blood cell will shrink (a process called *crenation,* fig. 2.17) as a result of the efflux of water.

Red blood cells can thus be used as *osmometers* to determine the osmolality of plasma because RBCs will neither expand nor shrink in an isotonic solution.

Procedure

1. Measure 2.0 mL of the solutions indicated in part C of the laboratory report into each of five numbered test tubes.
2. Wipe the tip of a finger with alcohol and, using a sterile lancet, prick the finger to draw a small drop of blood. Alternatively, dog or cat blood may be provided by the instructor.

   Note: Caution must be exercised when handling blood to guard against contracting infectious agents. Handle only your own blood and discard all objects containing blood into the receptacles provided by the instructor.

3. Allow the drop of blood to drain down the side of test tube 1. Mix the blood with the saline (salt) solution by inverting the test tube a few times.
4. Repeat the above procedure for test tubes 2–5. Additional drops of blood can be obtained by milking the finger.
5. Using a transfer pipette, place a drop of solution 1 on a slide, and cover it with a coverslip. Observe the cells using the 45× objective.
6. Repeat step 5 for the other solutions and record your observations in the laboratory report.

Clinical Applications

Intravenous fluids must be isotonic to blood in order to maintain the correct osmotic pressure and prevent cells from either expanding or shrinking from the gain or loss of water. Common isotonic fluids include *normal saline* (0.9% NaCl), *5% dextrose* in water *(D5W)*, and *Ringer's lactate* (with glucose, lactic acid, and different salts). In contrast to isotonic fluids, hypertonic solutions of *mannitol* (an osmotically active solute) are given intravenously to promote osmosis and thereby reduce the swelling of cerebral edema, a significant cause of mortality in people with brain trauma or stroke.

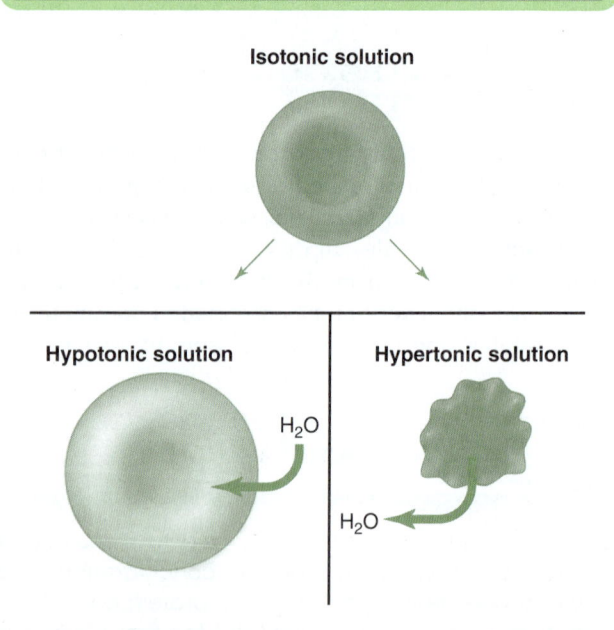

Figure 2.17 **Red blood cells in isotonic, hypotonic, and hypertonic solutions.** In each case, the external solution has an equal, lower, or higher osmotic pressure, respectively, than the intracellular fluid. As a result, water moves by osmosis into the red blood cells placed in hypotonic solutions, causing them to swell and even to burst. Similarly, water moves out of red blood cells placed in a hypertonic solution, causing them to shrink and become crenated.

(For a full-color version of this figure, see fig. 6.13 in *Human Physiology,* thirteenth edition, by Stuart I. Fox.)

Laboratory Report 2.6

Name _____
Date _____
Section _____

DATA FROM EXERCISE 2.6

A. Solubility of Compounds in Polar and Nonpolar Solvents

1. Describe the appearance of the solutions after completing steps 2 and 3. What is your conclusion regarding solubility and solvents?

2. Describe the appearance of the solutions after completing steps 4 and 5. What is your conclusion regarding solubility, solvents, and the detergent?

B. Osmosis Across an Artificial Semipermeable Membrane

1. Enter your data in this table.
   Time when meniscus level was marked: _____

Time	20% Molasses Solution, Distance Meniscus Moved	25% Molasses Solution, Distance Meniscus Moved
15 minutes		
30 minutes		
45 minutes		
60 minutes		

2. Was there any change in the solution level with time? Explain the forces involved.

3. Compare the movement of the 20% molasses solution to that of the 25% solution. Was the distance traveled by the two solutions predictable? Explain.

C. **Concentration and Tonicity**

1. Enter your data in this table.

Tube and Contents	Molality	Milliosmolality	Visual Appearance of RBCs	Estimated RBC Diameter (µm)
1 10 g/dL NaCl				
2 3.5 g/dL NaCl*				
3 0.85 g/dL NaCl	0.145 m	290 mOsm		
4 0.45 g/dL NaCl				
5 0.20 g/dL NaCl				

*Approximately the concentration of seawater.
Note: dL = 100 mL.

2. Which solution is isotonic? _____

3. Which solutions are hypotonic? _____

4. Which solutions are hypertonic? _____

82

REVIEW ACTIVITIES FOR EXERCISE 2.6

Test Your Knowledge

1. Define *osmosis*. _____

2. Describe what is meant by *osmotic pressure*. _____

3. Define *isotonic*. _____

4. Red blood cells in a hypertonic solution will _____

5. A 0.10 M NaCl solution is _____ (iso/hypo/hypertonic) to a 0.10 M glucose solution.

Test Your Understanding

6. When the body needs to conserve water, the kidneys excrete a hypertonic urine. What do the terms *isotonic* and *hypertonic* mean? Since the fluid that is to become urine begins as plasma (an isotonic solution), what must happen to change it to a hypertonic urine? Explain.

7. What component of the molasses solution was osmotically active? Explain why this is true.

Test Your Analytical Ability

8. The receptors for thirst are located in a part of the brain called the *hypothalamus*. These receptors, called *osmoreceptors*, are stimulated by an increase in blood osmolality. Imagine a man who has just landed on a desert island. Trace the course of events leading to his sensation of thirst. Can he satisfy his thirst by drinking seawater? Explain.

9. Before the invention of refrigerators, pioneers preserved meat by salting it. Explain how meat can be preserved by this procedure. (Hint: Think about what salting the meat would do to decomposer organisms, such as bacteria and fungi.)

10. Graph your data from exercise 2.6B, with height on the Y-axis and time on the X-axis. Is the graph linear? If not, what might cause it to have the shape it does?

Test Your Quantitative Ability

11. Suppose a salt and a glucose solution are separated by a membrane that is permeable to water but not to the solutes. The NaCl solution has a concentration of 1.95 g per 250 mL of water (molecular weight = 58.5). The glucose solution has a concentration of 9.0 g per 250 mL of water (molecular weight = 180).
    Calculate the molality, millimolality, and milliosmolality of both solutions. State whether osmosis will occur and, if it will, in which direction. Explain your answer.

Clinical Investigation Questions

12. What is the osmotic role of plasma albumin, and how would a low plasma albumin level promote edema?

13. Why are intravenous solutions generally isotonic and not hypotonic or hypertonic? What is the significance of normal saline?

Section 3

The Nervous System and Sensory Physiology

Despite changes in the external temperature, the availability of foods, the presence of toxic and threatening agents, and other influences, the internal environment of the body remains remarkably constant. The science of physiology is largely a study of the regulatory mechanisms that maintain this internal constancy *(homeostasis)*. This regulation is a major function of the nervous system and endocrine system.

To maintain homeostasis, an organism must be able to recognize specific environmental features and make appropriate responses. At its simplest level, recognition is achieved through the stimulation of specific types of **sensory receptors,** such as cutaneous (skin) receptors. A given receptor will usually be responsive to only one *modality* (specific type) of stimulus. The rod and cone photoreceptors of the eye, for example, are stimulated by light. Each receptor transduces the particular environmental stimulus into electrochemical nerve impulses that then go to the specific part of the brain where the sensation is identified. This principle is known as the *law of specific nerve energies.*

The appropriate response to the environmental stimulus occurs through the neural activation of **effector organs,** which are *muscles* and *glands.* The voluntary activation of skeletal muscles by somatic motor nerves emerging from the spinal cord may produce a simple reflex action or involve more complex nervous system interaction. Autonomic motor nerves are involuntary and stimulate cardiac muscles, smooth muscles, exocrine glands (such as sweat glands and gastric glands), and some endocrine glands (the adrenal medulla, for example). Many other endocrine glands are indirectly regulated by the hypothalamus through its control of the anterior pituitary, thus wedding the endocrine system to the nervous system. Interposed between these *afferent* (sensory) and *efferent* (motor) pathways are millions of association neurons within the brain and spinal cord that integrate these activities while promoting learning and memory. The brain is the ultimate interpretation center for all sensations, including touch, pain, vision, hearing, equilibrium, and taste.

Exercise 3.1	Recording the Nerve Action Potential
Exercise 3.2	Electroencephalogram (EEG)
Exercise 3.3	Reflex Arc
Exercise 3.4	Cutaneous Receptors and Referred Pain
Exercise 3.5	Eyes and Vision
Exercise 3.6	Ears: Cochlea and Hearing
Exercise 3.7	Ears: Vestibular Apparatus—Balance and Equilibrium
Exercise 3.8	Taste Perception

EXERCISE 3.1

Recording the Nerve Action Potential

MATERIALS

1. Frogs
2. Dissecting equipment and trays, glass probes, thread
3. Oscilloscope, or computer, and nerve chamber
4. Frog Ringer's solution (see Materials, exercise 5.1)
5. Alternatively, labs with more advanced equipment can measure the action potential in a single axon, using other animals, such as crayfish (crawdads). For example, see the Crawdad Project at www.crawdad.cornell.edu.

The potential difference across axon membranes undergoes changes during the production of action potentials. As the axons of a nerve produce action potentials, the surface of the nerve at this region has a difference in potential in relation to the surface of an unstimulated region. The polarity of this potential difference and its magnitude in millivolts can be seen in a computer or oscilloscope screen.

LEARNING OUTCOMES

You should be able to:

1. Describe the resting membrane potential and the distribution of Na^+ and K^+ across the axon membrane.
2. Describe the events that occur during the production of an action potential.
3. Demonstrate the recording of action potentials in the sciatic nerve of a frog, and explain how this recording is produced.

Textbook/Multimedia Correlations

Before performing this exercise, you should study the introductory material presented here. Further information relating to this exercise can be found in these pages of *Human Physiology*, thirteenth edition, by Stuart I. Fox:

- *Action Potentials.* Chapter 7, p. 174.
- *Conduction of Nerve Impulses.* Chapter 7, p. 178.
- Multimedia correlations are provided in *MediaPhys 3.0:* Topics 3.27–3.34 and 4.4–4.22.

Clinical Investigation

A patient accidentally cut himself on a piece of broken glass, and the physician applied lidocaine before suturing the laceration.

- Describe the role of plasma membrane Na^+ and K^+ channels in the production of action potentials.
- Explain how lidocaine and other local anesthetics work.

A **neuron** (nerve cell) consists of three regions specialized for different functions: (1) the **dendrites** receive input from sensory receptors or from other neurons; (2) the **cell body** contains the nucleus and serves as the metabolic center of the cell; and (3) the **axon** conducts the nerve impulse to other neurons or to effector organs (fig. 3.1). The origin of the axon near the cell body is an expanded region called the *axon hillock*. Adjacent to the axon hillock is the *axon initial segment,* which is the region where the first action potentials (nerve impulses) are generated. Toward their ends, axons can produce up to 200 or more branches called *axon collaterals,* and each of these can divide to synapse with many other neurons. A **nerve** is a bundle of axons that leaves the *central nervous system* (brain and spinal cord) as part of the *peripheral nervous system.*

MEMBRANE POTENTIALS

If one lead of a voltmeter is placed on the surface of an axon and the other lead is placed inside the cytoplasm, a *potential difference* (or voltage) will be measured across the axon membrane. The inside of the cell is about 70 millivolts negative (−70 mV) with respect to the outside. (The outside of the axon is positive with respect to the cytoplasm.) This resting membrane potential is maintained by the unequal distribution of ions on the two sides of the membrane. Na^+ is present in higher concentrations outside the cell than inside, whereas K^+ is more concentrated inside the cell. These differences are maintained, in part, by active transport processes (the sodium-potassium pump).

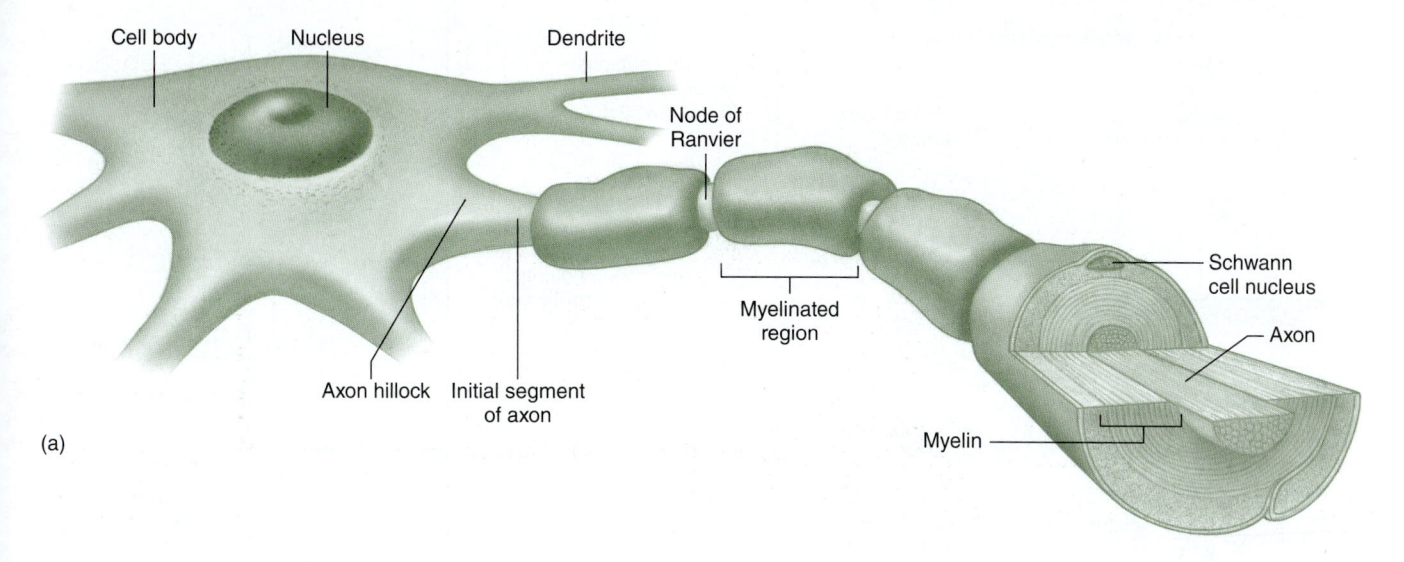

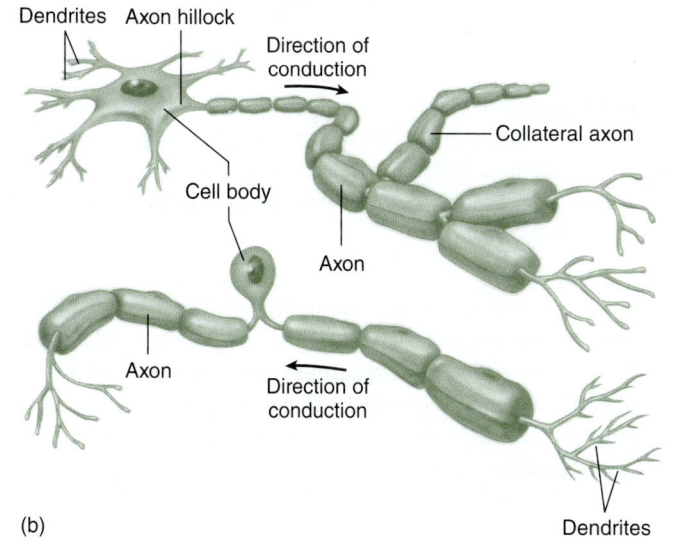

Figure 3.1 Neuron structure. (a) The components of a neuron. (b) A motor and a sensory neuron, showing the pathways of conduction (arrows).

(For a full-color version of this figure, see figs. 7.1 and 7.2 in *Human Physiology*, thirteenth edition, by Stuart I. Fox.)

When a neuron is appropriately stimulated, the barriers to Na⁺ are lifted (voltage-gated Na⁺ channels open), and the positively charged Na⁺ is allowed to diffuse into the cell along its concentration gradient. The flow of positive ions into the cell first eliminates the potential difference across the membrane and then continues until the polarity is actually reversed and the inside of the cell is positive with respect to the outside (about +30 mV). This phase is called **depolarization.** At this point, the barriers to K⁺ are lifted (voltage-gated K⁺ channels open), and the flow of this positively charged ion out of the cell along its concentration gradient helps to reestablish the resting potential **(repolarization).** During the time of repolarization, the diffusion of Na⁺ into the cell is blocked and the neuron is refractory (not capable of responding) to further stimulation; it is in a *refractory period.* The momentary reversal and reestablishment of the resting potential is known as the **action potential** (fig. 3.2).

In this exercise, a frog's sciatic nerve will be dissected and placed on two pairs of electrodes. One pair of electrodes (the stimulating electrodes) will deliver a measured pulse of electricity to one point on the nerve; the other pair (the recording electrodes) will be connected to a cathode-ray tube of an oscilloscope adjusted to sweep an electron beam horizontally across a screen when the nerve is unstimulated. The delivery of a small pulse of electricity through the stimulating electrodes produces a small vertical deflection at the beginning of the horizontal sweep. This initial vertical deflection (the *stimulus artifact*) increases in amplitude (height) as the strength of the stimulating voltage is increased (fig. 3.3).

When the stimulating voltage reaches a sufficient level (the *threshold potential*), the region of the nerve next to the stimulating electrodes becomes depolarized (the outside of the nerve becomes negative with respect to the inside). By the creation of "minicircuits," this region depolarizes the adjacent region of the nerve, while it (the region nearest the stimulating electrodes) is being repolarized (fig. 3.4). In this manner, a wave of "surface negativity" is conducted from the stimulating electrodes toward the two recording electrodes. When this wave reaches the first recording electrode, this electrode becomes electrically negative with respect to the second recording electrode (because both are on the

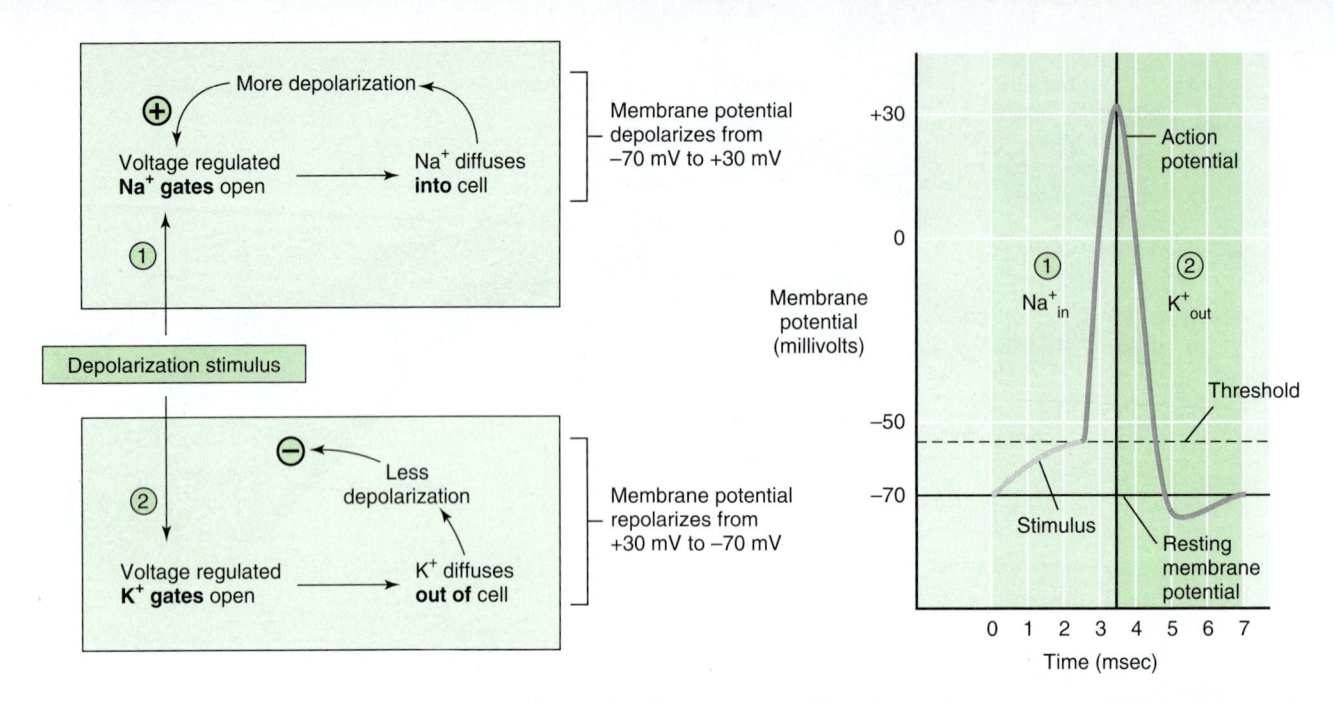

Figure 3.2 Depolarization of an axon affects Na⁺ and K⁺ diffusion in sequence. (1) Na⁺ gates open and Na⁺ diffuses into the cell. (2) After a brief period, K⁺ gates open and K⁺ diffuses out of the cell. An inward diffusion of Na⁺ causes further depolarization, which in turn causes further opening of Na⁺ gates in a positive feedback (+) fashion. The opening of K⁺ gates and outward diffusion of K⁺ makes the inside of the cell more negative, and thus has a negative feedback effect (−) on the initial depolarization.

(For a full-color version of this figure, see fig. 7.13 in *Human Physiology*, thirteenth edition, by Stuart I. Fox.)

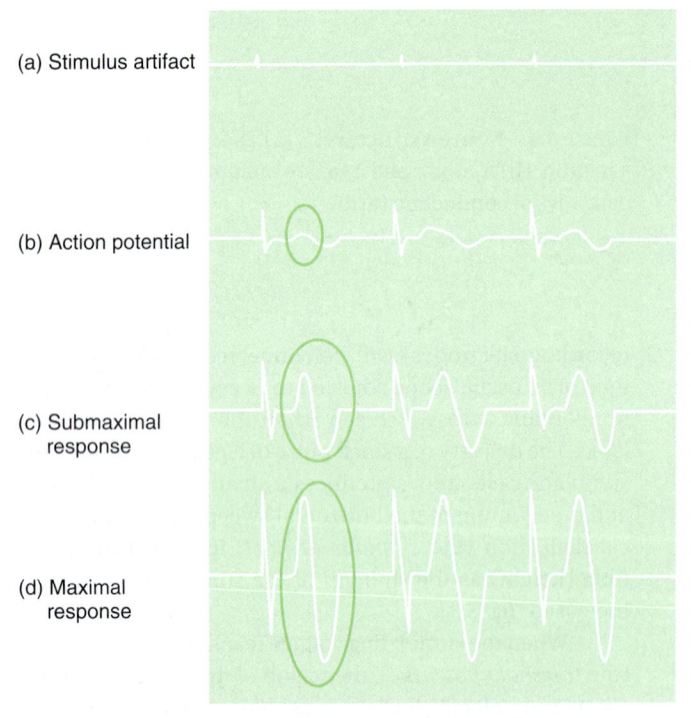

Figure 3.3 Recording from a frog sciatic nerve. As the stimulus voltage is increased from (a) through (d), the amplitude of the stimulus artifact and nerve action potential (circled areas) are also increased, but only to a maximum value, shown in (d). This effect is because the sciatic nerve contains hundreds of nerve fibers. The number of individual nerve fibers stimulated to produce all-or-none action potentials increases with increasing stimulus intensity because some fibers are located closer to the stimulating electrodes than others.

surface of the nerve and the depolarization wave has not yet reached this second electrode). The potential difference between these two recording electrodes produces a vertical deflection of the electron beam a few milliseconds (msec) after the stimulus artifact. This second event is the action potential of the nerve.

The electrical activity of the nerve can be intensified by increasing either the frequency or the strength of stimulation. Increasing the frequency of stimulation will increase the number of impulses conducted by the nerve in a given time, up to a maximum amount. This maximum (impulses about 2 msec apart) is due to the **refractory period** of the nerve and ensures that the action potentials will remain separate events even at high frequencies of stimulation. Similarly, increasing the strength of each stimulus will increase the amplitude of each action potential, up to a maximum level. This is because the action potential recorded from a nerve is the sum of the action potentials of all stimulated axons in that nerve. As the strength (voltage) of the stimulus is increased, the number of depolarized axons increases, increasing the amplitude of the overall nerve response (fig. 3.3).

However, when impulses are recorded from individual axons, the amplitude of the action potentials does *not* increase with increasing strength of stimulation. A neuron either does not "fire" (to any subthreshold stimulus) or it "fires" maximally (to any suprathreshold stimulus). This is the **all-or-none law** of nerve physiology. The stronger the stimulus, the greater the number (not the size) of action potentials carried by a nerve fiber in a given time. The strength of a stimulus, therefore, is coded in the nervous system by the *frequency* (not the amplitude) of action potentials.

voltage changes with time as vertical line deflections (fig. 3.3). Alternatively, the voltage changes detected by the recording electrodes can be used by an amplifier to produce a similar display on a computer monitor.

The beam is made to sweep from left to right across the screen at a particular rate. The image on the screen is a plot of voltage (y axis) against time (x axis). If the sweep of the beam is triggered by a stimulus to the nerve preparation, the electrical response of the nerve will always appear at the same time after the sweep has begun and at the same location on the screen. The oscilloscope or computerized display allows an action potential produced in response to a second stimulus to be superimposed on the one produced in response to the previous stimulus. An observer will see an apparently stable image of an action potential, even though each action potential lasts for only about 3 msec.

Vertical deflections of the beam are produced when a **potential difference,** or voltage, develops between the two *recording electrodes* that touch the nerve some distance away from a pair of *stimulating electrodes* (fig. 3.5). The first action potential, produced near the stimulating electrodes, results from a depolarizing current whose voltage, duration, and frequency (number of "shocks" per second) can be varied by the operator adjusting the stimulator module. Conducted by the nerve, the action potential is recreated in the region of the nerve in contact with the recording electrodes, and a vertical line deflection is observed on the screen.

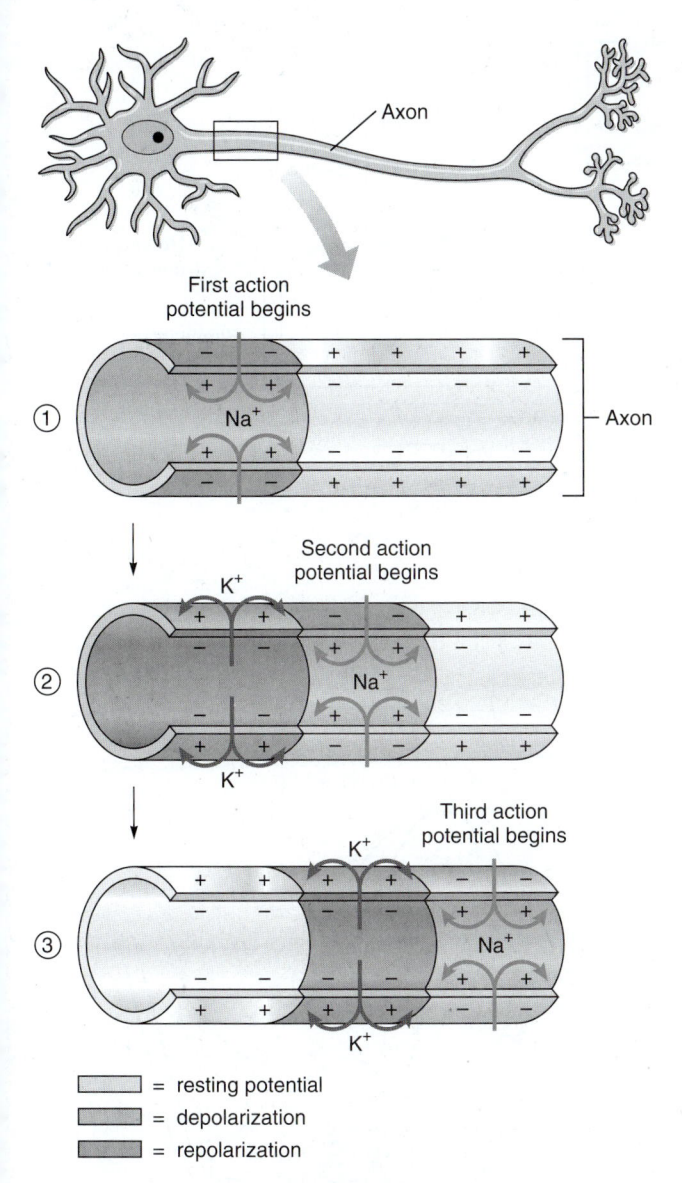

Figure 3.4 The conduction of action potentials in an unmyelinated axon. Each action potential "injects" positive charges that spread to adjacent regions. The region that has just produced an action potential is refractory. The next region, not having been stimulated previously, is partially depolarized. As a result, its voltage-regulated Na$^+$ gates open and the process is repeated. Successive segments of the axon thereby regenerate, or "conduct," the action potential.

(For a full-color version of this figure, see fig. 7.19 in Human Physiology, thirteenth edition, by Stuart I. Fox.)

Observing with an Oscilloscope or Computer Monitor

Electrical activity in nerves is frequently observed by using an oscilloscope. In an oscilloscope, electrons from a cathode-ray "gun" are sprayed across a fluorescent screen, producing a line of light. Changes in the potential difference between the two recording electrodes cause this line to deflect. Movement of the line upward or downward is proportional to the incoming voltage from the nerve preparation. The oscilloscope can thus function as a fast-responding voltmeter, displaying

Clinical Applications

Local anesthetics block the conduction of action potentials in axons. They do this by reversibly binding to specific sites within the voltage-gated Na$^+$ channels, reducing the ability of membrane depolarization to produce action potentials. *Cocaine* was the first local anesthetic to be used, but because of its potential for abuse and toxicity alternatives were developed. The first synthetic analog of cocaine used for local anesthesia, *procaine,* was produced in 1905. Other local anesthetics of this type include *lidocaine* and *tetracaine.*

Procedure

1. Decapitate or double-pith a frog (see section 5, pages 178–179 and figs. 5.7 and 5.8, for this procedure), skin its legs and the lower portion of its back, and place it in a prone position in a dissecting tray (fig. 3.6).
2. Make a 1-inch incision on both sides of its spine from the anal region toward the head. Lift the portion of spine free from surrounding muscle, and excise it to expose the right and left sciatic nerves, which run lateral and parallel to the spine.

Figure 3.5 Setup for recording from an isolated nerve. A nerve is laid across a pair of stimulating electrodes and a pair of recording electrodes. A stimulator delivers a square-wave pulse of a given voltage and duration to the nerve via the stimulating electrodes. This can be seen at the recording electrodes as the stimulus artifact (not shown). The action potential observed after the stimulus artifact represents the response of the nerve to the stimulus.

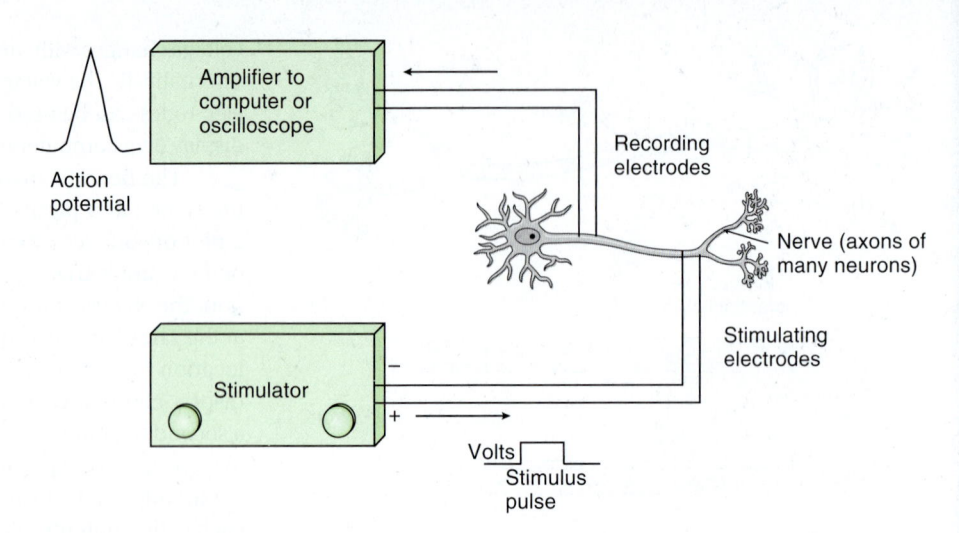

Note: *Be careful not to touch the nerve with your fingers or metal tools and not to let the nerve touch the surface of the frog's skin or cut muscle. Keep the nerve moist with Ringer's solution.*

3. Lift one of the sciatic nerves with a glass probe; tie it with a few inches of thread; and cut the nerve closer to the head, beyond the tie.

4. Separate the large posterior muscles of the thigh to expose the distal portion of the sciatic nerve, and lift it with a glass probe.

Note: *The sciatic nerve is easily identified because it runs in the same connective tissue sheath as the sciatic artery. Tie the nerve with a few inches of thread, and cut the nerve distally beyond the tie.*

5. Lift the two ends of the nerve with the two lengths of thread (fig. 3.6) and carefully free it from attached muscle and fascia.

6. Lay the nerve across the stimulating and recording electrodes of the nerve chamber, always keeping the nerve moist with Ringer's solution.

7. Connect the stimulating electrodes to the stimulator and the recording electrodes to the preamplifier of the oscilloscope.

8. Adjust the horizontal sweep; set the stimulus frequency, duration, and amplitude at their lowest values.

9. Slowly increase the stimulus strength until the threshold voltage is obtained (the lowest stimulus that produces an observed action potential). Threshold stimulus: _____ V

Note: *The first deflection on the screen is the stimulus artifact; the deflection to the right of the stimulus artifact is the action potential.*

10. Slowly increase the strength of the stimulating voltage until the action potential is at its maximum amplitude. Stimulus producing maximum response: _____ V

11. With the stimulating voltage set at a slightly suprathreshold level, gradually increase the frequency of stimulation and note this effect on the amplitude of the action potential.

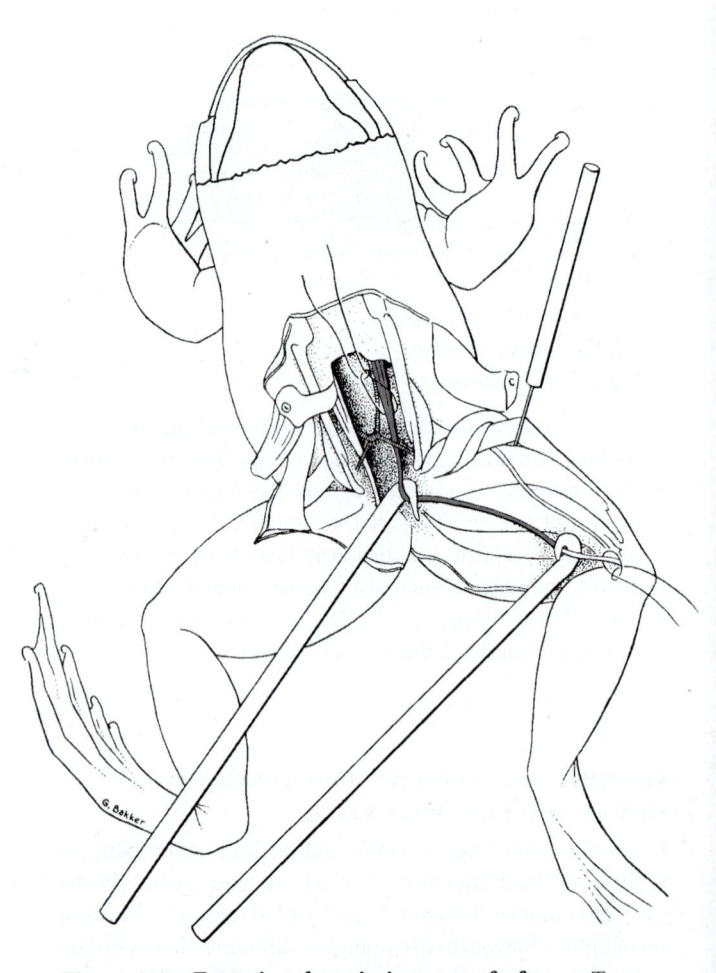

Figure 3.6 Exposing the sciatic nerve of a frog. Two glass probes are used to handle the nerve. The two ends have been tied with thread just beyond where they will be cut.

Laboratory Report 3.1

Name _____
Date _____
Section _____

REVIEW ACTIVITIES FOR EXERCISE 3.1

Test Your Knowledge

1. The parts of a neuron that receive input from other neurons or from sensory receptors are the _____.

2. The voltage across the membrane of a particular axon may be −70 mV. This is known as its _____.

3. During the action potential, the membrane polarity momentarily reverses.
   (a) This process is called _____.
   (b) It is caused by the diffusion of _____ into the axon.

4. The number of action potentials produced per unit time is known as its _____; this is proportional to the _____ of the stimulus.

5. What is the all-or-none law regarding action potentials? _____

Test Your Understanding

6. Both recording electrodes in this exercise were placed on the surface of the nerve—that is, both were extracellular. What must be done to record the resting membrane potential of an axon?

7. Describe the permeability properties of the axon membrane when the nerve is at rest and the permeability properties during the production of an action potential. Describe the direction and nature of the ion movements during the production of an action potential.

91

Test Your Analytical Ability

8. How is the strength of a sensory stimulus coded by a single axon? What additional way could the strength of a stimulus be coded by a collection of axons (a nerve) in the intact nervous system?

9. In this exercise, the amplitude of the action potential produced by the frog sciatic nerve increased when the stimulating voltage increased. How is this possible if the action potentials of individual axons obey the all-or-none law?

Clinical Investigation Questions

10. What function do the Na^+ channels serve in the production of action potentials? What function do K^+ channels serve?

11. How do local anesthetics block the perception of pain?

EXERCISE 3.2

Electroencephalogram (EEG)

MATERIALS

1. Oscilloscope and EEG selector box (Phipps and Bird), physiograph and high-gain coupler (Narco), or electroencephalograph recorder (Lafayette Instrument Company)
2. EEG electrodes and surface electrode
3. Long ECG elastic band and ECG electrolyte gel
4. Alternatively, the Biopac equipment can be used for visualizing the EEG on a computer (Biopac lessons 3 and 4), available in their basic, advanced, and ultimate systems.
5. In addition, the PowerLab system with its flat EEG electrodes may be used for visualizing the EEG on a computer screen.

Chemical neurotransmitters released by presynaptic axons produce excitatory or inhibitory postsynaptic potentials. These postsynaptic potentials account for most of the electrical activity of the brain and contribute to the electroencephalogram recorded by surface electrodes positioned over the brain.

LEARNING OUTCOMES

You should be able to:

1. Describe the structure of a chemical synapse.
2. Describe EPSPs and IPSPs, and explain their significance.
3. Demonstrate the recording of an electroencephalogram, and explain how an EEG is produced.

Textbook/Multimedia Correlations

Before performing this exercise, you should study the introductory material presented here. Further information relating to this exercise can be found in these pages of *Human Physiology*, thirteenth edition, by Stuart I. Fox:

- *The Synapse.* Chapter 7, p. 180.
- *Acetylcholine as a Neurotransmitter.* Chapter 7, p. 185.
- *Cerebral Cortex.* Chapter 8, p. 210.

The basic unit of neural integration is the **synapse,** the functional connection between the axon of one neuron and the dendrites or cell body (occasionally even the axon) of another neuron. Action potentials produced by the first neuron are conducted to the axon terminals, where they stimulate the release of chemical *neurotransmitters* from synaptic vesicles that are docked at the plasma membrane (fig. 3.7). Release (by exocytosis) of the neurotransmitters, such as **acetylcholine (ACh),** allows them to diffuse across the *synaptic cleft.* This is a small space separating the first neuron (the *presynaptic neuron*) from the next neuron (the *postsynaptic neuron*).

The interaction of these chemical transmitters with their specific receptors may stimulate a depolarization in the postsynaptic membrane. Unlike the all-or-none action potential, this **excitatory postsynaptic potential, or EPSP,** is a *graded* response—the larger the number of synaptic vesicles releasing neurotransmitters, the greater the depolarization. When the EPSP reaches a critical level of depolarization (the *threshold*) at the axon initial segment, it generates an action potential that can then be conducted down the axon of the second neuron to the next synapse. Chemical neurotransmitters released by other neurons that synapse with the second cell may produce the opposite response—a *hyperpolarization* of the postsynaptic membrane (the inside of the cell becomes even more negative with respect to the outside). This is called an **inhibitory postsynaptic potential, or IPSP.** The production of action potentials by the second cell, as well as their frequency, will be determined by the algebraic sum of these EPSPs and IPSPs produced by the convergence (fig. 3.8) of multiple neurons on the second cell.

These electrical activities—action potentials, EPSPs and IPSPs—are not equal in two different regions of the brain at the same time. Therefore, an extracellular potential difference (which fluctuates between 50 and 100 millionths of a volt) can be measured by placing two electrode leads on the scalp over two different regions of the brain. The recording of these "brain waves" is known as an **electroencephalogram (EEG).**

In practice, nineteen electrodes are placed at various standard positions on the scalp, with each pair of electrodes connected to a different recording pen. The record obtained reflects periodic waxing and waning of synchronous neuronal activity, producing complex waveforms characteristic of the regions of the brain sampled and the state of the subject (fig. 3.9).

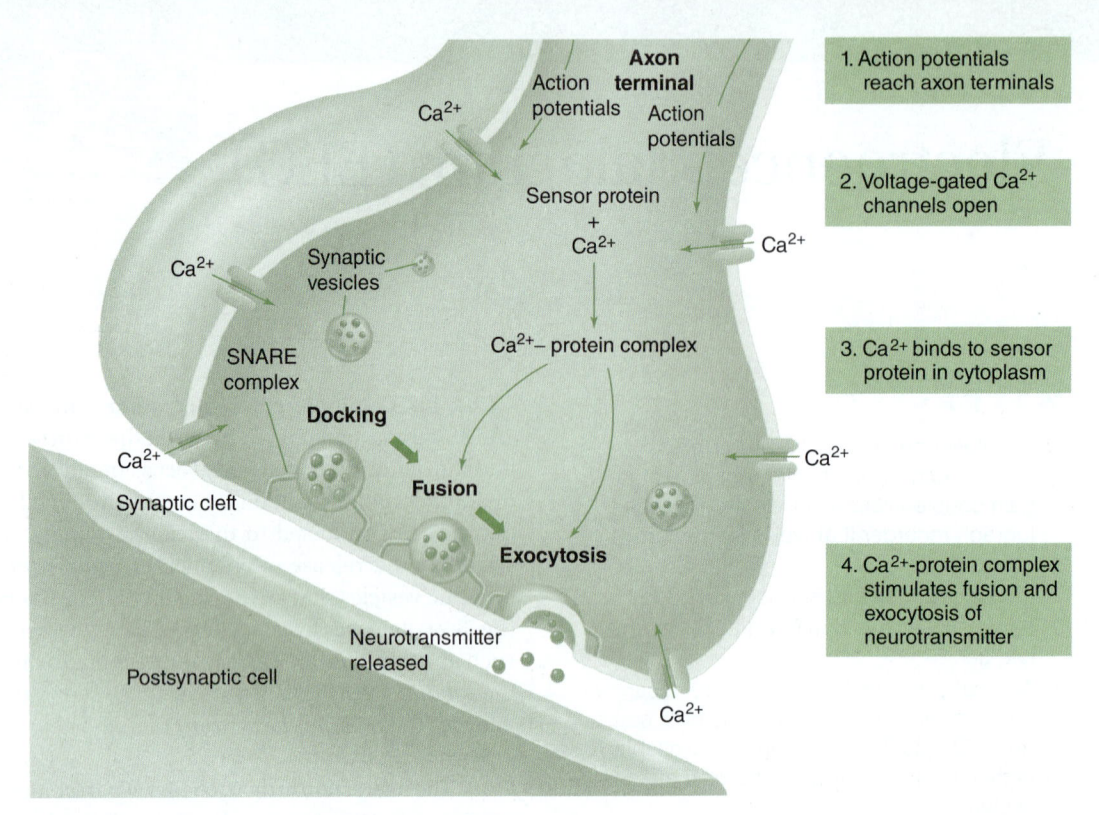

Figure 3.7 **The release of neurotransmitter.** Steps 1–4 summarize how action potentials stimulate the exocytosis of synaptic vesicles. Action potentials open channels for Ca^{2+}, which enters the cytoplasm and binds to a sensor protein, believed to be synaptotagmin. Meanwhile, docked vesicles are held to the plasma membrane of the axon terminals by a complex of SNARE proteins. The Ca^{2+}-sensor protein complex alters the SNARE complex to allow the complete fusion of the synaptic vesicles with the plasma membrane, so that neurotransmitters are released by exocytosis from the axon terminal.

(For a full-color version of this figure, see fig. 7.23 in *Human Physiology,* thirteenth edition, by Stuart I. Fox.)

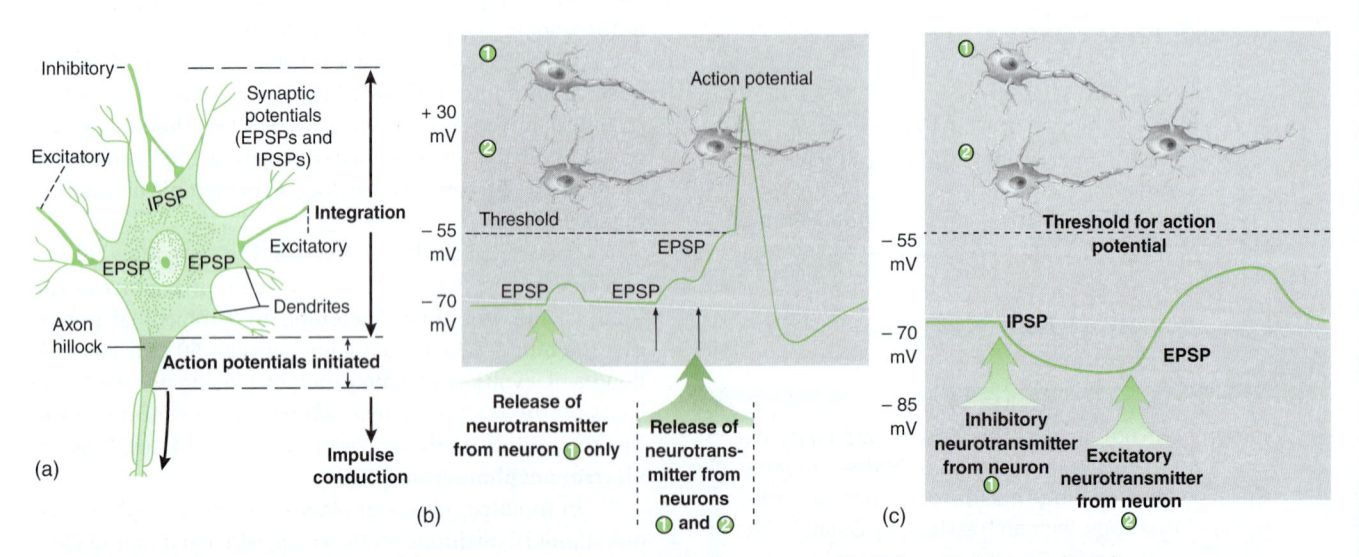

Figure 3.8 **Synaptic integration.** (a) Many presynaptic inputs, both excitatory (EPSP) and inhibitory (IPSP), can converge on a single neuron. (b) Excitatory input—the depolarizations of EPSPs—from different presynaptic neurons can summate in the postsynaptic neuron. (c) Inhibitory input—the hyperpolarization of an IPSP—can also summate with EPSPs, taking the membrane potential farther from the threshold required for action potentials.

(For a full-color version of this figure, see figs. 7.33 and 7.34 in *Human Physiology,* thirteenth edition, by Stuart I. Fox.)

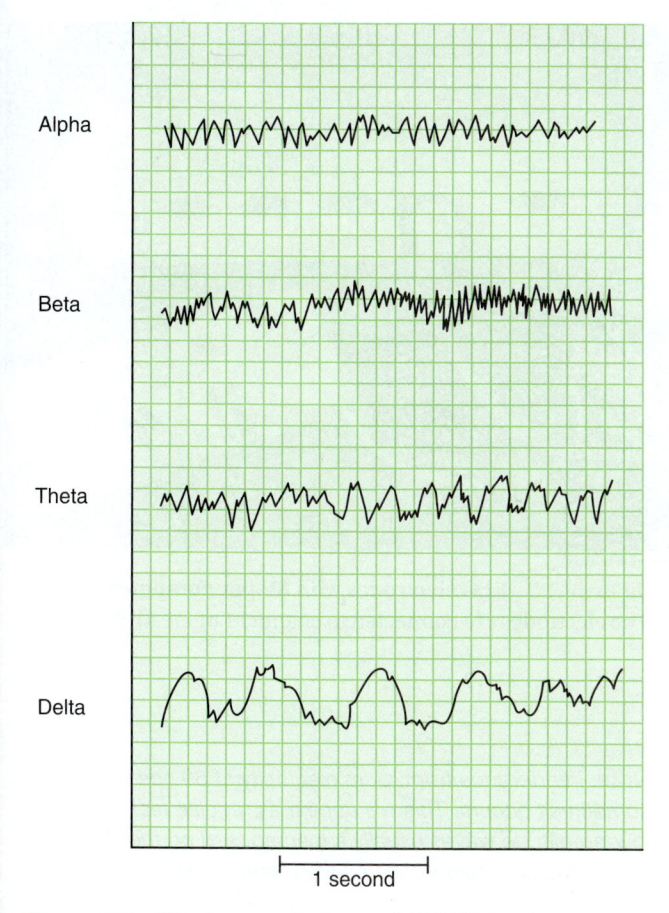

Figure 3.9 Electroencephalograph (EEG) rhythms.

Clinical Applications

Use of the *electroencephalograph* may help to diagnose a number of brain lesions, including epilepsy, intracranial infections, and encephalitis. It has been discovered that meditation, as performed by Zen monks and yogis, results in the production of slow alpha waves (7 cps) of increased amplitude and regularity (even with the eyes partially open), and that this change is associated with a decrease in metabolism and sympathetic nerve activity. Many people, through biofeedback techniques, attempt to enhance their ability to produce alpha rhythms so they can relax and lower their sympathetic nerve effects.

Procedure

1. If an **oscilloscope** will be used,
   (a) connect the EEG selector box to the preamplifier of the oscilloscope, and adjust the horizontal sweep and sensitivity;
   (b) plug the lead for the forehead into the EEG selector box outlet labeled L. Frontal and the lead for the earlobe into the outlet for the ground.
2. If a **physiograph** will be used,
   (a) insert a high-gain coupler into the physiograph;
   (b) set the gain on ×100, the sensitivity on 2 or 5, and the time constant on 0.03 or 0.3;
   (c) connect two EEG electrodes and one surface electrode to the high-gain coupler.
3. If an **electroencephalograph** will be used,
   (a) insert the cable from the electrode box/lead selector into the EEG amplifier within the recorder;
   (b) insert the electrodes into the selector box, and follow the instructions for the particular equipment model used.
4. Obtain a long elastic ECG strap and tie it around the forehead (with the knot at the back of the head) to form a snug headband.
   (a) If an **oscilloscope** will be used for recording, dab a little electrolyte gel onto the single electrode from the left frontal outlet and place it under the headband in the middle of the forehead.
   (b) If a **physiograph** will be used for recording, dab electrolyte gel onto both EEG electrodes and place them under the headband on the right and left sides of the forehead.
5. Dab a little electrolyte gel onto the ground electrode (the surface electrode for the physiograph), and with

There are four characteristic types of EEG wave patterns. **Alpha waves,** consisting of rhythmic oscillations with a frequency of 8 to 12 cycles per second (cps), were the first patterns to be characterized (fig. 3.9). These waves, seen with a single pair of electrodes, are produced by the visual association areas of the parietal and occipital lobes and predominate when the subject is relaxed and awake but has eyes closed. Alpha waves can be suppressed by opening the eyes or by doing mental arithmetic and are normally absent in a significant number of people. The alpha rhythm of children under the age of 8 occurs at a lower frequency (4–7 cps).

Beta waves (13–25 cps) are strongest from the frontal lobes and reflect the evoked activity produced by visual stimuli and mental activity; they are enhanced by barbiturate drugs. **Theta waves** (5–8 cps) are emitted from the temporal and occipital lobes and are common in newborn infants and sleeping adults. In awake adults, theta wave recordings generally indicate severe emotional stress. **Delta waves** (1–5 cps) seem to be emitted from the cerebral cortex. Delta waves are seen in awake infants. Common in adults during deep sleep, the presence of delta waves in an awake adult indicates brain damage. During a *petit mal epileptic seizure,* the EEG pattern may show regular spikes and waves at 3 cps. Abnormal patterns not observed in the "resting" EEG can often be revealed by stimulation of the subject through hyperventilation or flashing of a strobe light at different frequencies.

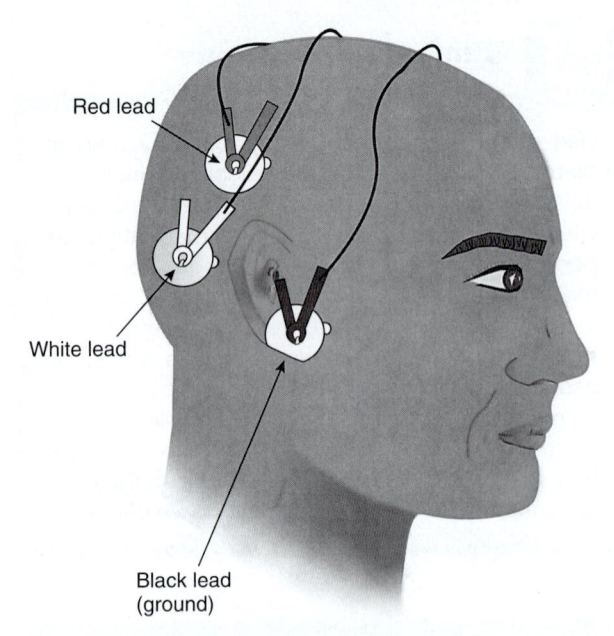

Figure 3.10 Biopac EEG electrode placement. Be sure that the EEG leads are in firm contact with the scalp.

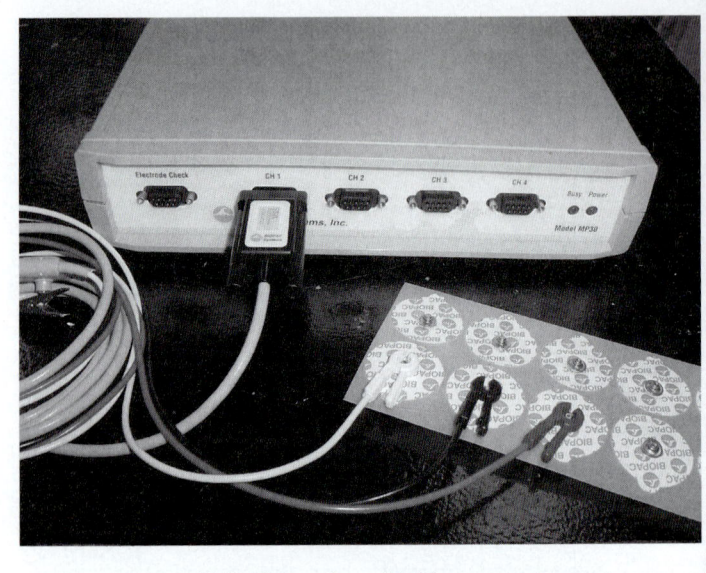

Figure 3.11 EEG electrodes with Biopac equipment. The electrodes are connected to channel 1 of the MP30.

your fingers, press this electrode against the skin behind the ear.

6. If the **Biopac** system is used (figs 3.10 and 3.11), follow the procedural steps provided on the computer screen. You can complete their lab exercises 3 and 4 once the electrodes are in place.

Note: *Alpha, beta, theta, and delta waves will be recorded and stored in your student folder.*

7. With the subject in a relaxed position (no muscular movements), with his or her eyes closed, observe the EEG pattern, checking particularly for the presence of alpha waves. Many people do not produce alpha waves in this situation, so test a number of subjects.

Note: *Interference from room electricity at 60 cps sometimes occurs. This will appear as regular, fast, low-amplitude waves usually superimposed on the slower, more irregular brain waves of larger amplitude.*

8. Ask the subject to open his or her eyes, and observe the effect this has on the EEG patterns.
9. Ask the subject to sing, and then to hyperventilate (for only a few seconds), and observe the effect on the EEG patterns.
10. If the subject has any experience in meditation, have the subject try this for a few minutes and observe if this has an effect on the EEG patterns.

Laboratory Report 3.2

Name _____
Date _____
Section _____

REVIEW ACTIVITIES FOR EXERCISE 3.2

Test Your Knowledge

1. A chemical released by an axon is known as a _____.

2. Binding of the previously-named chemical to its receptor in the postsynaptic membrane may produce a depolarization called a(n) _____.

3. Different chemicals released by axons may produce a hyperpolarization of the postsynaptic membrane; such a hyperpolarization is called a(n) _____.

4. Action potentials are all-or-none; synaptic potentials, by contrast, are _____.

Match these terms:
____ 5. alpha rhythm   (a) common in awake, tense subjects
____ 6. beta rhythm   (b) observed in some relaxed subjects
____ 7. theta rhythm   (c) observed in deep sleep
____ 8. delta rhythm   (d) seen in some children

Test Your Understanding

9. Describe where an EPSP is produced, and explain how it is produced. What is its significance?

10. Compare the properties of an EPSP with those of an action potential.

11. What is the significance of IPSPs? How are they produced?

Test Your Analytical Ability

12. Propose a rationale by which the observation of a person's brain waves might be used to help that person lower his or her blood pressure.

13. Which type of neural activity, action potentials, or synaptic potentials, produces the EEG? Explain your answer. Speculate as to why the EEG may be influenced by hyperventilation.

EXERCISE 3.3

Reflex Arc

MATERIALS
1. Rubber mallets
2. Blunt probes

In a reflex, specific sensory stimuli evoke characteristic motor responses very rapidly because few synapses are involved. Tests for simple reflex arcs are very useful in diagnosing neurological disorders because a specific simple reflex arc occurs at a specific spinal cord segment and involves particular nerves.

LEARNING OUTCOMES

You should be able to:

1. Describe the neurological pathways involved in a simple reflex arc.
2. Describe the structure and function of muscle spindles.
3. Demonstrate muscle stretch reflexes, and explain the clinical significance of these tests.
4. Demonstrate a Babinski reflex (Babinski's sign), and explain the clinical significance of this test.

Textbook/Multimedia Correlations

Before performing this exercise, you should study the introductory material presented here. Further information relating to this exercise can be found in these pages of *Human Physiology,* thirteenth edition, by Stuart I. Fox:

- *Spinal Cord Tracts.* Chapter 8, p. 232.
- *Cranial and Spinal Nerves.* Chapter 8, p. 236.
- *Neural Control of Skeletal Muscles.* Chapter 12, p. 385.

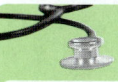

Clinical Investigation

A patient undergoing a physical exam is asked to remove her shoes and sit on the edge of a table with her feet dangling. The physician rubs the back end of a rubber mallet along the sole of her foot and uses the rubber mallet to tap just below her patella.

- Identify these tests.
- Explain what responses would be considered abnormal and what these could indicate.

The speed of a motor response to an environmental stimulus depends, in part, on the number of synapses to be crossed between the afferent flow of impulses and the activation of efferent nerves. A reflex is a relatively simple motor response made without the involvement of large numbers of association neurons.

The simplest reflex requires only one synapse between the sensory and motor neurons. These monosynaptic reflexes are muscle stretch reflexes; the knee-jerk, or patellar reflex, is the best-known example, although all skeletal muscles exhibit similar stretch reflexes. Impulses traveling on the sensory axons enter the CNS in the *dorsal root* of the peripheral nerve, make a single synapse with a motor neuron (an alpha motoneuron) within the central gray matter, and then leave the CNS in the *ventral root* of the spinal nerve (fig. 3.12). In this example, the stimulated muscles (quadriceps femoris) are *extensors*—they increase the angle of a joint (the knee joint in this case). *Flexors,* by contrast, act antagonistically to decrease the angle of a joint. The flexor muscles of the knee joint are known as the hamstrings.

In more complicated reflexes, the sensory impulses may travel longitudinally and transversely within the gray matter, stimulating other motor neurons. This may lead to the contraction of other flexor muscles on the same side (*ipsilateral* muscles) and the contraction of extensor muscles on the opposite side (*contralateral* muscles), while inhibiting the contraction of antagonistic muscles (ipsilateral extensors and contralateral flexors). A reflex of this type is known as a **crossed-extensor reflex** (fig. 3.13).

A. TESTS FOR SPINAL NERVE STRETCH REFLEXES

In this exercise, a number of reflex arcs will be tested that are initiated by distinctive *stretch receptors* within muscles. These receptors, called **muscle spindles,** are embedded within the connective tissue of the muscle and consist of specialized thin muscle fibers *(intrafusal fibers)* innervated by sensory neurons (see fig. 3.12). The intrafusal fibers are arranged in parallel with the normal muscle cells *(extrafusal fibers),* so that stretch of the muscle also places tension on the intrafusal fibers. Located within the spindles, the intrafusal fibers respond to the tension by stimulating (depolarizing) the sensory neuron. The sensory neuron arising from the intrafusal fiber synapses with the motor neuron in the spinal cord that, in turn, innervates the extrafusal fibers. The

99

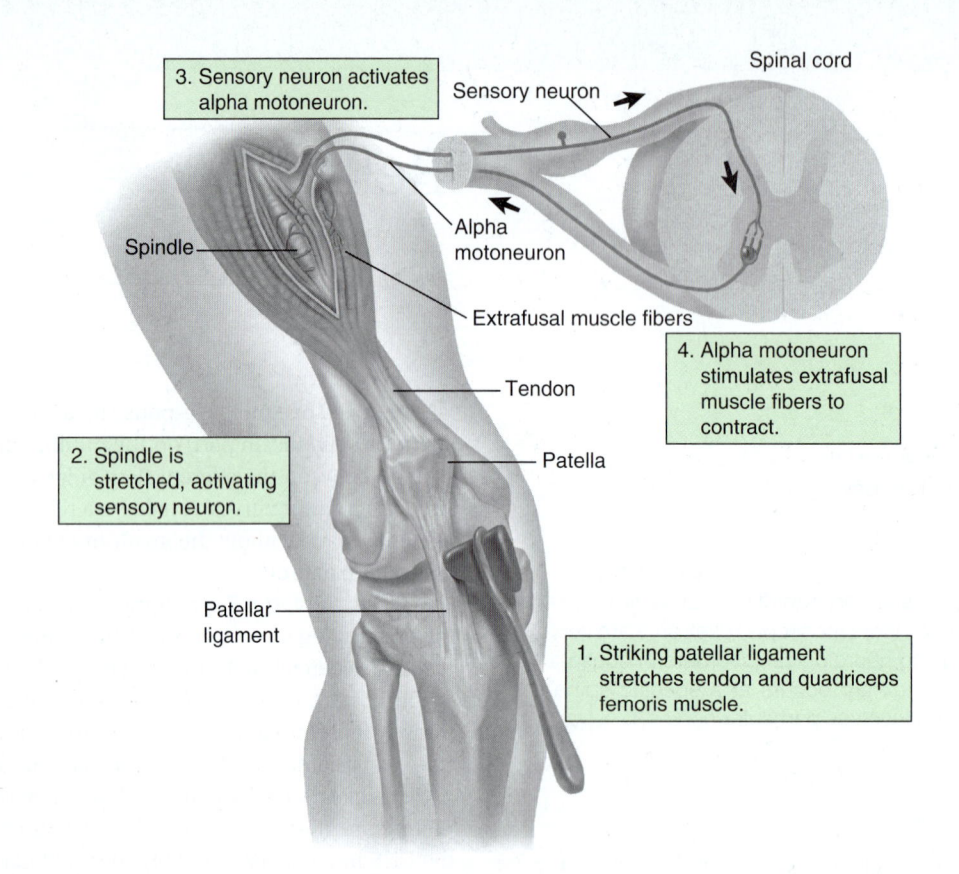

Figure 3.12 The knee-jerk reflex. This is an example of a monosynaptic stretch reflex.
(For a full-color version of this figure, see fig. 12.28 in *Human Physiology,* thirteenth edition, by Stuart I. Fox.)

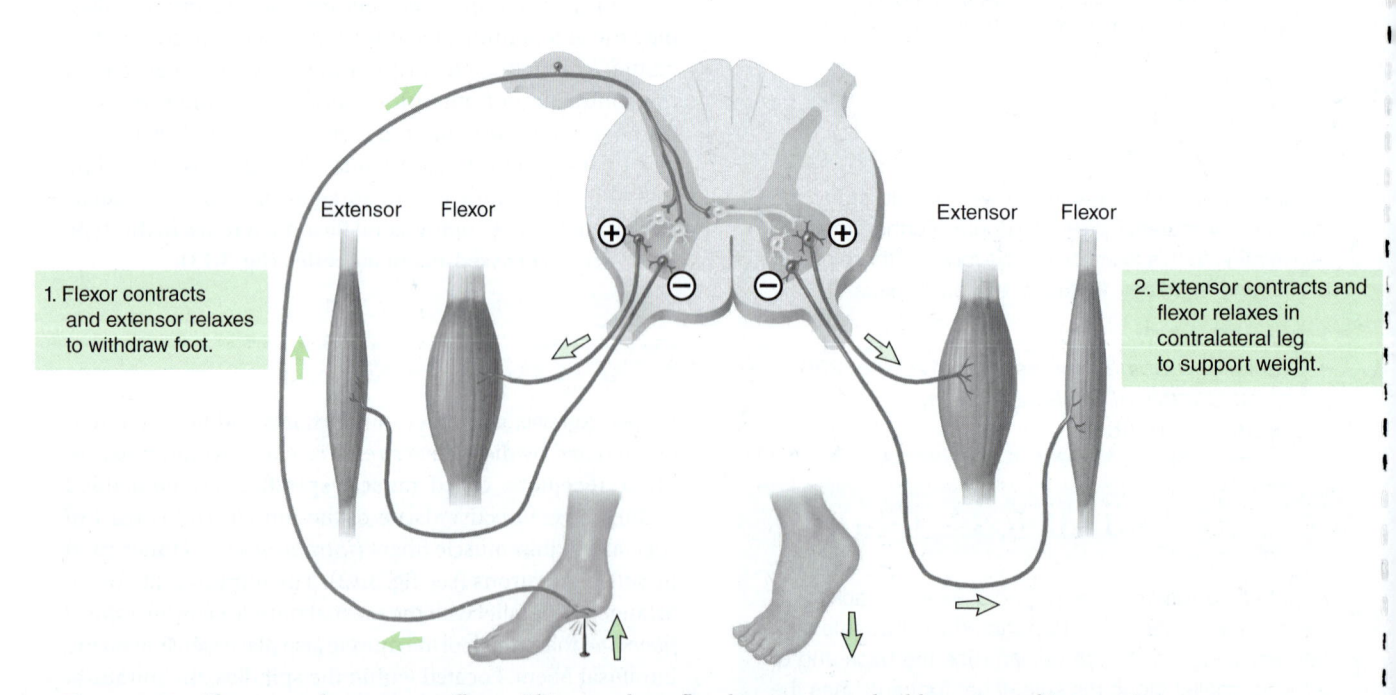

Figure 3.13 The crossed-extensor reflex. This complex reflex demonstrates double reciprocal innervation.
(For a full-color version of this figure, see fig. 12.31 in *Human Physiology,* thirteenth edition, by Stuart I. Fox.)

resultant contraction of the extrafusal fibers of the muscle then releases tension on the intrafusal fibers and decreases stimulation of the stretch receptors. Reflexes of this type are *monosynaptic stretch reflexes.* In a typical clinical examination, this reflex is elicited by striking the muscle tendon with a rubber mallet, creating a momentary stretch.

When the extrafusal muscle fibers contract during the stretch reflex, they produce a short, rapid movement of the limb (the jerk). This is very obvious for the knee-jerk reflex, but can be quite subtle for the biceps- and triceps-jerk reflexes.

Use of the *Flexicomp* allows the limb movement to be seen as a tracing on the computer screen. This allows measurements to be made, including the latent period between application of the stimulus and the contraction and the strength of the contractions under different conditions.

Procedure

Knee-Jerk Reflex (Fig. 3.14a) — Tests Femoral Nerve

1. Allow the subject to sit comfortably with his or her legs free.
2. Strike the ligament portion of the patellar tendon just below the patella (kneecap), and observe the resulting contraction of the quadriceps muscles and extension of the lower leg.

Procedure

Ankle-Jerk Reflex (Fig. 3.14b) — Tests Medial Popliteal Nerve

1. Have the subject kneel on a chair with his or her back to you and with feet (shoes and socks off) projecting over the edge.
2. While holding the subject's foot with one hand, strike the Achilles (calcaneal) tendon at the level of the ankle and observe the resulting plantar flexion of the foot. Plantar flexion is seen as a downward movement of the foot, as occurs when a person points their toes.

Procedure

Biceps-Jerk Reflex (Fig. 3.14c) — Tests Musculocutaneous Nerve

1. With the subject's arm relaxed but fully extended on the desk, gently press his or her biceps tendon in the antecubital fossa with your thumb or forefinger and strike this finger with the mallet.
2. If this procedure is performed correctly, the biceps muscle will twitch but usually will not contract strongly enough to produce arm movement.

Procedure

Triceps-Jerk Reflex (Fig. 3.14d) — Tests Radial Nerve

1. Have the subject lie on his or her back with the elbow bent, so that the arm lies loosely across the abdomen.
2. Strike the triceps tendon about 2 inches above the elbow. If there is no response, repeat this procedure, striking to either side of the original point.
3. If this procedure is correctly performed, the triceps muscle will twitch but usually will not contract strongly enough to produce arm movement.

B. A Cutaneous Reflex: The Plantar Reflex and Babinski's Sign

The **plantar reflex** is elicited by cutaneous (skin) receptors of the foot and is one of the most important neurological tests. In normal individuals, proper stimulation of these receptors located in the sole of the foot results in the flexion (downward movement) of the great toe, while the other toes flex and come together. The normal plantar reflex requires the uninterrupted conduction of nerve impulses along the *pyramidal motor tracts,* which descend directly from the cerebral cortex to lower motor neurons in the spinal cord. Damage anywhere along the pyramidal motor tracts produces a *Babinski reflex,* or **Babinski's sign,** to this stimulation, in which the great toe extends (moves upward) and the other toes fan laterally, as shown in figure 3.14*e*. Infants exhibit Babinski's sign normally because neural control is not yet fully developed.

Clinical Applications

Tests for simple muscle reflexes are basic to any physical examination when motor nerve or spinal damage is suspected. If there is spinal cord damage, these easily performed tests can help locate the level of the spinal cord that is damaged: motor nerves that exit the spinal cord above the damaged level may not be affected, whereas nerves that originate at or below the damaged level may not be able to produce their normal reflexes. The Babinski test is particularly useful in this regard because damage to the pyramidal motor (corticospinal) tract at any level may be detected by a positive Babinski reflex.

> **Procedure**

1. Have the subject lie on his or her back with knees slightly bent and with the thigh rotated so that the lateral (outer) side of the foot is resting on the couch.

2. Applying firm (but not painful) pressure, draw the tip of a blunt probe (such as the handle of the rubber mallet) along the lateral border of the sole, starting at the heel and ending at the base of the big toe (fig. 3.14*e*). Observe the response of the toes to this procedure.

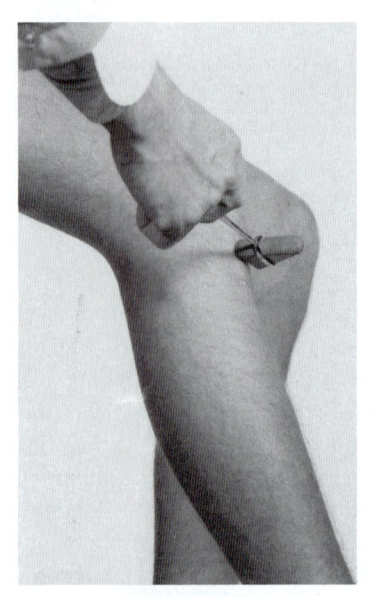

(a) Knee (patellar) reflex

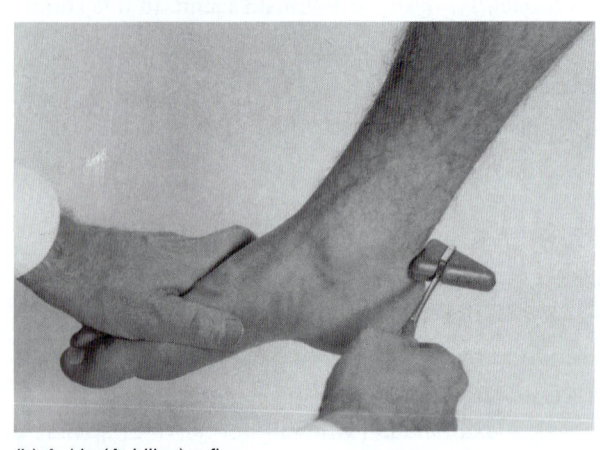

(b) Ankle (Achilles) reflex

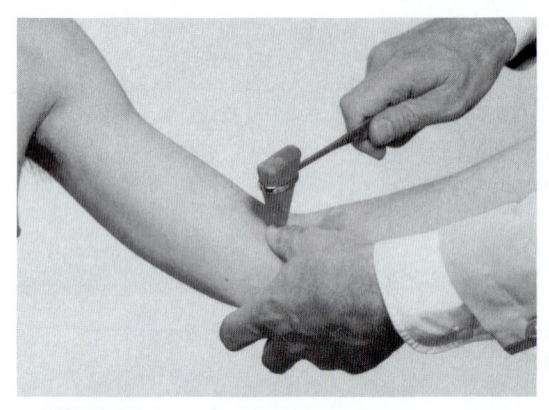

(c) Biceps reflex

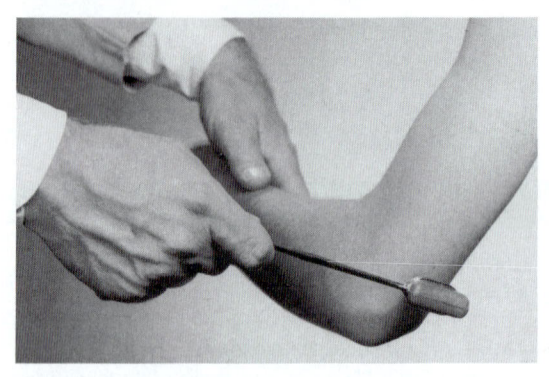

(d) Triceps reflex

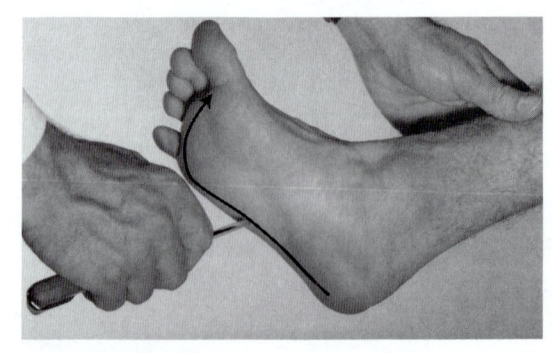

(e) Plantar reflex

Figure 3.14 **Some reflexes of clinical importance.**

Laboratory Report 3.3

Name _____

Date _____

Section _____

REVIEW ACTIVITIES FOR EXERCISE 3.3

Test Your Knowledge

1. Afferent neurons are _____.
2. Efferent neurons are _____.
3. Muscle spindles are receptors sensitive to _____.
4. A muscle spindle is composed of several _____.
5. How many synapses are crossed in a single reflex arc during a muscle stretch reflex? _____
6. The ventral root of spinal nerves contain _____ neurons, whereas the dorsal root contains _____ neurons.
7. The test you performed that involves the stimulation of ascending and descending spinal cord tracts: _____.

Match these tests with the nerve involved:

___ 8. biceps jerk   (a) femoral nerve
___ 9. triceps jerk   (b) musculocutaneous nerve
___ 10. knee jerk   (c) medial popliteal nerve
___ 11. ankle jerk   (d) radial nerve

Test Your Understanding

12. Describe the sequence of events that occurs from the time the patellar tendon is stretched to the time the leg is extended (knee-jerk reflex).

13. Compare the neural pathway involved in a muscle stretch reflex with that of the plantar (Babinski) reflex.

103

Test Your Analytical Ability

14. Suppose a person has spinal cord damage at the cervical level. Would this stop the knee-jerk reflex? How would it affect the plantar (Babinski) reflex? Explain.

15. Spinal cord damage interrupts descending motor tracts that act to suppress the activity of spinal motor neurons. Use this information, and the muscle stretch reflex pathway, to explain why someone with a spinal cord injury can have muscle spasms below the level of the injury.

Clinical Investigation Questions

16. What are the names of the neurological tests performed by the physician and what anatomical structures are being tested?

17. Describe the appearance of abnormal responses and explain what these could indicate.

EXERCISE 3.4

Cutaneous Receptors and Referred Pain

MATERIALS

1. Thin bristles, cold and warm metal rods
2. Calipers, cold and warm water baths

Specialized sensory organs and free nerve endings in the skin provide four modalities of cutaneous sensation. The modality and location of each sensation is determined by the specific sensory pathway in the brain; the acuteness of sensation depends on the density of the cutaneous receptors. Damage to an internal organ may elicit a perception of pain in a somatic location, such as a limb or a region of the body wall. This type of pain, called referred pain, is very important in clinical diagnosis.

LEARNING OUTCOMES

You should be able to:

1. Describe the punctate distribution of cutaneous receptors.
2. Describe the structures of cutaneous receptors and the modality of sensations they mediate.
3. Determine the two-point touch threshold in different areas of the skin, and explain the physiological significance of the differences obtained.
4. Define and demonstrate sensory adaptation, and explain its significance.
5. Define referred pain, and explain how this pain is produced.
6. Demonstrate the referred pain produced by striking the ulnar nerve with a mallet, and explain the clinical significance of other referred pains in the body.

Textbook/Multimedia Correlations

Before performing this exercise, you should study the introductory material presented here. Further information relating to this exercise can be found in these pages of *Human Physiology*, thirteenth edition, by Stuart I. Fox:

- *Cutaneous Sensations.* Chapter 10, p. 270.
- Multimedia correlations are provided in *MediaPhys 3.0:* Topics 6.38–6.40.

Clinical Investigation

A patient reported a pain in his back that became acute when he ate oily or fatty food. An examination performed after the patient ate peanut butter revealed the pain to be localized to below the right scapula. A stool sample revealed that the patient had fatty stools.

- Explain referred pains in general, and identify the possible source of the back pain experienced by this patient.
- Explain how pain below the right scapula might relate to fatty stools.

Sensory receptors in the skin transduce mechanical or chemical stimuli into nerve impulses (action potentials). These impulses are conducted by sensory neurons into the central nervous system and ultimately to the region of the brain that interprets these impulses as a particular sensation. Different receptors are most sensitive to different stimuli and thus activate different sensory neurons. It is the brain's interpretation of action potentials along particular sensory neurons that creates the sensory perception.

A. Mapping the Temperature and Touch Receptors of the Skin

Four independent modalities of cutaneous sensation have traditionally been recognized: *warmth, cold, touch,* and *pain.* (Pressure is excluded because it is mediated by receptors deep in the dermis, and the sensations of itch and tickle are usually excluded because of their more complex origins.) Mapping of these sensations, such as temperature and touch, on the surface of the skin has revealed that the receptors are not generalized throughout the skin but are clustered at different points (have a *punctate distribution*).

The punctate distribution is different for each of the four sensory modalities, so physiologists believed that each sensation was mediated by a different sensory receptor. This view is supported by the identification of different cutaneous receptors (table 3.1 and fig. 3.15) and by physiological studies of these receptors. However, because most natural stimuli

Table 3.1 Cutaneous Receptors

Receptor	Structure	Sensation	Location
Free nerve endings	Unmyelinated dendrites of sensory neurons	Light touch; hot; cold; nociception (pain)	Around hair follicles; throughout skin
Merkel's discs	Expanded dendritic endings associated with 50–70 specialized cells	Sustained touch and indented depth	Base of epidermis (stratum basale)
Ruffini corpuscle (endings)	Enlarged dendritic endings within open, elongated capsule	Skin stretch	Deep in dermis and hypodermis
Meissner's corpuscles	Dendrites encapsulated in connective tissue	Changes in texture; slow vibrations	Upper dermis (papillary layer)
Pacinian corpuscles	Dendrites encapsulated by concentric lamellae of connective tissue structures	Deep pressure; fast vibrations	Deep in dermis

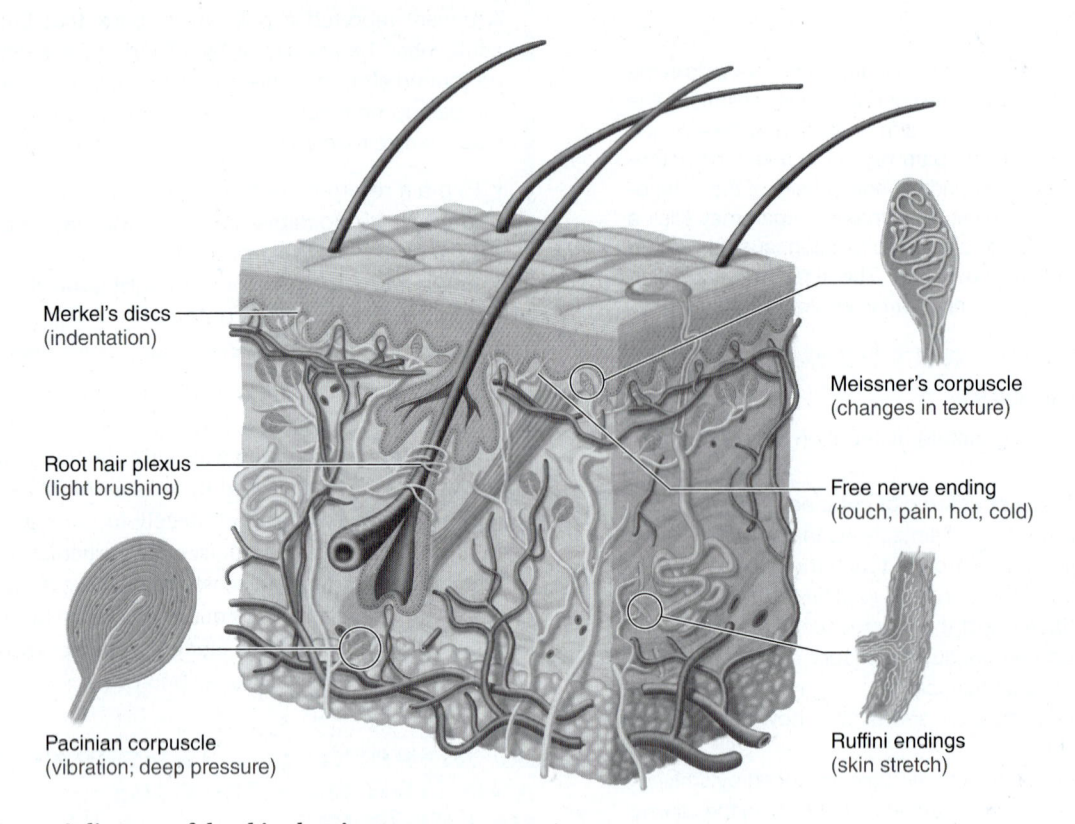

Figure 3.15 A diagram of the skin showing cutaneous receptors.
(For a full-color version of this figure, see fig. 10.4 in *Human Physiology*, thirteenth edition, by Stuart I. Fox.)

will activate more than one type of cutaneous receptor to a greater or lesser degree, the actual perception of the stimulus created in the brain will usually be more nuanced than each receptor type alone could provide.

Procedure

1. Using a pen with washable ink, draw a square (2 cm on a side) on the ventral surface of the subject's forearm. Alternatively, you can use a stamp with a square grid and ink it with washable ink. Stamp that square on the ventral surface of the subject's forearm and also in the "Data" section of the laboratory report. This is where you will draw your sensory map of that area of skin.

2. As you perform steps 3, 4, and 5, mark the square in the "Data" section of your laboratory report, rather than on the subject's forearm. This will produce a map of the cutaneous receptors in that area of skin.

3. With the subject's eyes closed, gently touch a dry, ice-cold, metal rod to different points in the square. Mark the points of cold sensation with a dark dot on the map in your laboratory report.

4. With the subject's eyes closed, gently touch a dry, warm, metal rod (heated to about 45°C in a water-bath) to different points in the square. Mark the

points of warm sensation with an open circle on the map in your laboratory report.

5. Gently touch a thin bristle to different areas of the square, and indicate the points of touch sensation with small x's on the map in your laboratory report.

B. THE TWO-POINT THRESHOLD IN TOUCH PERCEPTION

The density of touch receptors in some parts of the body is greater than in other parts. Therefore, the areas of the **somatosensory cortex** (*postcentral gyrus* of the central fissure, fig. 3.16b) that correspond to different regions of the body are of different sizes. Those areas of the body that have the largest density of touch receptors also receive the greatest motor innervation; the areas of the **motor cortex** (*precentral gyrus* of the central fissure, fig. 3.16a) that serve these regions are correspondingly larger than other areas. Therefore, a map of the sensory and motor areas of the brain reveals that large areas are devoted to the touch perception and motor activity of the face (particularly the tongue and lips) and hands, whereas relatively small areas are devoted to the trunk, hips, and legs (table 3.2).

The density of touch receptors is measured by the **two-point threshold test.** The two points of a pair of adjustable calipers are simultaneously placed on the subject's skin with equal pressure, and the subject is asked whether two separate points of contact are felt. If the answer is yes, the points of the caliper are brought closer together and the test is repeated until only one point of contact is felt. The minimum distance at which two points of contact can be felt is the two-point threshold.

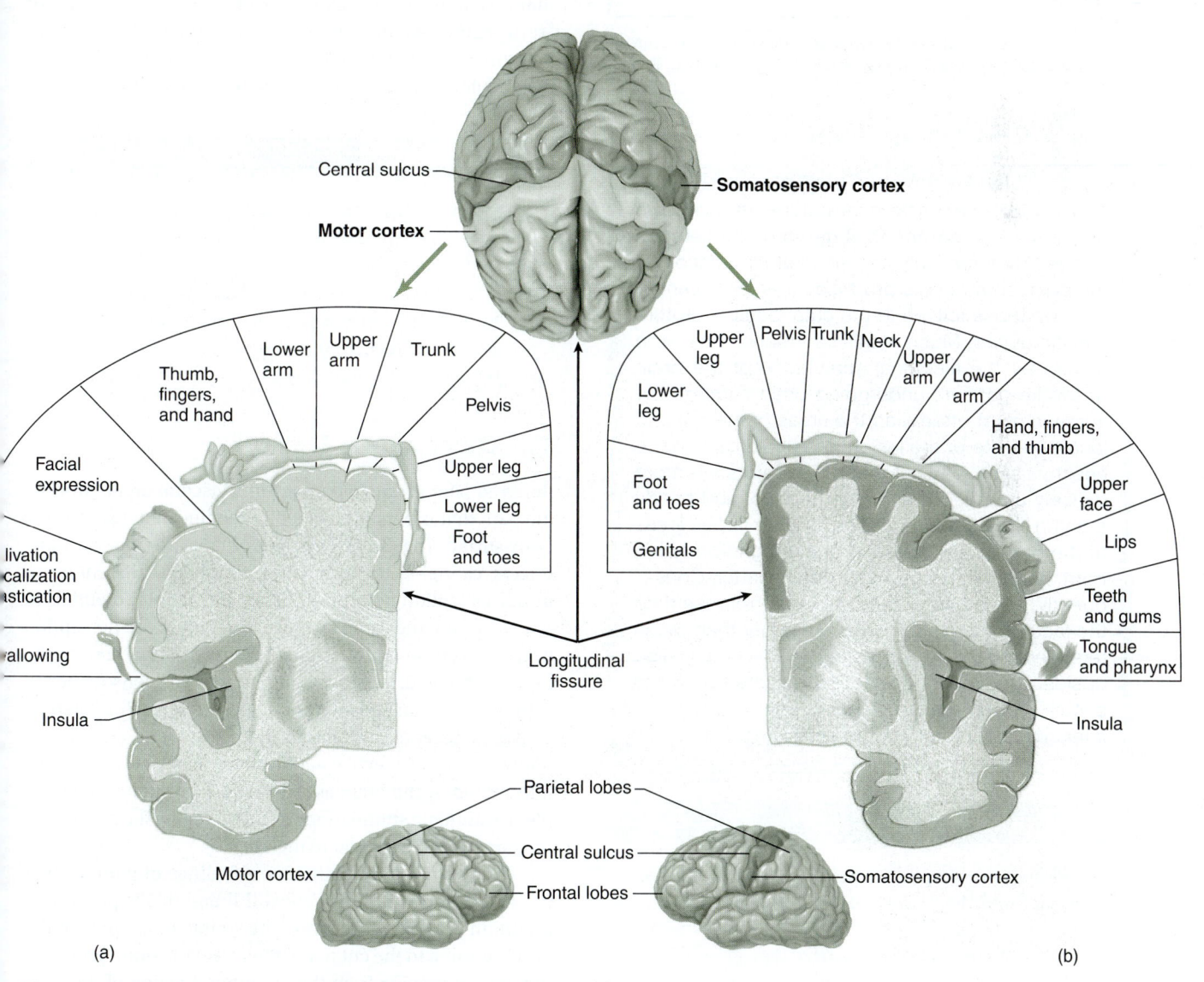

Figure 3.16   Motor and sensory areas of the cerebral cortex.   (a) Motor areas that control skeletal muscles and (b) sensory areas that receive somatesthetic sensations. Artistic liberties have been taken in (b), because the left hemisphere actually receives sensory input primarily from the right side of the body.

(For a full-color version of this figure, see fig. 8.7 in *Human Physiology*, thirteenth edition, by Stuart I. Fox.)

Table 3.2 The Two-Point Touch Threshold for Different Regions of the Body

Body Region	Two-Point Touch Threshold (mm)
Big toe	10
Sole of foot	22
Calf	48
Thigh	46
Back	42
Abdomen	36
Upper arm	47
Forehead	18
Palm of hand	13
Thumb	3
First finger	2

Source: From S. Weinstein and D.R. Kenshalo, editors, *The Skin Senses*, © 1968. Courtesy of Charles C. Thomas, Publisher, Ltd., Springfield, Illinois.

Clinical Applications

Sensory information from the cutaneous receptors projects to the **postcentral gyrus** of the cerebral cortex. Therefore, direct electrical stimulation of the postcentral gyrus produces the same sensations as those felt when the cutaneous receptors are stimulated. Much of this information has been gained by the electrical stimulation of the brain of awake patients undergoing brain surgery; the surgeon must often map the areas of the brain to locate the site of the lesion and avoid damage to healthy tissue. The cutaneous receptors are more densely arranged in the face, tongue, and hands than on the back and thighs, and so larger areas of the brain are involved in analyzing information from the former areas than from the latter. Consequently, the areas of the brain map representing the face, tongue, and hands are larger than those representing the back and thighs. The map is also upside down, with the feet represented near the superior surface and the head represented more inferiorly and laterally in the cortex (fig. 3.16).

Procedure

1. Starting with the calipers wide apart and the subject's eyes closed, determine the two-point threshold on the back of the hand. (Randomly alternate the two-point touch with one-point contacts, so that the subject cannot anticipate you.)
2. Repeat this procedure with the palm of the hand, fingertip, and back of the neck.
3. Write the minimum distance (in mm) in the data table provided in your laboratory report.

C. Adaptation of Temperature Receptors

Many of our sense receptors respond strongly to acute changes in our environment and then stop responding when these stimuli become constant. This phenomenon is known as **sensory adaptation.** Our sense of smell, for example, quickly adapts to the odors of the laboratory; and our touch receptors soon cease to inform us of our clothing, until these stimuli change. Sensations of pain, by contrast, adapt little if at all.

When one hand is placed in warm water and another in cold water, the strength of stimulation gradually diminishes until both types of temperature receptors have adapted to their new environmental temperature. If the two hands are then placed in water at an intermediate temperature, the hand that was in the cold water will feel warm, and the hand that was in the warm water will feel cold. The "baseline," or "zero," of the receptors has changed. The sensations of temperature are therefore not absolute but relative to the baseline previously established by sensory adaptation.

Procedure

1. Place one hand in warm water (about 40°C) and the other in cold water, and leave them in the water for a minute or two (but remove them as soon as the water becomes too uncomfortable).
2. Now place both hands in lukewarm water (about 22°C), and record your observations and conclusions regarding your sensations in the laboratory report.

D. Referred Pain

Referred pain is due to damage in a visceral organ producing pain that is perceived at a different location towards the body surface (fig. 3.17). Receptor organs are sensory transducers, changing environmental stimuli into afferent nerve impulses (action potentials). Since the action potentials in one nerve are the same as those in another, the perception of the sensation is determined entirely by the area of the brain stimulated, which is different for each sensory nerve. Although a given sensory nerve is normally stimulated by a specific receptor, trauma to the nerve along the afferent pathway may also evoke action potentials, and this will be interpreted by the brain as the normal sensation. An example of traumatic stimulation would be seeing flashes of light or "stars" when punched in the eye.

Amputees frequently report feelings of pain in their missing limbs as if they were still there; this is part of the phantom limb phenomenon. The source of nerve stimulation is trauma to the cut nerve fibers, yet the brain perceives the pain as coming from the amputated region of the body that had originally produced the action potentials along these nerves. This is a referred pain because the source of nerve stimulation is different from the perceived location of the stimulus.

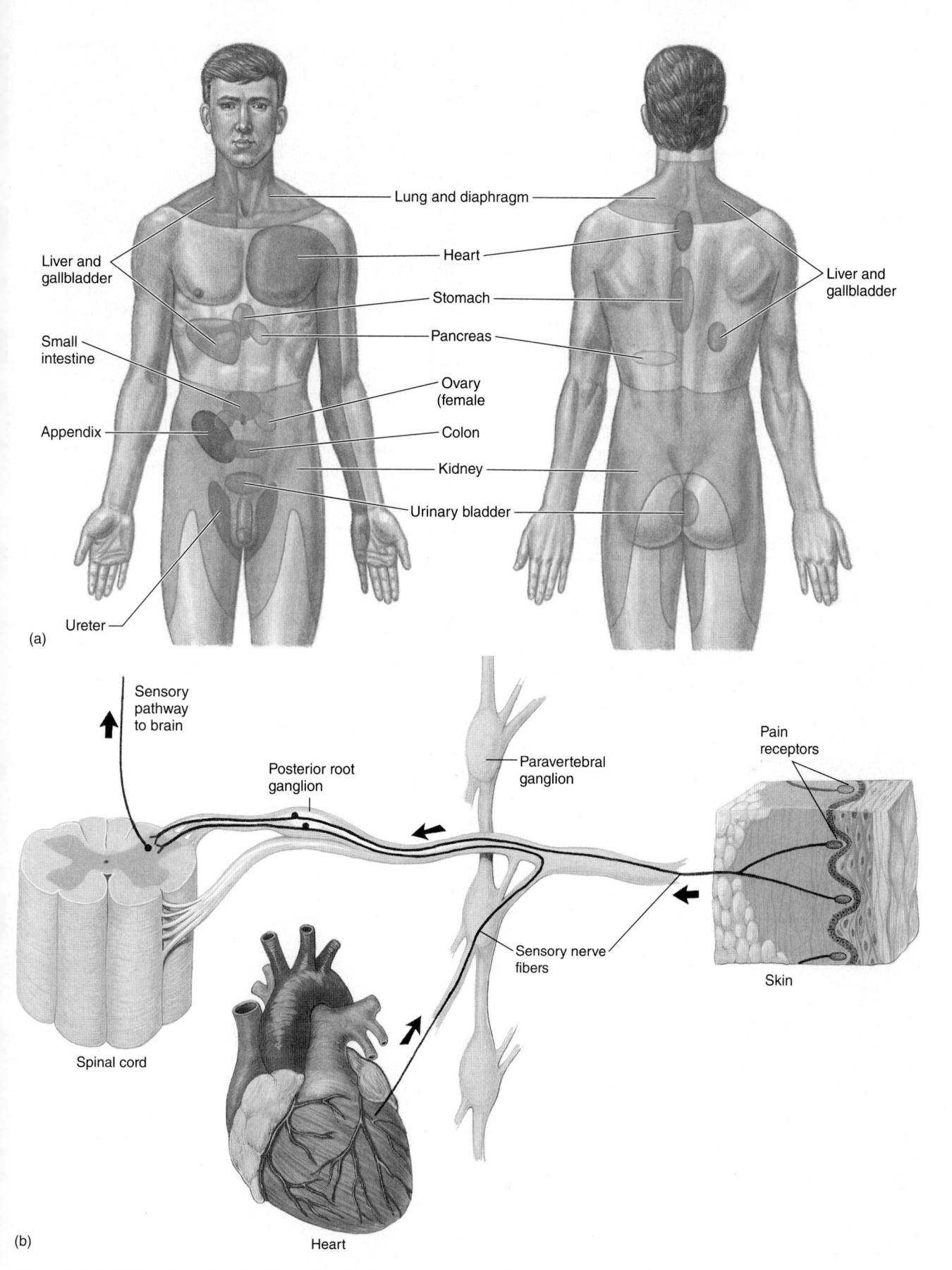

Figure 3.17 **Referred pain.** (a) Sites of referred pain are perceived cutaneously but actually originate from specific visceral organs. (b) One explanation why pain from a visceral organ, such as the heart, might be perceived over a particular area of skin. In this scenario, sensations from the two regions share a common nerve pathway.

Clinical Applications

Referred pains are important clinically, particularly for pains in the deep viscera (organs within the abdominal or thoracic cavities), which are characteristically dull and poorly localized. In ischemic heart disease, for example, the pain is referred to the left pectoral region and left arm and shoulder areas (see fig. 3.17a); this is called **angina pectoris.** In many patients with stomach ulcers, the pain is referred to the region between the scapulae of the back. A referred pain under the right scapula may be caused by a gallstone when the gallbladder contracts. Referred pains are believed to result because both visceral sensory and somatic sensory neurons can synapse on the same interneurons in the spinal cord (see fig. 3.17b). These, in turn, project to the same area of the somatosensory cortex (postcentral gyrus) devoted to a particular body location (such as the left arm).

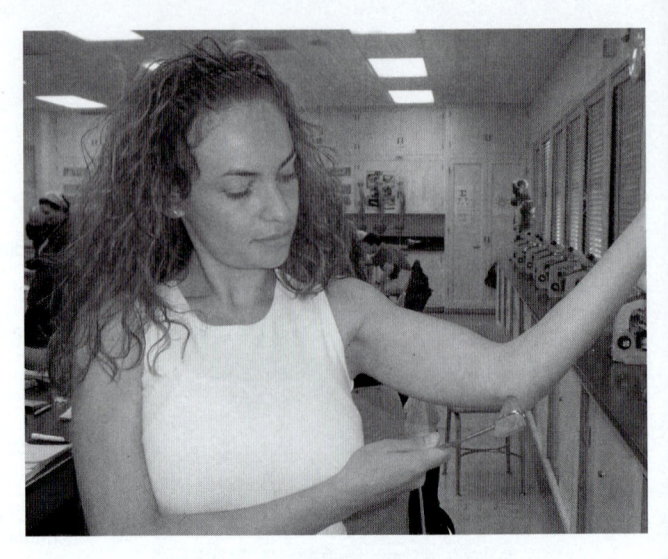

Figure 3.18 Position of the rubber mallet to test for referred pain. The mallet strikes the ulnar nerve in the groove lateral to the median epicondyle of the humerus.

Procedure

1. Gently tap the ulnar nerve where it crosses the median epicondyle of the elbow (fig. 3.18).
2. Describe the locations where you perceive tingling or pain.

Laboratory Report 3.4

Name _____
Date _____
Section _____

DATA FROM EXERCISE 3.4

A. Mapping the Temperature and Touch Receptors of the Skin

1. In this box, reproduce the map of the hot, cold, and touch receptors from the square on your forearm. If you used an ink stamp on your forearm, stamp this space as well, so you can reproduce your map here.

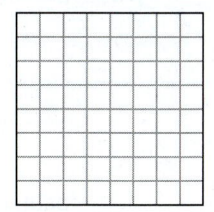

B. The Two-Point Threshold in Touch Perception

1. Write your results in this data table.

Location	Two-Point Threshold (mm)
Back of hand	
Palm of hand	
Fingertip	
Back of the neck	

2. Compare your results to the average values provided in table 3.2.

C. Adaptation of Temperature Receptors

1. Record your observations and conclusions here.

D. Referred Pain

1. Describe the locations where tingling or pain was felt. Was this feeling perceived to be in a different location than where the mallet was struck? If so, where?

111

REVIEW ACTIVITIES FOR EXERCISE 3.4

Test Your Knowledge

Match the receptor with the sensation with which it is most associated:
_____ 1. free nerve endings   (a) light touch; hot and cold
_____ 2. Ruffini corpuscle   (b) sustained touch and pressure
_____ 3. Pacinian corpuscle   (c) sustained pressure
_____ 4. Meissner's corpuscle   (d) changes in texture; slow vibrations
_____ 5. Merkel's discs   (e) deep pressure; fast vibrations

6. The motor cortex is the _____ gyrus of the cerebral cortex; the sensory cortex is the _____ gyrus.

7. Define *sensory adaptation*. _____

8. Name a sensory modality that adapts quickly: _____; name one that adapts slowly, if at all: _____.

9. Angina pectoris is an example of a(n) _____ pain.

10. Pain perceived in an amputated limb is known as the _____.

Test Your Understanding

11. Which parts of your body have the highest density of touch receptors? What benefits may be derived from that fact?

12. What does the map of the sensory cortex reveal about the density of touch receptors? Explain.

13. How did your right and left hands feel when placed in the same lukewarm water-bath? Explain how this occurred.

14. Describe the importance of referred pain in the diagnosis of deep visceral pain and give examples.

Test Your Analytical Ability

15. Describe the map of the motor cortex. How does it compare with the map of the sensory cortex? What does the map of the motor cortex reveal about motor control?

16. "Our perceptions of the external world are created by our brains." Discuss this concept, using the phantom limb phenomenon to support your argument.

Clinical Investigation Questions

17. What are referred pains, and how are they produced? Which organ or organs are most likely affected in this patient?

18. How might the pain below the right scapula in this patient relate to the presence of fatty stools? Explain.

EXERCISE 3.5

Eyes and Vision

MATERIALS

1. Snellen eye chart and astigmatism chart
2. Wire screen and meter stick
3. Ophthalmoscope
4. Lamp
5. Red, blue, and yellow squares on larger sheets of black paper or cardboard
6. Ishihara color blindness cards

The elastic properties of the eye's lens allow its refractive power to be varied so that the image of an object from almost any distance can be focused properly on the retina. Photoreceptors—rods and cones—are located in the retina. The refractive abilities of the eye and the functions of its inner structures are routinely tested in eye examinations.

LEARNING OUTCOMES

You should be able to:

1. Describe the structure of the eye and the functions of its component parts.
2. Test for visual acuity and accommodation, and describe common refractive problems.
3. Identify the extrinsic eye muscles, and describe their functions.
4. Describe the optic disc and fovea centralis, and explain their significance.
5. Demonstrate the presence of a blind spot and explain why light focused on this spot cannot be seen.

Textbook/Multimedia Correlations

Before performing this exercise, you should study the introductory material presented here. Further information relating to this exercise can be found in these pages of *Human Physiology*, thirteenth edition, by Stuart I. Fox:

- *The Eyes and Vision.* Chapter 10, p. 290.
- *Retina.* Chapter 10, p. 298.
- Multimedia correlations are provided in *MediaPhys 3.0:* Topics 6.42–6.47.

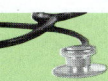

Clinical Investigation

A man who is color-blind with 20/300 vision has LASIK surgery that corrects his vision to 20/20 in each eye. Years later, however, he experiences difficulty reading and must get reading glasses. A few years later he develops cloudy patches in his right eye that the physician says will require surgery to replace the lens. A few years after his vision is restored by that surgery, the clarity of his central vision deteriorates and it is dangerous for him to drive at night.

- Explain the most likely cause of his color blindness, the visual defect indicated by 20/300 vision, and how the LASIK surgery changed his vision to 20/20.
- Explain why he needed reading glasses years after his LASIK surgery.
- Identify the most likely cause of the cloudy patches in his vision that was corrected by surgical replacement of his lens, and the likely cause of the later deterioration of his central vision.

You should be familiar with the gross structure of the eye (fig. 3.19). The eye has three walls, or tunics, that form an *outer fibrous layer* (the **sclera** and **cornea**), a middle vascular layer (the **choroid**), and an inner layer (the **retina**). The **lens,** suspended by a *suspensory ligament* attached to the **ciliary body,** divides the eye into anterior and posterior *cavities.* Filled with *aqueous humor,* the anterior cavity is further divided into anterior and posterior *chambers* by the colored **iris.** The iris is a muscular diaphragm that regulates the entry of light into the eye through its aperture *(pupil).* The semi-gelatinous *vitreous humor* (or *vitreous body*) occupies the posterior chamber and lends structural support to the eye.

A. REFRACTION: TEST FOR VISUAL ACUITY AND ASTIGMATISM

Light rays are bent *(refracted)* when they pass from air to a medium of greater density, where their rate of transmission is slower. The light rays that diverge from an object in the

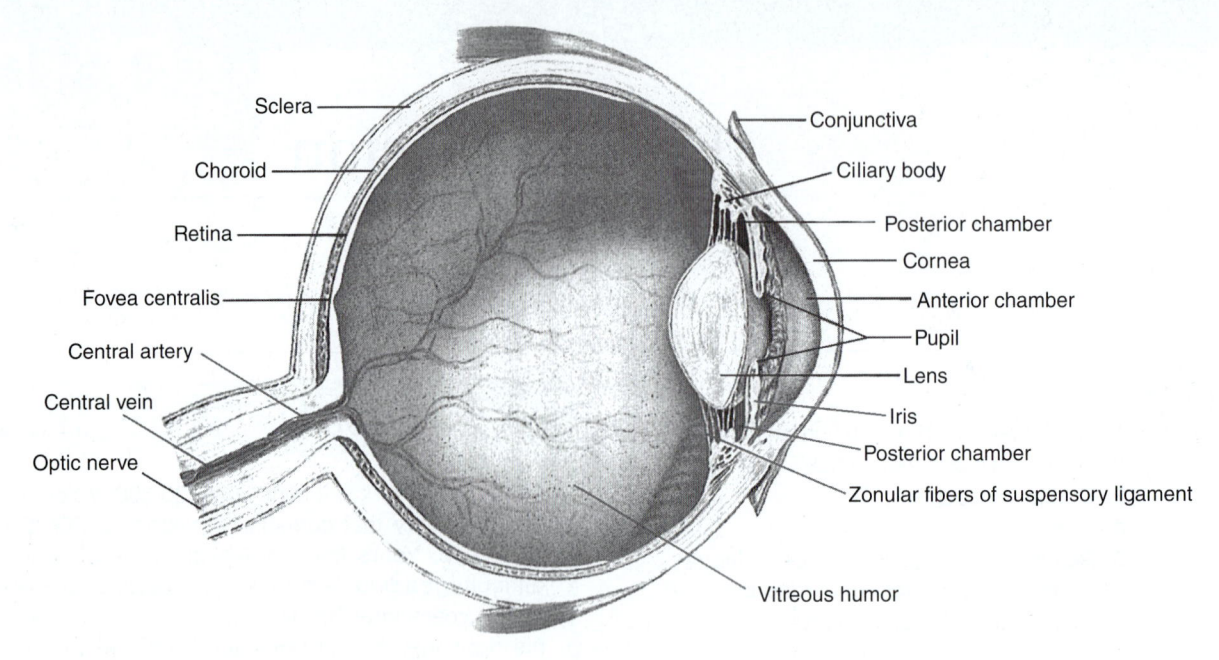

Figure 3.19 **The gross structure of the eye.**
(For a full-color version of this figure, see fig. 10.27 in *Human Physiology*, thirteenth edition, by Stuart I. Fox.)

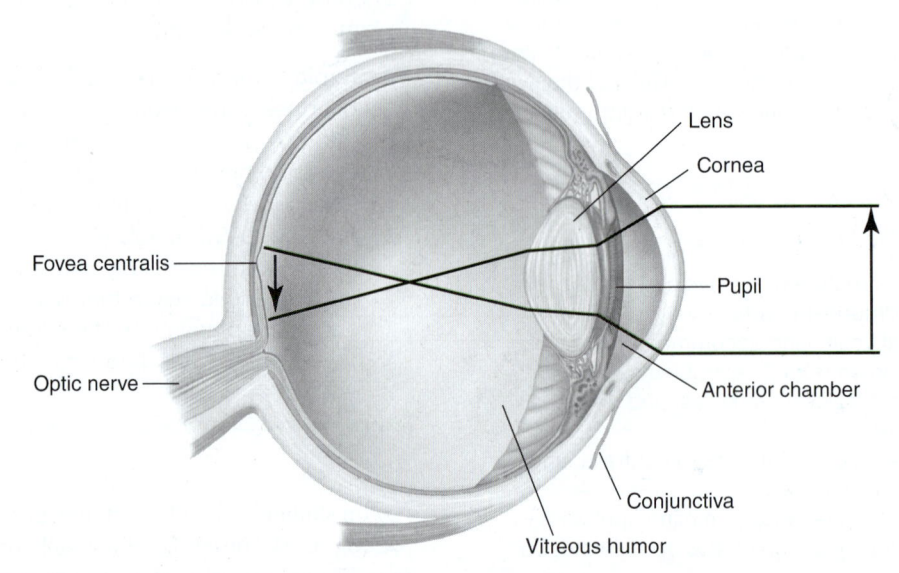

Figure 3.20 **The image is inverted on the retina.** Refraction of light, which causes the image to be inverted, occurs to the greatest degree at the air-cornea interface. Changes in the curvature of the lens, however, provide the required fine focusing adjustments.
(For a full-color version of this figure, see fig. 10.31 in *Human Physiology*, thirteenth edition, by Stuart I. Fox.)

visual field are refracted by the cornea, aqueous humor, lens, and vitreous humor of the eye so that the rays converge (are focused) on the retina and form an inverted image, reversed from left to right (fig. 3.20).

The refractive power of the cornea and vitreous humor is constant. The strength of the lens (i.e., its ability to refract light) can be varied by making it more or less convex. The greater the degree of convexity, the greater the strength of the lens (i.e., the greater the ability to bring parallel rays of light to a focus). A lens that brings light to a focus 0.25 m from its center is stronger (more convex) than a lens that brings light to a focus 1 m from its center. The strength of a lens is expressed in **diopters.**

$$\text{Strength (diopters)} = \frac{1}{\text{focal length (meters)}}$$

The lens that brings parallel waves of light to a focus 0.25 m from its center has a *focal length* of 0.25 m and a strength of 4 diopters, whereas the lens that brings light to a focus 2 m from its center has a focal length of 2 m and a strength of 0.5 diopters. The refractive power of the normal eye when an object is 20 feet or more away is 67 diopters.

Emmetropia (normal vision)
Rays focus on retina
(a)

No correction necessary

Myopia (nearsightedness)
Rays focus in front of retina
(b)

Concave lens corrects nearsightedness

Hyperopia (farsightedness)
Rays focus behind retina
(c)

Convex lens corrects farsightedness

Astigmatism
Rays do not focus
(d)

Uneven lens corrects astigmatism

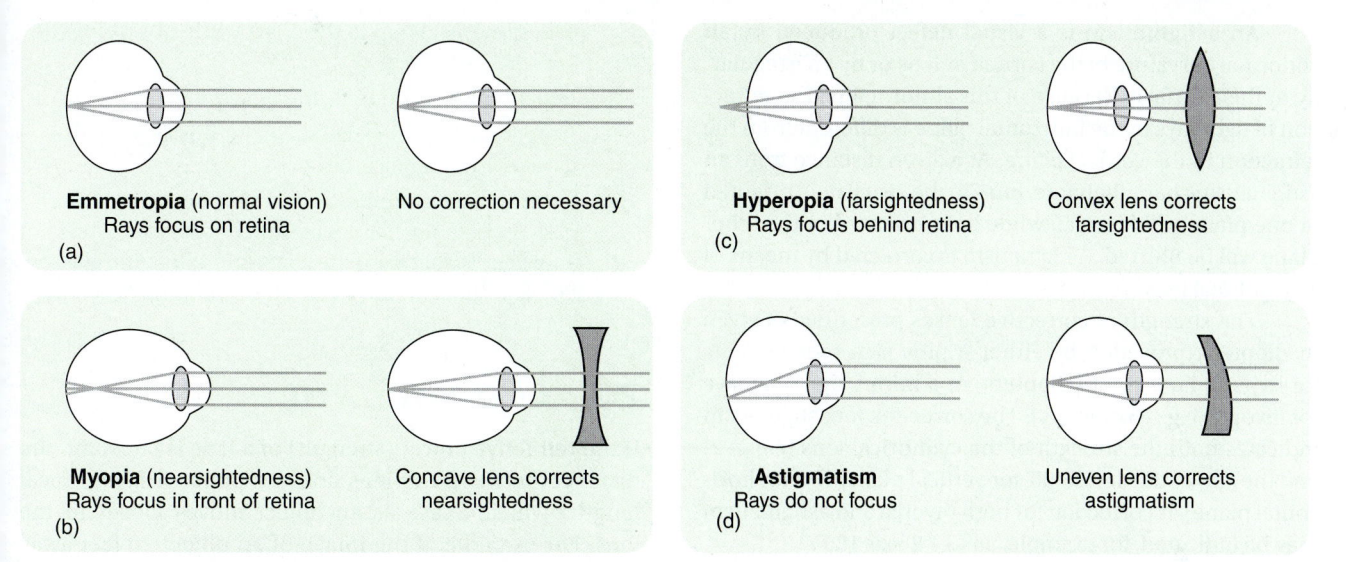

Figure 3.21 **Problems of refraction and how they are corrected.** In a normal eye (a), parallel rays of light are brought to a focus on the retina by refraction in the cornea and lens. If the eye is too long, as in myopia (b), the focus is in front of the retina. This can be corrected by a concave lens. If the eye is too short, as in hyperopia (c), the focus is behind the retina. This is corrected by a convex lens. In astigmatism (d), light refraction is uneven because of irregularities in the shape of the cornea or lens.

(For a full-color version of this figure, see fig. 10.35 in *Human Physiology*, thirteenth edition, by Stuart I. Fox.)

When the light rays that diverge from two adjacent points in the visual field are each brought to a perfect focus on the retina, two points will clearly be perceived. If, however, the light rays converge on a point in front of or behind the retina, the two points in the visual field will be perceived as one fuzzy or blurred point. To correct this defect in **visual acuity** (sharpness of vision), the individual must either adjust the distance between the eye and the object or wear corrective lenses that change the degree of refraction.

When a distant object (20 feet or more) is brought to a focus in front of the retina, the individual is said to have **myopia** *(nearsightedness)*. Myopia is usually due to an elongated eyeball (excessive distance from lens to retina). As a result, the object must be moved closer to the eyes for the image to be in focus on the retina and clearly seen. Glasses with concave lenses may be used to correct myopia. In the opposite condition, **hyperopia** (hypermetropia, or *farsightedness*), the image is brought to a focus behind the retina. This condition is usually due to an eyeball that is too short. As a result, the object must be moved farther from the eyes for the image to be in focus on the retina. Glasses with convex lenses may be used to correct hyperopia. These conditions, and the ways that glasses correct for them, are illustrated in figure 3.21. Normal visual acuity is called **emmetropia**.

Myopia is frequently tested by means of the **Snellen eye chart.** A person with normal visual acuity can read the line marked 20/20 from a distance of 20 feet. An individual with 20/40 visual acuity must stand 20 feet away from a line that a normal person can read at 40 feet. An individual with 20/15 visual acuity can read a line at a distance of 20 feet that the average, normal young adult could not read at a distance greater than 15 feet. The person with 20/40 vision has myopia, but the person with 20/15 vision does not have hyperopia. The farsighted person has a decreased ability to see near objects but cannot see distant objects any better than a person with normal vision. A person with 20/15 vision has better than the average normal distance vision.

Clinical Applications

The lens is normally completely clear, due to its unique structure. It is composed of about one thousand layers of cells aligned in parallel and tightly joined together, so that gaps don't form as the shape of the lens is changed. The lens is transparent because (1) it is avascular; (2) its cell organelles have been destroyed in a controlled process that halts before the cells die; and (3) the cell cytoplasm is filled with proteins called *crystallin*. Because of this structure, every region of the lens normally has the same refractive index. However, damage from ultraviolet light, dehydration, or oxidation may cause the crystallin proteins to change shape and aggregate to produce the cloudy patches in a person's visual field known as **cataracts.** Cataracts interfere with vision in more than half of people over the age of 65. This is generally treated by surgically replacing the lens with an artificial lens.

An **astigmatism** is a visual defect produced by an abnormal curvature of the cornea or lens or by an irregularity in their surface. Because of this abnormality, the refraction of light rays in the horizontal plane is different from the refraction in the vertical plane. At a given distance from an astigmatism chart, therefore, lines in the visual field oriented in one plane will be clear, while lines oriented in the other plane will be blurred. Astigmatism is corrected by means of a cylindrical lens.

The strength of corrective lenses prescribed is given in diopters, preceded by either a plus sign (convex lens for hyperopia; e.g., +14 diopters) or a minus sign (concave for myopia; e.g., –5 diopters). The correction for astigmatism indicates both the strength of the cylindrical lens (e.g., +2) and the axis of the defect (90° for vertical plane, 180° for horizontal plane). A correction for both myopia and astigmatism may be indicated, for example, as –3 + 2 axis 180°.

Clinical Applications

Many people with refractive problems choose to have a surgical procedure known as **LASIK** *(laser-assisted in situ keratomileusis)*. A computer-guided laser reduces the curve of the cornea for myopia so that the focus is moved back to the retina. It makes the cornea more steeply curved for hyperopia, and more spherical to correct for astigmatism. This does not correct for presbyopia, so reading glasses may be required. Alternatively, if the person has myopia, one eye can be deliberately undercorrected for reading while the other eye is corrected for distance vision. This is called *monovision,* and is tolerated better by some people than others.

Procedure

1. Stand 20 feet (6 m) from the *Snellen eye chart.* Covering one eye, attempt to read the line with the smallest letters you can see (with glasses off, if applicable). Walk up to the chart and determine the visual acuity of that eye.
2. Repeat this procedure using the other eye (with glasses off, if applicable).
3. Repeat this procedure for each eye with glasses on (if applicable).
4. Stand about 20 feet away from an astigmatism chart, and cover one eye (glasses off). This chart consists of a number of dark lines radiating from a central point, like spokes on a wheel. If astigmatism is present, some of the spokes will appear sharp and dark, whereas others will appear blurred and lighter because they are coming to a focus either in front of or behind the retina. Still covering the same eye, slowly walk up to the chart while observing the spokes.
5. Repeat this procedure using the other eye.
6. Repeat the test for astigmatism for both eyes with glasses on (if applicable).
7. To verify that astigmatism has been corrected with glasses, hold the glasses in front of your face while standing 10 feet from the chart and rotate the glasses 90°. The shape of the wheel should change when the glasses are rotated.

B. Accommodation

If the refractive power (strength) of a lens is constant, the distance between the lens and the point of focus (focal length) will increase as an object moves closer to the lens. For example, if the image of an object 20 feet away is in focus on the retina (or on the photosensitive film of a camera), the image of an object 10 feet away will be focused *behind* the retina (or the camera film) and will appear blurred. A camera can be adjusted to focus on an object 10 feet away by moving the lens outward until the focal length of the image equals the distance between the lens and the film. The object 10 feet away will now be in focus, but the object 20 feet away will be blurred because its image will now come to a focus in front of the film.

The human eye differs from the camera in that the distance between the retina and the lens of the eye cannot be changed to bring objects into focus. The human lens is elastic, however, so its degree of convexity (and therefore its refractive power) can be altered by changing the tension placed on it by the suspensory ligament; this, in turn, is regulated by the degree of contraction of the ciliary muscle (fig. 3.22). When the ciliary muscle is relaxed, the suspensory ligament pulls on the lens, thereby decreasing its convexity and power; distant objects (more than 20 feet away) are thus brought to a focus on the retina.

Near objects are brought to a focus on the retina by contraction of the ciliary muscle. The contraction reduces the tension on the suspensory ligament, allowing the lens to assume a more convex shape. This ability of the eye to focus the images of objects at different distances from the lens is called **accommodation.**

The convexity of the normal lens can be adjusted to give it a range of power from 67 diopters (for distant vision; least convex) to 79 diopters (for near vision; most convex). The elasticity of the lens and the degree of convexity it can assume for near vision decreases with age, a condition called **presbyopia** (meaning "old eyes"). Lens elasticity can be tested by measuring the near point of vision (the closest an object can be brought to the eyes while still maintaining visual acuity). The near point of vision changes dramatically with age, averaging about 8 cm at age 10 and 100 cm at age 70. Presbyopia may be corrected with *bifocals,* which contain two lenses of different refractive strengths.

(a)

Ciliary muscle fibers relaxed
Suspensory ligament taut
Lens thin and focused for distant vision

(b)

Ciliary muscle fibers contracted
Suspensory ligament relaxed
Lens thick and focused for close vision

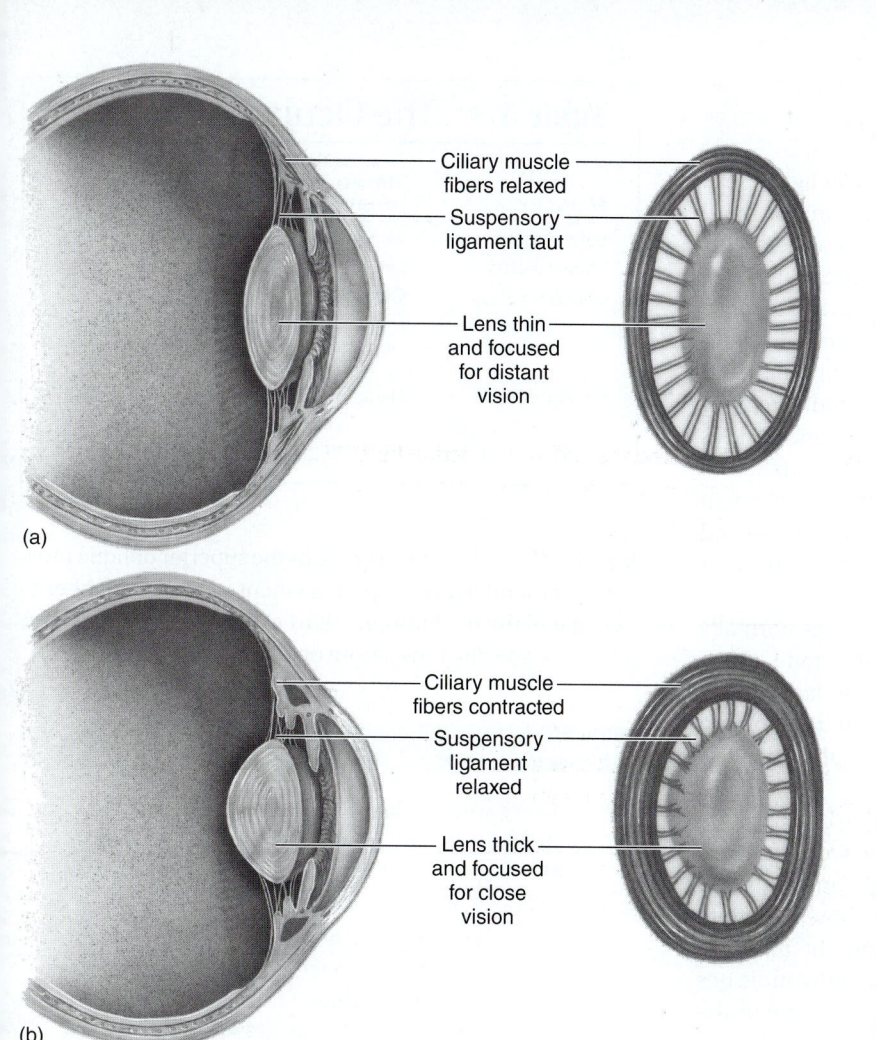

Figure 3.22 Changes in the shape of the lens permit accommodation. (a) The lens is flattened for distant vision when the ciliary muscle fibers are relaxed and the suspensory ligament is taut. (b) The lens is more spherical for close-up vision when the ciliary muscle fibers are contracted and the suspensory ligament is relaxed.

(For a full-color version of this figure, see fig. 10.34 in *Human Physiology*, thirteenth edition, by Stuart I. Fox.)

Clinical Applications

Glaucoma occurs when there is loss of retinal ganglion cell axons (see fig. 3.25), along with blood vessels and glia, in the optic nerve that produces characteristic changes in the appearance of the retina through an ophthalmoscope (see fig. 3.24). Glaucoma is generally caused by a problem in the flow of aqueous humor and can be classified as *closed-angle* or *open-angle*. Closed-angle glaucoma occurs when the drainage pathway for aqueous humor is blocked (by a tumor or inflammation); open-angle glaucoma occurs when the angle formed by the iris and cornea (see fig. 3.19) is unobstructed but the drainage (through the *canal of Schlemm*) is still inadequate. Primary open-angle glaucoma is the second leading cause of blindness in the United States, and can occur when the aqueous humor pressure is either normal (10 to 21 mm Hg) or elevated. Elevated intraocular pressure due to inadequate drainage of aqueous humor is the most important risk factor, and the only risk factor that can be treated to prevent this disease or slow its progression.

Procedure

1. Place a square of wire screen (or stretched cheesecloth) about 10 inches in front of your eyes, and observe a distant object through the screen.
2. After closing your eyes momentarily, open them and note whether the screen or the distant object is in focus.
3. Repeat this procedure, this time focusing the eyes on the screen before closing and reopening them.
4. To measure the near point of vision, place one end of a meter stick under one eye and extend it outward. Holding a pin at arm's length, gradually bring the pin toward the eye.
5. Record the distance at which the pin first appears blurred or doubled.

   Near point of vision _____ cm

6. Repeat this procedure, determining the near point of vision for the other eye.

   Near point of vision _____ cm

Note: *If the average near point of vision at age 10 is 8 cm, and at age 70 is 100 cm, what is your expected near point of vision?*

C. Extrinsic Muscles of the Eye and Nystagmus

The six extrinsic muscles of the eye are shown in figure 3.23. The cranial nerve innervations and the actions of these muscles are summarized in table 3.3. These muscles allow the eyes to follow a moving object by maintaining the image on the same location of the retina of each eye, the *fovea centralis*, which provides maximum visual acuity. These muscles also allow the visual field of each eye to maintain the correct amount of central overlap. (The medial regions of each visual field overlap, while the more lateral regions are different for each eye; this *retinal disparity* helps in three-dimensional vision and depth perception.) When an object is brought closer, the correct amount of overlap and retinal disparity is maintained by the medial movement, or *convergence*, of the eyes.

The actions of antagonistic ocular muscles normally maintain the eyes in a midline position. If the tone of one muscle is weak as a result of muscle or nerve damage, the eyes will drift slowly in one direction followed by a rapid movement back to the correct position. This phenomenon is known as **nystagmus.**

In a typical examination of the ocular muscles, the subject is asked to follow an object (such as a pencil) with his or her eyes as it is moved up and down, right and left. Continued oscillations of the eye (slow phase in one direction, fast phase in the opposite direction) indicate the presence of nystagmus. Inability to move the eye laterally indicates damage either to the abducens (sixth cranial) nerve or the lateral rectus muscle (table 3.3). Inability to move the eye downward when it is moved inward indicates damage to the trochlear (fourth cranial) nerve or the superior oblique muscle. All other defects in eye movement may be due either to damage of the oculomotor (third cranial) nerve or to damage of the specific muscles involved.

Table 3.3 The Ocular Muscles

Muscle	Cranial Nerve Innervation	Movement of Eyeball
Lateral rectus	Abducens	Lateral
Medial rectus	Oculomotor	Medial
Superior rectus	Oculomotor	Superior and medial
Inferior rectus	Oculomotor	Inferior and medial
Inferior oblique	Oculomotor	Superior and lateral
Superior oblique	Trochlear	Inferior and lateral

Procedure

1. Observe retinal disparity by holding a pencil in front of your face with one eye closed and then quickly changing eyes and noting the apparent position of the pencil.
2. Observe convergence by asking a subject to focus on the tip of a pencil as it is slowly brought from a distance of 2 feet in front of the face to the bridge of the nose. Notice the change in the diameter of the subject's pupil during this procedure.
3. Hold a pencil about 2 feet away from the bridge of the subject's nose. Then move the pencil to the left, to the right, and up and down, leaving the pencil in each position at least 10 seconds. Observe the movement of the eyes and note the presence or absence of nystagmus.

D. Pupillary Reflex

The correct amount of light is admitted into the eye through an adjustable aperture, the pupil, surrounded by the iris. The iris consists of two groups of smooth muscles with opposing actions. Operating like sphincters, the *circular muscles* constrict the pupil in bright light, whereas the *radial muscles* work to dilate the pupil in dim light. These responses are mediated by the autonomic nervous system. Sympathetic nerves stimulate the radial muscles to dilate the pupil, and parasympathetic nerves stimulate the circular muscles to constrict the pupil.

1. Trochlea
2. Superior oblique
3. Levator palpebrae superioris (cut)
4. Superior rectus
5. Medial rectus
6. Lateral rectus
7. Inferior rectus
8. Inferior oblique

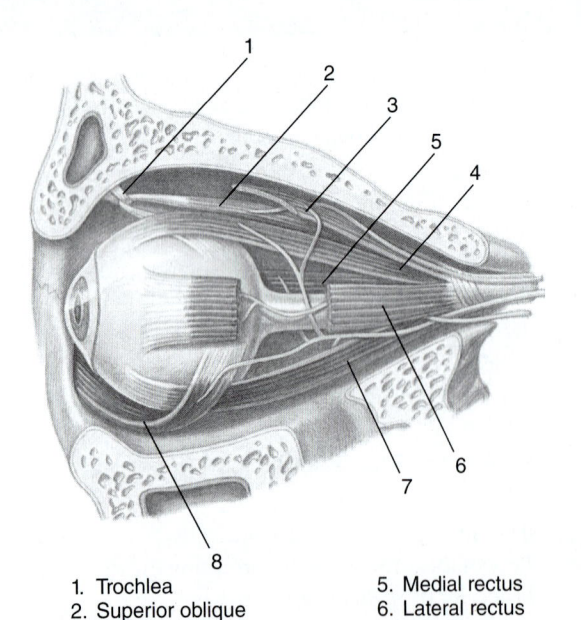

Figure 3.23 A lateral view of ocular muscles of the left eyeball.

Procedure

1. Stay with the subject in a darkened room for at least 1 minute, allowing his or her eyes to adjust to the dim light. This adjustment is known as *dark adaptation*.

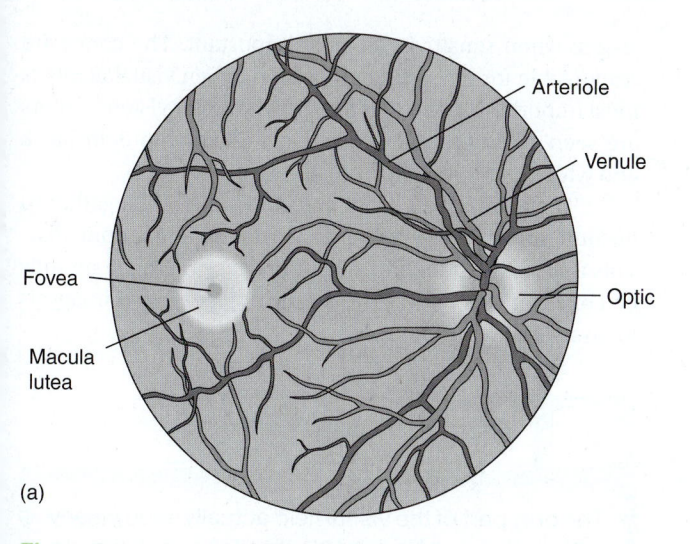

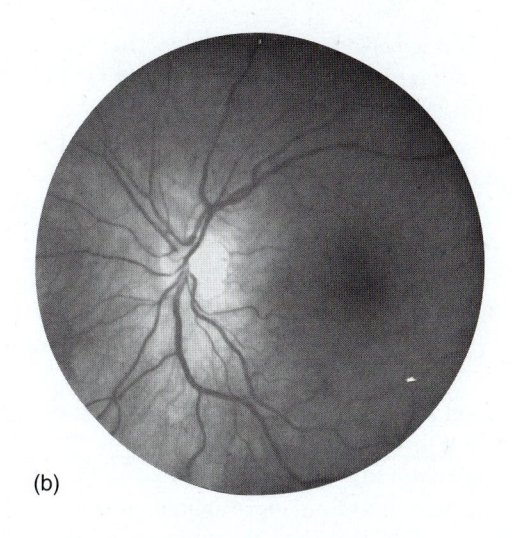

Figure 3.24 **A view of the retina as seen with an ophthalmoscope.** Optic nerve fibers leave the eyeball at the optic disc to form the optic nerve. (Note the blood vessels that can be seen entering the eyeball at the optic disc.)

(For a full-color version of this figure, see fig. 10.30 in *Human Physiology*, thirteenth edition, by Stuart I. Fox.)

2. Shine a narrow beam of light (from a pen flashlight or an ophthalmoscope, for example) from the right side into the subject's right eye. Observe the pupillary reflex in the right eye and also in the left eye. The pupillary reflex in the other (left) eye is called the *consensual reaction.*

3. Repeat this procedure (first dark-adapting the eyes again) from the left side with the left eye.

E. Examination of the Eye with an Ophthalmoscope

An **ophthalmoscope** is a device used to observe the posterior inner part of the eye (the *fundus*). A mirror positioned at the top of the instrument deflects light at a right angle into the eye, enabling an observer to see the interior of the eye through a small slit in the mirror. Different depths of focus are attained by changing the lenses positioned in the slit. The lenses are carried on a wheel in regular order according to their focal lengths. The strength of each lens is given in diopters preceded by a plus (+) for a convex lens or a minus (−) for a concave lens, with 0 indicating no lens.

In this exercise an opthalmoscope will be used to observe the arteries and veins of the fundus and two regions of the retina: the **optic disc** (the region where the optic nerve exits the eye, otherwise known as the *blind spot*) and the **macula lutea** (fig. 3.24). The macula lutea is a yellowish region containing a central pit, the **fovea centralis,** where the highest concentration of the photoreceptors responsible for visual acuity (the cones) is located. When the eyes are looking directly at an object, the image is focused on the fovea.

Clinical Applications

Clinical examination of the fundus **(ophthalmoscopy)** can aid the diagnosis of a number of ocular and systemic (body) diseases. The features noted in these examinations include the condition of the blood vessels; the color and shape of the disc; the presence of particles, exudates, or hemorrhage; the presence of edema and inflammation of the optic nerve *(papilledema);* and myopia and hyperopia.

Procedure

1. Have the subject sit in a darkened room and look at a distant object (blinking as needed).

2. Position your chair so that you are close to and facing the subject. Hold the ophthalmoscope with your right hand and use your right eye when observing the subject's right eye. (The situation is reversed when viewing the left eye.)

3. With your forefinger on the lens adjustment wheel and your eye as close as possible to the small hole in the ophthalmoscope (glasses off, if applicable), bring the instrument as close as possible to the subject's eye. (Steady your hand by resting it on the subject's cheek.)

4. Examine the subject's eye from the front to the back. Looking slightly from the side of the eye (not directly in front), examine the iris and lens using a +20 to +15 lens. The lens selected will vary among examiners who normally wear glasses. If both your eyes and the subject's eyes are normal, you will be

able to see clearly without a lens. (On the 0 setting, the refractive strength of the subject's eye will be sufficient to focus the light on the eye.)

5. Rotate the wheel counterclockwise to examine the fundus.
   (a) If a positive (convex) lens is necessary to focus on the fundus, and your eyes are normal, the subject has hyperopia (hypermetropia).
   (b) If a negative (concave) lens is necessary to focus on the fundus, and your eyes are normal, the subject has myopia.
6. Observe the arteries and veins of the fundus and follow them to their point of convergence. This will enable you to see the optic disc (fig. 3.24).
7. Finally, at the end of the examination, observe the macula lutea by asking the subject to look directly into the light of the ophthalmoscope.

F. The Blind Spot

The retina contains two types of *photoreceptors*, **rods** and **cones.** Rods and cones synapse with *bipolar neurons*, which in turn synapse with *ganglion cells* whose axons form the optic nerve transmitting sensory information out of the eye to the brain. In the fovea, only one cone will synapse with one bipolar cell, whereas several rods may converge on a given bipolar cell (fig. 3.25). Summation allows the rods to be more sensitive to low levels of illumination, whereas the cones require more light but provide greater visual acuity. The rods, therefore, are responsible for night *(scotopic)* vision, when sensitivity is most important. The cones are responsible for day *(photopic)* vision, when visual acuity is most important. The cones also provide color vision—colors are seen during the day, whereas night vision is in black and white.

The axons of all ganglion cells in the retina gather to become the optic nerve that exits the eye at the optic disc. This is also called the **blind spot** because there are no rods or cones in the optic disc, so an object whose image is focused here will not be seen.

Clinical Applications

The only part of the visual field actually seen clearly is that tiny part (about 1%) that falls on the fovea centralis. We are unaware of this because rapid eye movements shift different parts of the visual field onto the fovea. A common visual impairment, particularly in older people, is **macular degeneration**—degeneration of the macula lutea and its central fovea. People with macular degeneration lose the clarity of vision provided by the fovea in the central region of the visual field. In most cases, the damage is believed to be related to the loss of retinal pigment epithelium in this region. In some cases, the damage is made worse by growth of new blood vessels *(neovascularization)* from the choroid into the retina.

Figure 3.25 **Covergence in the retina and light sensitivity.** Bipolar cells receive input from the convergence of many rods (a) and a number of such bipolar cells converge on a single ganglion cell, so rods maximize sensitivity to low levels of light at the expense of visual acuity. By contrast, the 1:1:1 ratio of cones to bipolar cells to ganglion cells in the fovea (b) provides high visual acuity, but sensitivity to light is reduced.

(For a full-color version of this figure, see fig. 10.44 in *Human Physiology*, thirteenth edition, by Stuart I. Fox.)

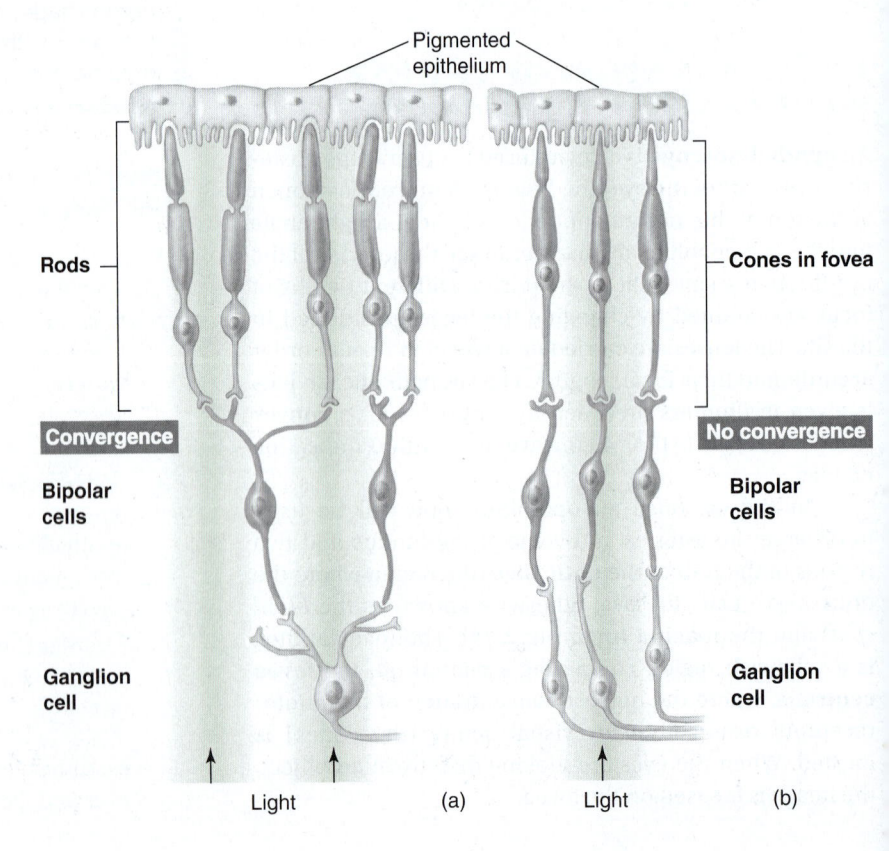

●   +

Figure 3.26 A diagram for demonstrating the blind spot.

Procedure

1. Hold the drawing of the circle and the cross (fig. 3.26) about 20 inches from your face with the left eye covered or closed. Focus on the circle; this is most easily done if the circle is positioned in line with the right eye.
2. Keeping the right eye focused on the circle, slowly bring the drawing closer to your face until the cross disappears. Continue to move the drawing slowly toward your face until the cross reappears.
3. Repeat this procedure with the right eye closed or covered and the left eye focused on the cross. Observe the disappearance of the circle as the drawing is brought closer to your face.

G. THE AFTERIMAGE

The light that strikes the receptors of the eye stimulates a photochemical reaction in which the pigment **rhodopsin** (within the rods) dissociates to form the pigment *retinene* and the protein *opsin*. This chemical dissociation produces electrical changes in the photoreceptors, which trigger a train of action potentials in the axons of the optic nerve. These events cannot be repeated in a given rod receptor until the rhodopsin is regenerated. This requires a series of chemical reactions in which one isomer of retinene is converted to another through the intermediate compound *vitamin A_1*. In other words, after the rhodopsin visual pigment in the rod has been "bleached" or dissociated by light from an object, some time is required before that receptor can again be stimulated.

When an eye adapted to a bright light, such as a lightbulb, is closed or quickly turned toward a wall, the bright image of the lightbulb will still be seen. This is called a **positive afterimage** and is caused by the continued "firing" of the photoreceptors. After a short period, the dark image of the lightbulb, called the **negative afterimage,** will appear against a lighter background due to the "bleaching" of the visual pigment of the affected receptors.

According to the **Young-Helmholtz** theory of color vision, there are three systems of cones that respond respectively to *blue, green,* and *red* (or violet) light, and all other colors are seen by the brain's interpretation of mixtures of impulses from these three systems. Color discrimination will be impaired if one system of cones is defective (color blindness) or if one system of cones has been "bleached" by the continued viewing of an object. In the latter case, the positive afterimage of the object will appear in the complementary color.

Clinical Applications

The eye is a receptor, transducing light into electrical nerve impulses. We actually see with our brain. Impulses from the retina pass, via the *lateral geniculate bodies*, to the *visual cortex* of the occipital lobe where the patterns of impulses are integrated to produce an image. The importance of the visual cortex in vision is illustrated by **strabismus,** a condition in which weak extrinsic eye muscles prevent the two eyes from converging on an object and fusing the images. To avoid confusion the cortical cells eventually stop responding to information from one eye, making that eye functionally blind. Visual information is integrated with input from the other senses in the cortex of the *inferior temporal lobe.* If this area is damaged (the **Klüver-Bucy syndrome),** visual recognition is impaired. Although the image is seen, it lacks meaning and emotional content.

Procedure

1. After staring at a lightbulb, suddenly shift your gaze to a blank wall. Observe the appearance of the *negative afterimage.*
2. For 1 minute, stare at a dot made in the center of a small red square pasted on a larger sheet of black paper.
3. Suddenly shift your gaze to a sheet of white paper, and note the color of the *positive afterimage.*
   Positive afterimage (red) is _____.
4. Repeat this procedure using blue squares and yellow squares.
   Positive afterimage (blue) is _____.
   Positive afterimage (yellow) is _____.

H. COLOR VISION AND COLOR BLINDNESS

Humans have **trichromatic color vision.** As stated in exercise 3.5G, color vision is provided by the stimulation of three types of cones, designated *blue, green,* and *red.* These terms do not refer to the color of the cones; rather, they refer to the region of the visible spectrum in which each cone's pigment absorbs light best (fig. 3.27). This is the cone's *absorption maximum.* On this basis, the blue cones are also called *S cones* (for short wavelength); the green cones are also called *M cones* (for medium wavelength); and the red cones are called *L cones* (for long wavelength). Each of these cones has the same retinene pigment present

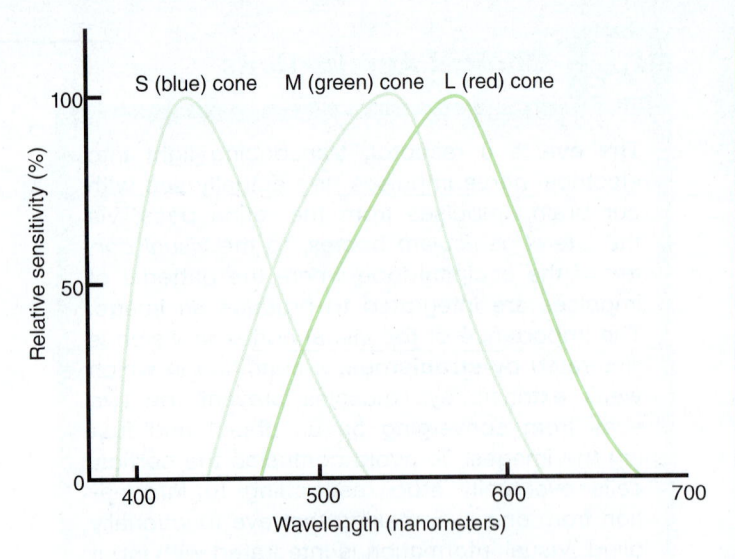

Figure 3.27 The three types of cones. Each type contains retinene, but the protein with which the retinene is combined is different in each case. Thus, each different pigment absorbs light maximally at a different wavelength. Color vision is produced by the activity of these blue (short wavelength, or S) cones, green (medium wavelength, or M) cones, and red (long wavelength, or L) cones.

(For a full-color version of this figure, see fig. 10.42 in *Human Physiology*, thirteenth edition, by Stuart I. Fox.)

in the rods, but each has a different protein (known as the *photopsins*) associated with the retinene to give the cone its distinct absorption maximum.

Examining figure 3.27, you can understand why a green (M) cone, for example, can be stimulated as much by a weak green light as by a more intense red light. The color we perceive actually depends on neural computations that compare the effects of a particular light wavelength on the different types of cones. In this way, the brain can distinguish both color and light intensity. This information is preserved as it is sent to the lateral geniculate nuclei of the thalamus and then to the primary visual cortex.

Clinical Applications

The gene for the S cone pigment is located on autosomal chromosome number 7, whereas the genes for the M and L cones are located on the X chromosome. **Color blindness** is caused by an inherited lack of one or more type of cones (see exercise 11.3C), usually the M (green) or L (red) cones. These people have only two functioning types of cones, and their vision is thus dichromatic. People who have only one cone in the middle to long wavelength region (M or L) have difficulty distinguishing reds from greens. Because the photopsins of the M and L cone pigments are coded on the X chromosome, and because men have only one X chromosome (and so cannot carry the trait in a recessive state), such red-green color blindness is far more common in men than in women.

Procedure

You can test for red-green color blindness using the *Ishihara test*. In this test, colored dots are arranged in a series of circles in such a way that a person with normal vision can see a number embedded within each circle. By contrast, a color-blind person will see only an apparently random array of colored dots.

Laboratory Report 3.5

Name _____
Date _____
Section _____

REVIEW ACTIVITIES FOR EXERCISE 3.5

Test Your Knowledge

1. The photoreceptors responsible for color vision are the _____; of these, there are _____ different types.

2. The region of the retina in which there are no photoreceptors is called the _____; this is also known as the _____.

3. Retinene (retinaldehyde) is derived from which vitamin? _____

4. When light enters the retina, it first passes through the _____ cell layer, then the _____ cell layer, before reaching the photoreceptors.

5. The axons of _____ cells gather to produce the optic nerve.

6. Define:
   (a) *visual acuity* _____
   (b) *accommodation* _____

Match these terms with the appropriate description:

____ 7. myopia   (a) abnormal curvature of the cornea or lens
____ 8. hyperopia   (b) eye too long
____ 9. presbyopia   (c) abnormally high intraocular pressure
____10. astigmatism   (d) eye too short
____11. glaucoma   (e) loss of lens elasticity

Test Your Understanding

12. Describe the muscle layers of the iris and the innervation to these muscles. Use this information to explain how pupils constrict in bright light and dilate in dim light.

13. Describe how the curvature of the lens is regulated, and use this information to explain how images are kept in focus on the retina as a distant object is brought closer to the eyes.

14. Explain the reason for blurred vision in a person with myopia, and describe how this person's vision is improved by the lenses in a pair of glasses.

15. Explain why vision is clearest where the image falls on the fovea centralis. Also, explain why vision in dim light is better out of the "corners of the eyes" than when the eyes look directly at an object.

Test Your Analytical Ability

16. Carrots contain large amounts of *carotene,* a precursor of vitamin A. Can eating carrots help improve blurred vision from myopia or hyperopia? Explain. Also, explain how eating carrots may influence scotopic (night) vision.

17. "You see with your brain, not with your eyes." Defend this statement with reference to the Young-Helmholtz theory of color vision and to the optic disc.

Clinical Investigation Questions

18. What is the most likely cause of this person's color blindness? What does 20/300 vision indicate, and how is that corrected by LASIK surgery?

19. What caused his need for reading glasses some years after his LASIK surgery? What caused the cloudy patches in his vision afterward? What is a likely cause of the later deterioration of his central vision? Explain.

EXERCISE 3.6

Ears: Cochlea and Hearing

MATERIALS
1. Tuning forks
2. Rubber mallets

Sound is conducted by the middle ear to the inner ear, where events within the cochlea result in the production of nerve impulses. Clinical tests of middle ear (conductive) and inner ear (sensory) function aid in the diagnosis of hearing disorders.

LEARNING OUTCOMES

You should be able to:

1. Describe the structure of the middle ear, and explain how the ossicles function.
2. Describe the structure of the inner ear, and explain how the cochlea functions.
3. Demonstrate Rinne's test and Weber's test, and explain the significance of each.
4. Explain how the source of a sound is localized.

Textbook/Multimedia Correlations

Before performing this exercise, you should study the introductory material presented here. Further information relating to this exercise can be found in these pages of *Human Physiology*, thirteenth edition, by Stuart I. Fox:

- *The Ears and Hearing.* Chapter 10, p. 282.
- Multimedia correlations are provided in *MediaPhys 3.0:* Topics 6.49–6.57.

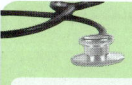

Clinical Investigation

A patient receiving Rinne's test reported that the sound reappeared in his left ear but did not reappear in the right ear. When Weber's test was performed, the patient reported that the sound was louder in the left ear.

- Describe Rinne's and Weber's tests and explain their significance.
- Distinguish between conduction and sensorineural deafness, and identify which type of deafness the tests suggest the patient may have.

You should be familiar with the gross structure of the ear (fig. 3.28). Sound waves are conducted through the **outer ear** (the pinna and the external auditory meatus) to the tympanic membrane (eardrum), causing it to vibrate. The vibration of the tympanic membrane causes the three ossicles of the **middle ear**—the *malleus* (hammer), *incus* (anvil), and *stapes* (stirrup)—to vibrate, thus pushing the footplate of the stapes against a flexible membrane, the *oval window* (fig. 3.29). Vibration of the oval window produces compression waves in the fluid-filled cochlea of the **inner ear.**

The compression waves of cochlear fluid flow over a thin, flexible membrane within the cochlea, the *basilar membrane,* causing it to vibrate. Within the **organ of Corti,** the basilar membrane is coated with sensory *hair cells,* which are displaced upward by this vibration into a stiff overhanging structure called the *tectorial membrane* (fig. 3.30). The distortion of the hair cells produced by this action stimulates a train of action potentials that travels along the cochlear branch of the vestibulocochlear (eighth cranial) nerve to the brain. Here, the action potentials are interpreted as the sound of a specific **pitch,** determined by the location of the stimulated hair cells on the basilar membrane. The basilar membrane has a *tonotopic organization,* meaning that different pitches of sound stimulate different locations of the basilar membrane and the different sensory neurons that serve those locations. The loudness of a sound is coded by the frequency of action potentials conducted by the neurons of the cochlear nerve.

Sensory neurons transmitting different pitches of sound from the basilar membrane send this information to the cochlear nuclei of the brain stem, which, like the basilar membrane, preserve a tonotopic organization. That is, different regions of the cochlear nuclei represent different pitches. This separation of neurons by pitch is preserved in the tonotopic organization of the auditory cortex, which allows the perception of different pitches of sound.

A. CONDUCTION OF SOUND WAVES THROUGH BONE: RINNE'S AND WEBER'S TESTS

Although hearing is normally produced by the vibration of the oval window in response to sound waves conducted through the movements of the middle-ear ossicles, the *endolymph* fluid of the cochlea can also be made to vibrate in response to sound waves conducted through the skull bones directly,

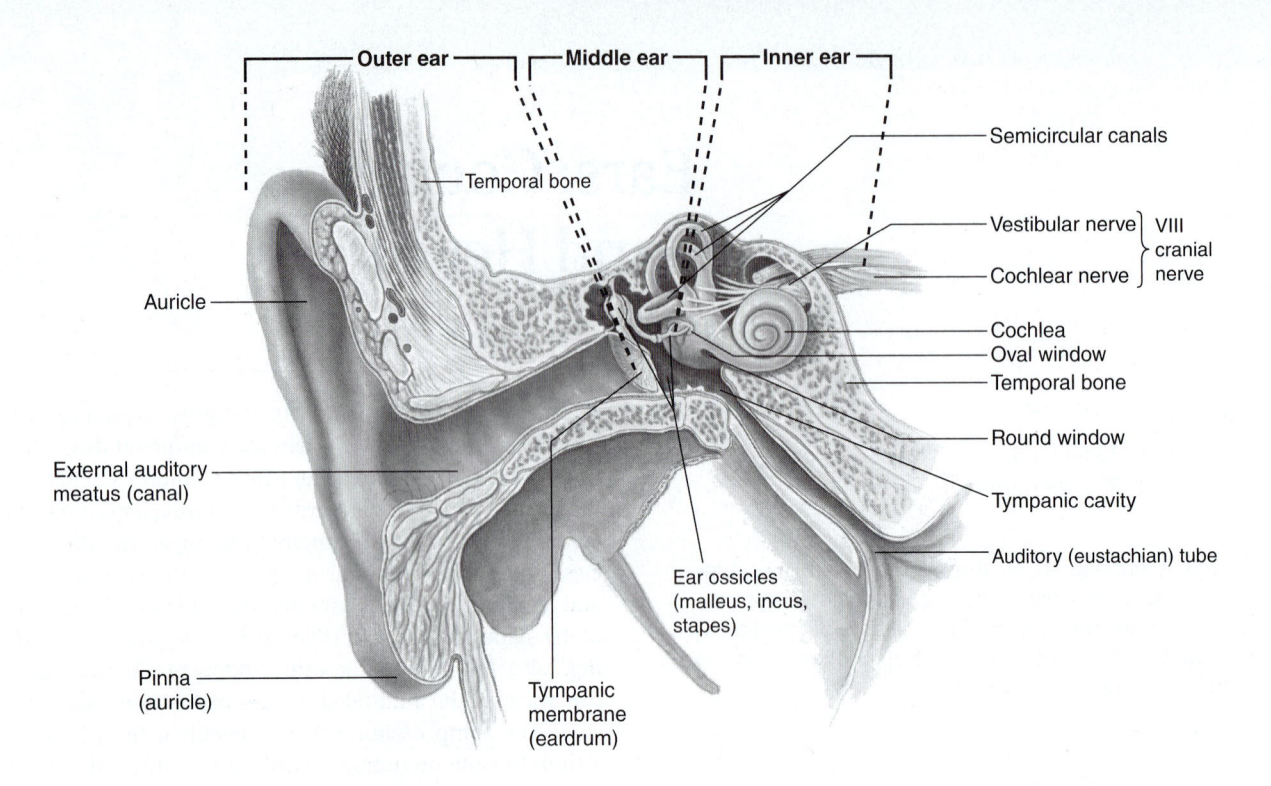

Figure 3.28 **The outer, middle, and inner ear.**
(For a full-color version of this figure, see fig. 10.18 in *Human Physiology*, thirteenth edition, by Stuart I. Fox.)

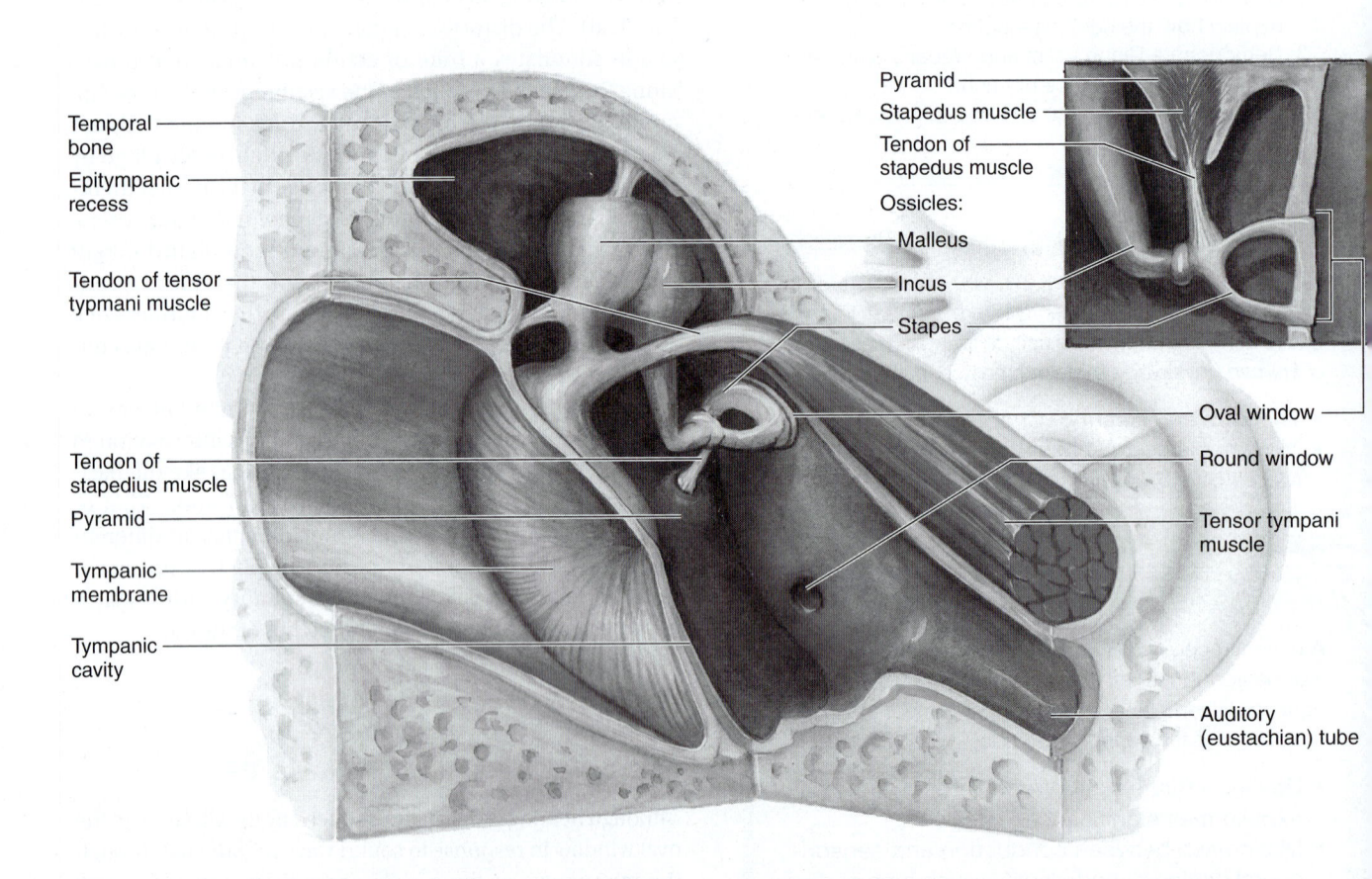

Figure 3.29 **A medial view of the middle ear.** The locations of auditory muscles, attached to the middle-ear ossicles, are indicated.
(For a full-color version of this figure, see fig. 10.19 in *Human Physiology*, thirteenth edition, by Stuart I. Fox.)

Figure 3.30 The organ of Corti. (a) The organ of Corti within the cochlea and (b) in greater detail.

(For a full-color version of this figure, see fig. 10.22 in *Human Physiology*, thirteenth edition, by Stuart I. Fox.)

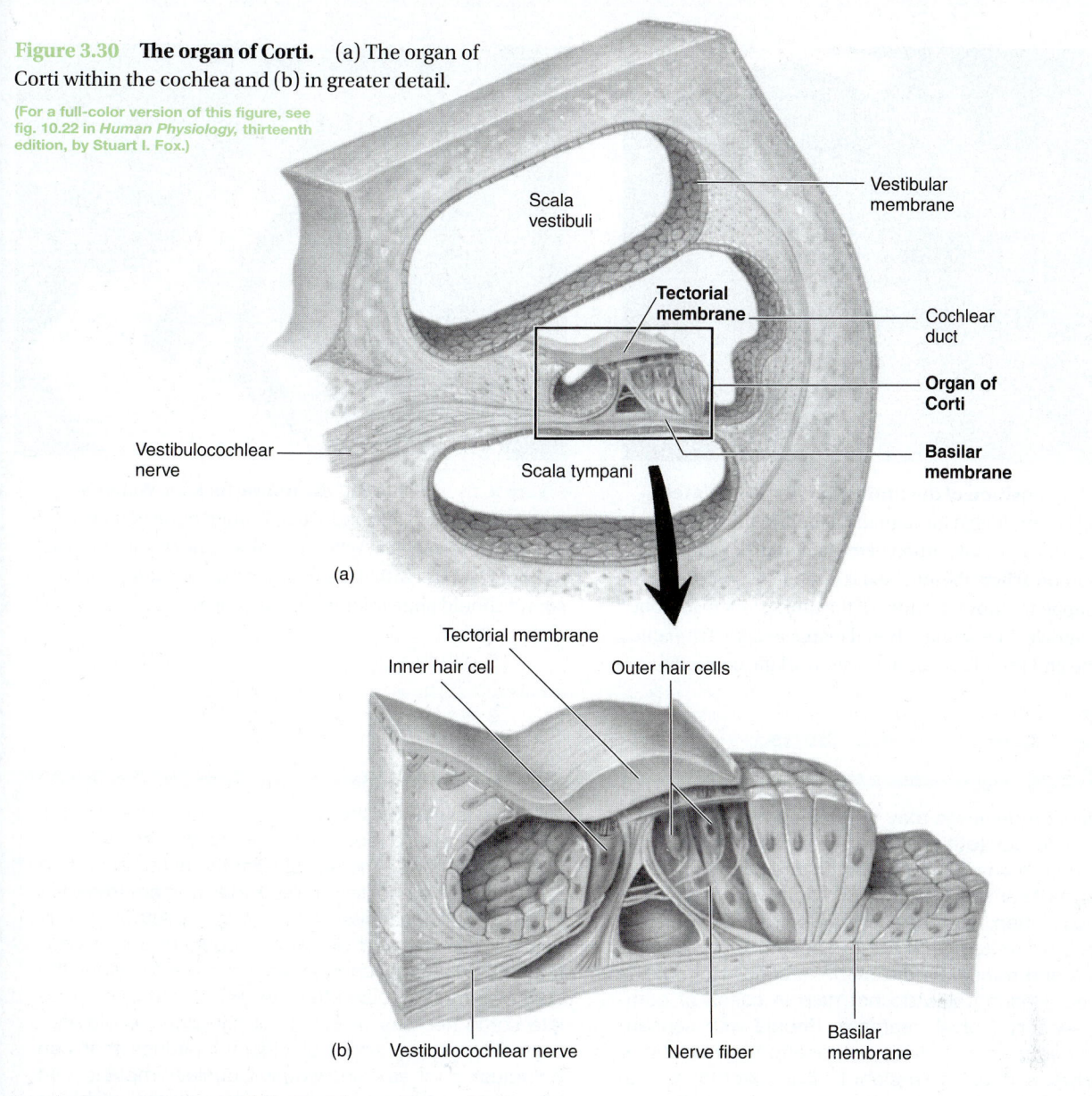

Labels (a): Scala vestibuli; Vestibular membrane; Tectorial membrane; Cochlear duct; Organ of Corti; Basilar membrane; Scala tympani; Vestibulocochlear nerve

Labels (b): Tectorial membrane; Inner hair cell; Outer hair cells; Vestibulocochlear nerve; Nerve fiber; Basilar membrane

thereby bypassing the middle ear. This makes it possible to differentiate between deafness resulting from middle-ear damage (**conduction deafness,** such as from damage to the ossicles in *otitis media*—middle-ear infection—or immobilization of the stapes in *otosclerosis*) and deafness resulting from damage to the cochlea or vestibulocochlear nerve (**sensory deafness,** such as from infections, streptomycin toxicity, or prolonged exposure to loud sounds).

Procedure

Rinne's Test

1. Strike a tuning fork with a rubber mallet to produce vibrations.
2. Perform Rinne's test by placing the handle of the vibrating tuning fork against the mastoid process of the temporal bone (the bony prominence behind the ear), with the tuning fork pointed down and behind the ear (fig. 3.31). When the sound has almost died away, move the tuning fork (by the handle) near the external auditory meatus. If there is no damage to the middle ear, the sound will reappear.
3. Simulate conduction deafness by repeating Rinne's test with a plug of cotton in the ear. In conductive deafness, conduction by bone (via the mastoid process) is more effective than conduction by air.

Weber's Test

1. Perform Weber's test by placing the handle of the vibrating tuning fork on the midsagittal line of the head (fig. 3.32), and listen. In conduction deafness, the sound will seem louder in the affected ear (room noise is excluded but bone conduction continues), whereas in sensory deafness, the cochlea is defective and the sound will be louder in the normal ear.
2. Repeat Weber's test with one ear plugged with your finger. The sound will appear louder in the plugged ear because external room noise is excluded.

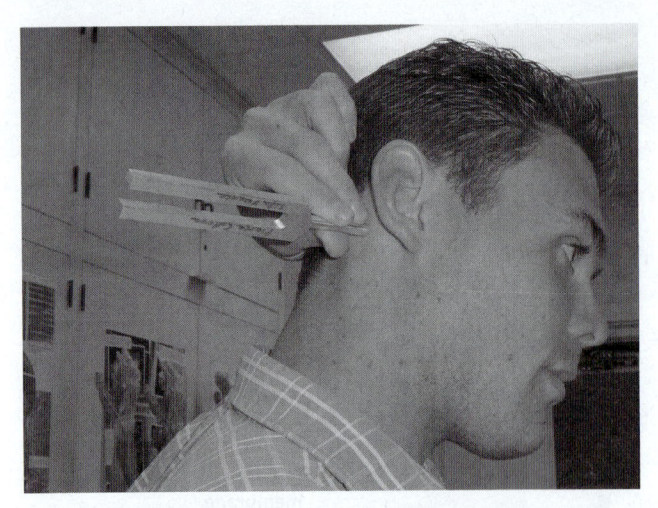

Figure 3.31 Position of the tuning fork for Rinne's test. Sound will be conducted through the handle of the tuning fork and through the mastoid portion of the temporal bone into the petrous portion, where the inner ear is located. As soon as the sound disappears, move the tines of the tuning fork to just outside your auricle. The sounds should reappear if the tympanic membrane and middle ear ossicles are working correctly.

Figure 3.32 Position of the tuning fork for Weber's test. Sound will be conducted through bone equally to the right and left ears. When you block one of your external auditory canals (with a finger in the ear, for example), the sound should appear louder in the plugged ear.

Clinical Applications

Conduction deafness may be caused by infections of the middle ear **(otitis media),** infections of the tympanic membrane **(tympanitis),** or an excessive accumulation of ear wax *(cerumen).* People with conduction deafness often wear **hearing aids** over the mastoid process of the temporal bone. These devices amplify sounds and transmit them by bone conduction to the cochlea. Hearing aids do not help in cases of complete sensory (nerve) deafness. People with sensorineural deafness sometimes choose to have **cochlear implants**. A cochlear implant has an external and an internal portion. The external portion consists of microphones, a speech processor that converts sound into electrical signals, and a transmitter that sends the electrical signal to the internal portion. The internal portion contains a receiver implanted into the temporal bone that receives the signal from the external portion and activates a number of electrodes that are threaded into different locations of the cochlea. Although hair cells and associated dendrites degenerate in sensorineural deafness, enough can survive to allow the cochlear implant to be effective. The electrodes stimulate some neurons of the spiral ganglion, providing a limited tonotopic signal of different pitches that can be adjusted for each individual. Cochlear implants can help young children acquire language skills and help adults who have lost their hearing understand spoken words, although this may take training and practice.

B. Binaural Localization of Sound

Just as binocular vision provides valuable clues for viewing scenes in three dimensions, binaural hearing helps to localize sounds (people have stereoscopic vision and stereophonic hearing). The ability to localize the source of a sound depends partly on the difference in loudness of the sound that reaches the two ears and partly on the difference in the time of arrival of the sound at the two ears.

There is an interaural intensity difference when one ear is closer to the sound than the other and the sound frequency is above 2000 Hz. At these higher frequencies, the wavelengths of sound are shorter than the distance between the ears. This information is supplemented by the interaural time difference if sound arrives at one ear before the other. This time difference is particularly important for localizing low-frequency sounds (below 2000 Hz). Humans can detect an interaural intensity difference as little as 1–2 decibels and an interaural time difference as short as 10 microseconds.

Procedure

1. With both eyes closed, the subject is asked to locate the source of a sound (e.g., a vibrating tuning fork).
2. The vibrating tuning fork is placed at various positions (front, back, and sides about a foot from the subject's head), and the subject is asked to describe the location of the tuning fork.
3. Repeat the procedures with one of the subject's ears plugged.

Laboratory Report 3.6

Name _____
Date _____
Section _____

REVIEW ACTIVITIES FOR EXERCISE 3.6

Test Your Knowledge

1. List the three ossicles of the middle ear in sequence, from the outermost to the innermost.

    Outermost _____

    Middle _____

    Innermost _____

2. The scientific name for the eardrum is the _____.
3. The middle chamber of the cochlea is the _____.
4. Sensory hair cells in the cochlea are located on a membrane called the _____.
5. The innermost middle ear ossicle presses against a flexible membrane called the _____.
6. The sensory structure of the inner ear responsible for transducing vibrations into nerve impulses is known as the _____.

Test Your Understanding

7. In sequence, describe the events that occur between the arrival of sound waves at the tympanic membrane and the production of nerve impulses.

8. Explain how different pitches of sound affect the basilar membrane and how this relates to the information sent to the brain.

Test Your Analytical Ability

9. Explain the result that might be obtained by performing the Rinne's and Weber's test on a patient with otosclerosis. How might these results compare with those obtained from a patient with sensory deafness? Explain.

10. Explain how the binaural localization of high-pitched sounds differs from low-pitched sounds and why this difference exists.

11. Hair cells are mechanoreceptors employed in the inner ear of mammals and also in the lateral line organs on the body of fish. How might the bending of hair cell processes lead to depolarization? Propose a general description of the function of hair cells in the animal kingdom.

Clinical Investigation Questions

12. How is Rinne's test performed, and what is the significance of the sound reappearing in the left ear but not the right in this patient? How is Weber's test performed, and what is the significance of the sound being louder in the left ear in this patient?

13. What type of deafness is suggested by these tests, and would this deafness be best treated with a hearing aid or cochlear implant? Explain.

Ears: Vestibular Apparatus—Balance and Equilibrium

EXERCISE 3.7

MATERIALS

1. Swivel chair

The vestibular apparatus provides a sense of balance and equilibrium. As a result of inertia acting on the structures within the vestibular apparatus, changes in the position of the head result in the production of afferent nerve impulses conducted to the brain along the eighth cranial nerve. This information results in eye movements and other motor activities that help to orient the body in space.

LEARNING OUTCOMES

You should be able to:

1. Describe the structure of the semicircular canals, and explain how movements of the head result in the production of nerve impulses.
2. Describe vestibular nystagmus, and explain how it is produced.

Textbook/Multimedia Correlations

Before performing this exercise, you should study the introductory material presented here. Further information relating to this exercise can also be found in these pages of *Human Physiology,* thirteenth edition, by Stuart I. Fox:

- *Vestibular Apparatus and Equilibrium.* Chapter 10, p. 278.
- Multimedia correlations are provided in *MediaPhys 3.0:* Topics 6.59–6.64.

Clinical Investigation

A woman spent a week on a cruise ship with periodic feelings of vertigo, but she successfully treated her initial nausea with drugs for motion sickness. A week after she disembarked, her feeling of vertigo returned. However, this time it seemed like the room was spinning and she now had tinnitus.

- Distinguish between the nature and causes of vertigo and nausea, and how motion sickness medicine can treat one but not the other.
- Explain the causes of vertigo on a sea cruise and speculate on what might be causing the vertigo and tinnitus afterward.

The **vestibular apparatus,** located in the inner ear above the cochlea, consists of three *semicircular canals* (oriented in three planes), the *utricle,* and the *saccule* (fig. 3.33). The utricle and saccule are together called *otolith organs* and provide a sense of linear acceleration. These structures, like the cochlea, are filled with *endolymph* and contain sensory cells activated by bending. These sensory hair cells of the semicircular canals support numerous hairlike extensions, embedded in a gelatinous "sail" (the *cupula*) that projects into the endolymph (fig. 3.34). Movement of the endolymph fluid, induced by acceleration or deceleration, bends the extensions of the hair cells. This sends a train of impulses to the brain along the *vestibulocochlear* (eighth cranial) *nerve.* The sensory hair cells of the utricle and the saccule serve to orient the head with respect to the gravitational pull of the earth.

Afferent impulses from the vestibular apparatus help make us aware of our position in space and affect a variety of efferent somatic motor nerves (such as those regulating the voluntary extrinsic eye muscles). Under intense vestibular activity, autonomic motor nerves may also become stimulated, producing involuntary responses that can include vomiting, perspiration, and hypotension. This exercise will

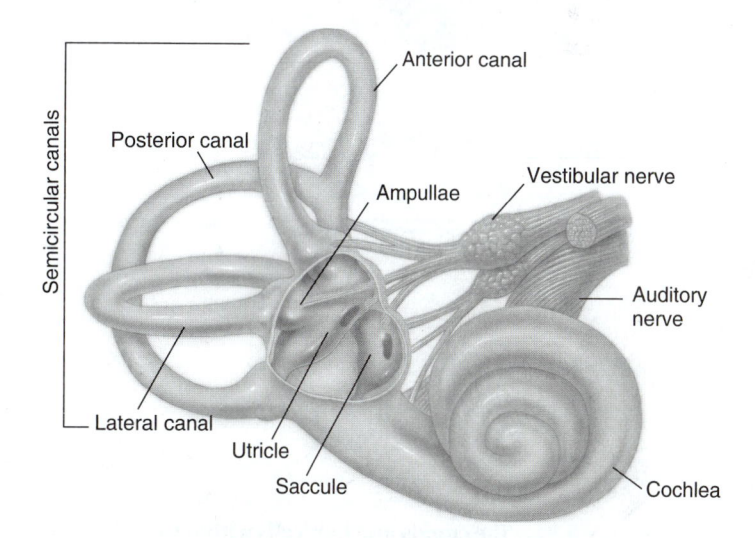

Figure 3.33 The inner ear. The vestibular apparatus (semicircular canals, utricle, and saccule), required for a sense of equilibrium, and the cochlea, required for hearing, make up the inner ear.

(For a full-color version of this figure, see fig. 10.12 in *Human Physiology,* thirteenth edition, by Stuart I. Fox.)

test the effect of vestibular activity on the extrinsic muscles of the eye by producing involuntary eye oscillations known as **vestibular nystagmus.**

Clinical Applications

Vestibular nystagmus is one of the symptoms of an inner-ear disease called **Ménière's disease.** The early symptom of this disease is often "ringing in the ears," or *tinnitus.* Because the endolymph of the cochlea and the endolymph of the vestibular apparatus are continuous through a tiny canal, vestibular symptoms of vertigo and nystagmus often accompany hearing problems in this disease.

The semicircular canals can be stimulated by rotating a subject in a chair. Stimulation of the posterior canal can occur when the head is flexed 30° forward, almost touching the chest. When the subject is first rotated to the right, the cupula will be bent to the left because of the inertial lag of the endolymph. This will cause nystagmus in which the eyes drift slowly to the left followed by a quick movement to the right (midline position). Nystagmus will continue until the inertia of the endolymph has been overcome and the cupula returns to its initial unbent position. At this time, the endolymph and the cupula are moving in the same direction and at the same speed. When the rotation of the subject is abruptly stopped, the greater inertia of the endolymph will cause bending of the cupula in the previous direction of spin (to the right), producing nystagmus with a slow drift phase to the right and a rapid phase to the left. This activation of the vestibular apparatus is often accompanied by **vertigo** (an illusion of movement, or spinning) and a tendency to fall to the right.

Clinical Applications

Vertigo may be accompanied by dizziness, but the two sensations are not the same. A person may be dizzy without experiencing the more severe effects of vertigo. In the following procedure, activation of the vestibular apparatus by rotation of the subject produces oscillatory eye movements (nystagmus). The eyes, in turn, can activate the vestibular apparatus and produce vertigo, as occurs in motion sickness (seasickness or car sickness, for example). Vertigo can also accompany diseases unrelated to the special senses, such as cardiovascular disease. Many of the unpleasant symptoms associated with vertigo, such as nausea and vomiting, are the result of activation of the autonomic motor system. Drugs taken for motion sickness (e.g., Dramamine or scopolamine) act by suppressing these autonomic responses.

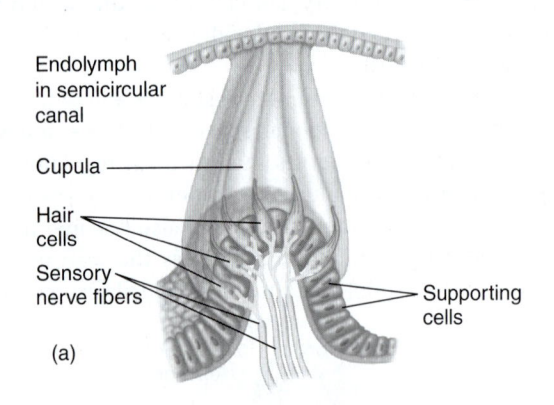

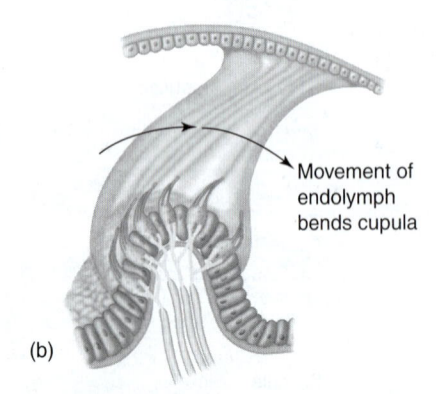

Figure 3.34 The cupula and hair cells within the semicircular canals. (a) Shown here, the structures are at rest or at a constant velocity. (b) Here, movement of the endolymph during rotation causes the cupula to bend, thus stimulating the hair cells.

(For a full-color version of this figure, see fig. 10.16 in *Human Physiology,* thirteenth edition, by Stuart I. Fox.)

Procedure

1. Have the subject sit in a swivel chair with the eyes open and the head flexed 30° forward (chin almost touching the chest). Rotate the chair quickly to the right for 20 seconds (about 10 revolutions).

    Note: *Only subjects not subject to motion sickness should be used. The exercise should be stopped immediately if the subject feels sick. Carefully observe the spinning student; dizziness could cause a student to fall off of the chair.*

2. Abruptly stop the chair and have the subject open his or her eyes as wide as possible. Note the direction of nystagmus (left to right or right to left, circle one). The subject should then close her or his eyes until nystagmus and any sense of dizziness has passed.

3. Repeat this procedure (using different subjects), alternating with the head resting on the right shoulder and the left shoulder (this stimulates the vertical canals), and note the direction of the postrotational nystagmus.

Laboratory Report 3.7

Name _____
Date _____
Section _____

REVIEW ACTIVITIES FOR EXERCISE 3.7

Test Your Knowledge

1. The organ of equilibrium is called the _____.
2. The structures sensitive to angular acceleration in three planes are the _____.
3. The structures sensitive to linear acceleration are the _____ and _____. Together, they are called the _____ organs.
4. The fluid within the organs of equilibrium is known as _____.
5. The sense of equilibrium is transmitted by the _____ (its number) cranial nerve, also known as the _____ (its name) nerve.
6. An illusion of movement or spinning is called _____.

Test Your Understanding

7. State the meaning of *inertia,* and explain how inertia affects the hair cell processes of the inner ear during acceleration and deceleration.

8. Including all of the relevant anatomical structures (and ending with the activation of the cranial nerve), explain how we sense spinning in a chair. What happens when the chair is suddenly stopped?

9. Which structures of the inner ear would be activated by a somersault? By a cartwheel? By fast acceleration in a race car?

Test Your Analytical Ability

10. Does nystagmus occur after a person in a rotating chair has achieved constant velocity? Explain.

11. Propose amusement park rides that would selectively activate the otolith organs and the different semicircular canals.

Clinical Investigation Questions

12. Which organs of the vestibular apparatus would be activated by movements of the ship at sea? What role does vision play in vertigo caused by this movement? Explain.

13. What might cause the nausea of seasickness, and how does medication help to relieve this nausea?

14. What is tinnitus, and how might it be produced when a person has vertigo?

EXERCISE 3.8

Taste Perception

MATERIALS

1. Cotton-tipped applicator sticks
2. Solutions of 5% sucrose, 1% acetic acid, 5% NaCl, and 0.5% quinine sulfate
3. Fruit (apples, grapes, or others)

There are five modalities of taste perception: sweet, sour, bitter, salty, and a more recently recognized umami (savory) taste. The receptors are taste cells within taste buds that are distributed on the tongue. Each taste cell is sensitive to only one modality of taste, and transmits that information on a taste-specific sensory neuron into the brain.

LEARNING OUTCOMES

You should be able to:

1. Describe the structure of a taste bud and the location of taste buds on the tongue.
2. List the four primary taste modalities, and describe how they are stimulated.

Textbook/Multimedia Correlations

Before performing this exercise, you should study the introductory material presented here. Further information relating to this exercise can be found in these pages of *Human Physiology*, thirteenth edition, by Stuart I. Fox:

- *Taste and Smell.* Chapter 10, p. 274.

The taste buds consist of specialized epithelial cells arranged in the form of barrel-shaped receptors (fig. 3.35c) associated with sensory (afferent) nerves. Long microvilli extend through a pore at the external surface of the taste bud and are bathed in saliva. The specialized epithelial **taste cells** can be depolarized to release chemical neurotransmitters that, in turn, stimulate associated sensory neurons leading to the brain. In adults these receptors are located primarily on the surface of the tongue, with a lesser number on the soft palate and epiglottis. In children the sense of taste is more diffuse, with additional receptors located on the inside of the cheeks.

The different categories of taste are produced by different chemicals that come into contact with the microvilli of taste cells. Four different categories of taste are traditionally recognized: *sour, salty, sweet,* and *bitter.* There is also a more recently discovered fifth category of taste, termed *umami* (a Japanese term related to a meaty flavor), for the amino acid glutamate (and stimulated by the flavor-enhancer monosodium glutamate). Although scientists long believed that different regions of the tongue were strictly specialized for different tastes, this is no longer thought to be true. Scientists now believe that all areas of the tongue are sensitive to all five tastes; indeed, a given taste bud may contain cells sensitive to different taste modalities. However, a particular taste cell is only sensitive to one category of taste and activates a sensory neuron that transmits information regarding that specific taste to the brain.

For example, the sweet taste evoked by sugar is carried to the brain on neurons devoted only to the sweet taste. Saccharin in low concentrations stimulates only the sweet receptors but at higher concentrations stimulates bitter receptors and gives saccharine an "aftertaste." The brain interprets the pattern of stimulation of these sensory neurons, together with the nuances provided by the sense of smell, as the complex tastes that we are capable of perceiving.

The *sour* taste of solutions is due to the presence of acid, or hydrogen ions (H^+), and the *salty* taste is due to the presence of Na^+ ions (fig. 3.36), but modified by the anion—sodium chloride tastes saltier than other sodium salts, such as sodium acetate. The remaining three taste modalities—sweet, bitter, and umami—involve the interaction of the taste molecules with membrane receptors coupled to G-proteins (fig. 3.36). Evidence demonstrates that each taste cell responds to only one of these taste modalities. Most organic molecules, but particularly sugars, taste sweet to varying degrees when they bind to the G-protein-coupled receptors on the taste cells "tuned" to detect a sweet taste. Bitter taste is evoked by quinine and seemingly unrelated molecules when they stimulate the taste cells that convey the bitter sense. It is the most acute sensation and is generally associated with toxic molecules (although not all toxins taste bitter).

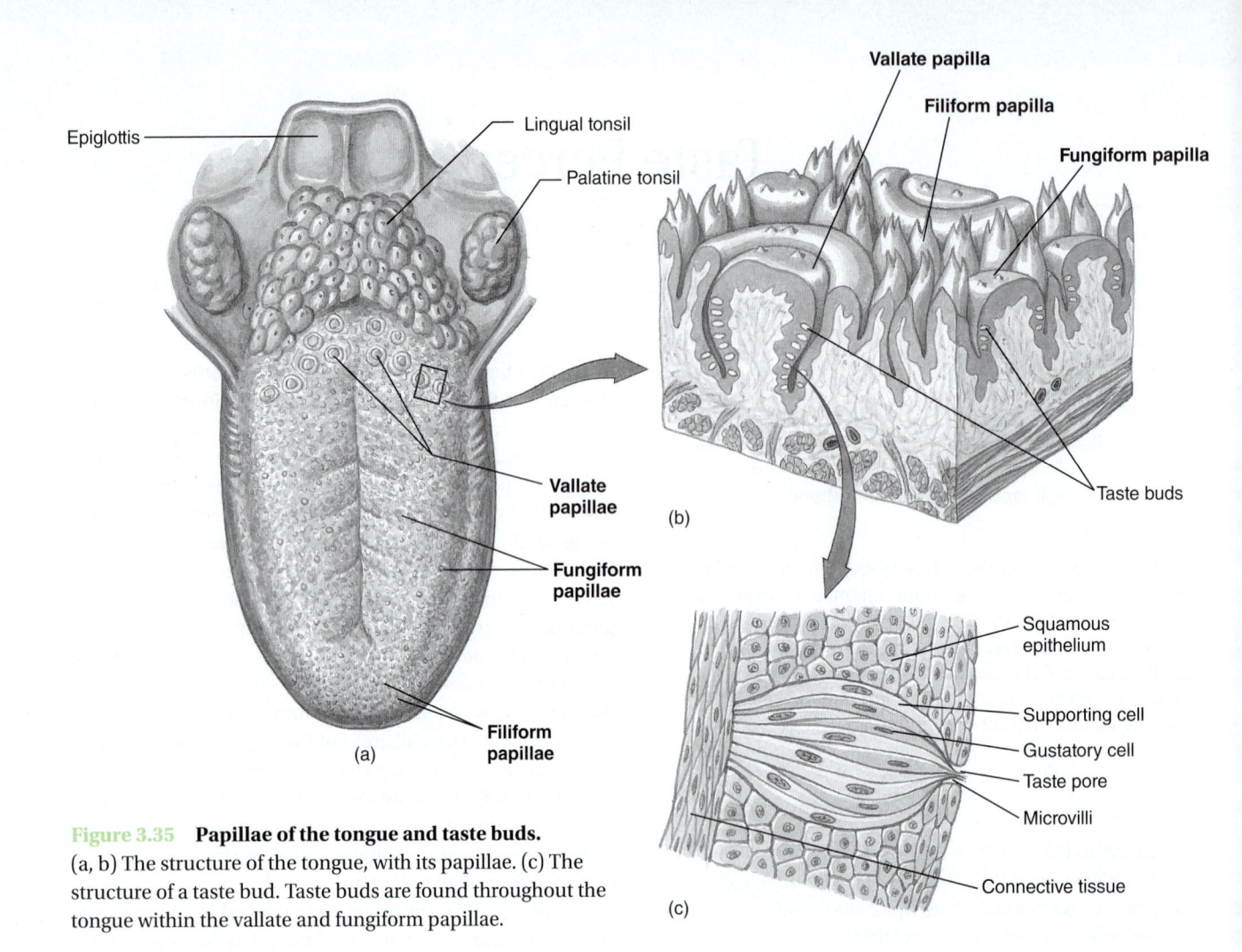

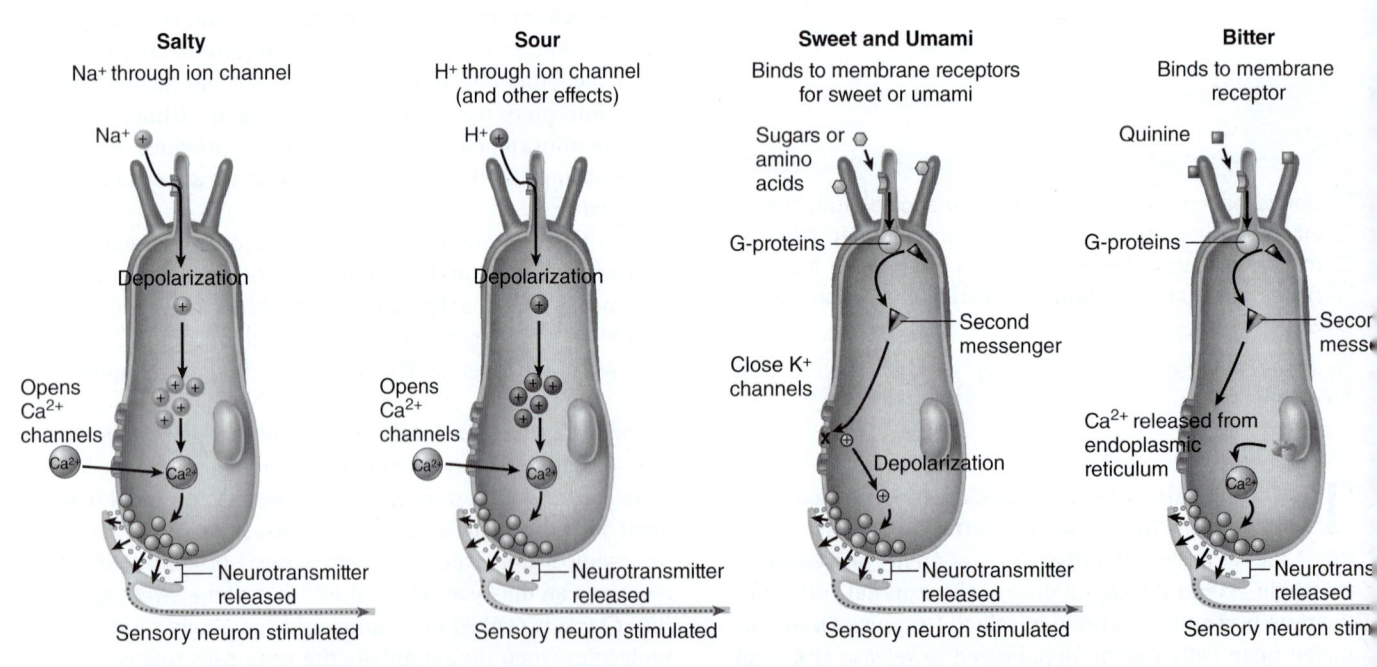

Figure 3.35 Papillae of the tongue and taste buds. (a, b) The structure of the tongue, with its papillae. (c) The structure of a taste bud. Taste buds are found throughout the tongue within the vallate and fungiform papillae.

Figure 3.36 The four major categories of taste. Each category of taste activates specific taste cells by different means. Taste cells for salty and sour are depolarized by ions (Na⁺ and H⁺, respectively) in the food, whereas taste cells for sweet and bitter are depolarized by sugars and quinine, respectively, by means of G-protein-coupled receptors and the actions of second messengers.

(For a full-color version of this figure, see fig. 10.8 in *Human Physiology,* thirteenth edition, by Stuart I. Fox.)

Figure 3.37 The innervation of the tongue.

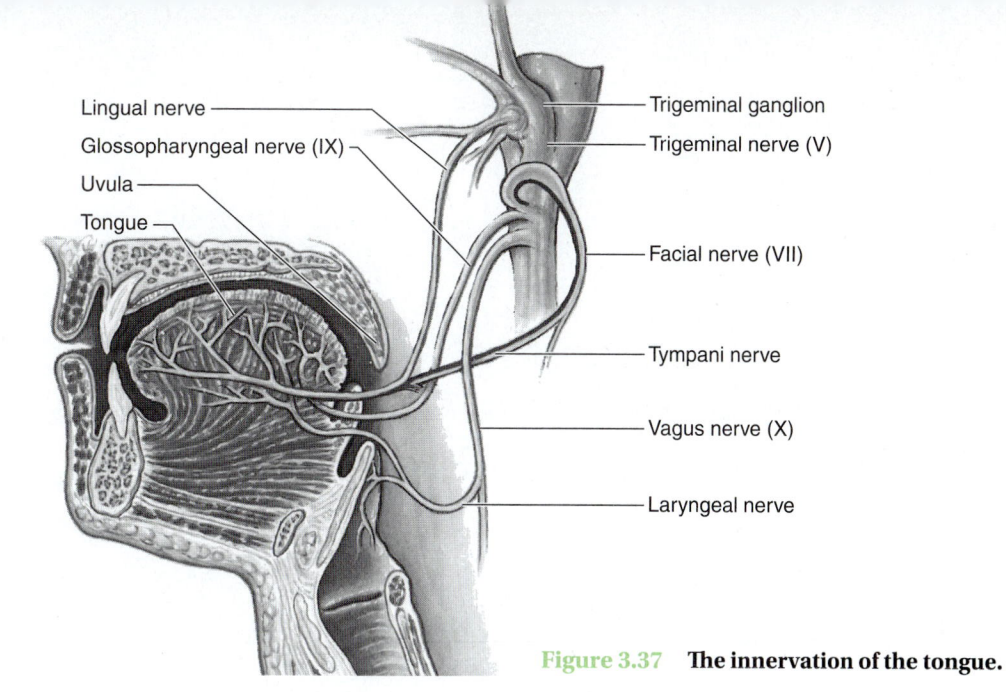

The particular type of G-proteins involved in taste have been termed **gustducin**. This term is used to emphasize the similarity to a related group of G-proteins called *transducin* that is associated with the photoreceptors in the eye. Dissociation of the gustducin G-protein subunit activates second-messenger systems, leading to depolarization of the receptor cell (fig. 3.36). The stimulated receptor cell, in turn, activates an associated sensory neuron that transmits impulses to the brain where they are interpreted as the corresponding taste perception.

The afferent pathway from the taste buds to the brain involves primarily two cranial nerves (fig. 3.37). Taste buds on the anterior two-thirds of the tongue are served by the *facial* (seventh cranial) *nerve*, whereas those on the posterior third of the tongue have a sensory pathway along the *glossopharyngeal* (ninth cranial) *nerve*. These nerves carry taste information to a nucleus in the medulla oblongata. From there, second-order neurons project to the thalamus and third-order neurons project from the thalamus to areas of the cerebral cortex. The cortical areas analyzing taste information include the somatosensory cortex of the post central gyrus devoted to the tongue, the insula, and the prefrontal (orbitofrontal) cortex, which is important in taste associations and the perception of flavor.

Clinical Applications

The sensations of *gustation* (taste) and *olfaction* (smell) are often grouped together in a single category—the *chemical senses*. The molecular basis of taste and smell is complex. Both sweet and bitter tastes (as well as perception of particular odorant molecules) are mediated by receptors coupled to membrane G-proteins. Dissociation of the G-protein subunit activates second-messenger systems that lead to depolarization of the receptor cell (see fig. 3.36). Together, these chemoreceptors function to provide the proper nuances of taste, which are extremely well developed in some people, such as wine tasters. Humans can distinguish up to 10,000 different odors, with a sensitivity capable of detecting a billionth of an ounce of perfume in the air. Because the sense of smell is so important in tasting, a stuffy nose from a cold or allergy can greatly affect the taste of foods.

Procedure

1. Dry the tongue with a paper towel and, using an applicator stick, apply a dab of 5% sucrose solution to the tip, sides, and back of the tongue.
2. Repeat this procedure using 1% acetic acid, 5% NaCl, and 0.5% quinine sulfate, being sure to rinse the mouth and dry the tongue between applications.

Note: *Apply quinine sulfate last; the effect is dominant and often lingering.*

3. Using the sketch provided in the laboratory report, record the location where you tasted each solution. Use the symbols sw for sweet, sl for salty, sr for sour, and b for bitter.
4. Pinch your nostrils closed with your fingers and bite into an apple, grape, or other fruit. Chew the food, and try to describe its flavor.
5. Now, remove your fingers from your nostrils so that you can also smell the food, and continue to bite and chew the fruit. Describe how its taste compares to what you described in step 4.

Laboratory Report 3.8

Name _____

Date _____

Section _____

DATA FROM EXERCISE 3.8

1. Map the areas of the tongue that seem to be sensitive to sweet, salty, sour, and bitter.

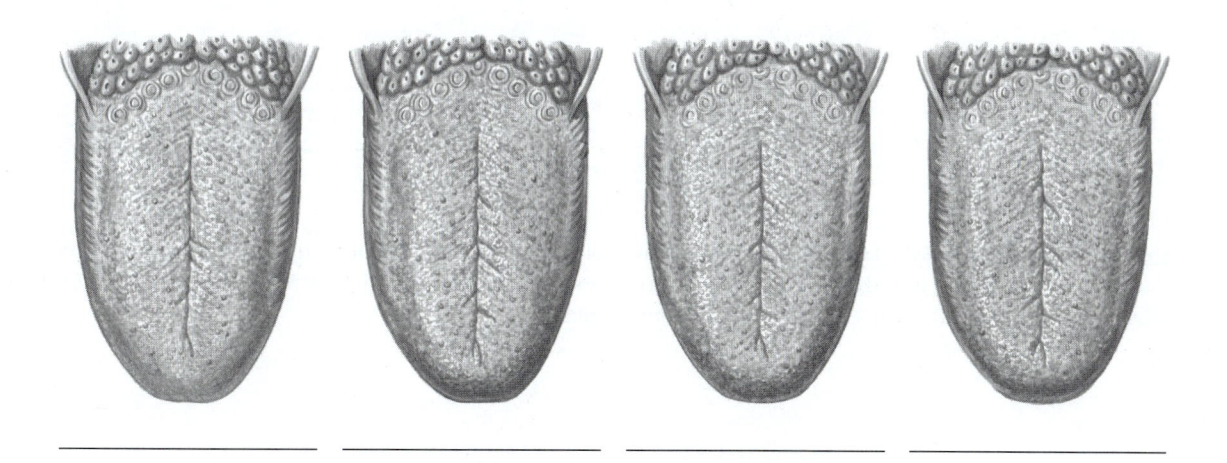

_____ _____ _____ _____

REVIEW ACTIVITIES FOR EXERCISE 3.8

Test Your Knowledge

1. What aspect of a solution causes it to taste sour? _____
2. What aspect of a solution causes it to taste salty? _____
3. Were tastes of sweet, salty, sour, or bitter restricted to any particular area of the tongue? Did any area of the tongue seem particularly sensitive to one modality of taste? If so, describe this distribution. If not, what can you conclude?

Test Your Understanding

4. Describe the common mechanism of action of molecules that taste sweet, umami, and bitter.

5. What does *chemical senses* mean? Explain how the sense of olfaction and gustation interact.

143

Test Your Analytical Ability

6. Did your map of taste perception support the traditional concept that different tastes could be mapped to specific locations on the tongue? Propose scientific tests to resolve this issue.

7. Tastes that are attractive (sweet and umami) or aversive (bitter) use G-protein-coupled receptors, as compared to sour and salty tastes. What are the advantages of G-protein-coupled receptors that might explain this distinction? What attributes of sour and salty tastes make them different? Explain.

8. Humans are said to be able to distinguish 10,000 different odors. Therefore, do you think we have 10,000 different receptors, each specialized for a different odor? Why or why not? Propose another mechanism that might account for the ability to distinguish so many odors.

Section 4

The Endocrine System

Glands are clusters of secretory cells derived from glandular epithelial membranes that invaginate into the underlying connective tissues (fig. 4.1). When the invagination persists, a duct is formed leading from the secretory cells to the epithelial membrane surface and to the outside of the body. Ducts may thus lead directly to an external body surface or indirectly to the luminal lining of tubes within the digestive, respiratory, urinary, or reproductive tract, which ultimately lead to the outside. These glands are called **exocrine glands,** and include the sebaceous (oil) glands, sweat glands, mammary glands, salivary glands, and the pancreas. In contrast, if the invagination disappears the chemical product of the gland, a **hormone,** is secreted internally into the blood capillaries. Ductless glands producing hormones are called **endocrine glands.**

Hormones secreted by endocrine glands regulate the activities of other organs (table 4.1). This regulation complements that of the nervous system and serves to direct the metabolism of the hormone's target organs and cells along paths that benefit the body as a whole. By regulating the activity of enzymes within their target cells, hormones primarily regulate total body metabolism and the function of the reproductive system.

Exercise 4.1	Histology of the Endocrine Glands
Exercise 4.2	Thin-Layer Chromatography of Steroid Hormones
Exercise 4.3	Insulin Shock

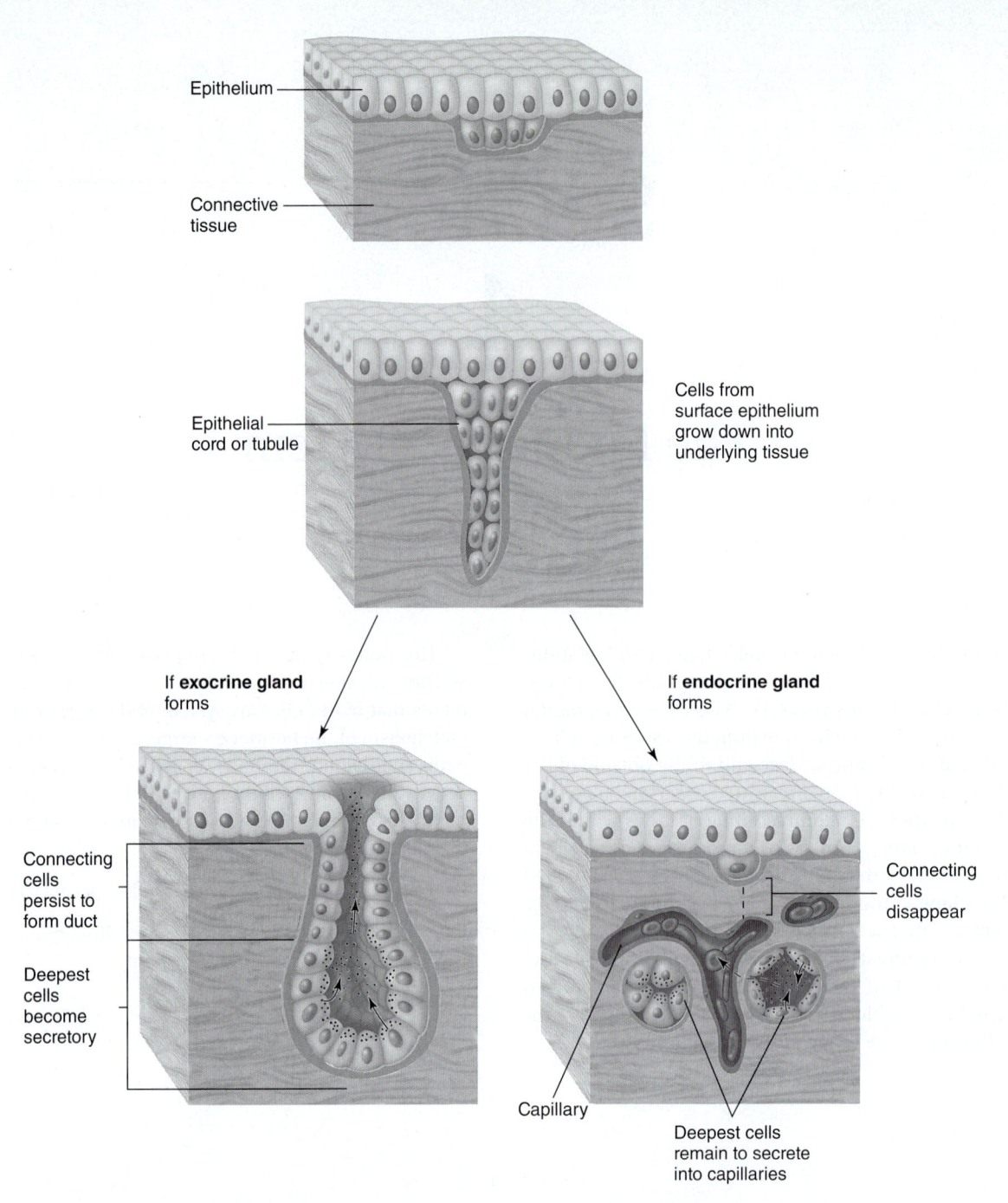

Figure 4.1 The formation of exocrine and endocrine glands from epithelial membranes. Exocrine glands retain a duct that can carry their secretion to the surface of the epithelial membrane, whereas endocrine glands are ductless.

(For a full-color version of this figure, see fig. 1.15 in *Human Physiology*, thirteenth edition, by Stuart I. Fox.)

Table 4.1 A Partial Listing of the Endocrine Glands

Endocrine Gland	Major Hormones	Primary Target Organs	Primary Effects
Adipose tissue	Leptin	Hypothalamus	Suppresses appetite
Adrenal cortex	Glucocorticoids (mainly cortisol) Mineralocorticoids (mainly aldosterone)	Liver and muscles Kidneys	Glucocorticoids influence glucose metabolism; mineralocorticoids promote Na^+ retention, K^+ excretion
Adrenal medulla	Epinephrine	Heart bronchioles and blood vessels	Causes adrenergic stimulation
Heart	Atrial natriuretic hormone	Kidneys	Promotes excretion of Na^+ and water in the urine
Hypothalamus	Releasing and inhibiting hormones	Anterior pituitary	Regulates secretion of anterior pituitary hormones
Small intestine	Secretin and cholecystokinin	Stomach, liver, and pancreas	Inhibits gastric motility and stimulates bile and pancreatic juice secretion
Islets of Langerhans (pancreas)	Insulin Glucagon	Many organs Liver and adipose tissue	Insulin promotes cellular uptake of glucose and formation of glycogen and fat; glucagon stimulates hydrolysis of glycogen and fat
Kidneys	Erythropoietin	Bone marrow	Stimulates red blood cell production
Liver	Somatomedins	Cartilage	Stimulates cell division and growth
Ovaries	Estradiol-17β and progesterone	Female reproductive tract and mammary glands	Maintains structure of reproductive tract and promotes secondary sex characteristics
Parathyroid glands	Parathyroid hormone	Bone, small intestine, and kidneys	Increases Ca^{2+} concentration in blood
Pineal gland	Melatonin	Hypothalamus and anterior pituitary	Affects secretion of gonadotrophic hormones
Pituitary, anterior	Trophic hormones (ACTH, TSH, FSH, LH, and prolactin)	Endocrine glands and other organs	Stimulates growth and development of target organs; stimulates secretion of other hormones
Pituitary, posterior	Antidiuretic hormone oxytocin (produced in hypothalamus)	Kidneys and blood vessels Uterus and mammary glands	Antidiuretic hormone promotes water retention and vasoconstriction; oxytocin stimulates contraction of uterus and mammary secretory units, promoting milk ejection
Skin	1,25-Dihydroxyvitamin D_3	Small intestine	Stimulates absorption of Ca^{2+}
Stomach	Gastrin	Stomach	Stimulates acid secretion
Testes	Testosterone	Prostate seminal vesicles, testes, and other organs	Stimulates secondary sexual development, spermatogenesis, other effects
Thymus	Thymopoietin	Lymph nodes	Stimulates white blood cell production
Thyroid gland	Thyroxine (T_4) and triiodothyronine (T_3); calcitonin	Most organs	Thyroxine and triiodothyronine promote growth and development and stimulate basal rate of cell respiration (basal metabolic rate or BMR); calcitonin may participate in the regulation of blood Ca^{2+} levels

EXERCISE 4.1

Histology of the Endocrine Glands

MATERIALS

1. Microscopes
2. Prepared slides

Endocrine glands vary greatly in structure but have some features in common because they all secrete hormones into the blood. The histological structure of the endocrine glands shows how these glands function and how they are related to surrounding tissues.

LEARNING OUTCOMES

You should be able to:

1. Describe the structure of the ovaries and testes and the functions performed by their component parts.
2. Describe the histological structure of the pancreas and identify both its endocrine and exocrine structures.
3. Describe the histological structure of the adrenal and thyroid glands and the functions of their component parts.
4. Describe the embryological origin and structure of the anterior pituitary and posterior pituitary and their relationships to the hypothalamus.
5. List the hormones secreted by the anterior pituitary and posterior pituitary, and explain how the secretion of these hormones is regulated.

Textbook/Multimedia Correlations

Before performing this exercise, you should study the introductory material presented here. Further information relating to this exercise can be found in these pages of *Human Physiology*, thirteenth edition, by Stuart I. Fox:

- *Pituitary Gland.* Chapter 11, p. 331.
- *Adrenal Glands.* Chapter 11, p. 338.
- *Thyroid and Parathyroid Glands.* Chapter 11, p. 342.
- *Pancreas and Other Endocrine Glands.* Chapter 11, p. 346.
- Multimedia correlations are provided in *MediaPhys 3.0:* Topics 13.17–13.62.

Clinical Investigation

A biopsy of a patient's thyroid gland was performed, and the biopsy report stated that there was both hypertrophy and hyperplasia of the thyroid follicles. The physician told the patient that he suspected she had Graves' disease, an autoimmune disorder in which antibodies act like TSH to stimulate the thyroid.

- Describe the microscopic structure of the thyroid gland and how hypertrophy and hyperplasia would affect this structure.
- Explain the origin and function of TSH and how autoantibodies that act like TSH might cause the observed structural changes of the thyroid.

Endocrine glands may be independent organs, or they may be part of an organ that also performs nonendocrine functions (table 4.1). Organs that perform both endocrine and nonendocrine functions include the adipose tissue, brain, stomach, pancreas, liver, and skin.

Hormones are carried by the blood to all organs of the body; only certain organs, however, can respond to a given hormone. These are called the **target organs** for the hormone. Hormones affect the metabolism of their target organs and in so doing help to regulate growth and development, total body metabolism, and reproduction.

A. OVARY

The ovary is an endocrine gland as well as the producer of female gametes *(ova)*. The ovum (egg cell) can be thought of as an exocrine secretion because it enters a duct—the uterine tube—after leaving the ovary. The primary hormones of the ovary, *estrogen* and *progesterone,* are secreted directly into the blood of the circulatory system.

The **ovarian follicles** are brought to maturity under the influence of the *gonadotropic hormones* (FSH and LH, mentioned later) secreted by the anterior pituitary. In every cycle one of the mature follicles, or **graafian follicle,** eventually ruptures through the surface of the ovary to release its ovum in a process called *ovulation*. The empty follicle is then converted into a new endocrine structure, the **corpus luteum.** If fertilization does not occur the corpus luteum regresses and the cycle is ready to begin again.

The microscopic appearance of the ovary is thus continuously changing as the cycle progresses. A single slide of the ovary will reveal many follicles at different stages of maturation, including primary, secondary, and graafian follicles (fig. 4.2).

Procedure

Using the low-power objective, scan the slide of the ovary, and try to locate a circular-to-elliptical structure, the follicle, that encloses a space filled with fluid and scattered cells. Identify the following parts of the follicle (fig. 4.2):

1. **Ovum.** The ovum (egg cell) is the largest cell in the follicle and, at this stage of development, is called a *secondary oocyte.*
2. **Granulosa cells.** Granulosa cells are the numerous small cells found within the follicle.
3. **Antrum.** The antrum is the central fluid-filled cavity of the follicle.
4. **Cumulus oophorus.** Cumulus oophorus means "egg-bearing hill." This is the mound of granulosa cells that supports the ovum.
5. **Corona radiata.** The corona radiata is the layer of granulosa cells that surrounds the ovum. The ovum continues to be surrounded by its corona radiata after ovulation, and this layer of cells presents the first barrier to sperm penetration during fertilization.
6. **Zona pellucida.** The zona pellucida is a clear region containing glycoproteins between the plasma membrane of the ovum and the corona radiata.

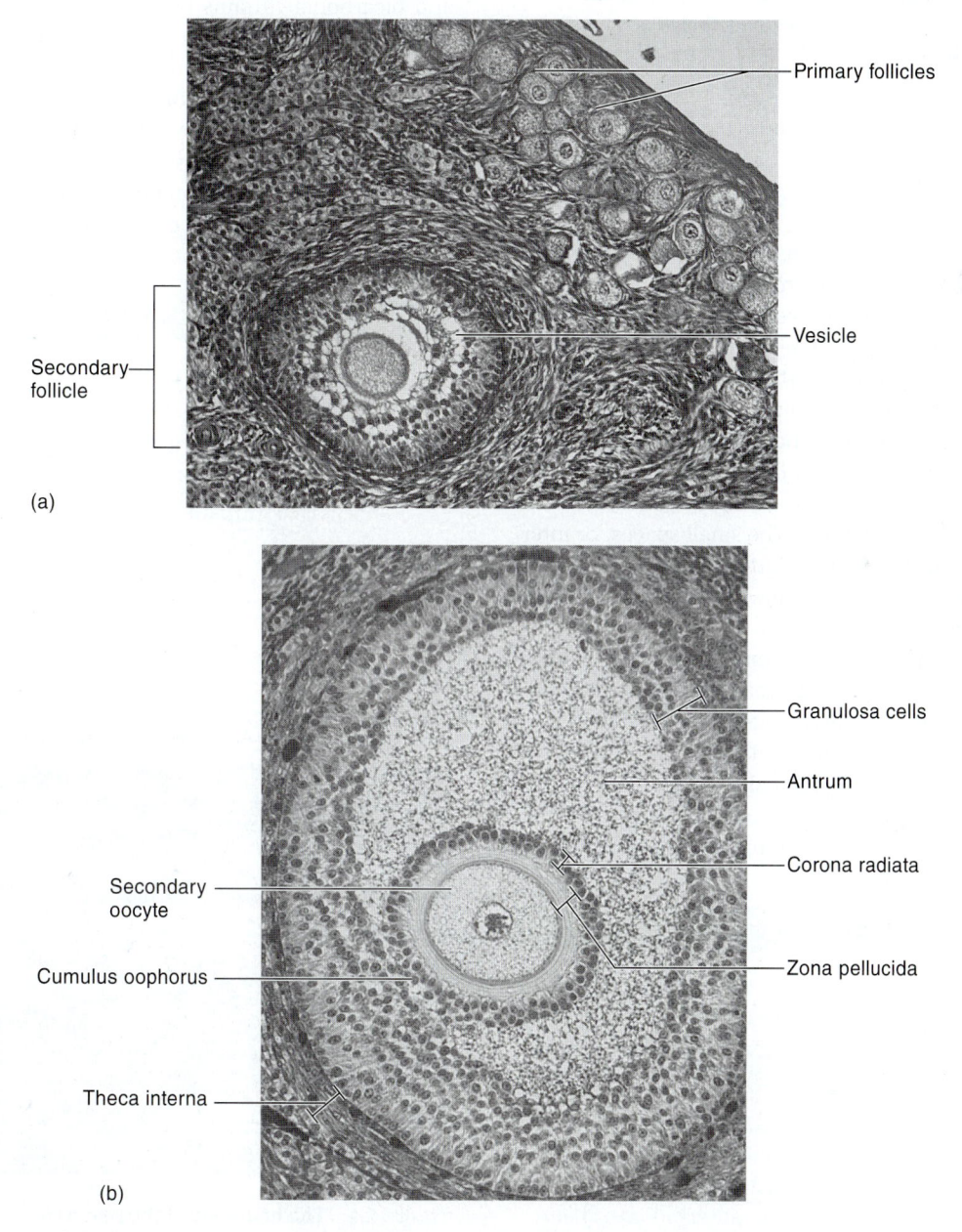

Figure 4.2 Ovarian follicles. Photomicrographs of (a) a secondary follicle and (b) a graafian follicle within an ovary (450×).

(For a full-color version of this figure, see fig. 20.27 in *Human Physiology*, thirteenth edition, by Stuart I. Fox.)

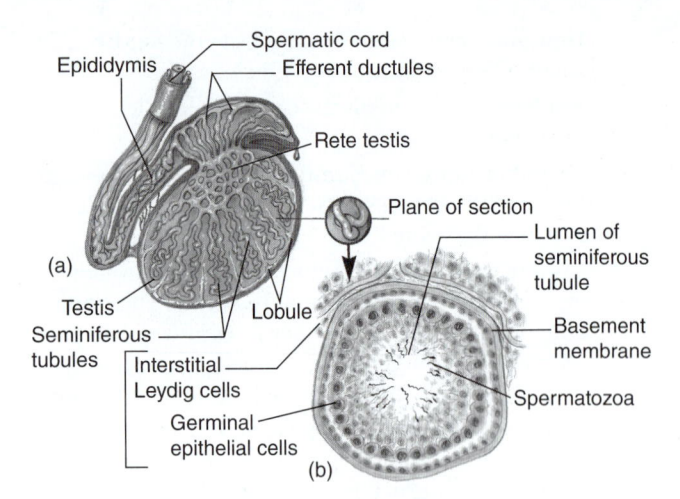

Figure 4.3 **The structure of the testis.** (a) A sagittal section of a testis and (b) a transverse section of a seminiferous tubule.

(For a full-color version of this figure, see fig. 20.11 in *Human Physiology*, thirteenth edition, by Stuart I. Fox.)

B. Testis

The testis produces both the male gametes *(sperm)* and the male sex hormone *(testosterone)*. Sperm are produced within the **seminiferous tubules** and travel through these tubules to the **epididymis,** where the sperm are passed into a single tubule that becomes the **ductus (vas) deferens** (fig. 4.3). The ductus deferens picks up fluid from the *seminal vesicles* and the *prostate* and passes its contents, now called **semen,** to the *ejaculatory duct.*

The seminiferous tubules are highly convoluted and tightly packed within the testis. The small spaces, or interstices, between adjacent convolutions of the tubules are filled with connective tissue known as *interstitial tissue*. Within this connective tissue are the interstitial **Leydig cells,** endocrine cells that produce the *androgens* (male sex steroid hormones). The major androgen secreted by the Leydig cells of sexually mature males is testosterone.

Because the seminiferous tubules are highly convoluted, the chances of seeing a longitudinal section of a tubule are remote. Most of the tubules will be cut more or less in cross section, giving a circular or oblong appearance (fig. 4.3*b*).

Procedure

Using the low-power objective, scan the slide of the testis and locate one seminiferous tubule within a section, or lobule. Switch to a high-power objective and observe these structures (fig. 4.3):

1. **Spermatogenic cells.** Spermatogenic cells form the *germinal epithelium* found along the outer wall of the tubules. These cells divide by **meiosis** to produce the sperm. The outermost cells are *diploid* (46 chromosomes), whereas the cells toward the lumen have completed meiotic division and are *haploid* (23 chromosomes). Within the tubular epithelium, chromosomes at various stages of meiosis can be seen as darkened structures. Haploid **spermatozoa** can often be seen within the tubular lumen.

2. **Leydig cells.** These endocrine cells can be seen in the interstitial connective tissue between adjacent convolutions of the tubules.

C. Pancreatic Islets (of Langerhans)

The pancreas has both an exocrine and an endocrine function. The exocrine secretion *(pancreatic juice)* is produced by pancreatic cells arranged in clusters called **acini** around a central duct. Pancreatic juice, containing digestive enzymes and bicarbonate, drains into *interlobular ducts* located in bands of connective tissue. From here the secretion flows into the pancreatic duct and empties into the duodenum.

The endocrine secretions of the pancreas, **insulin** and **glucagon,** are produced by scattered groups of cells called the **pancreatic islets,** or **islets of Langerhans.** These hormones do not enter the interlobular ducts but rather leave the pancreas by way of the circulatory system.

Procedure

Using the low-power objective, scan the slide of the pancreas and attempt to identify these structures (fig. 4.4):

1. **Acini.** The pancreatic acini are dark-staining clusters of cells that form most of the body of the pancreas.

2. **Interlobular ducts.** The interlobular ducts may be mistaken for veins because of their large size; thin

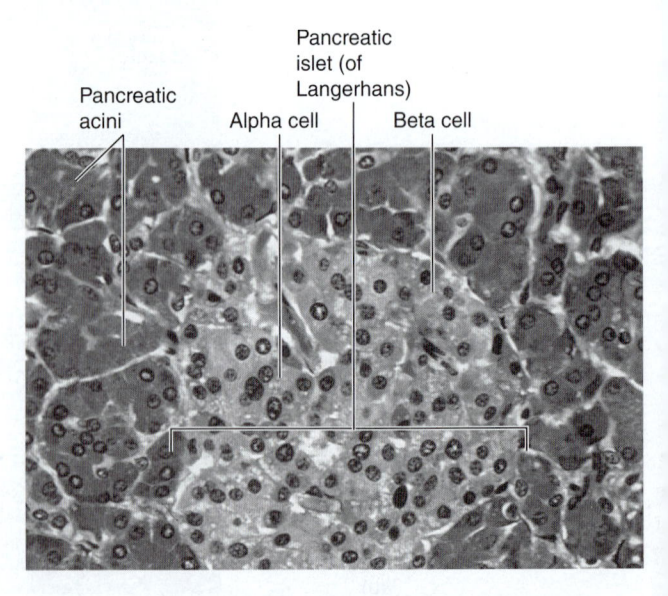

Figure 4.4 **The histology of the pancreas.** The exocrine pancreatic acinar cells (that form secretory structures called acini) and endocrine pancreatic islet (of Langerhans) are seen.

walls; and flattened, irregular shape. Unlike veins, however, their walls are composed of only a single layer of columnar epithelial cells, and no red blood cells will be seen in the lumina.

3. **Pancreatic islets (of Langerhans).** Under low power, the pancreatic islets will appear as light patches, circular in shape, against the dark background of the acini. Under high power, the *alpha cells* (which secrete glucagon) can easily be distinguished from the *beta cells* (which secrete insulin). The alpha cells are smaller and contain pink-staining granules, whereas the beta cells are larger and stain blue.

D. Adrenal Gland

The **adrenal gland** is actually two glands located in the same organ. In lower organisms, these glands are separated; in higher organisms (including humans), they are closely associated as the *adrenal cortex* (outer part) and the *adrenal medulla* (inner part) (fig. 4.5).

The **adrenal cortex** secretes **corticosteroid hormones** and weak androgens (male sex hormones). The corticosteroid hormones include those that regulate salt balance *(mineralocorticoids)* and hormones that regulate glucose homeostasis *(glucocorticoids)*. The **adrenal medulla** secretes two hormones, **epinephrine** and **norepinephrine,** that act together with sympathetic nerve stimulation to enhance the response of the cardiovascular system to increased physical demand.

The cells of the adrenal medulla are derived from the same embryonic tissue (neural crest ectoderm) as postganglionic sympathetic neurons, whereas the adrenal cortex is derived from a different embryonic tissue (mesoderm). Therefore, these two regions of the adrenal gland are different both physiologically and histologically. The adrenal cortex is stimulated to secrete its corticosteroid hormones by ACTH, a hormone secreted by the anterior pituitary gland (discussed shortly). The adrenal medulla is stimulated to secret epinephrine and norepinephrine by axons of preganglionic sympathetic neurons.

Procedure

Before observing the slide of the adrenal gland under the microscope, hold it up to the light and note the clear distinction between the adrenal cortex and adrenal medulla. Using the low-power objective, focus on the outer edge of the gland. Scan from this point inward and identify these structures (fig. 4.5):

1. **Capsule.** The capsule is a thin, tough layer of connective tissue that surrounds the gland.
2. **Zona glomerulosa.** The zona glomerulosa is the outer layer of the adrenal cortex; its cells are tightly packed in an irregular arrangement. The z. glomerulosa secretes the mineralocorticoids (mainly *aldosterone* and *deoxycorticosterone*). The secretion of these hormones is largely under the control of angiotensin II.

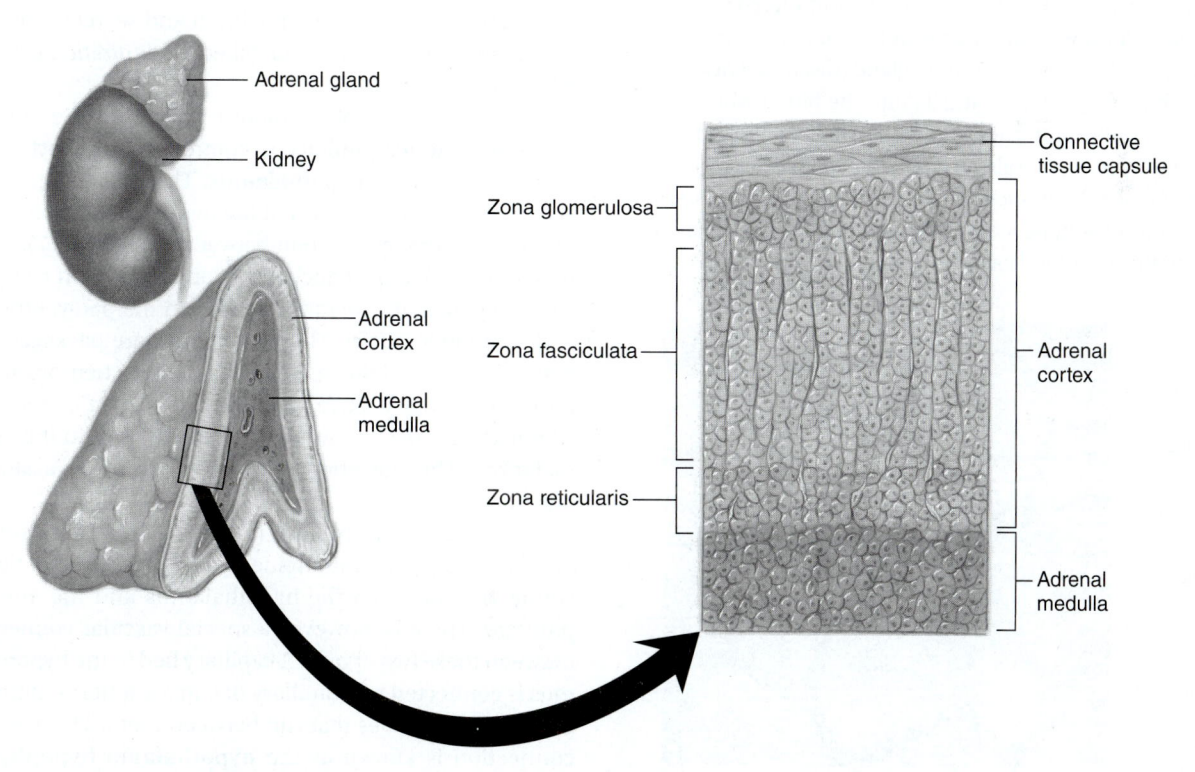

Figure 4.5 The adrenal gland.   The adrenal medulla and adrenal cortex are shown in the sectioned gland, and the histology of the adrenal cortex, showing its three zones, is illustrated as it would appear at a magnification of 450×.

(For a full-color version of this figure, see fig. 11.18 in *Human Physiology*, thirteenth edition, by Stuart I. Fox.)

3. **Zona fasciculata.** The zona fasciculata is the second layer of the adrenal cortex, located below the zona glomerulosa. This layer is the thickest part of the adrenal cortex, with its cells arranged in columns. They secrete the glucocorticoids when stimulated by adrenocorticotropic hormone (ACTH) secreted by the anterior pituitary. The most important glucocorticoids are *hydrocortisone (cortisol)* and *corticosterone.*
4. **Zona reticularis.** The z. reticularis is the innermost layer of the adrenal cortex and primarily secretes weak androgens (principally androstenedione). The epithelial cells in this layer form interconnections (anastomoses) with one another and stain a darker color than those of the z. fasciculata.
5. **Adrenal medulla.** The adrenal medulla forms the distinctive central region of the gland that stains a lighter color than the surrounding z. reticularis. It is composed of tightly packed clusters of *chromaffin cells.* The adrenal medulla secretes epinephrine and lesser amounts of norepinephrine.

E. Thyroid

Like the ovary, the functional units of the thyroid are called **follicles** (fig. 4.6). Each thyroid follicle is composed of a single layer of epithelial cells surrounding a homogeneous protein-rich fluid, the *colloid.*

The protein in the colloid is *thyroglobulin.* The thyroid follicles transport iodine into the colloid and attach it to the thyroglobulin. This serves as the precursor for the thyroid hormones: **thyroxine,** also called **tetraiodothyronine** and abbreviated T_4; and **triiodothyronine (T_3).** When stimulated by TSH from the anterior pituitary gland (discussed next), the thyroid follicles secrete T_4 and T_3 into the blood where they function as important regulators of metabolism and growth.

The thyroid gland also contains **parafollicular cells** that secrete the hormone *calcitonin* (also called *thyrocalcitonin*). This hormone is believed to play a relatively minor role in the regulation of blood calcium concentrations.

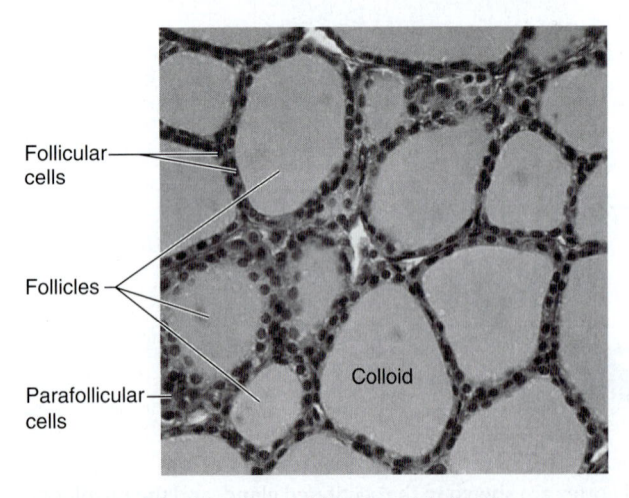

Figure 4.6 The histology of the thyroid gland.

(For a full-color version of this figure, see fig. 11.22 in *Human Physiology,* thirteenth edition, by Stuart I. Fox.)

Procedure

Scan the slide under *low* power and observe the follicles (fig. 4.6). Note the clear space between the lighter, inner colloid and the darker, surrounding epithelial cells. This space is an artifact (produced by manipulation of the tissue when the slide was prepared) and is not present *in vivo.*

F. Pituitary Gland

The **pituitary,** or *hypophysis,* like the adrenal gland, is derived from two distinct embryonic tissues. The **anterior pituitary,** also known as the **adenohypophysis** (*adeno* means "glandular"), originates in the embryo from a dorsal outpouching *(Rathke's pouch)* of oral epithelium.

Sometimes referred to as a master gland, the anterior pituitary secretes hormones that control other glands. These hormones include *adrenocorticotropic hormone (ACTH),* which stimulates the adrenal cortex; *thyroid-stimulating hormone (TSH),* which stimulates the thyroid; *gonadotropic hormones (FSH* and *LH—follicle-stimulating hormone* and *luteinizing hormone),* which stimulate the gonads; and *prolactin,* which stimulates the mammary glands. In addition, the anterior pituitary secretes *somatotropic hormone,* or *growth hormone (GH),* which stimulates growth in children.

In contrast, the **posterior pituitary,** also known as the **neurohypophysis,** is derived in the embryo from a ventral outpouching of the floor of the brain and secretes only two hormones: *vasopressin* (also called *antidiuretic hormone, ADH*) and *oxytocin.*

The secretion of hormones from both the anterior and the posterior pituitary is controlled by a part of the brain known as the **hypothalamus.** The posterior pituitary is derived as a downgrowth of the hypothalamus, providing the direct neural connection between them (fig. 4.7). Vasopressin (ADH) and oxytocin are manufactured in cell bodies of hypothalamic neurons located in the paraventricular and supraoptic nuclei. These hormones are packaged into vesicles that travel down the axons of these neurons to the posterior pituitary. Here, these hormones are stored until stimulated for release by axons of the neurons in the hypothalamus. The posterior pituitary is therefore a storage organ.

The anterior pituitary is derived from oral epithelium and not from brain tissue, so there is no direct neural connection between the hypothalamus and the anterior pituitary. There is, however, a special vascular connection between these two organs. A capillary bed in the hypothalamus is connected to a capillary bed in the anterior pituitary by means of venules that run between them. This vascular connection is known as the **hypothalamo-hypophyseal portal system** (fig. 4.8).

Unlike the posterior pituitary, the anterior pituitary manufactures its own hormones. These hormones are

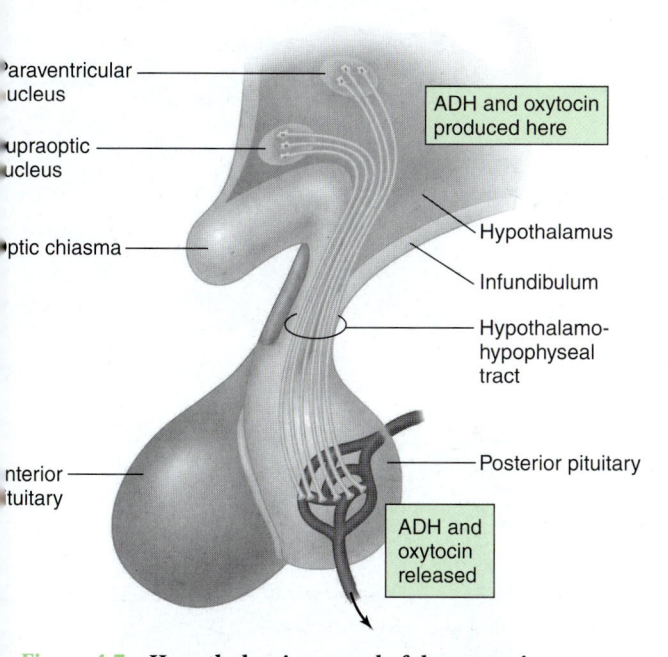

Figure 4.7 Hypothalamic control of the posterior pituitary. The posterior pituitary, or neurohypophysis, stores and releases hormones—vasopressin and oxytocin—produced in neurons within the supraoptic and paraventricular nuclei of the hypothalamus. These hormones are transported to the posterior pituitary by axons in the hypothalamo-hypophyseal tract.

(For a full-color version of this figure, see fig. 11.13 in *Human Physiology*, thirteenth edition, by Stuart I. Fox.)

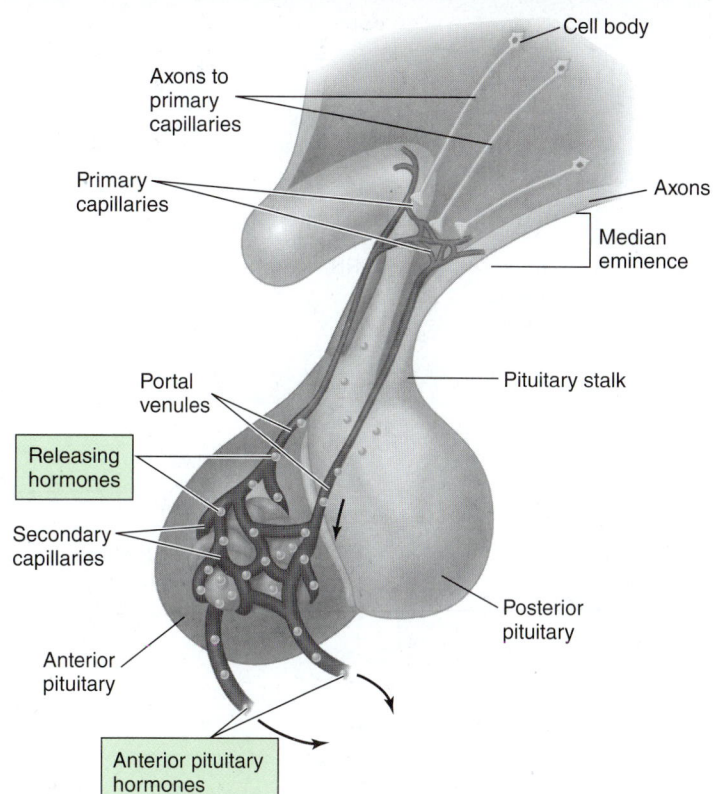

Figure 4.8 Hypothalamic control of the anterior pituitary. Neurons in the hypothalamus secrete releasing hormones (shown as spheres) into the blood vessels of the hypothalamo-hypophyseal portal system. These releasing hormones stimulate the anterior pituitary to secrete its hormones into the general circulation.

(For a full-color version of this figure, see fig. 11.15 in *Human Physiology*, thirteenth edition, by Stuart I. Fox.)

released upon the arrival of specific chemical messengers called *releasing hormones,* which are secreted into the hypothalamo-hypophyseal portal system by the hypothalamus, as illustrated in figure 4.8. The anterior pituitary, then, is not the "master gland" because the secretion of its hormones is controlled by releasing hormones secreted by the hypothalamus. So, is the hypothalamus the "master gland?" Not exactly, because both the hypothalamus and anterior pituitary are regulated by hormones from the target glands. For example, sex steroids secreted by the gonads can inhibit the secretion of the releasing hormone from the hypothalamus (*gonadotropin-releasing hormone,* or *GnRH*) and the anterior pituitary's responsiveness to GnRH (fig. 4.9). This type of regulation is termed **negative feedback** inhibition. This is the more common type of regulation. However, there is one instance of **positive feedback** regulation—when estrogen from the ovaries stimulates the anterior pituitary to secrete LH, thereby causing ovulation.

There isn't a linear chain of command among these endocrine glands as implied by the term *master gland* because the anterior pituitary is regulated by the hypothalamus, and the hypothalamus and anterior pituitary are regulated by the target glands' hormones. In this sense, the master gland concept can be a misleading simplification.

Clinical Applications

Knowledge of the normal histology of the endocrine glands helps in understanding their normal physiology and in diagnosing various pathological states. Endocrine glands may *atrophy* (lose structure) or develop hormone-secreting tumors known as *adenomas*.

In the testes, various diseases, such as those resulting from *Klinefelter's syndrome* (XXY genotype) or from the *mumps*, are associated with atrophy of the seminiferous (sperm-producing) epithelium. In the ovaries, granulosa cell tumors may secrete excessive estrogen. In *diabetes mellitus*, the beta cells within the islets of Langerhans in the pancreas may decrease in number and have decreased numbers of insulin-containing granules per cell. *Pheochromocytoma*, a tumor of the adrenal medulla, secretes excess epinephrine. Tumors of the anterior pituitary may result in *gigantism* and *acromegaly* (from elevated levels of growth hormone), hyperpigmentation (from excessive adrenocorticotropic hormone), or persistent lactation (from elevated prolactin secretion). These examples are only a few of the disorders associated with an abnormal histology of the various endocrine glands.

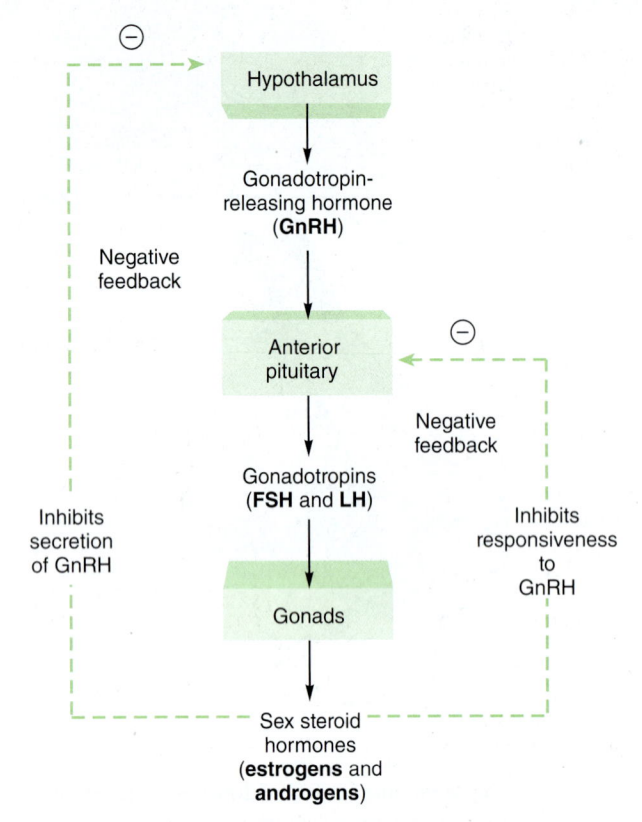

Figure 4.9 **The hypothalamus-pituitary-gonad (control system).** The hypothalamus secretes GnRH, which stimulates the anterior pituitary to secrete the gonadotropins (FSH and LH). These, in turn, stimulate the gonads to secrete the sex steroids. The secretions of the hypothalamus and anterior pituitary are regulated by negative feedback inhibition from the sex steroids.

(For a full-color version of this figure, see fig. 11.17 in *Human Physiology*, thirteenth edition, by Stuart I. Fox.)

Procedure

Scan the slide of the pituitary gland under *low* power (fig. 4.10). Distinguish the anterior pituitary (observe capillaries and darkly stained blood cells) from the posterior pituitary (characteristic nerve tissue, lightly stained). Next, switch to the *high*-power objective, and identify these structures:

1. **Anterior pituitary**
   (a) **Sinusoids.** Sinusoids are modified capillaries that lack an endothelial wall. They can easily be identified by the presence of red blood cells.

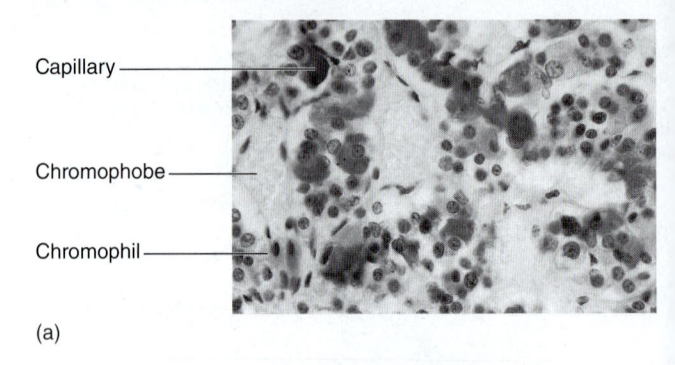

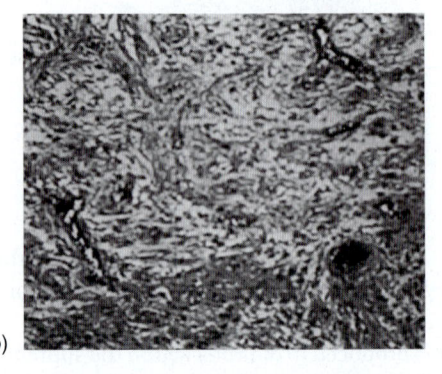

Figure 4.10 **Histology of the pituitary gland.** (a) The anterior pituitary (160×); and (b) the posterior pituitary (100×).

   (b) **Chromophils.** "Color-loving" chromophils are pituitary cells that readily take up stain into granules present in the cytoplasm. They are divided into two general categories on the basis of their staining properties: acidophils contain red-staining granules; basophils contain blue-staining granules. These two categories of cells produce different hormones.
   (c) **Chromophobes.** The cytoplasm of chromophobes ("color-fearing" cells) does not pick up stain, and hence these cells appear dull next to the chromophils. It is believed that the chromophobes are not involved in hormone production.

2. **Posterior pituitary**
   (a) **Nerve fibers.** Nerve fibers are the axons of neurons extending from the hypothalamus and compose most of the mass of the gland.
   (b) **Pituicytes.** Pituicytes are randomly distributed among the nerve fibers and lack the bright color of the anterior pituitary cells.

Laboratory Report 4.1

Name _____
Date _____
Section _____

REVIEW ACTIVITIES FOR EXERCISE 4.1

Test Your Knowledge

Match the gland with the hormone it secretes:

____ 1. ovarian follicle (a) growth hormone
____ 2. interstitial cells of Leydig (b) glucagon
____ 3. alpha cells of islets of Langerhans (c) hydrocortisone
____ 4. beta cells of islets of Langerhans (d) insulin
____ 5. zona glomerulosa of the adrenal cortex (e) testosterone
____ 6. zona fasciculata of the adrenal cortex (f) aldosterone
____ 7. adrenal medulla (g) estrogen
____ 8. posterior pituitary (h) oxytocin
____ 9. anterior pituitary (i) epinephrine
　　　　　　　　　　　　　　　　(j) thyroxine

10. The fluid-filled central cavity of an ovarian follicle is called the _____.

11. Most of the mass of the testis is composed of the _____.

12. Another name for tetraiodothyronine is _____; it is secreted by the _____ gland.

13. Two glands derived from neural tissue are the _____ and the _____.

14. Another name for vasopressin is _____; it is secreted by the _____.

Test Your Understanding

15. Distinguish between exocrine and endocrine glands.

16. Describe the structures involved in the production, transport, and secretion of oxytocin and vasopressin.

17. Where is ACTH produced? What does it do? Explain how its secretion is regulated.

18. The anterior pituitary has sometimes been called a "master gland." Why was this term used? Why is this term an oversimplification?

Test Your Analytical Ability

19. Do you think that ligation (tying) of the vas deferens will affect the blood concentration of testosterone? Explain.

20. In *Cushing's syndrome,* the adrenal cortex secretes excessive amounts of glucocorticoids. Which hormones are these? Propose three different possible causes of excessive secretion of glucocorticoids.

Clinical Investigation Questions

21. What are hypertrophy and hyperplasia? How would these processes affect the appearance of the thyroid under the microscope? How would the gross structure of the thyroid be affected?

22. What is the origin and function of TSH? How would too much TSH, or antibodies that act like TSH (in Graves' disease), affect the structure of the thyroid? Given your answer, how do you think inadequate TSH would affect thyroid structure?

EXERCISE 4.2

Thin-Layer Chromatography of Steroid Hormones

MATERIALS

1. Thin-layer plates (silica gel, F-254 [F = fluorescent]), chromatography developing chambers, capillary tubes
2. Driers (chromatography or hair driers), ultraviolet viewing box (short wavelength), rulers or spotting template (optional)
3. Steroid solutions: 1.0 mg/mL in absolute methanol of testosterone, hydrocortisone, cortisone, corticosterone, and deoxycorticosterone; 5 mg/mL of estradiol
4. Unknown steroid solution containing any two of the steroids previously described
5. Developing solvent: 60 mL toluene plus 10 mL ethyl acetate plus 10 mL acetone, or a volume containing a comparable 6:1:1 ratio of solvents
6. Fume hood, to prepare developing solution and to maintain developing chambers

Slight differences in steroid structure are responsible for significant differences in biological effects. Differences in structure and solubility can be used to separate a mixture of steroids and to identify unknown molecules.

LEARNING OUTCOMES

You should be able to:

1. Identify the major classes of steroid hormones and the glands that secrete them.
2. Describe the primary differences between different functional classes of steroid hormones.
3. Demonstrate the technique of thin-layer chromatography, and explain how this procedure works.

Textbook/Multimedia Correlations

Before performing this exercise, you should study the introductory material presented here. Further information relating to this exercise can be found in these pages of *Human Physiology*, thirteenth edition, by Stuart I. Fox:

- *Chemical Classification of Hormones.* Chapter 11, p. 319.
- *Mechanism of Steroid Hormone Action.* Chapter 11, p. 323.
- *Functions of the Adrenal Cortex.* Chapter 11, p. 338.
- *Gonads and Placenta.* Chapter 11, p. 350.

Clinical Investigation

A well-muscled male patient who complained of various symptoms was given a physical exam. He was found to have an enlarged prostate, abnormally small testicles, and the beginnings of gynocomastia. The patient admitted that he was a regular user of anabolic steroids.

- Identify the chemical and physiological nature of anabolic steroids, and explain how their use could cause an enlarged prostate and small testicles.
- Describe how the metabolism of anabolic steroids can result in gynecomastia.

A. STEROID HORMONES

The steroid hormones, secreted by the adrenal cortex and the gonads (fig. 4.11), are characterized by a common four-ring structure (with the rings designated *A*, *B*, *C*, and *D*). The carbon atoms in this structure are numbered as shown here:

Seemingly slight modifications in chemical structure result in very great differences in biological activity. On the basis of their activity and their structure, the steroid hormones can be grouped into the following functional categories: (1) androgenic hormones, (2) estrogenic hormones, (3) progestational hormones, and (4) corticosteroid hormones. The corticosteroid hormones can be divided further into the subcategories of glucocorticoids and mineralocorticoids.

The **androgenic hormones** have a characteristic nineteen-carbon steroid structure and function in the development of male secondary sex characteristics. The

157

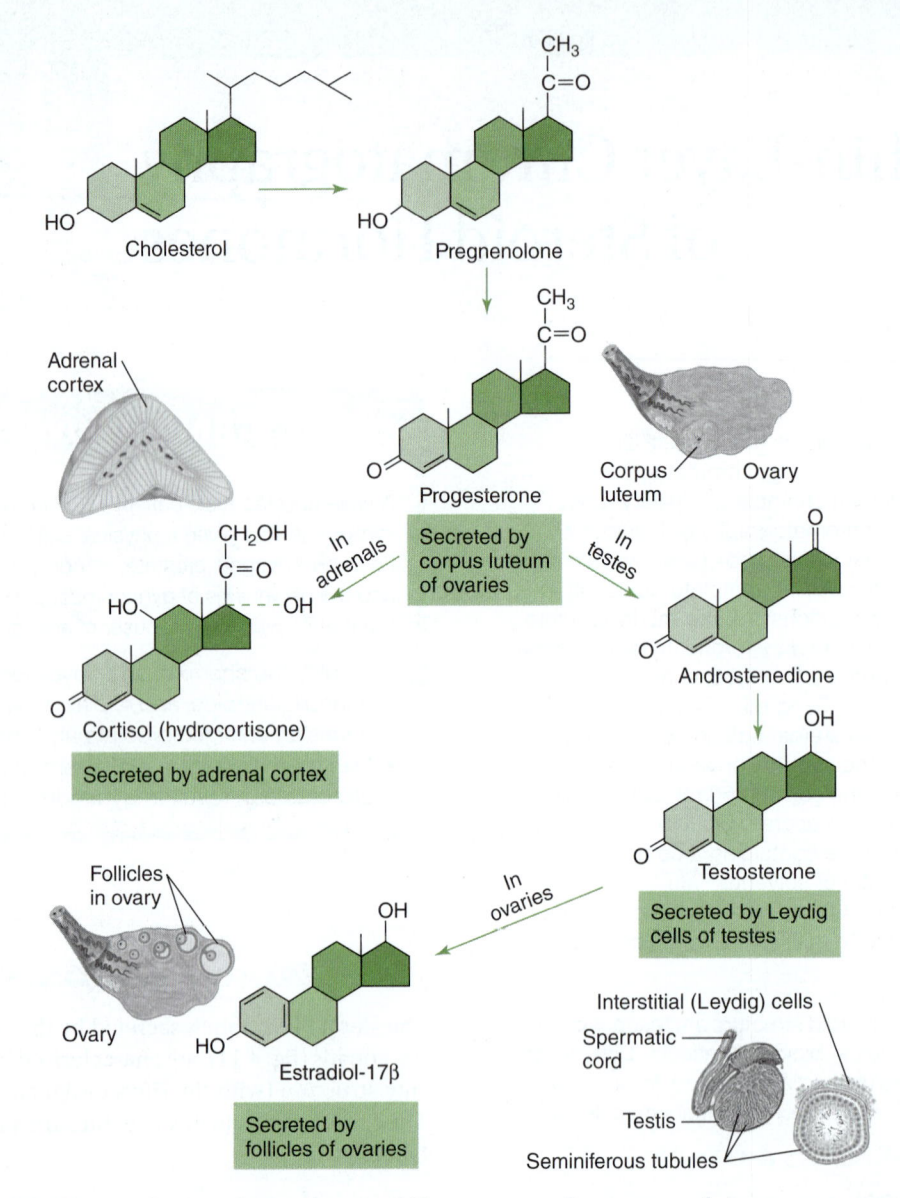

Figure 4.11 Simplified biosynthetic pathways for steroid hormones. Progesterone (a hormone secreted by the ovaries) is a common precursor of all other steroid hormones, and testosterone (the major androgen secreted by the testes) is a precursor of estradiol-17β, the major estrogen secreted by the ovaries.

(For a full-color version of this figure, see fig. 11.2 in *Human Physiology*, thirteenth edition, by Stuart I. Fox.)

most potent androgenic hormone is **testosterone,** secreted by the testes.

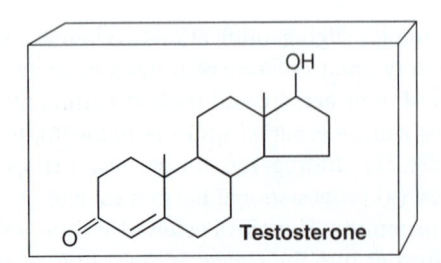

Although the primary source of androgens is the testes, the adrenal cortex also secretes small amounts (adrenal androgens are *dehydroepiandrosterone,* or *DHEA,* and *androstenedione*). Adrenal hyperplasia (Cushing's syndrome) and tumors of the adrenal cortex can also cause excessive levels of plasma androgen, which can have a masculinizing effect in females.

Testosterone and the other androgens are secreted by the interstitial Leydig cells of the testes. This secretion is stimulated by a gonadotropic hormone of the anterior pituitary, which is sometimes called *interstitial cell stimulating hormone* (*ICSH*) because it stimulates the interstitial Leydig cells of the testes. More commonly, it is referred to as *luteinizing hormone* (*LH*) because it is identical to the LH secreted by the anterior pituitary gland of females.

The structural difference between the androgens and **estrogens** is seemingly slight. The estrogens are eighteen-carbon steroids with three points of unsaturation (double

158

bonds, see appendix 1) in the A ring. These two categories of steroids, however, stimulate the development of markedly different (male and female) secondary sex characteristics. The chief estrogenic hormone is **estradiol.**

Estradiol

The estrogens are normally secreted in cyclically increasing and decreasing amounts by the ovaries, reaching a peak about the time of ovulation. The cyclical secretion of estrogens is stimulated, in turn, by the cyclical secretion of follicle-stimulating hormone (FSH), a gonadotropic hormone of the anterior pituitary. Abnormally high concentrations of circulating estrogenic hormones may be due to tumors of the adrenal cortex or the ovary. Excessive levels of these hormones can have a feminizing effect in males.

In the normal female cycle, estrogens stimulate growth and development of the endometrium (inner lining) of the uterus. The final maturation of the endometrium is under the control of the hormone **progesterone,** secreted by the corpus luteum of the ovaries during the luteal phase of the cycle (after ovulation). The cyclical secretion of progesterone is, in turn, stimulated by the cyclical secretion of luteinizing hormone (LH) from the anterior pituitary.

During pregnancy, the placenta secretes increasing amounts of progesterone in accordance with the development of the fetus. Progesterone is a twenty-one-carbon steroid.

Progesterone

The **corticosteroid hormones** are steroid hormones of the adrenal cortex. Also composed of twenty-one carbons, corticosteroids differ from progesterone by the presence of three or more oxygen groups. These hormones are divided into two functional classes, mineralocorticoids and glucocorticoids, secreted by two distinct regions (zona) of the adrenal cortex.

The **mineralocorticoids,** secreted by the zona glomerulosa of the adrenal cortex, are involved in the regulation of sodium and potassium balance. The most potent mineralocorticoids are **aldosterone** and, to a lesser degree, *deoxycorticosterone (DOC).* Secretion of aldosterone is stimulated by angiotensin II, which is regulated, in turn, by the secretion of renin from the kidneys.

Aldosterone

DOC

Clinical Applications

Anabolic steroids are synthetic androgens (male hormones) that promote protein synthesis in muscles and other organs. Use of these drugs by bodybuilders, weightlifters, and others is prohibited by most athletic organizations. Although administration of exogenous androgens does promote muscle growth, it can also cause a number of undesirable side effects. The liver and adipose tissue can change androgens into estrogens, so male athletes who take exogenous androgens often develop *gynecomastia*—an abnormal growth of female-like mammary tissue. High levels of exogenous androgens also inhibit the secretin of FSH and LH from the pituitary, causing atrophy of the testes and erectile dysfunction. The exogenous androgens also promote acne; aggressive behavior; male pattern baldness; and premature closure of the epiphyseal discs (growth plates in bones), stunting the growth of adolescents. Female users of exogenous androgens display masculinization and antisocial behavior. In both sexes, the anabolic steroids raise blood levels of LDL cholesterol (the "bad cholesterol") and triglycerides, while lowering the levels of HDL cholesterol (the "good cholesterol"), thus predisposing users to increased risk of heart disease and stroke.

An abnormal secretion of the mineralocorticoids is usually associated with hypertension and may be produced by primary aldosteronism or by secondary aldosteronism due to low blood sodium, high blood potassium, hypovolemia, cardiac failure, kidney failure, or cirrhosis of the liver.

The **glucocorticoids,** secreted by the zona fasciculata and the zona reticularis of the adrenal cortex, stimulate the breakdown of muscle proteins and the conversion of amino acids into glucose, a process termed *gluconeogenesis*. The secretions of the z. fasciculata and the z. reticularis are stimulated by adrenocorticotropin (ACTH) from the anterior pituitary. The most potent glucocorticoid in humans is **cortisol** (or **hydrocortisone**), whereas in many other mammals it is **corticosterone.** *Cortisone* and its analogues (such as Prednisone) are glucocorticoids used medically to inhibit inflammation and suppress the immune system.

An increased secretion of the glucocorticoids is associated with Cushing's syndrome (adrenal hyperplasia); pregnancy; and stress due to disease, surgery, and burns.

Each steroid has a different structure, so each will dissolve in a given solvent to a different degree. These differences in solubility will be used to separate and identify the steroids on a *thin-layer chromatography plate*.

The thin-layer plate consists of a thin layer of porous material (in this procedure, silica gel) coated on one side of a plastic, glass, or aluminum plate. The solutions of steroids are applied on different spots along a horizontal line near the bottom of the plate (a procedure called "spotting"), and the plate is placed on edge in a solvent bath *with the spots above the solvent.*

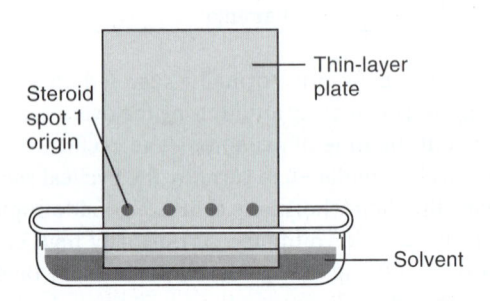

As the solvent creeps up the plate by capillary action, the steroids dissolve off their original spots (or *origins*) and are carried upward with the solvent toward the top end of the plate. Because the solubility of each steroid is different, the time required for the solvent to dissolve and transport the steroids will vary. The relative distance traveled by each steroid will also differ. If the process is halted before all the steroids have reached the top of the plate, some will have migrated farther from the origin than others.

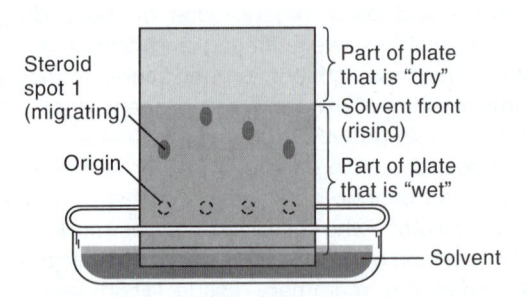

The final pattern (**chromatogram**) obtained using the same steroids, solvent, and conditions is predictable and reproducible. In other words, the distance that a given steroid migrates in a given solvent relative to the travel of the solvent front can be used as an identifying characteristic of that steroid. Each steroid can then be identified by a characteristic numerical value (R_f value) determined by calculating the distance the steroid traveled (D_s) relative to the distance traveled by the solvent front (D_f), as shown here:

$$R_f = \frac{\text{distance from origin to steroid spot } (D_s)}{\text{distance from origin to solvent front } (D_f)}$$

B. Thin-Layer Chromatography

In this exercise, an attempt will be made to identify two unknown steroids present in the same solution. To accomplish this task, you must (1) separate the two steroids and (2) identify these steroids by comparing their behavior with that of known steroids.

Using this method, an unknown steroid can be identified by comparing its R_f value in a given solvent with the R_f values of known steroids in the same solvent.

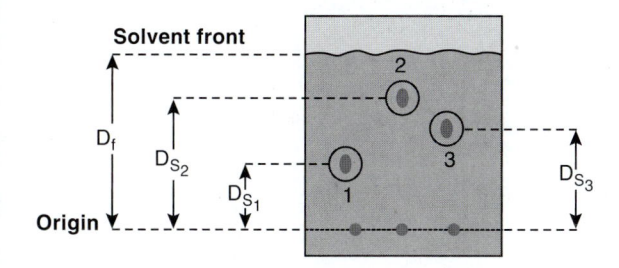

Clinical Applications

The chromatographic separation and identification of steroid hormones has revealed much about endocrine physiology. The placenta, for example, secretes estrogens more polar (more water-soluble) than estradiol, the predominant ovarian estrogen. These polar placental estrogens, *estriol* and *estetrol,* are now measured clinically during pregnancy to assess the health of the placenta.

Chromatograms of androgens recovered from their target tissues (such as the prostate) has revealed that these tissues convert testosterone into other active products. Further, these compounds appear to be more biologically active (more androgenic) than testosterone. Testosterone secreted by the testes, therefore, can be a *prehormone* that is enzymatically converted in target tissues into more active products, such as **dihydrotestosterone (DHT).** Conversion of testosterone to DHT is required for androgen stimulation of the prostate, for example; drugs used to treat an enlarged prostate work by blocking the enzyme (known as *5α-reductase*) that converts testosterone to DHT. Males who have a congenital deficiency of 5α-reductase exhibit some symptoms of androgen deficiency even though their testes secrete large amounts of testosterone.

Procedure

1. Using a pencil, make a tiny notch on the left margin of the thin-layer plate, approximately 1½ inches from the bottom. The origin of all the spots will lie on an imaginary line extending across the plate from this notch.

2. Using a capillary pipette, carefully spot steroid solution 1 (estradiol) about ½ inch in from the left-hand margin of the plate, along the imaginary line. Repeat this procedure, using the same steroid at the *same* spot, two more times. Allow the spot to dry between applications.

3. Repeat step 2 with each of the remaining steroid solutions (2, testosterone; 3, hydrocortisone; 4, cortisone; 5, corticosterone; 6, deoxycorticosterone; 7, unknown), spotting each steroid approximately ½ inch to the right of the previous steroid, along the imaginary line.

4. Observe the steroid spots at the origin under an ultraviolet lamp. (**Caution:** Do not look directly at the UV light.)

5. Place the thin-layer plates in a developing chamber filled with solvent (toluene/ethyl acetate/acetone, 6:1:1), and allow the chromatogram to develop for 1 hour. This should be done in a fume hood.

6. Remove the thin-layer plate, dry it, and observe it under the UV light. Using a pencil, outline the spots observed under the UV light.

7. In the laboratory report, record the R_f values of the known steroids and determine the steroids present in the unknown solution.

Laboratory Report 4.2

Name _____
Date _____
Section _____

DATA FROM EXERCISE 4.2

1. Record your data in this table and calculate the R_f value for each spot.

Steroid	Distance to Front	Distance to Spot	R_f Value
1. Estradiol			
2. Testosterone	same		
3. Hydrocortisone	same		
4. Cortisone	same		
5. Corticosterone	same		
6. Deoxycorticosterone	same		
7. **Unknown 1**	same		
8. **Unknown 2**	same		

2. The unknown solutions contained these steroid hormones:
   Unknown 1: _____.
   Unknown 2: _____.

REVIEW ACTIVITIES FOR EXERCISE 4.2

Test Your Knowledge

1. The chief estrogenic hormone is _____, secreted by the _____.
2. The chief androgenic hormone is _____, secreted by the _____.
3. Dehydroepiandrosterone is what kind of a hormone? _____ Which gland secretes it? _____.
4. Name two different endocrine glands that secrete progesterone: the _____ and the _____.
5. The major mineralocorticoid is _____. It is secreted by the _____ of the _____.
6. The major glucocorticoid in humans is _____. Another name for this hormone is _____. It is secreted by the _____ of the _____.
7. Name these steroids:
   (a) eighteen-carbon sex steroid _____.
   (b) nineteen-carbon sex steroid _____.
   (c) twenty-one-carbon sex steroid _____.

163

Test Your Understanding

8. Create an outline or flow chart of the categories and subcategories of the steroid hormones. Indicate the gland that secretes each hormone.

9. What are "anabolic steroids," and why do some athletes and bodybuilders illegally (and ill-advisedly) take them? Describe some of their side effects, and explain the mechanisms of how they can (a) stimulate growth of female-like breast tissue in men and (b) cause atrophy of the testes.

10. Which of the steroid hormones used in this exercise was most soluble in the solvent? Which was least soluble? Explain.

Test Your Analytical Ability

11. Suppose a man took a drug that acted as a 5α-reductase inhibitor. What effects might this drug have on the prostate? Explain.

12. Suppose the 5α-reductase inhibitor drug described in question 11 also caused hair to grow in a man with male-pattern baldness. (*Note:* There is such a drug—*Propecia.*) What could you conclude about the cause of male-pattern baldness? Explain.

13. The concentration of estriol and estetrol increases in the blood of pregnant women as the pregnancy progresses. How would the migration and R_f values of these two hormones compare with the migration and R_f value of estradiol in this chromatography exercise? Explain.

Test Your Quantitative Ability

14. Suppose that on a thin-layer chromatography plate the solvent front migrates a distance of 14 cm, and two steroid spots move distances of (a) 6 cm and (b) 12 cm. Calculate the R_f values for each steroid.

Clinical Investigation Questions

15. Which two natural hormones do anabolic steroids mimic? What physiological mechanisms could be responsible for an enlarged prostate and shrunken testes in men who use anabolic steroids?

16. What is gynecomastia? By what mechanisms could a user of anabolic steroids develop gynecomastia?

EXERCISE 4.3

Insulin Shock

MATERIALS

1. Large beaker filled with preaerated distilled water or spring water (dechlorinated)
2. Guppy, goldfish, or another fish of comparable size
3. Insulin (insulin, zinc—100 IU), glucose

Insulin stimulates the tissue uptake of blood glucose and thereby acts to lower the blood glucose concentration. Excessive insulin secretion can cause hypoglycemia, which, because of the brain's reliance on blood glucose, can affect brain function and even produce coma and death.

LEARNING OUTCOMES

You should be able to:

1. Describe the mechanism by which insulin regulates the blood glucose concentration.
2. Demonstrate the effects of excessive insulin on a small fish, and explain the clinical significance of these effects.

Textbook/Multimedia Correlations

Before performing this exercise, you should study the introductory material presented here. Further information relating to this exercise can be found in these pages of *Human Physiology*, thirteenth edition, by Stuart I. Fox:

- *Pancreatic Islets (Islets of Langerhans).* Chapter 11, p. 346.
- *Energy Regulation by the Pancreatic Islets.* Chapter 19, p. 677.
- *Diabetes Mellitus and Hypoglycemia.* Chapter 19, p. 681.

Clinical Investigation

A patient, in her fifties and significantly overweight, was told that she has diabetes due to a "relative insulin deficiency." When her blood tests revealed a high level of triglycerides, her physician stated that she was in danger of getting metabolic syndrome.

- Distinguish between type 1 and type 2 diabetes and determine which type the patient is likely to have.
- Describe the physiological function of insulin and what happens when there is an insulin deficiency.
- Describe metabolic syndrome and its dangers.

Insulin is a polypeptide hormone secreted by the *beta cells of the pancreatic islets (islets of Langerhans)*. Insulin stimulates the transport of glucose from the blood into the muscles, liver, and adipose tissue, thus lowering the blood glucose concentration. The effects of insulin injection on the blood glucose concentration of humans is depicted in figure 4.12.

Insulin acts on the skeletal muscles, liver, and adipose tissue to promote the uptake of blood glucose by stimulating the insertion of GLUT4 carriers into the plasma membrane (fig. 4.13). These carrier proteins are part of the membranes of intracellular vesicles, and when the carriers are inserted into the plasma membrane they permit the passive, facilitative diffusion of glucose from the plasma into the cytoplasm of these cells. In this way, an increased secretion of insulin causes a lowering of the blood glucose concentration.

When the islets are incapable of secreting an adequate amount of insulin, a condition known as *diabetes mellitus* develops in which the transport of glucose from the blood into the body tissues is impaired. This results in an increase in the blood sugar level *(hyperglycemia)* and the appearance of glucose in the urine *(glucosuria)*.

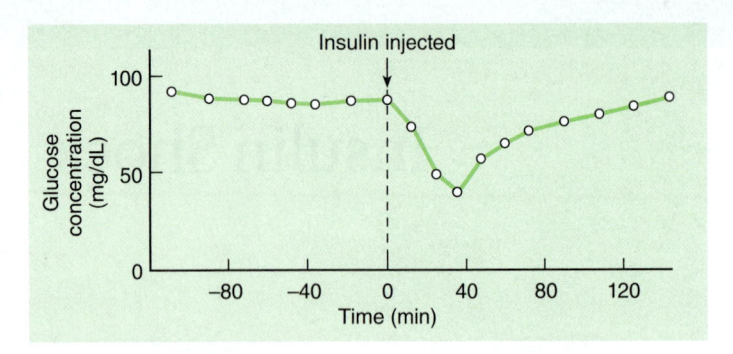

Figure 4.12 Homeostasis of the blood glucose concentration. Average blood glucose concentrations of five healthy individuals are graphed before and after a rapid intravenous injection of insulin. The "0" indicates the time of the injection. Following injection of insulin, the blood glucose is brought back up to the normal range. This occurs as a result of the action of hormones antagonistic to insulin, which cause the liver to secrete glucose into the blood. In this way, homeostasis is maintained.

(For a full-color version of this figure, see fig. 1.5 in *Human Physiology*, thirteenth edition, by Stuart I. Fox.)

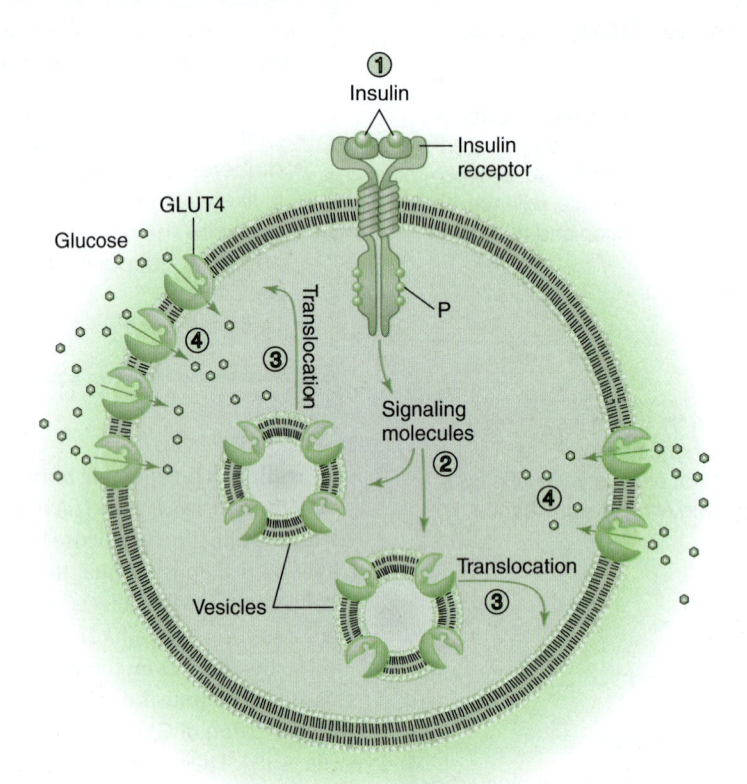

Figure 4.13 Insulin stimulates uptake of blood glucose. (1) Binding of insulin to its plasma membrane receptors causes the activation of cytoplasmic signaling molecules, which (2) act on intracellular vesicles that contain GLUT4 carrier proteins in the vesicle membrane. (3) This causes the intracellular vesicles to translocate and fuse with the plasma membrane, so that the vesicle membrane becomes part of the plasma membrane. (4) The GLUT4 proteins permit the facilitated diffusion of glucose from the extracellular fluid into the cell.

(For a full-color version of this figure, see fig. 11.30 in *Human Physiology*, thirteenth edition, by Stuart I. Fox.)

Under these conditions, the body tissues cannot obtain sufficient glucose for cellular respiration and increasingly rely on the metabolism of fat for energy. The intermediate products of fat metabolism (ketone bodies) accumulate in the blood, resulting in *ketoacidosis*. Because one of these products is a volatile compound called *acetone* (which has a fruity odor), a person with this condition has fruity-smelling breath. The excretion of ketone bodies and glucose in the urine is accompanied by the excretion of large amounts of water, causing dehydration. The combination of acidosis and dehydration that results from insufficient insulin may produce a diabetic coma.

People with **type 1 diabetes mellitus** must be given insulin injections to maintain blood glucose homeostasis. This is because there is a lack of insulin secretion due to the progressive destruction of the beta cells by the person's own immune system; type 1 diabetes is an autoimmune disease. If a person with this type of diabetes is given too much insulin, however, the blood sugar level will fall below normal (*hypoglycemia*). Because the central nervous system can only use plasma glucose for energy, the lowering of blood sugar essentially starves the brain. The ensuing condition is called *insulin shock*. The symptoms of insulin shock can be illustrated by the reaction of fish to insulin in this exercise.

Hypoglycemia can, however, have other causes and can result in symptoms less severe than those accompanying insulin shock. *Reactive hypoglycemia* may occur after a meal if the beta cells secrete excessive insulin in response to carbohydrates in the digested food. This type of hypoglycemia sometimes occurs in the beginning stages of diabetes mellitus and can often be controlled by eating smaller, more frequent meals lower in carbohydrates. For reasons not well understood, hypoglycemia can also occur as a result of alcohol ingestion. Other causes of hypoglycemia include tumors of the beta cells (insulinomas), which secrete excessive insulin, and liver diseases in which the ability to produce glucose from glycogen and noncarbohydrate molecules is impaired.

The symptoms of hypoglycemia appear when the blood glucose concentration is about 45 mg/dL (normal serum glucose concentration ranges from 70 to 100 mg/dL). Symptoms can appear, however, at higher glucose concentrations if the cerebral circulation is impaired, such as observed in elderly people with atherosclerosis. The symptoms of glucose deficiency—faintness, weakness, nervousness, hunger, muscular trembling, and tachycardia—are symptoms similar to those seen when the brain lacks sufficient oxygen. More prolonged hypoglycemia may damage other parts of the brain, resulting in behavior resembling neuroses and psychoses. Indeed, severe brain damage may result in coma and death.

Type 1 diabetes mellitus is caused by an insulin deficiency, whereas **type 2 diabetes mellitus** is associated with a decreased tissue sensitivity to insulin, called an *insulin resistance*. In people with type 2 diabetes, the islet beta cells fail to secrete adequate amounts of insulin to compensate for the insulin resistance. For this reason, there can be hyperglycemia despite even elevated levels of insulin secretion. As the disease progresses, there can be both insulin resistance and a reduced secretion of insulin. Type 2 diabetes is far more common than type 1 (table 4.2). It is generally associated with obesity, which increases insulin resistance, and can usually be controlled by moderate weight loss and exercise. If these are insufficient, medications are available that reduce the insulin resistance of the target tissue (muscle, liver, and adipose tissue) and have other actions that improve insulin's effectiveness.

Gestational diabetes occurs in about 4% of pregnancies due to insulin secretion that is inadequate to meet the increased demand imposed by the fetus. In women who do not develop gestational diabetes, the increased insulin requirements during pregnancy are normally met by an increased insulin secretion. This is believed to be due to the normal proliferation of beta cells in the mother's pancreatic islets during pregnancy.

Table 4.2 Comparison of Type 1 and Type 2 Diabetes Mellitus

Feature	Type 1	Type 2
Usual age at onset	Under 20 years	Over 40 years
Development of symptoms	Rapid	Slow
Percentage of diabetic population	About 10%	About 90%
Development of ketoacidosis	Common	Rare
Association with obesity	Rare	Common
Beta cells of islets (at onset of disease)	Destroyed	Not destroyed
Insulin secretion (at onset of disease)	Decreased	Normal or increased
Autoantibodies to islet cells	Present	Absent
Treatment	Insulin injections	Diet and exercise; oral stimulators of insulin sensitivity

Clinical Applications

Obesity, especially involving visceral fat, promotes insulin resistance and type 2 diabetes. These are also often associated with *hypertension* (high blood pressure) and *dyslipidemia:* high levels of blood triglycerides and low levels of HDL (the carrier proteins that transport cholesterol from arteries to the liver). This promotes a greater risk of atherosclerosis and cardiovascular disease. When a person has *central obesity* (defined by a waist circumference greater than specific values that vary by sex and ethnicity), hypertension, and other conditions described above, the condition is termed **metabolic syndrome.** Metabolic syndrome may raise the risk of coronary heart disease and stroke by a factor of three, and is a growing health concern in the United States.

Procedure

1. Place a small fish (guppy or goldfish) in a large beaker of water to which a few hundred units of insulin have been added.
2. Observe the effects of insulin overdose. (If no effects are seen in 30 minutes, repeat this step with another fish.)
3. Remove the fish to a second beaker of water containing 5% glucose. Observe the recovery.

Laboratory Report 4.3

Name _____

Date _____

Section _____

REVIEW ACTIVITIES FOR EXERCISE 4.3

Test Your Knowledge

1. Insulin is secreted by the _____ cells of the _____.
2. Insulin stimulates the _____.
3. Insulin acts to _____ the blood glucose concentration.
4. A deficiency of insulin causes _____ (term for blood glucose concentration); by contrast, an excess of insulin causes _____.
5. The disease in which there is inadequate action of insulin is called _____.
6. A person with the disease named in question 5 would, if untreated, have _____ (a term describing urine).

Test Your Understanding

7. Explain the meaning of *insulin shock* and why insulin shock is dangerous.

8. Explain why a person with type 1 diabetes mellitus may have a "fruity" breath. Why is this person in danger of dehydration?

Test Your Analytical Ability

9. Explain why a person with type 1 diabetes mellitus can lose weight rapidly.

10. The most common type of diabetes mellitus is type 2. In this condition, the person has beta cells that secrete insulin, sometimes even in high amounts. Despite this, the person can still have hyperglycemia. Explain how this can occur. Most people with type 2 diabetes mellitus are overweight. What does that suggest about large adipocytes compared to small adipocytes?

Clinical Investigation Questions

11. Which type of diabetes did the patient likely have? Explain.

12. What is the physiological function of insulin? How does a deficiency of insulin affect body function and health?

13 What are the symptoms of metabolic syndrome and what are its dangers?

Section 5

Skeletal Muscles

The basic mechanism of contraction for **striated muscles** (*skeletal* and *cardiac* muscles) can be divided into three phases: (1) electrical excitation of the muscle cell; (2) excitation-contraction coupling; and (3) sliding of the muscle filaments, or contraction.

At rest, there is a *potential difference* across the *muscle fiber* plasma membrane equal to approximately −80 mV (millivolts). The negative sign indicates that the inside of the membrane is negatively charged in comparison to the outside of the cell. When the muscle fiber is appropriately stimulated, either by a direct electric shock or by the motor nerve that innervates the muscle, the permeability of the membrane to cations changes. The diffusion of Na^+ into the cell *depolarizes* the membrane and its polarity momentarily reverses. This is immediately followed by the outward diffusion of K^+, which *repolarizes* the membrane and reestablishes the resting membrane potential. This rapid depolarization and repolarization of the membrane at the stimulated point is called an **action potential.**

Action potentials are conducted by the plasma membrane of skeletal muscle fibers and myocardial cells in much the same way that action potentials are conducted by unmyelinated axons. Action potentials produced by muscle cells stimulate a rise in the cytoplasmic concentration of Ca^{2+}. In skeletal muscles, this Ca^{2+} comes from a system of intracellular tubules called the *sarcoplasmic reticulum.* In the resting muscle, the absence of this calcium allows the protein *tropomyosin,* located in the thin filaments (described later), to inhibit contraction. As a result of action potentials produced by the muscle fiber, Ca^{2+} is released from the sarcoplasmic reticulum. This Ca^{2+} enters the cytoplasm and binds to a regulatory protein, *troponin,* in the thin filaments attached to tropomyosin. This causes the tropomyosin to change its position in the thin filaments so that it can no longer inhibit contraction. When this occurs the muscle fiber contracts. Because the binding of troponin to Ca^{2+} leads to the movement of tropomyosin and thereby contraction, and action potentials are needed for Ca^{2+} to be available in the muscle cytoplasm, Ca^{2+} is said to *couple electrical excitation to muscle contraction.*

Within the muscle fiber, there are numerous subunits *(fibrils)* oriented parallel to the long axis of the fiber (fig. 5.1*a*). Each fibril, in turn, is composed of numerous repeating subunits called **sarcomeres** (fig. 5.1*b*). The sarcomere is the functional unit of contraction. When contraction is stimulated by Ca^{2+}, the **thick** and **thin filaments** (composed of **myosin** and **actin,** respectively), within the sarcomeres slide over one another. This sliding of the filaments allows each sarcomere to shorten while its filaments remain the same length. As the sarcomeres become shorter, the fibrils, and thus the entire muscle fiber, shorten, resulting in muscle contraction (fig. 5.1*c*). This is known as the **sliding filament theory** of contraction.

Exercise 5.1   Neural Control of Muscle Contraction

Exercise 5.2   Summation, Tetanus, and Fatigue

Exercise 5.3   Electromyogram (EMG)

Figure 5.1 The sliding filament model of muscle contraction. (a) The structure of a muscle fiber. (b) The structure of a sarcomere, showing the banding patterns produced by the thick and thin filaments. (c) The changes in band patterns as the muscle contracts. Notice that the Z discs are brought closer together. The A bands remain the same length during contraction, but the I and H bands become progressively narrower and may eventually disappear.

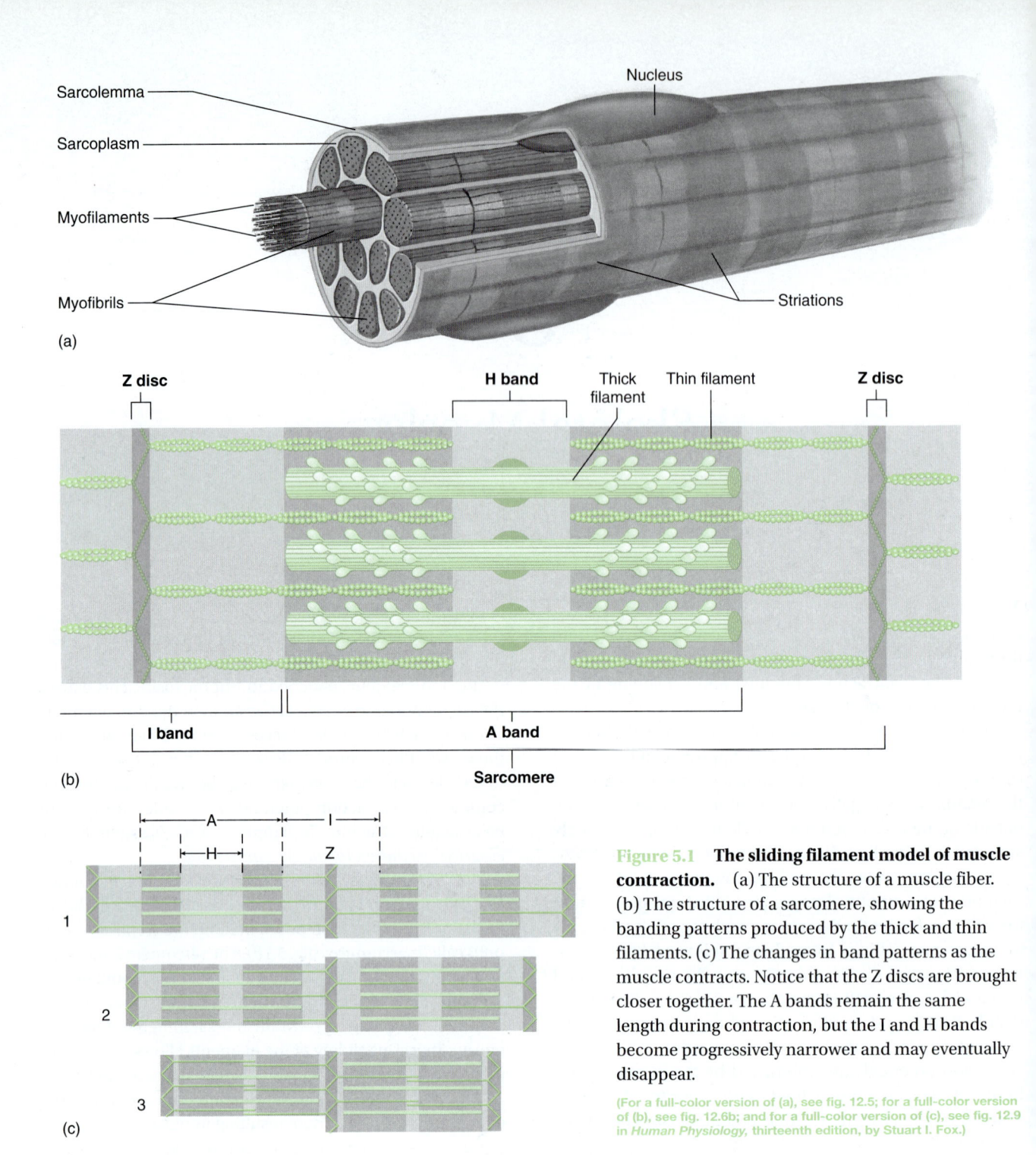

(For a full-color version of (a), see fig. 12.5; for a full-color version of (b), see fig. 12.6b; and for a full-color version of (c), see fig. 12.9 in *Human Physiology*, thirteenth edition, by Stuart I. Fox.)

EXERCISE 5.1

Neural Control of Muscle Contraction

MATERIALS

1. Frogs
2. Surgical scissors, forceps, sharp probes, dissecting trays, glass probes
3. Recording equipment (either kymograph or electrical recorder, such as physiograph) and electrical stimulators
4. As an alternative to pen-and-paper recording equipment, a computerized data acquisition and analysis system, such as those provided by Biopac, iWorx, and others, may be used to perform the exercises in this section
5. Straight pins (bent into a "Z" shape), or thin-wired fish hooks (barbless or debarbed) may be used instead; thread
6. Bone clamp (if kymograph is used) or myograph transducer (if physiograph is used)
7. Frog Ringer's solution (dissolve 6 g of NaCl, 0.075 g of KCl, 0.10 g of $CaCl_2$, and 0.10 g of $NaHCO_3$ in a liter of water)
8. Alternatively, simulated skeletal muscle exercises may be performed using the *Physiology Interactive Lab Simulations (Ph.I.L.S.):* Skeletal Muscle Function (Exercises 1, 2, and 3)

Isolated muscles from a pithed frog can be used to study the physiology of muscle contraction. Isolated frog muscles can be stimulated directly by an electric shock and indirectly through the activation of the appropriate motor nerve.

LEARNING OUTCOMES

You should be able to:

1. Prepare a pithed frog for the study of muscle physiology.
2. Describe how muscle contraction can be stimulated by a direct electric shock.
3. Explain how motor nerves stimulate the contraction of skeletal muscles.

Textbook/Multimedia Correlations

Before performing this exercise, you should study the introductory material presented here. Further information relating to this exercise can be found in these pages of *Human Physiology*, thirteenth edition, by Stuart I. Fox:

- *Skeletal Muscles.* Chapter 12, p. 360.
- *Regulation of Contraction.* Chapter 12, p. 370.
- *Neural Control of Skeletal Muscles.* Chapter 12, p. 385.
- Multimedia correlations are provided in *MediaPhys 3.0:* Topic 5.10; and *Physiology Interactive Lab Simulations (Ph.I.L.S.):* Skeletal Muscle Function (Exercises 1, 2, and 3).

Clinical Investigation

A patient complained of droopiness in her forehead and right eyelid. When questioned, she mentioned that she recently had Botox injections.

- Explain how nerves stimulate skeletal muscles to contract.
- Identify the nature of Botox, and explain how Botox injections reduce facial wrinkles but can also cause the side effects reported by this patient.

RECORDING PROCEDURES

The **physiograph** is a device for recording the mechanical aspects of muscular contraction (fig. 5.2). It is very sensitive because the mechanical movements of the muscle are first transduced (converted) into electrical current and then greatly amplified prior to recording. Mechanical events with different energies (from muscle contraction to sound waves) can also be recorded, as can nerve impulses and other primarily electrical activity, such as the *electrocardiograph* (ECG), *electromyograph* (EMG), or the *electroencephalograph* (EEG) recordings. Because a number of physiological

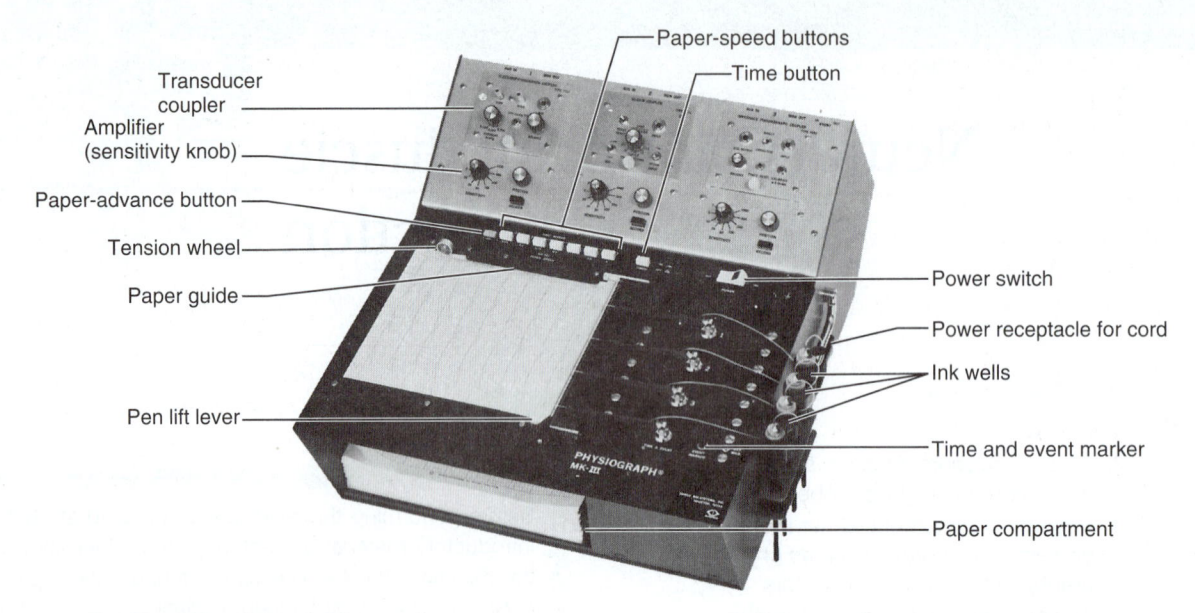

Figure 5.2 The Physiograph Mark III recorder.

parameters can be simultaneously recorded on different channels of the physiograph, the temporal (time) relationship between events can be studied.

The physiograph consists of four basic parts: (1) the **transducer** changes the original energy of the physiological event into electrical energy; (2) the **coupler** makes the input energy from the transducer compatible with the built-in amplifier (fig. 5.3); (3) the **amplifier** then increases the strength of the electrical current and forwards the signal to a galvanometer; (4) the **galvanometer** responds to the current generated by directing movement of a pen. The movement of the pen is proportional to the strength of the electrical current generated by the physiological event being measured. Because recording paper moves continuously at a known speed under the pen, both the *frequency* (number per unit time) and the *strength* (amplitude of the pen deflection from a baseline) of the physiological event can be continuously recorded (fig. 5.2).

Similar principles are involved when the physiological event (such as the contraction of a frog skeletal muscle or heart) is transduced into an electric current, and this is now fed into a computer rather than into a pen-and-paper recorder. Unlike the pen-and-paper recorder, the computer not only displays the data but also analyzes it. **Computerized data acquisition and analysis** is usually performed with particular commercial products, such as those produced by Biopac and iWorx. These companies provide exercises comparable to some in this laboratory guide, and these can be used in conjunction with this guide as determined by the instructor.

For Physiograph Recording

1. Insert the *transducer coupler* into the physiograph and plug the *myograph transducer* into the coupler (fig. 5.3).*
2. Raise the inkwells and lower the pens onto the paper by lowering the pen lifter (fig. 5.2). With an index finger covering the hole on the rubber bulb, squeeze to force ink into each pen.
3. Turn the physiograph on with the rocking power switch. Set the paper speed by depressing the *paperspeed* button marked 0.5 cm per second. Turn on the paper drive by depressing the *paper-advance* button and releasing it, allowing it to rise (the *up* position is on).
4. Move the *time switch* to "on." The bottom pen, labeled *time & event*, will make upward deflections every second.

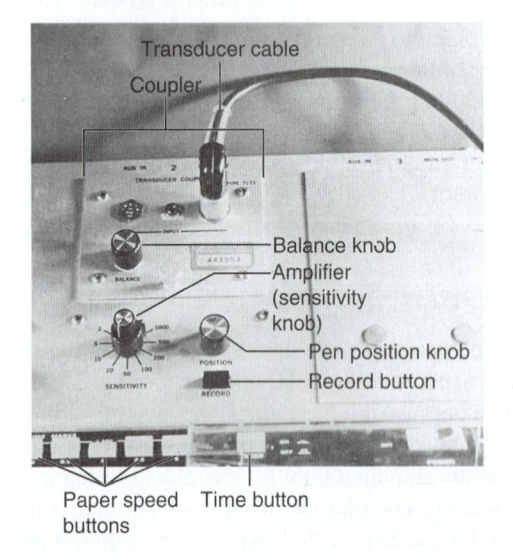

Figure 5.3 **The Narco Mark III Physiograph with an inserted transducer coupler.** The transducer coupler is connected by means of a cable to a myograph transducer.

---
*For Physiograph Mark III, Narco Bio-Systems.

Note: *At a paper speed of 0.5 cm/sec, these deflections will be separated by a distance equal to the width of one small box on physiograph recording paper. If the paper speed is increased to 1.0 cm/sec, the deflections of the time-and-event pen will be two small boxes apart.*

5. Turn the outer knob of the *amplifier sensitivity* control to its lowest number (this will be its greatest sensitivity) (fig. 5.3). With the *record* button off (in the up position), adjust the position of the recording pen for the appropriate channel with the *position* knob, so that the pen writes on the heavy horizontal line closest to the center of the channel being recorded. This position can later be lowered to allow the pen a greater excursion for a stronger signal.
6. Depress the *record* button (the *down* position is on), causing the pen to move away from the heavy horizontal line. Bring the pen back to the line by rotating the *balance* knob. The pen should now remain on the heavy line whether or not the record button is depressed and regardless of the setting of the sensitivity control.

For Computerized Recording

1. If the **Biopac** system is used, their force transducer (SS12LA), stimulator (BSLSTM), and amplifier (MP30 or MP35) can be connected to each other and to the computer as illustrated in figure 5.4. A close-up view of the MP30 connections is shown in figure 5.5.
   a. You can follow the Biopac Setup procedure as outlined in their BSL *PRO* Lesson A02 (available on the Internet).
   b. A different manual stimulator may be used instead of the Biopac stimulator; if a different stimulator is used, you will not get a recording of the stimulation, only of the muscle responses. In such a case, channel 1 will not be connected.
   c. Once the equipment is set up, the exercises can be performed separately or in conjunction with the Biopac *PRO* lesson.
2. If the **iWorx** system is used (fig. 5.6), the iWorx data acquisition unit, cable, and displacement transducer should be connected to each other and to the computer as shown in iWorx Experiments 7 and 8 (available on the Internet). The exercises in this section can then be performed separately or in conjunction with iWorx Experiments 7 and 8.

A. Frog Muscle Preparation and Stimulation

To study the physiology of frog muscle and nerve, the frog must be killed but its tissues kept alive. This can be accomplished by destroying or **pithing** the frog's central nervous system. The frog is clinically dead (clinical death is defined as the irreversible loss of higher brain function), but its muscles and peripheral nerves will continue to function as long as their cells remain alive. Under the proper conditions this state can be maintained for several hours.

There are two techniques for pithing a frog. One technique involves grasping the frog securely with one hand and

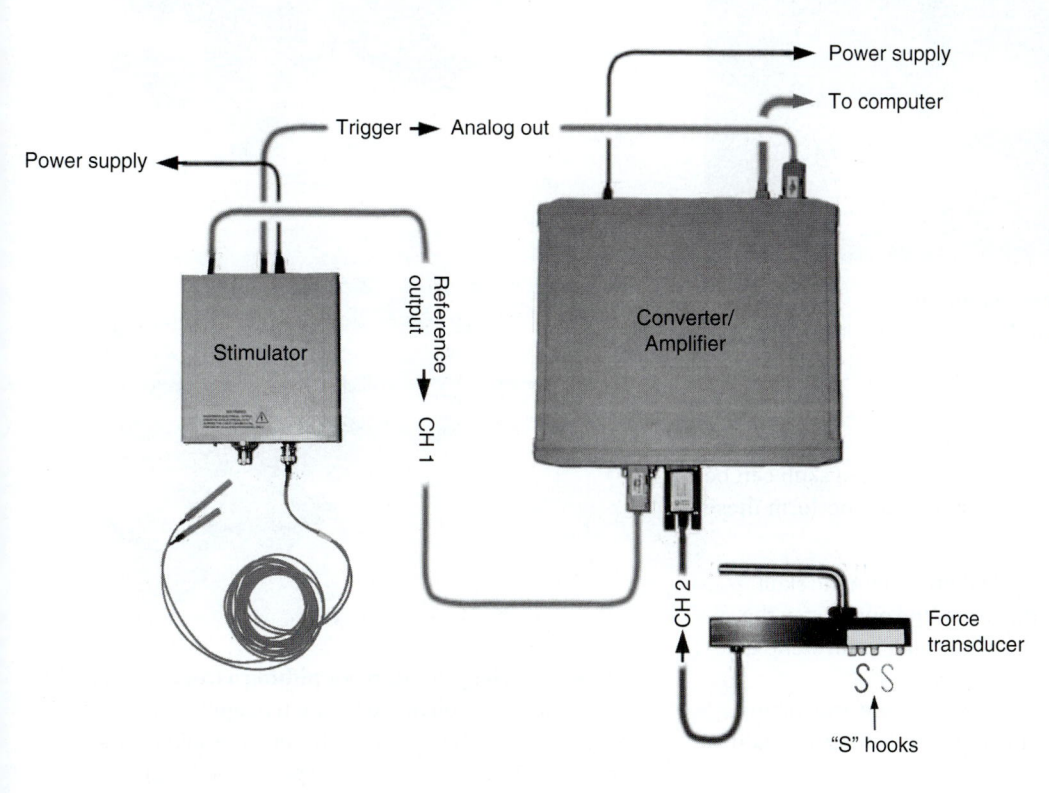

Figure 5.4 Connection of hardware for use of Biopac system with frog muscle exercises. The details of the setup are described in the Biopac *PRO* Lesson A02. Once the equipment is set up, it can be used to perform the exercises in this section.

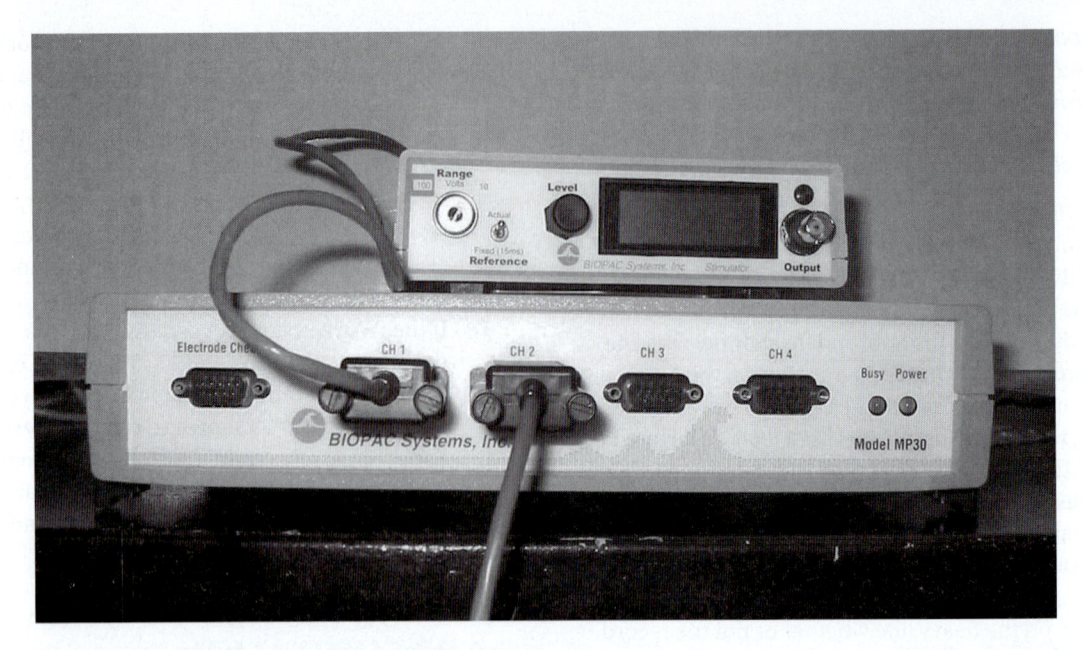

Figure 5.5 Connections to Biopac amplifier (MP30). The force transducer (not shown) is plugged into channel 2, and the stimulator is plugged into channel 1. (If a different stimulator is used, it need not be plugged into channel 1.) If the Biopac stimulator is used (shown on top of the MP30), its "trigger" connector (not shown) must be plugged into the back of the MP30 or MP35 (see the diagram in fig. 5.4). The output of the MP30 or MP35 is connected to a computer for data visualization and analysis.

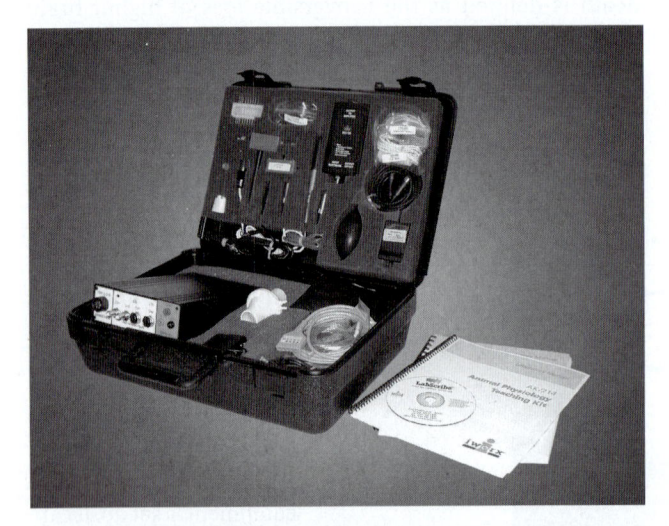

Figure 5.6 The iWorx system components. Together with their force transducer and stimulator, this system can be set up as described in iWorx Experiments 7 and 8 and used to perform the exercises in this section.

flexing its head forward so that the base of the skull can be felt with the fingers of the other hand. Then perform these steps:

1. Quickly insert a sturdy metal probe into the skull through the foramen magnum (the opening in the skull where the spinal cord joins the brain stem) as in figure 5.7a.
2. Move the probe around in the skull, destroying the brain and preventing the frog from feeling any pain (it is now clinically dead).

(a)

(b)

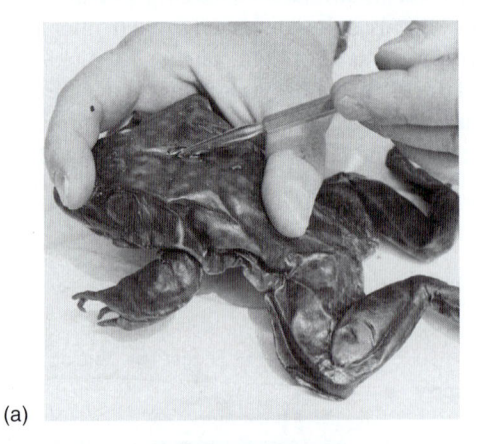

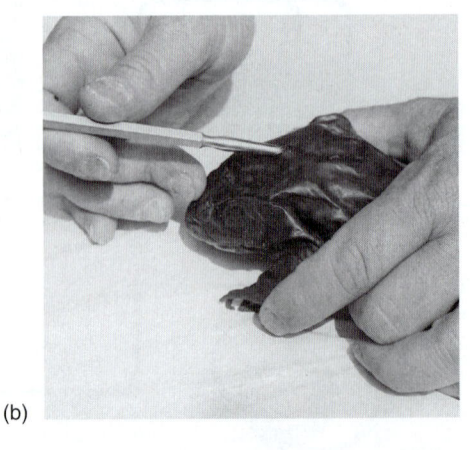

Figure 5.7 The procedure for pithing a frog. (a) A probe is first inserted through the foramen magnum into the skull. (b) The probe is then inserted through the spinal cord. (Procedure was simulated with a preserved frog.)

3. Keeping the head flexed, partially withdraw the probe and turn it so that it points toward the hind end of the frog. Insert the probe downward into the spinal canal (fig. 5.7b), destroying the spinal cord and its reflexes. The frog's legs will straighten out as the inserted probe causes reflex stimulation of the spinal nerves. When the spinal cord is destroyed, the frog will become limp.

Alternatively, the following procedure may be employed. Force one blade of a pair of sharp scissors into the frog's mouth as shown in figure 5.8a. Quickly decapitate the frog by cutting behind its eyes. The frog is dead as soon as its brain has been severed from its spinal cord. Insert a probe down into the exposed spinal cord, as previously described, to destroy the frog's spinal reflexes (fig. 5.8b).

After the frog has been pithed, completely skin one of its legs to expose the underlying muscle. Discard the skin. Then run one blade of a pair of scissors under the Achilles tendon and cut it, leaving part of the tendon still attached to the gastrocnemius muscle (fig. 5.9).

(a)

(b)

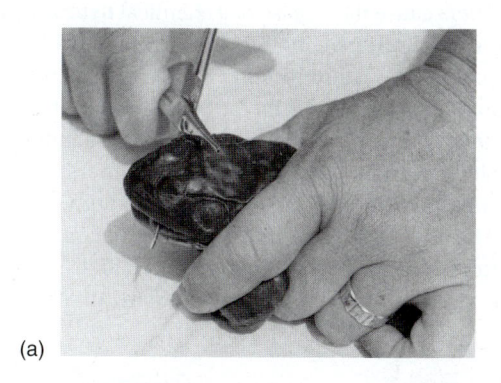

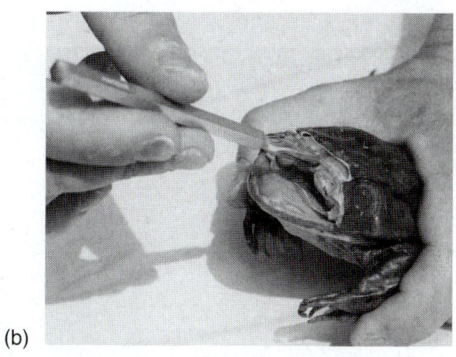

Figure 5.8 **An alternate pithing procedure.** (a) The frog is first decapitated. (b) A probe is then inserted into the spinal cord. (Procedure was simulated with a preserved frog.)

Procedure

SETUP FOR DIRECT STIMULATION OF THE MUSCLE

1. Secure the frog to a dissecting tray by inserting sharp pins through the arms and legs.
2. Push a Z-bent pin through the Achilles tendon. Tie one end of a cotton thread to the bent pin and the other end to the hook of a force transducer. Position the myograph so that it is directly above the muscle, and adjust the height of the force transducer so that the muscle is under tension. Insert two stimulating electrodes directly into the gastrocnemius muscle (fig. 5.9).
3. At this time, you can also set up the frog sciatic nerve for stimulation. To do this, follow steps 1 and 2 on page 182.
4. Determine the **threshold stimulus.** This is the minimum stimulus that will produce a response on the recorder (a muscle *twitch*). First, determine the threshold stimulus when both recording electrodes are placed in the muscle.

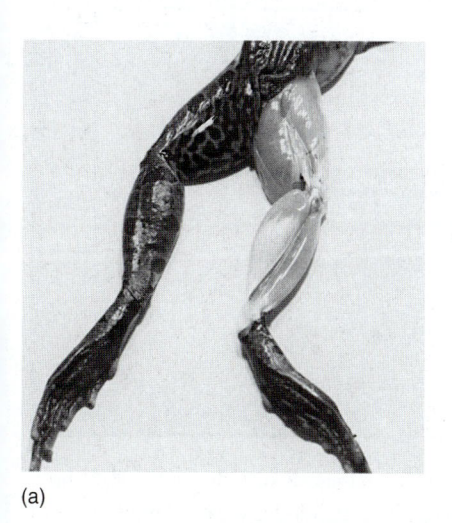

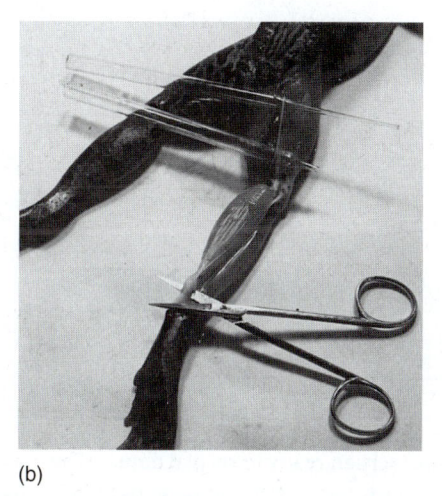

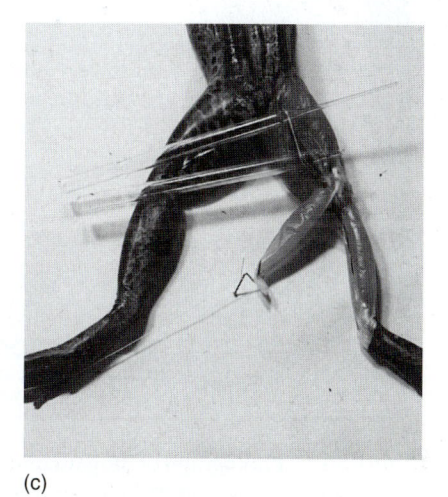

(a) (b) (c)

Figure 5.9 **Preparation of gastrocnemius.** After the frog's leg is skinned (a), the Achilles tendon is cut (b) and a bent pin is inserted into it (c). A length of thread attaches this pin to the hook of the myograph transducer (not shown). In (b), the sciatic nerve is shown between two glass probes in preparation for exercise 5.1.

a. Deliver shocks to the muscle at gradually increasing voltages, starting at a voltage too low to produce a contraction. When a contraction is first produced, record that voltage in your laboratory report. This can be performed with a stimulator produced by any manufacturer.
b. If the Biopac stimulator is used, the shock can be delivered by manually pressing a button or by using the stimulator dialog box (fig. 5.10). The contraction is recorded on the bottom half of the computer screen (fig. 5.10). This portion can be enlarged by clicking and dragging on the separation bar.
c. Recording with the Biopac system is started by pressing the start button on the bottom right portion of the computer screen.

5. Next, determine the threshold stimulus when both recording electrodes are placed on the sciatic nerve. This is described in the next section.

Note: Rinse the muscle periodically with Ringer's solution (a salt solution balanced to the extracellular fluid of the frog). Do not allow the muscle to dry.

B. STIMULATION OF A MOTOR NERVE

In the body *(in vivo),* skeletal muscles are stimulated to contract by somatic motor nerves. Action potentials in the motor nerve fibers cause the release of a chemical neurotransmitter called **acetylcholine (ACh)** from the axon endings. This transmitter combines with a receptor protein in the muscle cell membrane (a "nicotinic" subtype of ACh receptor) and stimulates the production of new action potentials in the muscle fiber. Electrical stimulation of the muscle fibers causes Ca^{2+} to be released from the sarcoplasmic reticulum (fig. 5.11).

The Ca^{2+} released into the sarcoplasm binds to a regulatory protein called **troponin,** which is together with a protein known as **tropomyosin** attached to the

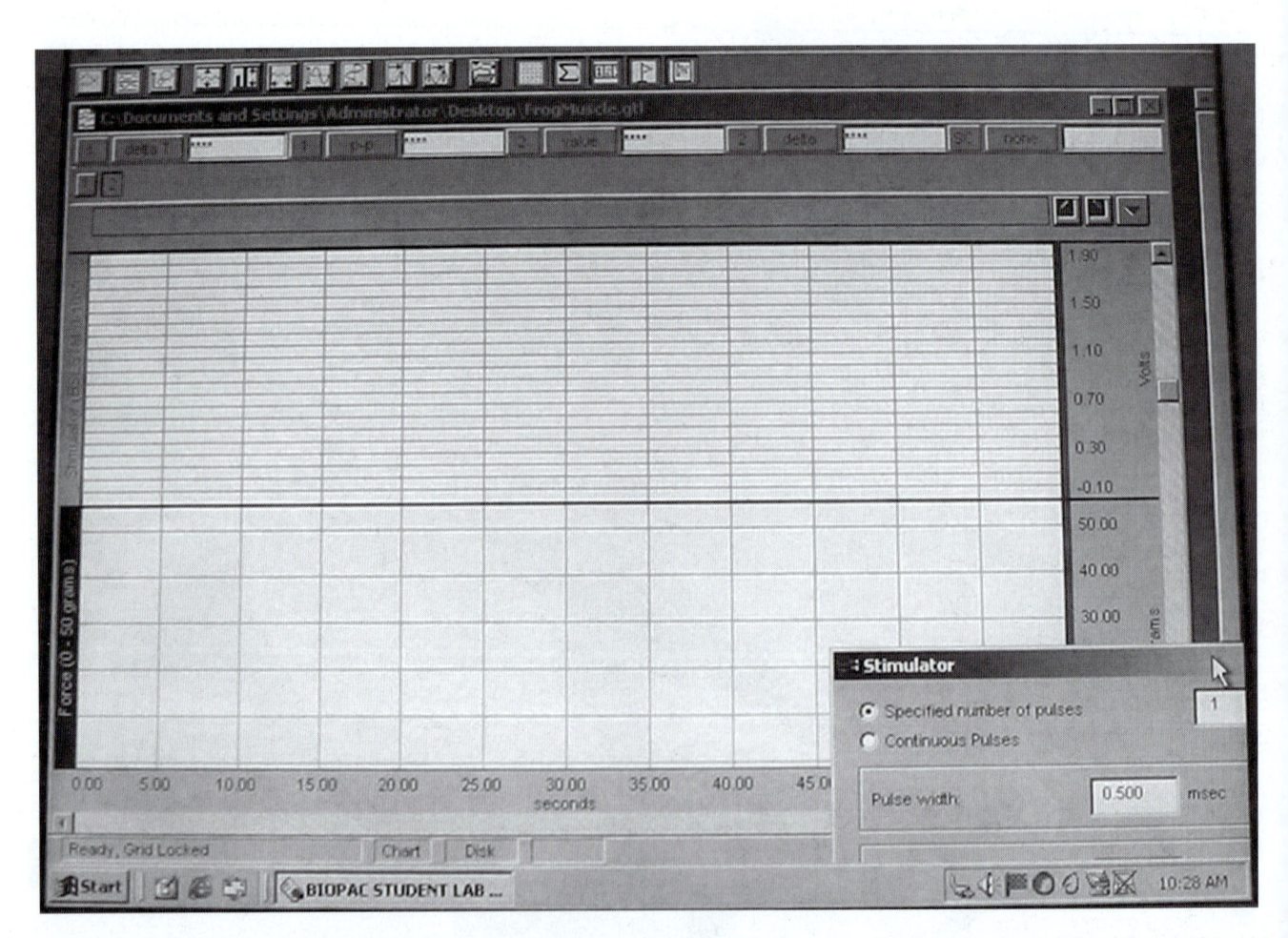

Figure 5.10 **Photograph of Biopac computer screen ready to record data.** The upper portion records stimulus voltages, if the Biopac stimulator is used. The lower portion records the muscle contractions (this portion can be expanded, if desired) regardless of the nature of the stimulator used. The pop-up screen in the lower right allows control of the Biopac stimulator. This screen should be dragged and dropped away from that position to access the "start" button for recording, hidden behind the stimulator pop-up screen.

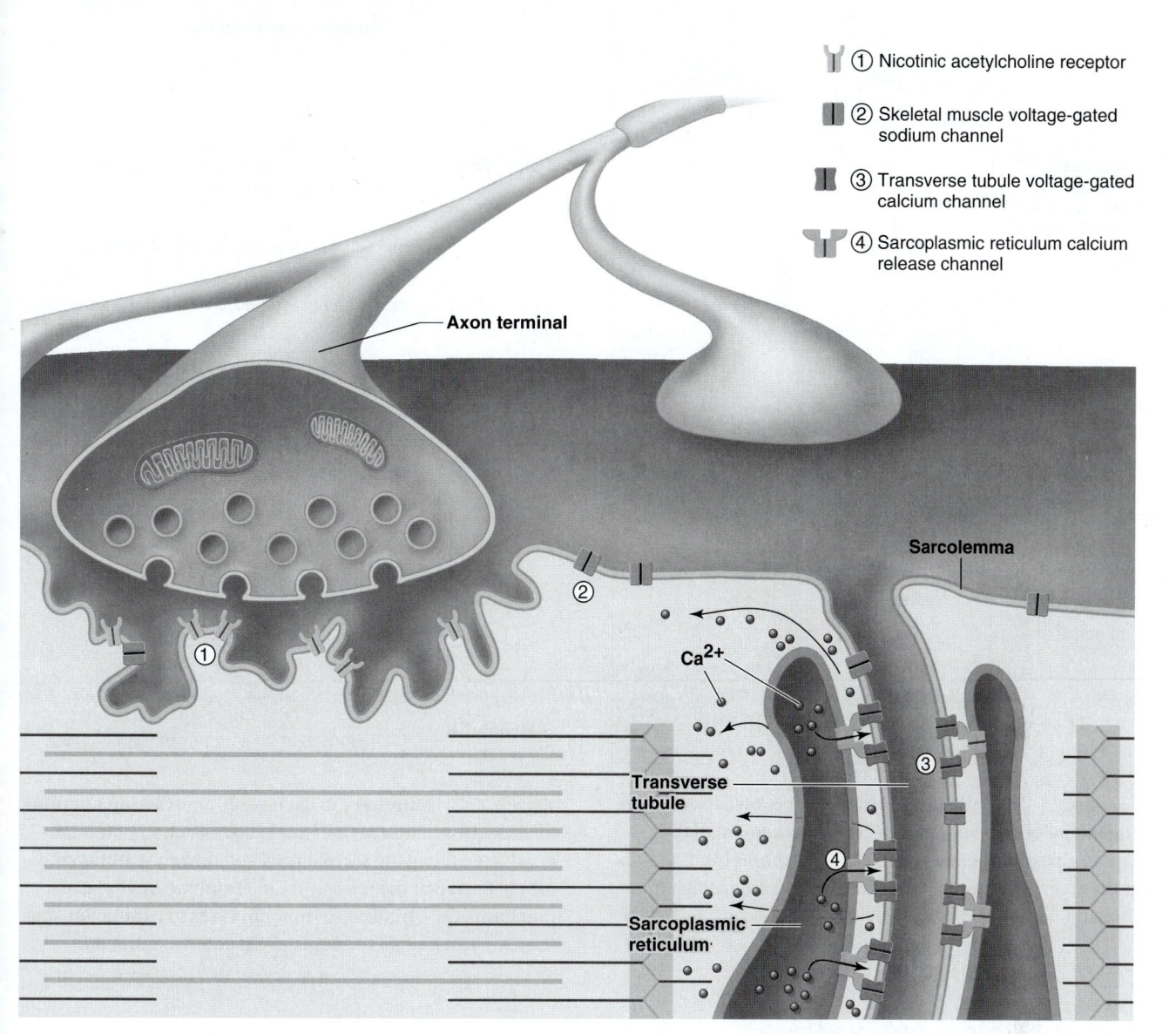

Figure 5.11 **Excitation-contraction coupling in skeletal muscles.** The numbered structures depict the molecules involved in *excitation-contraction coupling*, which refers to the release of Ca^{2+} from the sarcoplasmic reticulum in response to electrical excitation (action potentials), and the stimulation of muscle contraction in response to the released Ca^{2+}. Voltage-gated Ca^{2+} channels in the transverse tubules (activated by action potentials) interact with Ca^{2+} release channels in the sarcoplasmic reticulum, leading to the diffusion of Ca^{2+} into the sarcoplasm. The Ca^{2+} can then bind to troponin to stimulate contraction.

(For a full-color version of this figure, see fig. 12.16 in *Human Physiology*, thirteenth edition, by Stuart I. Fox.)

thin actin filaments. When troponin binds to Ca^{2+}, the troponin-tropomyosin complex shifts position on the thin filaments. This exposes binding sites on the actin for the cross-bridges of the myosin-thick filaments. Binding of the cross-bridges to actin causes power strokes, producing sliding of the filaments and contraction (fig. 5.12). The release of Ca^{2+} in response to action potentials thus causes the cross-bridge cycle that produces contraction. The story of how electrical excitation leads to contraction is known as **excitation-contraction coupling** (summarized in fig. 5.13).

The electrical activity of somatic motor neurons is normally stimulated in the spinal cord by synapses with other neurons. These other neurons may be association neurons located in the brain or spinal cord, or they may be sensory (afferent) neurons. Alternatively, action potentials in motor, or efferent, nerve fibers may be stimulated by damage to the fibers peripherally. This damage produces an *injury current* that stimulates action potentials and subsequent muscle contractions when, for example, a nerve is pinched.

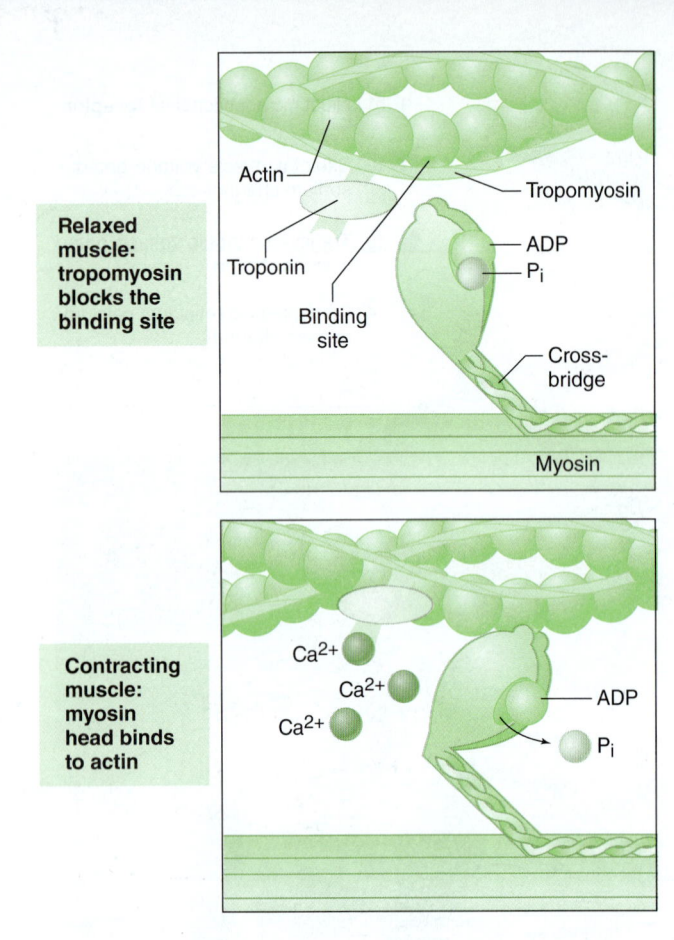

Figure 5.12 **The role of Ca^{2+} in muscle contraction.** The attachment of Ca^{2+} to troponin causes movement of the troponin-tropomyosin complex, which exposes binding sites on the actin. The myosin cross-bridges can then attach to actin and undergo a power stroke.

(For a full-color version of this figure, see fig. 12.14 in *Human Physiology*, thirteenth edition, by Stuart I. Fox.)

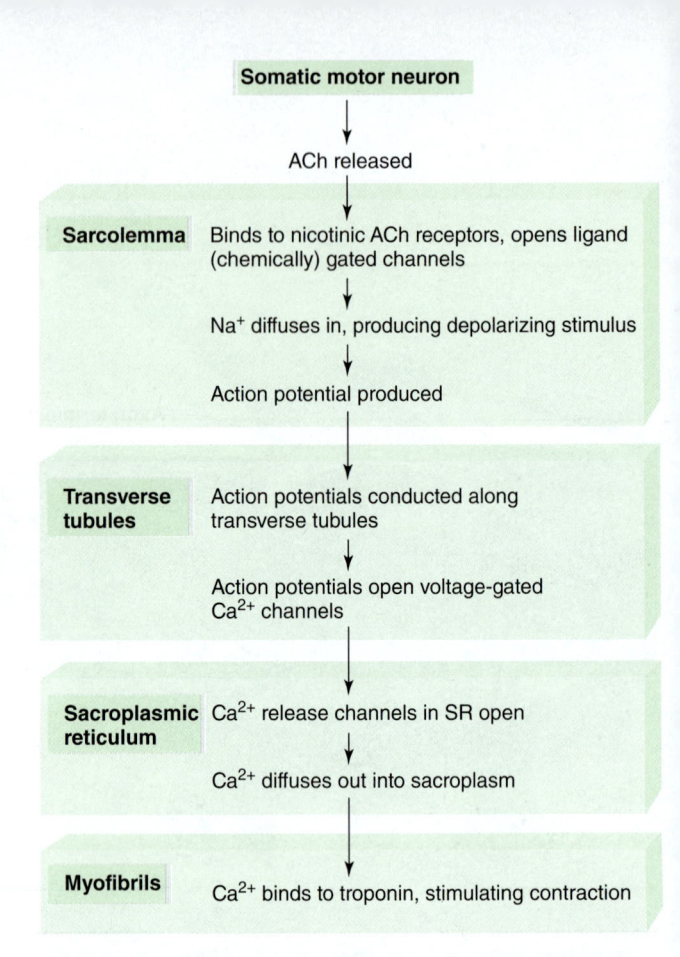

Figure 5.13 **Summary of excitation-contraction coupling.** Electrical excitation of the muscle fiber—action potentials conducted along the sarcolemma and down the transverse tubules—trigger the release of Ca^{2+} from the sarcoplasmic reticulum. Ca^{2+} binding to troponin leads to contractions, so the Ca^{2+} can be said to couple excitation to contraction.

(For a full-color version of this figure, see fig. 12.17 in *Human Physiology*, thirteenth edition, by Stuart I. Fox.)

Clinical Applications

Acetylocholine (ACh) released by axon terminals diffuses across the synaptic cleft and binds to ACh receptors in the plasma membrane of the muscle fiber, stimulating action potentials that lead to contraction. The potentially deadly **botulinum toxin**, produced by the bacteria *Clostridium botulinum*, is selectively taken into cholinergic nerve endings and cleaves proteins needed for the exocytosis of synaptic vesicles that releases ACh. This reduces the ACh in the synaptic cleft and thereby blocks nerve stimulation of the muscles, producing a flaccid (relaxed) paralysis. Botulinum toxin is used medically in certain cases to relieve muscle spasms due to excessive nerve stimulation; for example, it may be injected into an extrinsic eye muscle to help correct deviation of the eye (*strabismus*). Injections of **Botox** (a brand name for botulinum toxin) are also given to relax facial muscles for the temporary cosmetic treatment of skin wrinkles.

Procedure

1. With the frog on its belly, part the posterior muscles of the thigh around the femur to reveal the white sciatic nerve (fig. 5.9*a*).
2. Using glass probes, gently free the nerve from its attached connective tissue, and raise it between two glass probes (fig. 5.9*b*).
3. Place both stimulating electrodes under the sciatic nerve. Starting with the stimulator set at 0 V, gradually increase the stimulus voltage in small increments until the minimum stimulus that will produce a muscle twitch is attained. Record this threshold voltage in the space provided in your laboratory report.
4. Turn off the stimulator but leave the recorder running. Using a length of cotton thread, tie a knot in the nerve. Record and observe the response of the gastrocnemius muscle.
5. Push the stimulating electrodes again into the gastrocnemius muscle, in preparation for exercise 5.2.

Laboratory Report 5.1

Name _____

Date _____

Section _____

DATA FOR EXERCISE 5.1

1. Your threshold stimulus voltage when the electrodes were placed in the muscle: _____ V.
2. Your threshold stimulus voltage when the electrodes were placed on the nerve: _____ V.

REVIEW ACTIVITIES FOR EXERCISE 5.1

Test Your Knowledge

1. Arrange these structures in decreasing order of size: sarcomere, fibril, filaments, fiber:
   (a) _____
   (b) _____
   (c) _____
   (d) _____

2. The electrical events conducted along the cell (plasma) membrane that stimulate contraction are called _____.

3. Actin and myosin comprise the _____ and _____ filaments, respectively.

4. What substance couples electrical excitation to muscle contraction? _____

5. The substance named in question 4 is stored in which intracellular organelle? _____

6. The substance named in question 4 binds to a regulatory protein known as _____, which in turn is bound to an inhibitory protein called _____.

7. The neurotransmitter chemical that stimulates contraction of skeletal muscles: _____.

Test Your Understanding

8. Draw a sarcomere and label the parts and the bands. Then, describe and illustrate how the structure of a sarcomere changes during muscle contraction.

9. Trace the course of events starting from the moment ACh binds to its receptors in the sarcolemma and ending when Ca^{2+} enters the sarcoplasm.

183

10. Trace the course of events starting from the moment Ca^{2+} enters the sarcoplasm and ending when the cross-bridges have completed one power stroke.

Test Your Analytical Ability

11. Which had the lower threshold for stimulation of muscle contraction—stimulation of the muscle directly or stimulation of the nerve that innervates the muscle? Propose an explanation for these results.

12. Using your knowledge of the regulation of muscle contraction, predict what might happen to the beating of a heart if the blood concentration of Ca^{2+} were abnormally increased.

13. Predict the effects on muscles of a drug that blocks the action of acetylcholinesterase, an enzyme that breaks down acetylcholine. Compare that to the effects on muscles of a drug that blocks acetylcholine receptors.

Clinical Investigation Questions

14. How does a somatic motor nerve stimulate a skeletal muscle fiber to contract? List the sequence of events, and indicate where Botox would block that sequence.

15. How could Botox help to treat a person with strabismus? How could Botox injections produce smoother skin, but also cause a droopy forehead and eyelid?

EXERCISE 5.2

Summation, Tetanus, and Fatigue

MATERIALS

1. Frogs
2. Equipment and setup used in exercise 5.1
3. Electrocardiograph plates and electrolyte gel
4. Alternative equipment: Physiogrip, (Intellitool, Inc.); Biopac system with hand dynamometer; or the Biopac finger twitch transducer
5. Also, the iWorx equipment (with computers) can be used to perform these exercises by adapting exercise 6, per the laboratory manual available at www.iWorx.com
6. A PowerLab data acquisition system module is also available for performing the frog muscle exercises
7. Alternatively, simulated skeletal muscle exercises may be performed using the *Physiology Interactive Lab Simulations (Ph.I.L.S.):* Skeletal Muscle Function (Exercises 1, 2, and 3)

Twitch, summation, and tetanus can be produced by direct electrical stimulation of frog muscles *in vitro* (in the laboratory) and by stimulation of human muscles *in vivo* (within the living body). These procedures demonstrate how normal muscular movements are produced.

LEARNING OUTCOMES

You should be able to:

1. Define *twitch, summation, tetanus,* and *fatigue.*
2. Demonstrate twitch, summation, and tetanus in both frog and human muscles, and demonstrate fatigue in the frog muscle preparation.
3. Explain how a smooth, sustained contraction is normally produced.

Textbook/Multimedia Correlations

Before performing this exercise, you should study the introductory material presented here. Further information relating to this exercise can be found in these pages of *Human Physiology,* thirteenth edition, by Stuart I. Fox:

- *Motor Units.* Chapter 12, p. 363.
- *Contractions of Skeletal Muscles.* Chapter 12, p. 374.
- *Muscle Fatigue.* Chapter 12, p. 382.
- Multimedia correlations are provided in *MediaPhys 3.0:* Topics 5.16–5.18; and *Physiology Interactive Lab Simulations (Ph.I.L.S.):* Skeletal Muscle Function (Exercises 1, 2, and 3).

Clinical Investigation

A young man noticed that the middle finger of his right hand was twitching, and a friend mentioned that he might have a pinched nerve. Subsequently, when he developed a spasm in his leg muscle that awoke him from his sleep, his friend told him that he might have been hyperventilating or that he might need more calcium.

- Distinguish among twitch, a normal contraction, and tetany, and explain how a pinched nerve might cause a muscle to twitch.
- Explain how hyperventilation or a calcium deficiency might cause tetany.

A muscle fiber's action potential lasts approximately 10 milliseconds, whereas its contraction typically lasts about ten times that long. This is a **twitch**—a rapid contraction followed by relaxation. It is possible to stimulate a muscle fiber *in vitro* (outside of the body) with electrodes so quickly that successive action potentials occur before the muscle fiber relaxes. This keeps Ca^{2+} in the sarcoplasm for a longer time, so that the muscle fiber's contraction is more sustained. However, in the body (*in vivo*), muscle fibers are stimulated in a different manner by somatic motor neurons.

A motor neuron and the muscle fibers it innervates is known as a *motor unit* (see fig. 5.18). When a muscle is stimulated to contract *in vivo,* somatic motor neurons that innervate the muscle become activated rapidly one after another. The muscle fibers that each motor neuron innervates are stimulated to contract at the same time, generally with a twitch. However, because of the rapid, successive, *asynchronous activation* of many motor units, some will just be starting to twitch after other motor units have already begun and after others have started to relax. Consequently, the production of smooth, graded skeletal muscle contractions *in vivo* occurs by the **summation** of the fiber twitches of different motor units.

Summation of twitches can be demonstrated *in vitro* by stimulating an isolated muscle with a series of rapid, successive electrical shocks. Two successive shocks result in a second, stronger twitch that partially "rides piggyback" on the first in a recording (fig. 5.14*a*). Stronger muscle contractions are produced by the summation of the twitches of more muscle fibers. This is similar to the situation *in vivo*, where a stronger muscle contraction is produced by the recruitment of more motor units (and thus more muscle fibers) into the summated contraction.

If the activation of successive motor units occurs sufficiently rapidly, there will be no visible relaxation between the twitches of the different groups of muscle fibers.

(a)

(b)

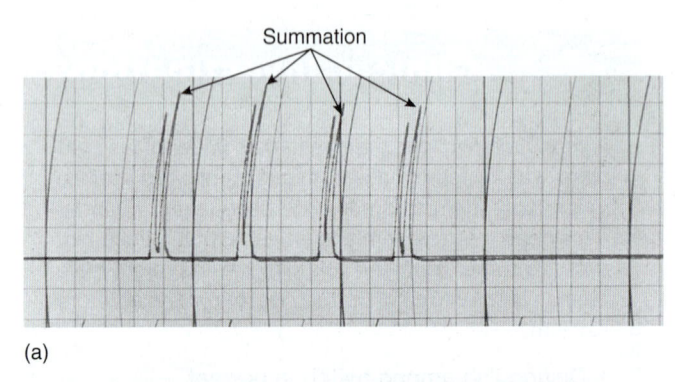

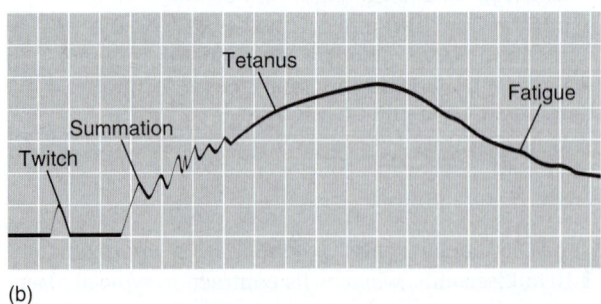

Figure 5.14 Recording of muscle contractions.
(a) A recording of the summation of two muscle twitches on a physiograph recorder. Contraction to the second stimulus is greater than contraction to the first stimulus (the intensity of the first and second stimuli is the same). (b) Diagram illustrating twitch, summation, tetanus, and fatigue.

Contraction will be smooth and sustained. Maintenance of a sustained muscle contraction is called **tetanus**. (The term *tetanus* should not be confused with the disease tetanus, which is accompanied by a painful state of muscle contracture, or *tetany*.)

Tetanus can be demonstrated in the laboratory by setting the stimulator to deliver shocks automatically to the muscle at an ever-increasing frequency until the twitches fuse into a smooth contraction (fig. 5.14*b*). This is similar to what occurs in the body when different motor neurons in the spinal cord are activated to stimulate a muscle at slightly different times.

If the stimulator is left on so that the muscle remains in tetanus, a gradual decrease in contraction strength will be observed. This is due to **muscle fatigue**. Fatigue during a sustained maximal contraction, as when lifting a very heavy weight, appears to be due to an accumulation of extracellular K^+. This depolarizes the membrane potentials and interferes with the ability of the muscle fiber to produce action potentials.

Muscle fatigue that occurs during most types of exercise, however, appears to have different causes. Chiefly, there is depletion of muscle glycogen and a reduced ability of the sarcoplasmic reticulum to release Ca^2, leading to failure of excitation-contraction coupling. Although failure of excitation-contraction coupling is known to produce muscle fatigue, the reasons for that failure—despite nearly a century of study—are incompletely understood.

It has been known from the early twentieth century that fatigue occurs when lactate accumulates, and that restoring aerobic respiration allows muscle glycogen and contractile ability to recover. This led to the widespread belief that lowered muscle pH caused by the H^+ released from lactic acid causes muscle fatigue. However, ongoing research suggests that lactate production may be more coincidental with muscle fatigue than a cause of it. Several other changes may produce muscle fatigue by interfering with excitation-contraction coupling. The relative contribution of each change to muscle fatigue depends on the type of exercise performed. The muscle changes that may contribute to fatigue include:

1. Increased concentration of PO_4^{3-}, derived from the breakdown of phosphocreatine, in the cytoplasm. This is currently believed to reduce the force developed by cross-bridges and serves as a major contributor to muscle fatigue.
2. A decline in ATP, particularly around the junction of the transverse tubules and sarcoplasmic reticulum, that may hinder the action of the Ca^{2+} pumps. ATP declines significantly in fast-twitch fibers during high-intensity exercise, but does not decline enough to produce rigor complexes (as in rigor mortis). ATP does not measurably decline in slow-twitch fibers during exercise.
3. Depletion of muscle glycogen. The mechanisms by which this contributes to fatigue are not fully understood, but the depletion of muscle glycogen

appears to decrease the release of Ca^{2+} from the sarcoplasmic reticulum.

4. Increased ADP in the cytoplasm. This causes a decrease in the velocity of muscle shortening during muscle fatigue.

A. Twitch, Summation, Tetanus, and Fatigue in the Gastrocnemius Muscle of the Frog

Summation, tetanus, and fatigue can be demonstrated with the frog gastrocnemius muscle preparation used in exercise 5.1.

Procedure

1. Set the stimulus voltage above threshold, and press down on the switch that delivers a single pulse to the muscle two or three times in rapid succession. If this is done with enough speed, successive twitches can be made to "ride piggyback" on preceding twitches, demonstrating *summation* (fig. 5.14*a*).
   a. If a manual stimulator is used, it must be set to deliver a single shock when a trigger is depressed. Remember that if such a stimulator is used with the Biopac system, the contractions will be recorded but the stimulating shocks will not be recorded.
   b. If the Biopac stimulator is used, it is best to use the manual trigger at the back of the stimulator to deliver the separate shocks in such a way as to produce summation of twitches.
2. Set the stimulus switch to deliver shocks to the muscle automatically at a frequency of about one per second. Gradually increase the frequency of stimulation until the twitches fuse into a smooth, sustained contraction, demonstrating *tetanus* (fig. 5.14*b*).
   a. If a manual stimulator is used, the stimulator must be set to "continuous" and the frequency increased by gradually turning a knob toward a frequency that produces a complete tetanus.
   b. If the Biopac stimulator is used, use the stimulator dialog box to set the stimulator to continuous pulses and gradually increase the pulse rate by 1 Hz.
   c. Regardless of which type of stimulator is used, if the Biopac system is used for recording, the "start" button on the lower right portion of the computer screen must be pressed to begin recording.
3. Maintain stimulation until the strength of contraction gradually diminishes as a result of muscle *fatigue* (fig. 5.14*b*).
4. Enter your recordings in the laboratory report.

B. Twitch, Summation, and Tetanus in Human Muscle

The properties of frog muscle contractions observed *in vitro* reflect the behavior of human muscle *in vivo* in many ways. A single pulse of electrical stimulation produces a single, short contraction, or twitch. Many single, electrical shocks delivered in rapid succession produce summation, where one twitch starts before the previous twitch has finished. As the frequency of stimulation is increased, the individual twitches are barely evident; this is *incomplete tetanus*. At a sufficiently high frequency of stimulation, the contraction becomes a smooth, sustained *complete tetanus*. This mimics the smooth contractions that we normally produce *in vivo* by the rapid, asynchronous activation of somatic motor neurons and the muscle fibers they innervate.

Clinical Applications

Sustained muscular spasm (*tetany*) may be produced by hypocalcemia and by alkalosis. (The most common cause of tetany is alkalosis produced by hyperventilation.) Cramps may be due to a variety of conditions, including salt depletion. General muscle weakness may be caused by alterations in plasma potassium levels (due, for example, to excessive diarrhea or vomiting).

Muscular dystrophy refers to any of a variety of diseases in which there is a progressive lack of support (dystrophy) and weakness of skeletal muscles (although heart muscle may also be involved) that does not seem to be caused by inflammation or neural disease. *Duchenne's muscular dystrophy* is the most severe of the muscular dystrophies, afflicting 1 out of 3,500 boys each year who inherit the disease as an X-linked recessive trait. The gene produces a defective protein known as *dystrophin* that makes the muscle fibers more easily damaged, resulting in fiber necrosis and replacement by fibrous connective and fatty tissues.

Procedure

1. Rub a small amount of electrolyte gel on the skin near the wrist and attach an ECG electrode plate to this area with an elastic band. Rub electrolyte gel on a second ECG electrode plate and place it on the anterior, medial area of the forearm, just below (distal to) the elbow. Do not attach this electrode to the arm, as this will be the *exploring electrode* (fig. 5.15).
2. Connect the electrode plates to a stimulator, making sure that the stimulator is *off* at this time.
3. Set the stimulus intensity at 15 V and deliver a single pulse of stimulation. If no twitch is observed or felt in the fingers, move the exploring electrode around the medial area of the forearm until an effect is seen or felt. (See fig. 5.15 for the approximate position of the electrode.)

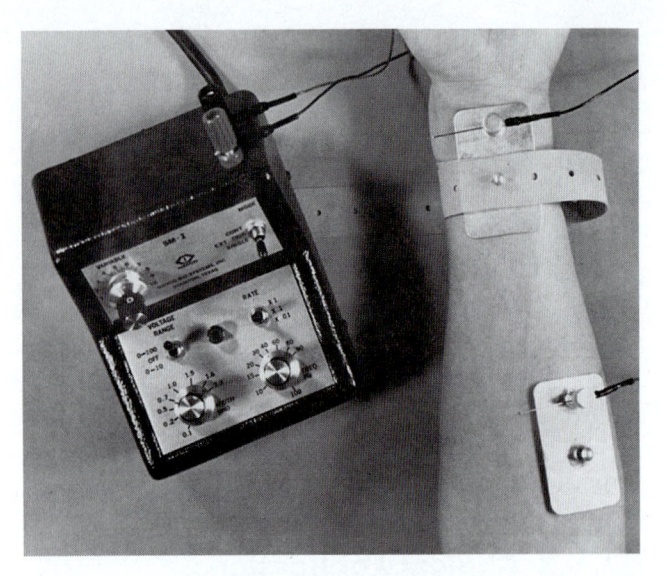

Figure 5.15 Placement of electrodes for eliciting finger twitches in response to electrical stimulation.

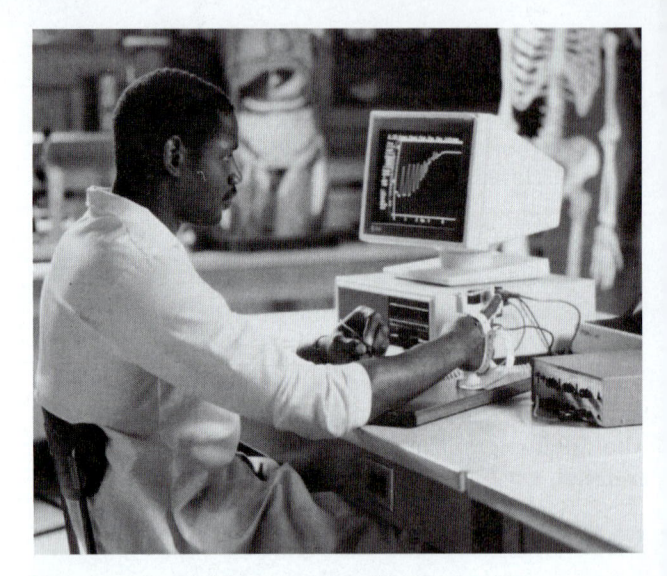

Figure 5.16 Student using the Physiogrip to demonstrate muscle twitches, summation, and tetanus.

⚠ **Caution:** *The stimulus intensity may have to be increased for some people, but do not exceed 30 V! An effect can generally be obtained at a lower voltage by moving the exploring electrode to a slightly different position or by adding electrolyte gel. A tingling sensation means that the stimulus intensity is adequate, although the position may have to be changed.*

4. Once a muscle twitch has been observed, set the stimulator so that it automatically delivers one pulse of stimulation per second. Adjust the exploratory plate so that only one finger twitches.
5. Keeping the stimulus intensity constant, gradually increase the frequency of stimulation until a maximum contraction is reached. Gradually decrease the stimulus frequency until the individual twitches are reproduced.

Procedure

ALTERNATIVE FOR INTELITOOL PHYSIOGRIP

1. As in the previous procedure, determine the correct points for placement of the electrodes on the forearm so that the flexor digitorum superficialis muscle is stimulated. This will result in flexion of the finger that grips the trigger of the Physiogrip.
2. Hold the Physiogrip (fig. 5.16) with light pressure while another student delivers electrical shocks using the stimulator. Start with about 15 V at a duration of 1 millisecond and gradually increase the voltage until threshold is observed.
3. Continue to increase the voltage in small increments, demonstrating the graded increase in contraction strength in response to stronger electrical stimuli.
4. With the voltage constant, gradually increase the frequency of stimulation to demonstrate tetany.

Procedure

ALTERNATIVE FOR BIOPAC SYSTEM

1. The hand dynamometer (fig. 5.17a) used in the Biopac Lesson 2 (EMG II) can be adapted for this exercise, once the stimulating electrodes are placed on the forearm to stimulate the digitorum superficialis muscle.
2. As per the previous instructions, deliver the electrical shocks using the stimulator. Start with about 14 V at a duration of 1 millisecond and gradually increase the voltage until a threshold is observed.
3. Continue to increase the voltage in small increments, demonstrating the graded increase in contraction strength in response to stronger electrical stimuli.
4. With the voltage constant, gradually increase the frequency of stimulation to demonstrate summation and tetany.
5. Alternatively, the Biopac human finger twitch equipment, using their finger twitch transducer (SS61L) (fig. 5.17b) may be adapted for this exercise. Refer to the Biopac PRO Lesson H06 on their Web site.

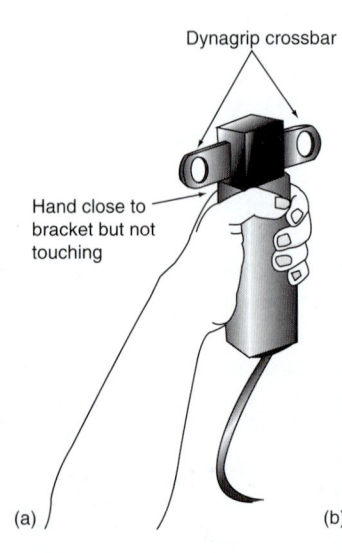

Figure 5.17 Biopac devices to record human muscle contractions: (a) the hand dynamometer; (b) the finger twitch transducer.

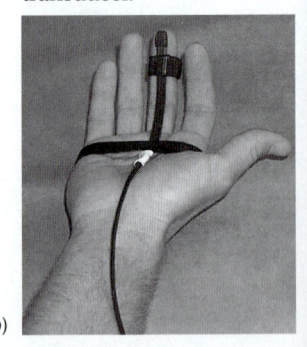

188

Laboratory Report 5.2

Name _____
Date _____
Section _____

DATA FROM EXERCISE 5.2

A. Twitch, Summation, Tetanus, and Fatigue in the Gastrocnemius Muscle of the Frog

1. In this space, tape your recording or draw a facsimile.

2. Label twitch, summation, tetanus, and fatigue in your recording and compare your results with that shown in figure 5.14.

B. Twitch, Summation, and Tetanus in Human Muscle

1. Describe the results of your procedure in this space. Alternatively, if you used the Intelitool or Biopac systems and have your data in a computer file, print a hard copy and paste it in the space provided. Indicate twitch, summation, tetanus, and fatigue (if obtained).

REVIEW ACTIVITIES FOR EXERCISE 5.2

Test Your Knowledge

1. Define:
   (a) twitch _____
   (b) summation _____
   (c) tetanus _____
2. Sustained muscular spasm is called _____. The two most common causes of this are _____ and _____.

Test Your Understanding

3. Describe how you produced summated contractions with the isolated muscle and how you produced a tetanus contraction. Explain how summation of twitches is accomplished *in vivo* and how a sustained, complete tetanus contraction is produced.

4. Describe how you could increase the strength of the contraction with the isolated muscle, and explain how this is accomplished *in vivo*.

5. Describe how you demonstrated muscle fatigue in the laboratory, and describe the changes within the muscle that may help to explain how fatigue is produced.

Test Your Analytical Ability

6. Suppose you hold a 10-pound weight steadily in your right hand with your elbow slightly bent, so that you maintain a contraction of your biceps brachii muscle. After some time, you experience pain and it becomes increasingly difficult to keep the weight at the same level. Explain why this is so.

7. Continuing with the situation described in question 6, suppose your hand starts to shake. Describe the changes in the state of your skeletal muscle fibers as your hand goes from being held steady to shaking. Relate these changes to the recordings obtained in this exercise.

Clinical Investigation Questions

8. How do twitch and tetany differ from a normal muscle contraction? How could a pinched nerve cause a muscle to twitch?

9. How does hyperventilation or a calcium deficiency cause muscle tetany? Which is the more likely cause? Explain?

EXERCISE 5.3

Electromyogram (EMG)

MATERIALS

1. Physiograph or another electrical recorder and high-gain coupler
2. EMG plates and disposable adhesive paper washers for EMG plates
3. Electrolyte (ECG) gel or paste and alcohol swabs
4. Alternative: Biopac system equipment for EMG I and II (Lessons 1 and 2)
5. As an alternative, the iWorx equipment can be used with their hand dynamometer. This is described in exercise 8 from the laboratory manual available at www.iWorx.com.
6. Another alternative is the PowerLab system with their flat EEG electrodes and hand dyanamometer.

The electrical activity produced by muscles can be recorded using surface electrodes. This recording may demonstrate the action of antagonistic muscles and may be useful in biofeedback training of muscles.

LEARNING OUTCOMES

You should be able to:

1. Demonstrate the antagonism between the action of the biceps and triceps muscles using the electromyogram (EMG).
2. Distinguish between isotonic and isometric muscle contractions, with examples of each.
3. Describe the EMG of the biceps and triceps during flexion and extension of the arm.
4. Explain the importance of antagonist inhibition in skeletal movements.
5. Explain how the EMG can be used in biofeedback techniques.

Textbook/Multimedia Correlations

Before performing this exercise, you should study the introductory material presented here. Further information relating to this exercise can be found in these pages of *Human Physiology*, thirteenth edition, by Stuart I. Fox:

- *Motor Units.* Chapter 12, p. 363.
- *Contractions of Skeletal Muscles.* Chapter 12, p. 374.
- *Alpha and Gamma Motoneurons.* Chapter 12, p. 387.
- Multimedia correlations are provided in *Physiology Interactive Lab Simulations (Ph.I.L.S.):* Skeletal Muscle Function (Exercises 1, 2, and 3).

Clinical Investigation

A patient who had fallen in a construction site accident had suffered trauma to his spinal cord that caused paraplegia. After a period of flaccid paralysis from spinal shock, his leg muscles became spastic. Dorsiflexion and release of his foot produced a flapping tremor.

- Describe flaccid and spastic paralysis and explain how damage to descending motor tracts can cause both.
- Identify the agonist and antagonist muscles involved in a flapping tremor of the foot and explain how this tremor was produced.

Skeletal muscle contraction occurs in response to the production of action potentials by the muscle fibers. In the previous exercises, following stimulation the mechanical response of the muscles was observed and recorded as tension exerted on a myograph transducer. Although the muscles were stimulated by electric shocks, the electrical activity of the individual muscle cells was not recorded. Electric shocks delivered to the muscle fibers induce the production of action potentials, and it is these action potentials

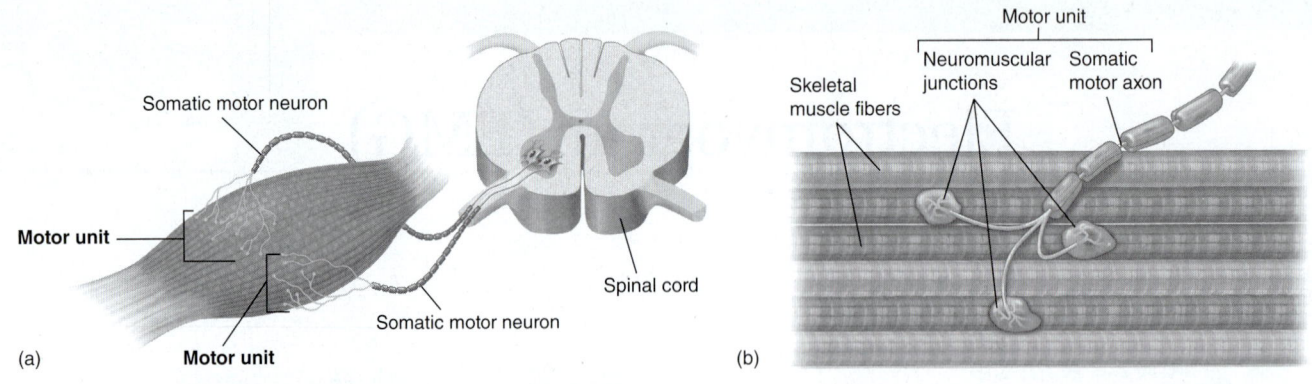

Figure 5.18 Motor units. A motor unit consists of a motor neuron and the muscle fibers it innervates. (a) Illustration of a muscle containing two motor units. In reality, a muscle would contain many hundreds of motor units, and each motor unit would contain many more muscle fibers than are shown here. (b) A single motor unit consisting of a branched motor axon and the three muscle fibers it innervates (the darker fibers) is depicted. The other muscle fibers would be part of different motor units and would be innervated by different neurons (not shown).

(For a full-color version of this figure, see fig. 12.4 in *Human Physiology*, thirteenth edition, by Stuart I. Fox.)

(acting via the release of Ca^{2+} from the sarcoplasmic reticulum) that lead to contraction of the muscles.

In the body *(in vivo)* muscle fibers are stimulated to contract by motor neurons. The axon of a motor neuron branches to innervate a number of muscle fibers, all of which contract when the axon is stimulated. The axon and the muscle fiber it stimulates is known as a **motor unit** (fig. 5.18). When more strength is required for a muscle contraction, more motor units are enlisted, or *recruited,* into the contraction.

The contraction that results in muscle shortening is called **isotonic** ("equal tension"), because the force of contraction remains relatively constant throughout the movement. Isotonic contractions are easily observed because the muscle becomes shorter and the corresponding limb or object attached is moved, such as movements when walking, lifting a chair, or doing push-ups (fig. 5.19*a*). In **isometric** ("equal measure") contractions, the length of a muscle remains constant and no movement is seen because the force in the muscle being contracted is opposed by an equal opposing force, such as gravity. For example, an isometric contraction occurs when a person supports an object in a fixed position (fig. 5.19*b*), such as in postural muscles when standing or sitting motionless. An isometric contraction can be converted to an isotonic contraction when an increased force generated within the muscle overcomes the opposing resistance and results in muscle movement. This occurs, for example, when a straining body successfully clears the floor during a push-up exercise.

If the tension produced by a muscle contraction exceeds the load on the muscle, the contraction will cause shortening of the muscle. This may be isotonic, but the contraction can be described more generally as a **concentric** (or **shortening**) **contraction.** When the force of the muscle load is greater than the tension produced by the muscle contraction, the force of the load will stretch the muscle *despite* its contraction. This is known as an **eccentric** (or **lengthening**) **contraction.** For example, when you do a "curl" with a dumbbell, your biceps brachii muscle produces a concentric contraction as you flex your forearm. When you gently lower the dumbbell back to the resting position, your biceps produce an eccentric contraction. In this case, the tension produced by the eccentric contraction of your biceps prevents the dumbbell from lowering too fast under the force of gravity.

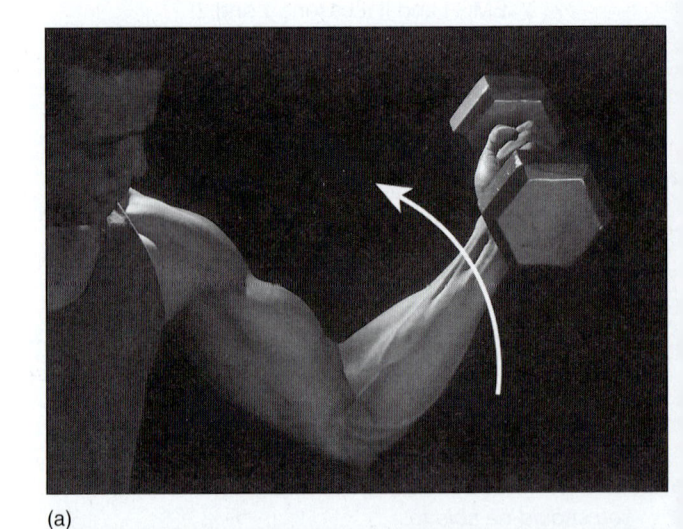

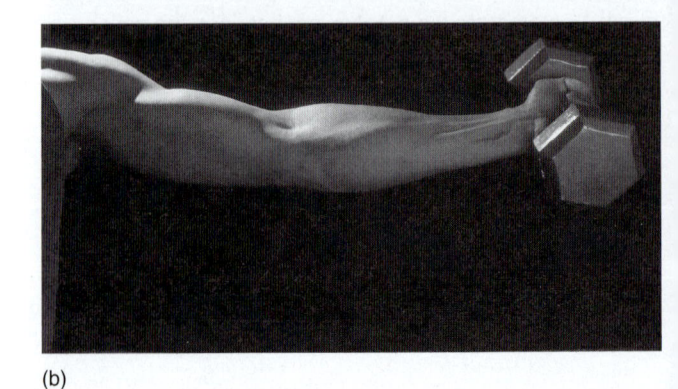

Figure 5.19 Isotonic (a) and isometric (b) muscle contractions.

A. Electromyogram Recording

When somatic motor nerves stimulate skeletal muscles to contract, the action potentials produced by the muscles transmit potential differences to the overlying skin that can be recorded by a pair of surface electrodes on the skin. The recording obtained is called an **electromyogram (EMG).** When the pair of electrodes are placed on the anterior surface of the upper arm, they record potentials generated by the *biceps brachii* muscle; and when they are placed over the posterior surface of the upper arm, they record the electrical activity of the *triceps brachii* muscle. Contraction of the biceps flexes the arm, whereas contraction of the triceps extends the arm. These two groups of muscles, therefore, are **antagonistic muscles.**

When the biceps muscle contracts and the arm flexes, the antagonistic triceps muscle is stretched; when the triceps muscle contracts and the arm extends, the biceps muscle is stretched. Stretching of a muscle elicits the muscle stretch reflex (exercise 3.3A), which would evoke contraction of the antagonistic muscles and interfere with the arm movements. However, during arm flexion the spinal motor neurons that stimulate the triceps are inhibited, and during arm extension the motor neurons that stimulate the biceps are inhibited. This inhibition occurs within the spinal cord by means of the inhibitory neurotransmitters glycine and GABA, released by association neurons.

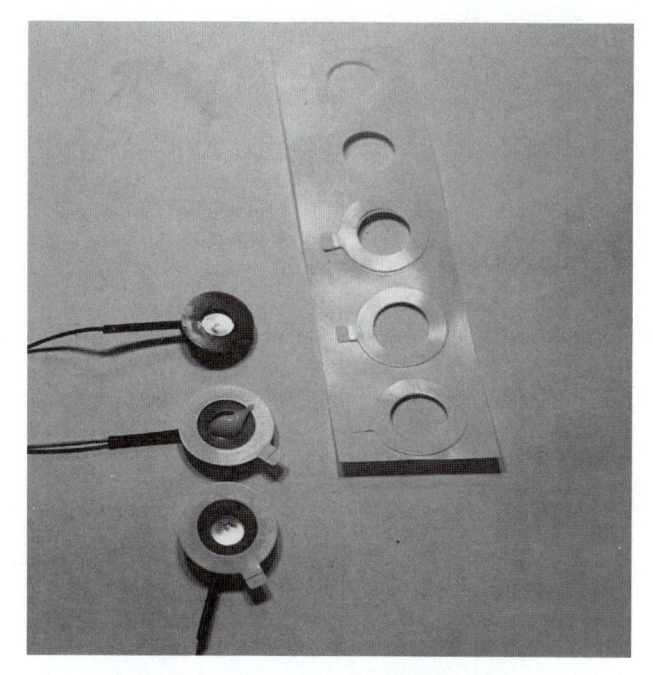

Figure 5.20 Electrode plates and adhesive washers needed for electromyograph (EMG) procedure.

Procedure

RECRUITMENT OF MOTOR UNITS

1. Using cotton or a paper towel soaked in alcohol, cleanse the skin over the biceps and triceps muscles.
2. Apply the self-sticking paper washers to the raised plastic area surrounding the electrode plates (fig. 5.20). Squeeze *electrolyte gel* onto the metal electrode plates. Use a paper towel to smooth the gel so that it completely fills the well between the electrode and the surrounding plastic.
3. Remove the paper coverings over the adhesive area of the washers and apply the electrodes to the skin over the biceps muscle. Apply one electrode to the skin over the proximal portion of the biceps and the other over the distal portion, aligned with the first (fig. 5.21). Apply the ground electrode over the triceps muscle.
4. Plug the electrodes into the *high-gain coupler* module of the physiograph. Set this module to a *gain* of ×100, a *time constant* of 0.03, and a *sensitivity* between 20 and 100.

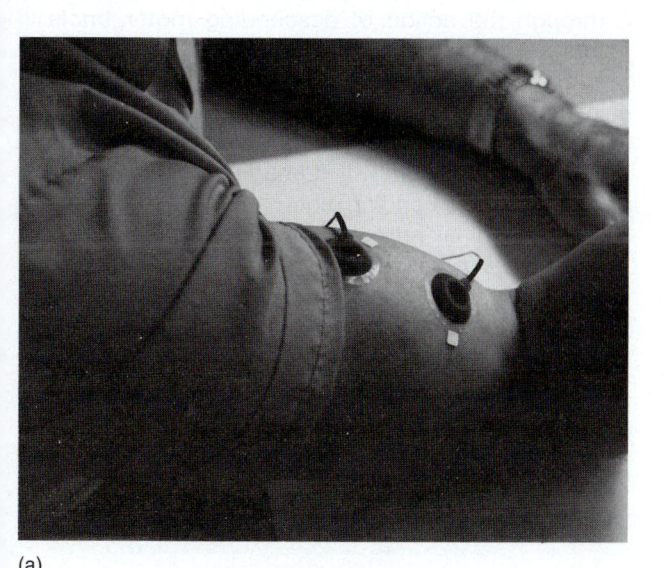

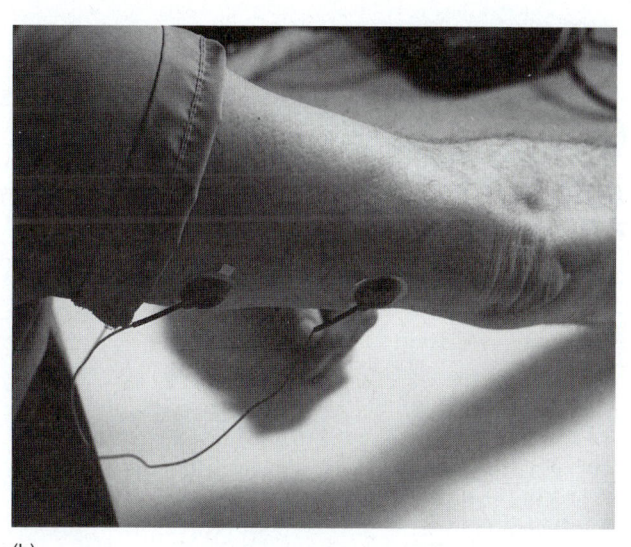

(a)   (b)

Figure 5.21 **Placement of the EMG electrodes.** The positions are shown for recording from (a) the biceps and (b) the triceps muscles.

5. Set the *chart speed* at 0.5 cm/sec. With the arm relaxed and hanging down, establish a baseline, or control, in the recording. Then flex the arm (bringing the hand upward), and observe the recording of an isotonic contraction. Extend the arm back to its previous position, and then flex it again so that the difference between flexion and extension can be seen in the recording.
6. Flex the arm again, this time lifting a chair or other weight. Observe the effect of this activity on the height (amplitude) of the recording. Now, "make a muscle" of increasing strength up to a maximum, demonstrating the recruitment of motor units into the isometric contraction.

ALTERNATIVE PROCEDURE USING THE BIOPAC SYSTEM

1. Stick three electrodes on the forearm in the positions indicated in figure 5.22, and attach the correctly colored electrodes as shown in that figure.
2. Proceed with the set-up instructions as per Lesson 1 (EMG I) for the Biopac student lab.
3. Demonstrate recruitment of motor units by clenching your fist tighter and tighter, up to your maximum strength.
4. Repeat this procedure after moving the electrodes to your other forearm.

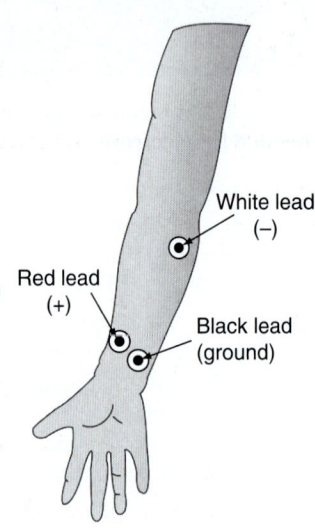

Figure 5.22 Biopac electrode positions for the EMG.

Procedure

EMG DURING ARM FLEXION AND EXTENSION

1. With the electrodes placed over the biceps brachii muscle as previously described, flex your arm as if you were trying to lift the table.
2. Change the position of the electrodes so that the two recording electrodes are lined up over the triceps muscle (one proximal and one distal) and the ground electrode is over the biceps muscle. Flex and extend the arm as before and observe the recording.
3. Place the hand on a table with the elbow bent and slowly extend the arm, as if doing a pushup. Observe the effect of this action on the EMG. Note the change in the recording as an isometric contraction becomes isotonic. Enter your recordings in the laboratory report.

B. BIOFEEDBACK AND THE ELECTROMYOGRAPH

Our behavior changes as a result of the pleasant or unpleasant consequences of our actions. That is, positive and negative reinforcements modify behavior; this represents a type of learning that experimental psychologists call *operant conditioning*. The pairing of a particular behavior, such as cigarette smoking, with unpleasant sensations has been used successfully to shape human behavior through *aversion conditioning*.

Biofeedback techniques similarly affect learning, usually through feedback provided by electronic monitors of specific physiological states. The electromyogram, for example, provides a visual display of muscle stimulation that can be used as a psychological reward to reinforce effort spent attempting to contract specific muscle groups. In this exercise, the EMG will be used to demonstrate biofeedback techniques involved in learning how to increase the strength of contraction of the triceps muscle.

Clinical Applications

The activity of antagonistic muscle groups is controlled in the central nervous system, so that when one group of muscles (the *agonist*) is stimulated to contract, the *antagonist* muscle group is inhibited and will be stretched. This inhibition of antagonistic muscle groups occurs largely through the action of descending motor tracts that originate in the brain. When a person has spinal cord damage that blocks these descending inhibitory influences, the antagonistic muscles may contract when they are stretched by the movement of a limb. If damage to upper motor neurons or descending motor tracts causes exaggerated spinal reflexes, *increased muscle tone* and *spastic paralysis* may be produced. When an examiner either dorsiflexes or plantarflexes and then releases the foot of someone with spinal cord damage, a **flapping tremor,** or **clonus,** may occur. This results when the spinal cord damage blocks the inhibitory effects of upper motor neurons on skeletal muscles reflexes, allowing antagonistic muscles to alternatively stretch and contract.

Procedure

1. Prepare the EMG electrodes as described in the previous procedure. Cleanse the skin with alcohol, and place the two recording electrodes over the triceps muscle and the ground electrode over the biceps muscle.
2. Set the *high-gain coupler* to a *gain* of ×100, a *time constant* of 0.03, and a *sensitivity* between 20 and 100. Set the *chart speed* of the recorder to 0.5 cm/sec.
3. Extend the arm and observe the highest amplitude of the recording. Attempt a forced extension and observe the amplitude (height) of the recording. Attempt to increase the amplitude of the EMG by various procedures. (*Hint:* Try to extend the arm with the back of the hand against a table.)
4. Alternatively, the Biopac system can be used with the same setup previously described. Demonstrate biofeedback by attempting to reproduce a particular muscle tension while watching the recording on the computer screen.

Clinical Applications

Biofeedback techniques serve a variety of clinical functions. The EMG is sometimes used to train people with neuromuscular disorders to regain use of affected limbs. Physiological monitoring of the heart rate and blood pressure has enabled patients with high blood pressure to lower their pulse; and the production of alpha rhythms on *electroencephalogram (EEG)* recordings has been used to teach people with stress techniques for relaxation.

Laboratory Report 5.3

Name _____
Date _____
Section _____

DATA FROM EXERCISE 5.3

A. **Electromyogram Recording**
   1. **Recruitment of Motor Units**
      In this space, tape your recordings or draw facsimiles. Label the parts of your recording.

   2. **EMG During Arm Flexion and Extension**
      In this space, tape your recordings or draw facsimiles.
      (a) Label flexion and extension for both the biceps and triceps muscles in your recording.
      (b) Compare your reading of the *isotonic* contraction to that of the *isometric* contraction.

B. **Biofeedback and the Electromyograph**
   In this space, tape your recordings or draw facsimiles. Label the region of your recording that demonstrates biofeedback.

REVIEW ACTIVITIES FOR EXERCISE 5.3

Test Your Knowledge

1. During arm flexion, the biceps brachii muscle is the _____ (agonist/antagonist) and the triceps brachii is the _____ (agonist/antagonist).

2. Define an *isotonic contraction*.

3. Define an *isometric contraction*.

4. Define *motor unit*.

5. The process of enlisting more and larger motor units to produce a stronger contraction is called

6. The recording of the electrical currents produced by contraction of skeletal muscle is called a(n):

Test Your Understanding

7. Draw two motor units, one smaller and one larger. With reference to your illustration, explain how these motor units could produce three different strengths of contraction.

8. Distinguish between isometric and isotonic contraction, and give examples of each.

9. Describe the activity of the biceps and triceps brachii during arm flexion and extension. Also, describe their activity when you "make a muscle."

10. Describe what is meant by concentric and eccentric contractions. Use these terms to describe the contractions of (a) your antagonistic arm muscles when you bench press a weight and gently bring it back to your chest and (b) your antagonistic leg muscles when you hold a weight on your shoulders as you do a squat and then stand up again.

Test Your Analytical Ability

11. Suppose you want to lift an object very slowly and carefully. Using the concepts of motor unit size and biofeedback, explain how you could do this.

12. Suppose a person had a stroke and could no longer write with their dominant hand. How could you use the equipment and techniques in this lab exercise to help train this person to write with the nondominant hand?

Clinical Investigation Questions

13. What are flaccid and spastic paralysis, and how would they be produced by spinal cord damage?

14. How do agonist and antagonist muscles produce a flapping tremor of the foot when there is damage to the spinal cord?

Section 6

Blood: Gas Transport, Immunity, and Clotting Functions

The various components of the blood serve different physiological functions. The **plasma,** or fluid portion of the blood (fig. 6.1), provides the major means for distributing chemicals between organs. For example, it transports oxygen from the lungs, food molecules absorbed through the small intestine, hormones secreted by the endocrine glands, and antibodies produced by certain white blood cells. The plasma also helps to eliminate metabolic wastes by carrying these unwanted molecules to the liver (for excretion in the bile); to the kidneys (for excretion in the urine); and, in the case of CO_2 gas, to the lungs (for excretion in the exhaled air).

In addition to plasma, the blood also contains two major types of cells: **red blood cells** (erythrocytes) and **white blood cells** (leukocytes) (fig. 6.1a). Red blood cells contribute to the respiratory function of the blood by providing transport for oxygen and carbon dioxide. Blood is "typed" based upon the presence or absence of specific molecules displayed by red blood cells (the ABO system and the Rh factor are examples). White blood cells and their products help to provide immunity from infection by recognizing and attacking foreign molecules and cells. The blood also contains **platelets** (thrombocytes), membrane-bound fragments derived from a bone marrow cell called a megakaryocyte. Platelets, together with proteins in the plasma, help to maintain the integrity of blood vessels by forming blood clots.

Exercise 6.1	Red Blood Cells and Oxygen Transport
Exercise 6.2	White Blood Cell Count, Differential Count, and Immunity
Exercise 6.3	Blood Types
Exercise 6.4	Blood Clotting System

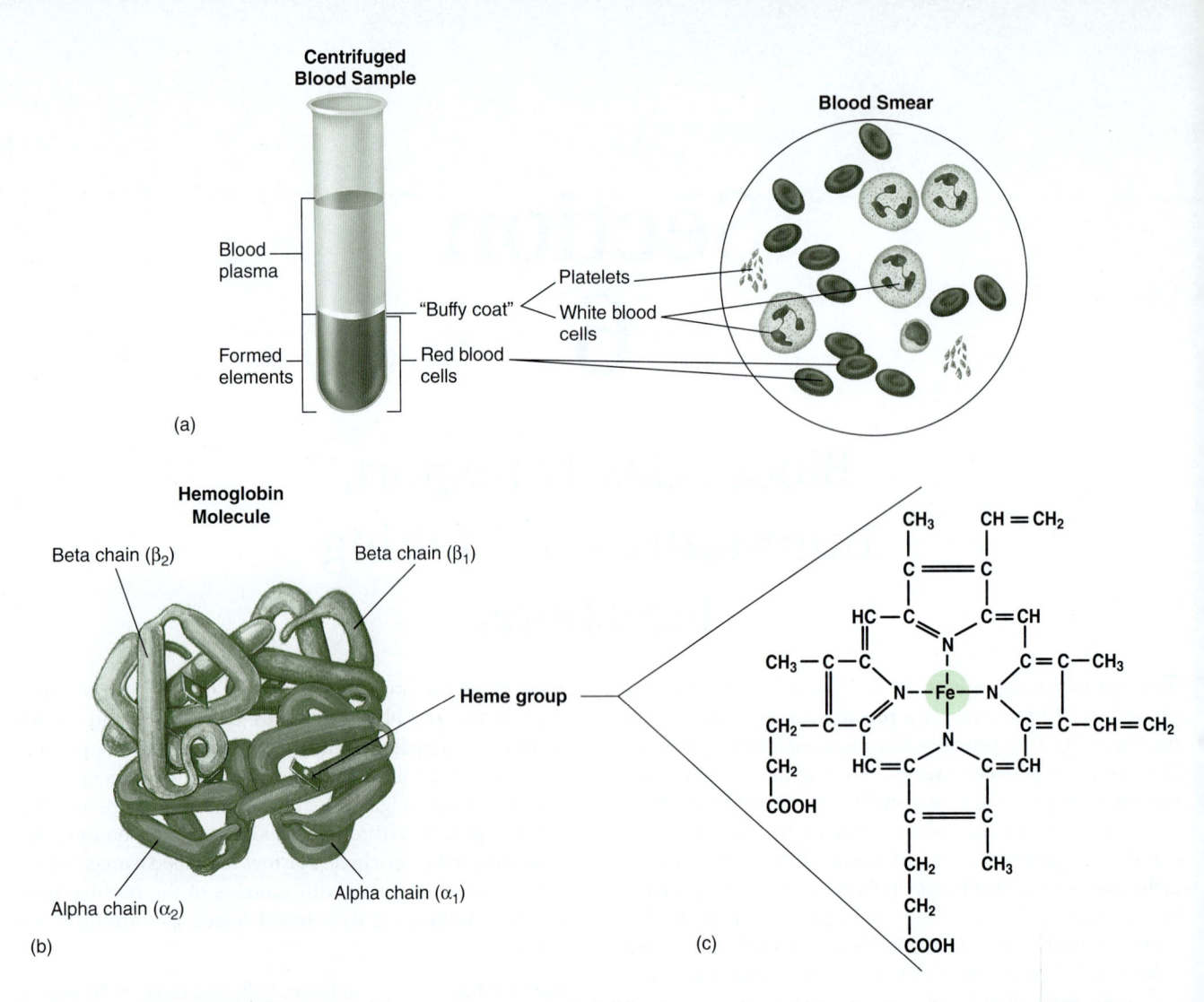

Figure 6.1 Composition of blood. (a) Centrifugation of whole blood causes the red blood cells to become packed at the bottom of the tube; this layer is covered by a thin, buffy coat of leukocytes and platelets, and all of these formed elements of blood are separated from the plasma. (b) The structure of hemoglobin within red blood cells, and (c) the structure of each heme group. There are four heme groups per hemoglobin molecule.

(For a full-color version of this figure, see figs. 13.1 for part (a) and 16.32 for part (b) in *Human Physiology*, thirteenth edition, by Stuart I. Fox.)

EXERCISE 6.1

Red Blood Cells and Oxygen Transport

MATERIALS

1. Hemocytometer
2. Unopettes (Becton-Dickinson) for manual red blood cell count and hemoglobin measurements.

   Note: *Unopettes are no longer sold, and so the procedures for this exercise that use Unopettes—the red blood cell count and hemoglobin concentration measurement—are provided only on the text website at www.mhhe.com/fox13.*

3. Heparinized capillary tubes, clay capillary tube sealant (Seal-ease), microcapillary centrifuge, hematocrit reader
4. Microscope
5. Sterile lancets and 70% alcohol for preparing fingertip blood. Alternatively, dog, cat, or sheep blood (obtained from a veterinarian) may be used.
6. Colorimeter and cuvettes
7. Container for disposal of blood-containing items

Almost all of the oxygen transported by the blood is carried within the red blood cells attached to hemoglobin. Measurements of the oxygen-carrying capacity of blood include the red blood cell count, hemoglobin concentration, and hematocrit. Anemia results when one or more of these measurements is abnormally low.

LEARNING OUTCOMES

You should be able to:

1. Describe the composition of blood.
2. Describe the composition of hemoglobin, and explain how hemoglobin participates in oxygen transport.
3. Demonstrate the procedures for taking the red blood cell count and hemoglobin and hematocrit measurements, and list the normal values for these measurements.
4. Explain how measurements of the oxygen-carrying capacity of blood can be used to diagnose anemia and polycythemia.

Textbook/Multimedia Correlations

Before performing this exercise, you should study the introductory material presented here. Further information relating to this exercise can be found in these pages of *Human Physiology*, thirteenth edition, by Stuart I. Fox:

- *The Formed Elements of Blood.* Chapter 13, p. 407.
- *Hemoglobin and Oxygen Transport.* Chapter 16, p. 558.
- Multimedia correlations are provided in *MediaPhys 3.0:* Topics 10.37–10.49.

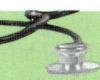

Clinical Investigation

A premenopausal female patient experiencing chronic fatigue was tested and found to have a red blood cell count of 3.9 million/mm^3, a hemoglobin concentration of 10.0 g/dL, and a hematocrit of 32. She did not appear jaundiced.

- Evaluate these measurements to determine if they are abnormal, what condition they may indicate, and how this relates to the patient's fatigue.
- Calculate the MCV and MCH and explain the most likely cause of any abnormalities you find.

Each ventilation cycle delivers a fresh supply of oxygen to the alveoli of the lungs. The amount of oxygen that leaves the lungs dissolved in plasma is equal to 0.3 mL of O_2 per 100 mL of blood. The amount of oxygen leaving the lungs in whole blood, however, is equal to 20 mL of O_2 per 100 mL of blood. Most of the oxygen (19.7 mL O_2 per 100 mL blood), therefore, must be carried within the cellular elements of the blood. This oxygen is carried by **hemoglobin** molecules within the red blood cells (fig. 6.1b). In this way the oxygen is transported to the body cells and used for aerobic cell respiration.

Each hemoglobin molecule consists of two pairs of polypeptide chains (one pair called the alpha chains and one pair called the beta chains) and four disc-shaped organic groups called heme groups. Each heme group contains one central ferrous ion (Fe^{2+}) capable of bonding with one molecule of oxygen (fig. 6.1b). Thus, one molecule of hemoglobin can combine with four molecules of oxygen.

The hemoglobin within the red blood cells load up with oxygen in the capillaries of the lungs and unload oxygen in the tissue capillaries. In both cases, oxygen moves according to its diffusion gradient. Red blood cells always metabolize anaerobically (so they do not consume the oxygen they carry), thereby maintaining a maximum diffusion gradient for oxygen between the red blood cells and the tissues.

The **oxygen-carrying capacity** of the blood is dependent on the total number of red blood cells and, consequently, on the total amount of hemoglobin. The total number of red blood cells is dependent on a balance between the rates of red blood cell production and destruction. The rate of red blood cell production by the bone marrow is regulated by the hormone **erythropoietin,** secreted by the kidneys. Erythropoietin secretion is increased when blood oxygen levels fall, such as when traveling in high-altitude environments. The rate of renal erythropoietin secretion is, therefore, regulated by the oxygen requirements of the body.

Older red blood cells (those approximately 120 days old) are routinely destroyed by the action of phagocytic cells fixed to the sides of blood channels (sinusoids) by a meshwork (reticulum) of fibers. Located in the spleen, liver, and bone marrow, these fixed phagocytes compose the **reticuloendothelial system.** These reticuloendothelial cells digest the hemoglobin within the old red blood cells into the component parts of protein, iron, and the heme pigment. The protein is hydrolyzed and returned to the general amino acid pool of the body, the iron is recycled to the bone marrow, and the heme is changed into a new pigment called **bilirubin.**

Bilirubin released by the spleen, liver, and bone marrow (reticuloendothelial system) enters the blood as free bilirubin, which is nonpolar and attached to albumin in the plasma. Because it is bound to albumin, it cannot be filtered by the kidneys or secreted by the liver into the bile. However, the liver combines some of the free bilirubin with another molecule (glucuronic acid)—producing conjugated bilirubin—that makes it more polar and water-soluble. The liver can secrete conjugated bilirubin into the bile.

The iron derived from the heme groups by the reticuloendothelial system of the liver and spleen travels in the blood attached to a protein carrier called *transferrin*. The transferrin is taken out of the blood by cells of the bone marrow, and it is this recycled iron that supplies most of the body's need for iron. The balance of the requirement for iron, though small, must be made up for in the diet. Dietary iron is absorbed from the intestine and is also transported in the blood bound to transferrin.

Clinical Applications

Jaundice is a yellow staining of the tissues produced by high blood concentrations of either free or conjugated bilirubin. In adults, jaundice caused by high blood levels of conjugated bilirubin may occur when bile excretion is blocked by gallstones. Jaundice caused by high blood levels of free bilirubin is usually caused by an excessively high rate of red blood cell destruction. *Physiological jaundice of the newborn* is due to high levels of free bilirubin in otherwise healthy neonates. In premature infants, it may be caused by inadequate amounts of the liver enzymes needed to conjugate bilirubin so that it can be excreted in the bile. These jaundiced babies are usually treated by exposing them to blue light (with wavelengths of 400 to 500 nm), which converts free bilirubin in cutaneous vessels into a more water-soluble form that can be excreted in the bile and urine.

A. Hematocrit

When whole blood is centrifuged the red blood cells become packed at the bottom of the tube, leaving the plasma at the top. The ratio of the volume of packed red blood cells to the total blood volume is called the **hematocrit.**

Normal Values Normal hematocrit values: For an adult male, is 47 ± 7%; and for an adult female is 42 ± 5% (of the total blood volume).

Procedure

1. Prick your finger with a sterile lancet to obtain a drop of blood as described in the previous procedure. Again, discard the first drop onto an alcohol swab, and dispose of this properly in a designated container.
2. Obtain a heparinized capillary tube (**heparin** is an *anticoagulant*). One end of the tube is marked with a red band. Touch the end of the capillary tube opposite the marked end to the drop of blood, allowing blood to enter the tube by capillary action and gravity (fig. 6.2). The tube does not have to be completely full (half full or more is adequate), and air bubbles are not important (they will disappear during centrifugation).
3. Seal the red-banded (fire-polished) end of the capillary tube by gently pushing it upright into clay capillary sealant. Carefully rotate and remove the tube.

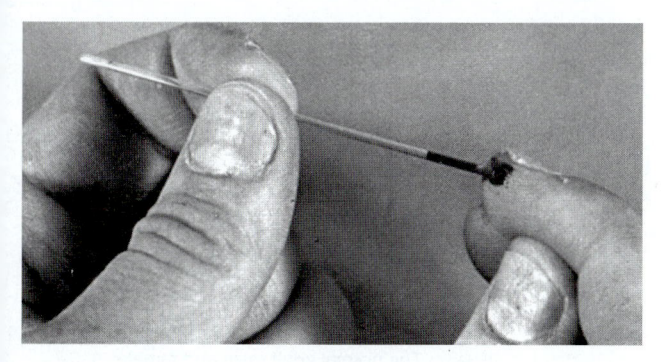

Figure 6.2 The method for filling a capillary tube with fingertip blood.

4. Place the sealed capillary tube in a numbered slot of the microcapillary centrifuge, with the plugged end of the capillary tube facing outward against the rubber gasket. Screw the top plate onto the centrifuge head and centrifuge for 3 minutes. At the end of the centrifugation, determine the hematocrit with the hematocrit reader provided, and enter this value in your laboratory report.

B. Red Blood Cell Count

> **Normal Values** Normal red blood cell counts: For an adult male is 4.5–6.0 million per cubic mm (mm^3); and for an adult female is 4.0–5.5 million/mm^3.

Unopette materials and procedures, and accompanying figures, are available on the text website at www.mhhe.com/fox13 because only certain laboratory courses have Unopettes to perform this exercise. This relates to exercises 6.1B (Red Blood Cell Count) and 6.1C (Hemoglobin Concentration). However, the information in these exercises is important for all physiology students and so is retained in this laboratory guide.

C. Hemoglobin Concentration

Hemoglobin absorbs light in the visible spectrum and hence is a pigment (a colored compound). It should therefore be possible to measure the concentration of hemoglobin in a hemolyzed sample of blood by measuring the intensity of its color. This procedure, however, is complicated in that red blood cells contain different types of hemoglobin, and each type absorbs light to different degrees in the different regions of the visible spectrum (i.e., has a slightly different color).

When the oxygen concentration of the blood is high, such as in the capillaries of the lungs, normal **deoxyhemoglobin** combines with oxygen to form the compound **oxyhemoglobin.** When the concentration of oxygen in the blood is low, such as in the capillaries of the tissues, the oxyhemoglobin dissociates to form reduced hemoglobin and oxygen.

$$\text{deoxyhemoglobin} + \text{oxygen} \underset{\text{tissues}}{\overset{\text{lungs}}{\rightleftarrows}} \text{oxyhemoglobin}$$

Arterial blood is bright red due to the predominance of the oxyhemoglobin pigment, whereas venous blood has the darker hue characteristic of deoxyhemoglobin. It should be emphasized, however, that venous blood, although darker in color, still contains a large amount of oxyhemoglobin; this functions as an oxygen reserve.

A less common though clinically important form of hemoglobin is **carboxyhemoglobin,** a complex of hemoglobin and carbon monoxide. This complex, unlike oxyhemoglobin, does not readily dissociate; thus, the hemoglobin bonded to carbon monoxide cannot participate in oxygen transport. The carboxyhemoglobin complex has a bright, cranberry red color.

A small percentage of hemoglobin contains iron oxidized to the ferric state (Fe^{3+}) instead of being in the normal ferrous state (Fe^{2+}). Hemoglobin in this oxidized state is called **methemoglobin** and is incapable of bonding with either oxygen or carbon monoxide. An increase in the amount of methemoglobin is associated with some genetic diseases, or it may result from the action of certain drugs, such as nitroglycerin.

The procedure for this exercise is provided on the text website at www.mhhe.com/fox13. In this procedure, all of the hemoglobin types in an unknown sample are converted into methemoglobin and the absorbance is measured in a colorimeter. The hemoglobin concentration of the blood can then be determined by comparison with the absorbance of a standard solution of known hemoglobin concentration.

> **Normal Values** Normal hemoglobin concentration for an adult male is 13–16 g/dL, and for an adult female is 12–15 g/dL.

D. Calculation of Mean Corpuscular Volume (MCV) and Mean Corpuscular Hemoglobin Concentration (MCHC)

An abnormally low hemoglobin, hematocrit, or red blood cell count may indicate a condition known as anemia. Anemia may be caused by iron deficiency, vitamin B_{12} and folic acid deficiencies, bone marrow disease, hemolytic disease (e.g., sickle-cell anemia), loss of blood through hemorrhage, or infections. Diagnosis of a specific type of anemia is aided by relating the measurements of hemoglobin, hematocrit, and red blood cell count to derive the **mean corpuscular volume (MCV)** and the **mean corpuscular hemoglobin concentration (MCHC).**

Clinical Applications

Anemia is subdivided into a number of categories on the basis of the **MCV** and **MCHC**. *Macrocytic anemia* (MCV greater than 94, MCHC within normal range) may be caused by folic acid deficiency or by vitamin B_{12} deficiency associated with the disease *pernicious anemia*. In this condition, a polypeptide "intrinsic factor" necessary for vitamin B_{12} absorption from the small intestine is not secreted as it normally is by the stomach. *Normocytic normochromic anemia* (normal MCV and MCHC) may be due to acute blood loss, hemolysis, aplastic anemia (damage to the bone marrow), or a variety of chronic diseases. *Microcytic hypochromic anemia* (abnormally low MCV and low MCHC), the most common type, is caused by inadequate amounts of iron.

Procedure

Note: *Use the values for hematocrit obtained in this exercise, and the values for red blood cell count and hemoglobin concentration obtained using the procedures on the text website, www.mhhe.com/fox13, or use hypothetical values, to calculate the MCV and MCHC.*

1. Calculate the mean corpuscular volume (MCV) according to the following formula:

$$\text{MCV} = \frac{\text{hematocrit} \times 10}{\text{RBC count (millions per mm}^3 \text{ blood)}}$$

Example
Hematocrit = 46
RBC count = 5.5 million

$$\text{MCV} = \frac{46 \times 10}{5.5} = \mathbf{84}$$

Calculate your mean corpuscular volume (MCV), and enter it in the laboratory report.

Normal Values The normal adult male and female mean corpuscular volume (MCV) ranges from 82–92 cubic micrometers.

2. Calculate your mean corpuscular hemoglobin concentration (MCHC) according to the following formula:

$$\text{MCHC} = \frac{\text{Hemoglobin (g/dL)} \times 100}{\text{Hematocrit}}$$

Example
Hematocrit = 46
Hemoglobin = 16 g/dL

$$\text{MCHC} = \frac{16 \times 100}{46} = \mathbf{35}$$

Calculate your mean corpuscular hemoglobin concentration (MCHC), and enter it in the laboratory report.

Normal Values The average normal adult male and female mean corpuscular hemoglobin concentration (MCHC) is 32–36 (in percent).

Laboratory Report 6.1

Name _____

Date _____

Section _____

DATA FROM EXERCISE 6.1

A. Hematocrit
1. Enter your hematocrit (in percent): _____.

B. Red Blood Cell Count
1. Enter your red blood cell count per cubic millimeter (mm^3) of blood: _____ per mm^3 of blood.

C. Hemoglobin Concentration
1. Enter your hemoglobin concentration in g per 100 mL (or deciliter, dL) of blood: _____ g/dL of blood.

D. MCV and MCHC
1. Calculate your mean corpuscular volume (MCV) and mean corpuscular hemoglobin concentration (MCHC) and enter these values here.

   MCV: _____ cubic micrometers (μm^3)
   MCHC: _____ percent (%)

2. Compare your values to the normal values, and write your conclusions here.

REVIEW ACTIVITIES FOR EXERCISE 6.1

Test Your Knowledge

1. One hemoglobin molecule contains _____ heme groups; each heme group normally combines with one molecule of _____.

2. The hormone _____ stimulates the bone marrow to produce red blood cells; this hormone is secreted by the _____.

3. Old red blood cells are destroyed by the _____ system, which includes these three organs:
   (a) _____
   (b) _____
   (c) _____

4. Heme derived from hemoglobin, minus the iron, is converted into a different pigment, known as _____; an accumulation of this pigment can cause a yellowing known as _____.

5. Define the term *hematocrit*. _____

6. The molecule formed by the binding of oxygen to deoxyhemoglobin: _____.

7. A hemoglobin molecule containing oxidized iron (Fe^{3+}) is called _____.

8. A molecule formed from the combination of hemoglobin and carbon monoxide is _____

9. A general term for an abnormally low red blood cell count or hemoglobin concentration: _____

10. The most common cause of the condition described in question 9 is _____.

Test Your Understanding

11. Describe some of the causes of anemia. Why is anemia dangerous?

12. Newborn babies, particularly premature ones, often have a rapid rate of red blood cell destruction and have jaundice. What is the relationship between these two conditions? How is this jaundice treated, and how does this treatment work?

13. Could a person have a low hematocrit yet have a normal red blood cell count? Explain what might cause this condition.

Test Your Analytical Ability

14. Results of blood tests performed in this exercise would be different for anemia and for carbon monoxide poisoning, yet in one respect these two conditions are similar. Explain why this statement is true.

15. People who live at high altitudes often have a high red blood cell count, a condition called polycythemia. Explain the cause of the polycythemia and its possible benefit. Do you think it could have any adverse effects? Explain.

16. Athletes sometimes use "blood doping" to enhance performance. They "bank" red blood cells prior to an athletic event and then return the banked red blood cells to their blood just before the competition. This is considered cheating and is banned by most athletic organizations. How could blood doping provide an advantage to the athlete? What could be its health disadvantage? Why do you think this practice could be difficult to detect?

Test Your Quantitative Ability

17. Calculate the hemoglobin concentration in an unknown blood sample if the unknown had an absorbance of 0.32 and a standard (with a hemoglobin concentration of 12 g/dL) had an absorbance of 0.24.

18. Calculate the mean corpuscular volume (MCV) given the following values: hematocrit = 52; rbc count = 4.6 million.

19. Calculate the mean corpuscular hemoglobin concentration (MCHC) given the following values: hematocrit = 52; hemoglobin concentration = 13 g/dL.

Clinical Investigation Questions

20. What do the patient's blood measurements indicate? How might these relate to her chronic fatigue? What is jaundice and how might it be produced?

21. What are the patient's MCV and MCHC, and what do they suggest? What do you think is the most likely cause of the patient's symptoms and how might this be treated?

EXERCISE 6.2

White Blood Cell Count, Differential Count, and Immunity

MATERIALS

1. Microscopes, hemocytometer slides
2. Thoma diluting pipettes
3. Lancets and alcohol swabs, for preparing fingertip blood. Alternatively, dog or cat blood (obtained from a veterinarian) may be used instead.
4. For total white blood cell count: Methylene blue in 1% acetic acid; and for differential count: Wright's stain (or Hardy Diagnostics One Step Wright's Stain)
5. Heparinized capillary tubes and glass slides

White blood cells—lymphocytes, monocytes, neutrophils, eosinophils, and basophils—are agents of the immune system. Lymphocytes provide immunity against specific antigens, whereas the other leukocytes are phagocytic. The total white blood cell count and the relative proportion of each type of white blood cell (differential count) change in a characteristic way in different disease states.

LEARNING OUTCOMES

You should be able to:

1. Distinguish the different types of leukocytes by the appearance of their nuclei and their cytoplasm.
2. Describe the origin and function of B and T lymphocytes.
3. List the phagocytic white blood cells, and explain their functions during local inflammation.
4. Perform a total and a differential white blood cell count, and explain the importance of this information in the diagnosis of diseases.

Textbook/Multimedia Correlations

Before performing this exercise, you should study the introductory material presented here. Further information relating to this exercise can be found in these pages of *Human Physiology*, thirteenth edition, by Stuart I. Fox:

- *The Formed Elements of Blood.* Chapter 13, p. 407.
- *Defense Mechanisms.* Chapter 15, p. 494.
- *Functions of B Lymphocytes.* Chapter 15, p. 503.
- *Functions of T Lymphocytes.* Chapter 15, p. 507.

Clinical Investigation

A patient had a white blood cell count of 12,000, with neutrophils comprising 85% of the total. Shortly thereafter, he had abdominal pain from appendicitis and the appendix was surgically removed. A year later, he was given corticosteroids to treat an inflammation and had an eosinophil differential count of 1%.

- Explain the significance of his total white blood cell count and differential count, and relate these to his subsequent appendectomy.
- Explain the significance of his eosinophil differential count and how that measurement might have been produced.

The **white blood cells (leukocytes)** are divided into two general categories on the basis of their histological appearance: granular (or polymorphonuclear) and agranular. Leukocytes in the granular category have granules in the cytoplasm and lobed or segmented nuclei, whereas those in the agranular category lack visible cytoplasmic granules and have unlobed nuclei (see **plate 1,** following p. 234).

The *granular leukocytes* are distinguished by their affinity for specific stains. The cytoplasmic granules of **eosinophils** stain bright red (the color of eosin stain), and the granules of **basophils** stain dark blue (the color of basic stain). The granules of **neutrophils** have a low affinity for stain; therefore, the cytoplasm of these cells appears relatively clear.

The *agranular leukocytes* include **lymphocytes** and **monocytes.** Lymphocytes are the smaller of these two cell types and are easily identified by their round nuclei and scant cytoplasm. Larger monocytes have kidney-bean-shaped nuclei, often with brainlike convolutions, and their cytoplasm has a ground glass appearance. Monocytes may also sometimes be identified by the appearance of short, blunt, cytoplasmic extensions (pseudopods).

Leukocytes can leave the vascular system and enter the connective tissues of the body by squeezing through capillaries (a process known as *diapedesis* or *extravasation*). During an inflammation response (fig. 6.3), the release of *histamine* from tissue mast cells and basophils increases the permeability of the capillaries and consequently promotes the process of diapedesis. This sequence of events also produces the local edema, redness, and pain associated with inflammation.

Figure 6.3 **The events in a local inflammation.** Antigens on the surface of bacterial cells (1) bind to antibodies, which coat the bacteria. This activates complement and (2) promotes phagocytosis by neutrophils and macrophages. Activation of complement also (3) stimulates mast cells to release histamine and other mediators of inflammation, including chemicals that promote capillary permeability and (4) extravasation (diapedesis) of leukocytes, which invade the inflamed site.

(For a full-color version of this figure, see fig. 15.5 in *Human Physiology*, thirteenth edition, by Stuart I. Fox.)

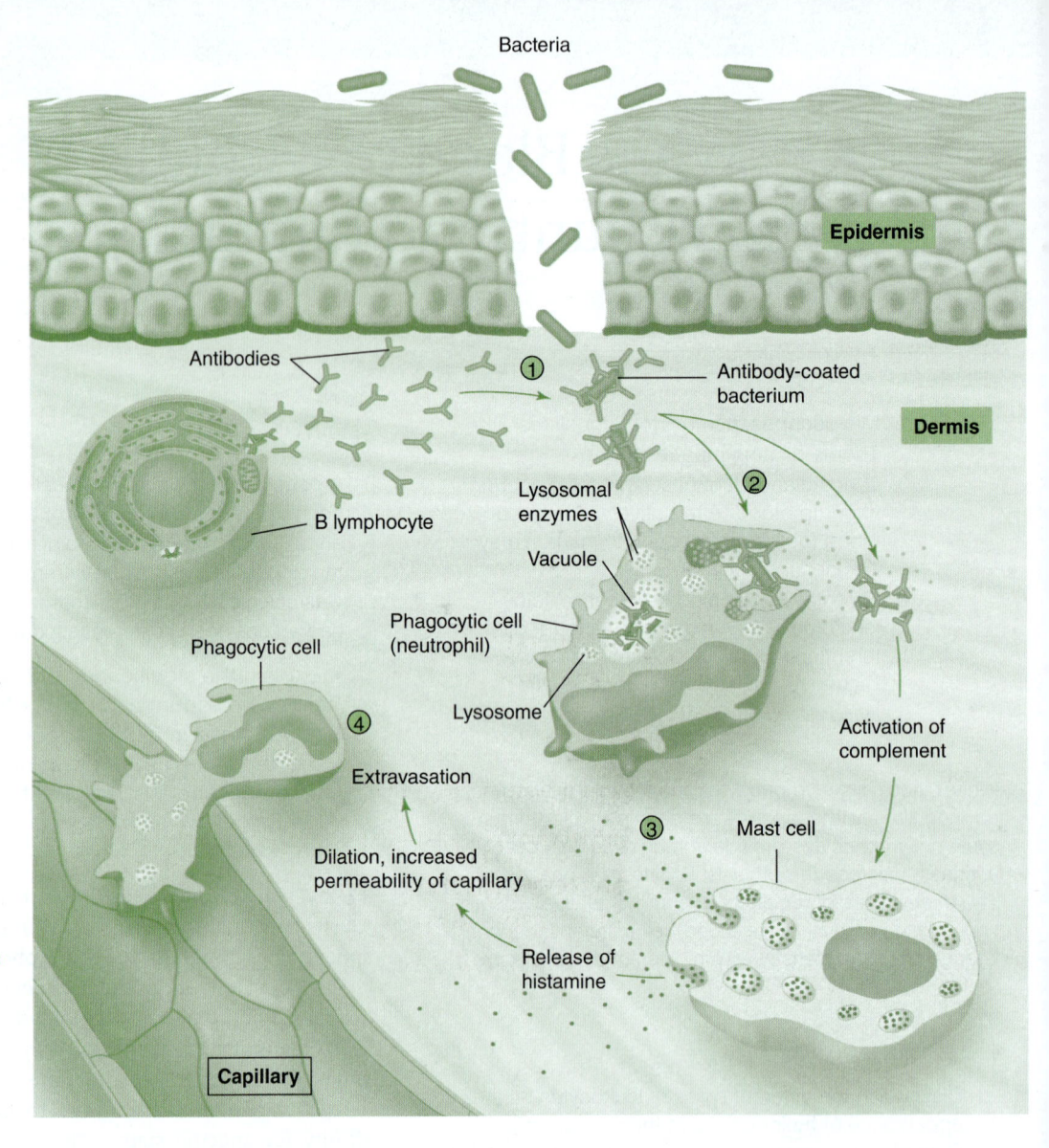

Neutrophils and, to a lesser degree, eosinophils, destroy the invading pathogens by phagocytosis. The battle is then joined by monocytes, which also enter the connective tissues and are transformed into voracious phagocytic cells known as *tissue macrophages*.

Neutrophils kill microorganisms by phagocytosis and the release of enzymes and antimicrobial peptides. They also release *NETS—neutrophil extracellular traps*—composed of extracellular fibers that trap invading pathogens. In the process, the neutrophils undergo programmed cell death where they release other peptidases (protein-digesting enzymes) that liquefy the surrounding tissues. This produces a viscous, protein-rich fluid that, together with the dead neutrophils, forms **pus**.

If these nonspecific immunological defenses are not sufficient to destroy the pathogens, lymphocytes may be recruited and their specific actions used to reinforce the nonspecific immune responses. Lymphocytes are first produced in the embryonic bone marrow, which then seeds the other lymphopoietic sites: the *thymus, lymph nodes,* and *spleen*. The thymus, in turn, sends cells to other locations and apparently regulates the general rate of lymphocyte production at all these sites through the release of a hormone. Therefore, all lymphocytes may be categorized in terms of their ancestry as either bone marrow-derived **B cells** or thymus-derived **T cells**.

Antigens are molecules that activate the immune system. A specific *receptor protein* displayed on the outer membrane of each lymphocyte is capable of recognizing and binding to a specific antigen. By means of this bonding, the antigen selects the lymphocyte capable of attacking it. Bonding of the antigen to its membrane receptor protein stimulates that lymphocyte to divide numerous times, until a large population of genetically identical cells (a *clone*) is produced. This **clonal selection theory** accounts for the fact that the immune response to a second and subsequent exposures to an antigen is greater than the immune response to the initial exposure to the antigen.

Neutrophils   Eosinophils   Basophils

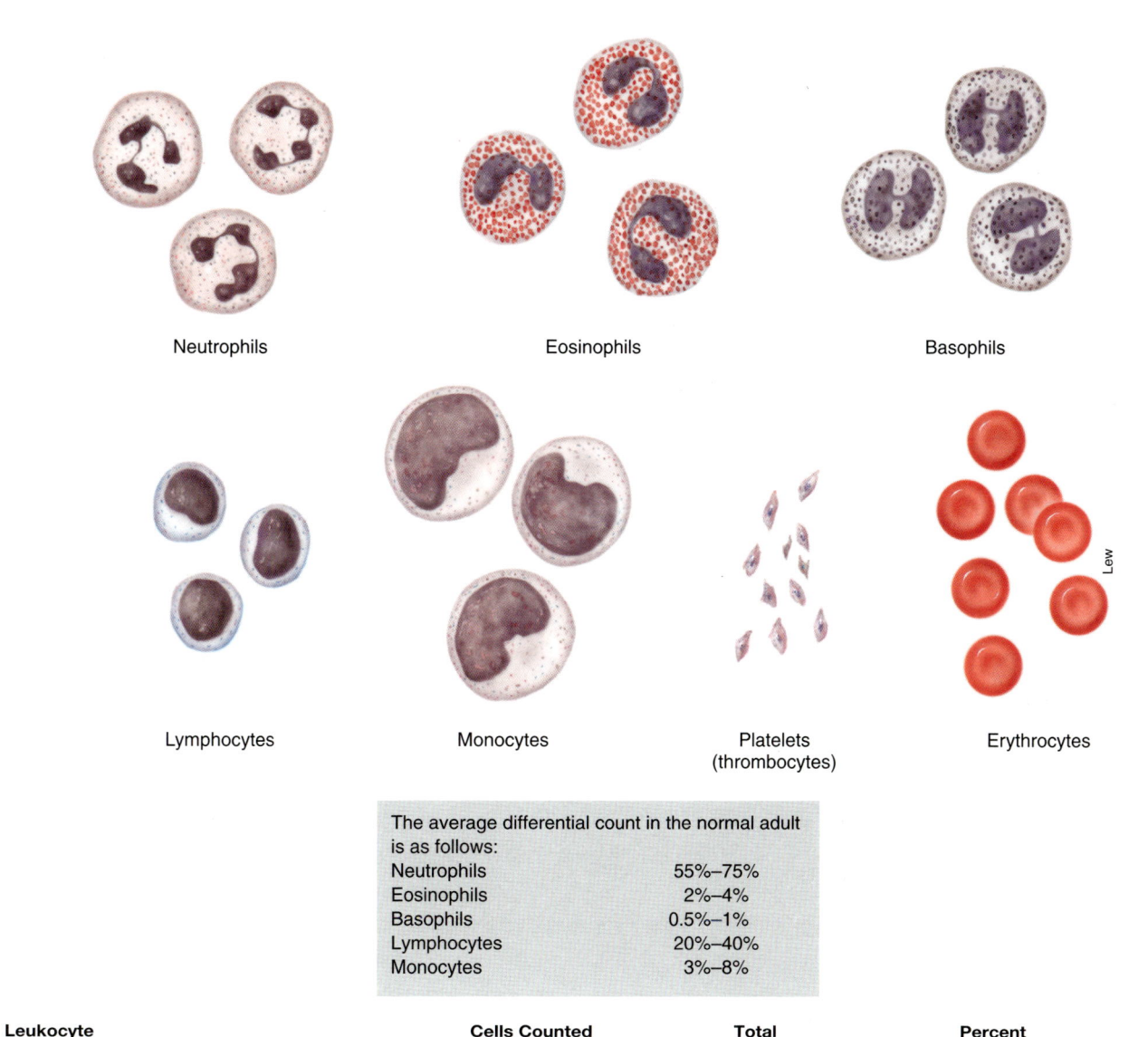

Lymphocytes   Monocytes   Platelets (thrombocytes)   Erythrocytes

The average differential count in the normal adult is as follows:
Neutrophils   55%–75%
Eosinophils   2%–4%
Basophils   0.5%–1%
Lymphocytes   20%–40%
Monocytes   3%–8%

Leukocyte	Cells Counted	Total	Percent
Neutrophils			
Eosinophils			
Basophils			
Lymphocytes			
Monocytes			

Plate 1 Formed elements of blood.

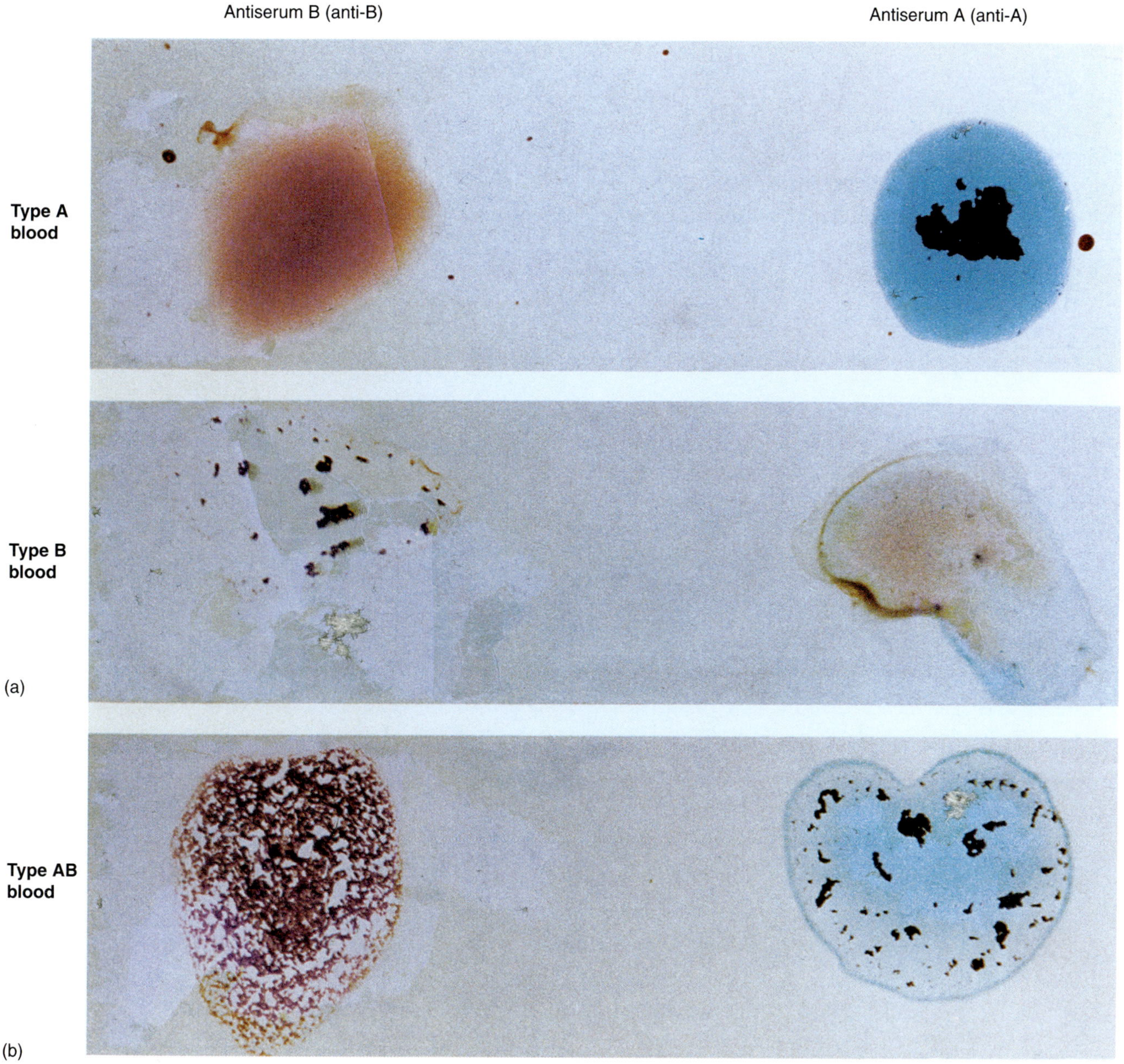

Plate 2 Results of blood typing. (a) Type A blood agglutinating (clumping) with antiserum A (above right), and type B blood agglutinating with antiserum B (below left). Type O blood (not shown) would not have agglutinated with either antiserum. (b) Type AB blood agglutinating with both antiserum B (left) and antiserum A (right).

When stimulated by antigens (generally bacterial), B lymphocytes (or B cells) develop into **plasma cells** that secrete large numbers of **antibody** molecules into the plasma, thus providing **humoral immunity.** These antibodies promote the destruction of bacteria in two ways: (1) the antibodies coat the bacterial cell, making it more easily attacked by the phagocytic neutrophils and tissue macrophages; (2) the attachment of antibody to antigen on the bacterial surface activates a system of plasma proteins—*complement*—that lyses the bacterial cell. These two systems work together, because a chemical released from complement attracts the phagocytic white blood cells and increases capillary permeability. Inflammation can later be suppressed by eosinophils, which engulf free antigen-antibody complexes, thus preventing the complement reaction.

T lymphocytes (or T cells) do not secrete antibodies. Instead, they must move into close proximity with their victim cells to destroy them. Consequently, T lymphocytes are said to provide **cell-mediated immunity,** often involving the secretion of chemicals, called *lymphokines*, released by some T cells. There are three major types of T cells: *killer* (or *cytotoxic*) *T cells,* which kill the victim cells; *helper T cells,* which promote the activity of killer T cells and B cells; and *regulatory* (previously called *suppressor*) *T cells,* which dampen immune responses. This cell-mediated immunity against cells infected with viruses, cancer cells, and cells of tissue transplants is again directed against specific antigens on the victim cell surface. Therefore, both T and B lymphocytes are specific in their immune attack and cooperate with each other in the immune defense against disease.

A. Total White Blood Cell Count

In this procedure, a small amount of blood is diluted with a solution that disintegrates the red blood cells (causes hemolysis) and lightly stains the white blood cells (WBC or leukocytes). The stained white blood cells are counted in the four large corner squares of a hemocytometer (see exercise 6.1A, Red Blood Cell Count).

Because the dilution factor is 20 and each of the four squares counted has a volume of 0.1 cubic millimeters (mm^3), the number of white blood cells per cubic millimeter of blood can be calculated as:

$$\text{WBC per mm}^3 = \frac{\text{white cells} \times 20}{4 \times 0.1 \text{ mm}^3}$$

or,

$$\text{WBC per mm}^3 = \text{white cells} = 50$$

> **Normal Values** The normal white blood cell count is 5,000–10,000 cells per cubic millimeter (mm^3) of blood.

Procedure

Caution: *Because of the danger of exposure to the AIDS virus and other harmful agents when handling blood, each student should perform this and other blood exercises with his or her blood only. All objects that have been in contact with blood must be discarded in a container indicated by the instructor.*

1. As described in exercise 6.1A, obtain a drop of blood after discarding the first drop, and fill the diluting pipette (the one with the white bead) to the *0.5 mark*. Avoid air bubbles; if too much blood is drawn into the pipette, remove it by touching the tip of the pipette to a filter paper.
2. Draw the diluting fluid to the *11 mark* on the pipette.
3. Shake or roll the pipette for 3 minutes.
4. Place a cover slip over one of the silvered areas of the hemocytometer.
5. Discard the first 4 drops from the diluting pipette, then direct the pipette tip to the edge of the cover slip that overlies the hemocytometer. Allow a drop from the pipette tip to be sucked under the cover slip by capillary action.
6. Allow the cells to settle for 1 minute; then using the low-power objective, count the number of white blood cells in the four large corner squares (labeled A, B, C, and D in fig. 6.4). If a cell is lying on the upper or left-hand line, include it in your count, but do not include cells touching the lower or right-hand line.
7. Calculate the number of white blood cells per cubic millimeter of blood, and enter this value in the laboratory report.

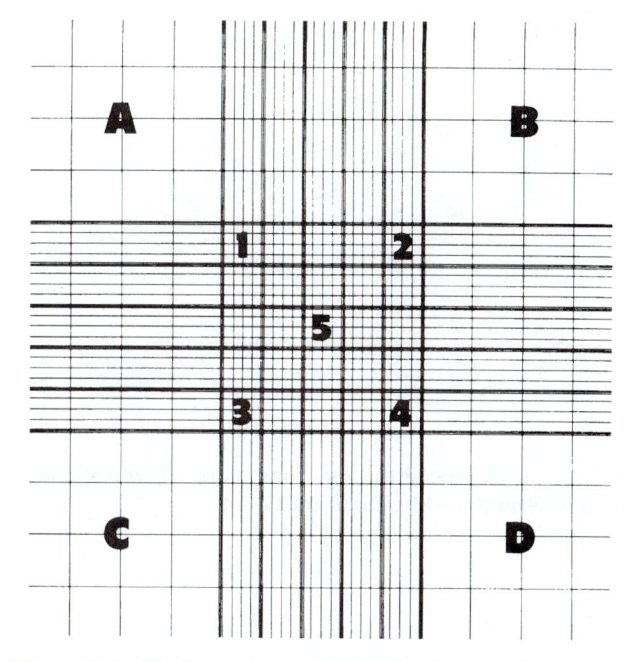

Figure 6.4 **The hemocytometer grid.** Squares 1–5 are used for red blood cell counts; squares A–D are used for white blood cell counts.

B. Differential White Blood Cell Count

Clinically, it is also important to determine the relative quantity (percentage) of each leukocyte type within a population of white blood cells. This percentage is obtained by microscopic identification of each leukocyte type out of a total count of 100 white blood cells (see plate 1).

Procedure

Making a Blood Smear

1. Fill a heparinized capillary tube with blood. This can serve as a reservoir of blood for making a number of slides.
2. Using the capillary tube, apply a small drop of blood on one end of an *absolutely clean* glass slide (fig. 6.5a). Place this slide flat on a laboratory bench.
3. Lower a second glass slide at an angle of 30° to the first slide, so that it is lightly touching the first slide *in front of* the drop of blood (fig. 6.5b).
4. Gently pull the second slide backward into the drop of blood, maintaining the pressure and angle that allows the blood to spread out along the edge of the second slide (fig. 6.5b).
5. Keeping the same angle and pressure, push the second slide across the first in a rapid, smooth motion. The blood should now be spread in a thin film across the first slide. Done correctly, the concentration of blood in the smear should diminish toward the distal end, producing a feathered appearance (fig. 6.5c,d).

Clinical Applications

An increase in the white blood cell count **(leukocytosis)** may be produced by an increase in any one of the leukocyte types. These include: (1) *neutrophil leukocytosis,* due to appendicitis, rheumatic fever, smallpox, diabetic acidosis, or hemorrhage; (2) *lymphocyte leukocytosis,* due to infectious mononucleosis or chronic infections (such as syphilis); (3) *eosinophil leukocytosis,* due to parasitic diseases (such as trichinosis), psoriasis, bronchial asthma, or hay fever; (4) *basophil leukocytosis,* due to hemolytic anemia, chicken pox, or smallpox; and (5) *monocyte leukocytosis,* due to malaria, Rocky Mountain spotted fever, bacterial endocarditis, or typhoid fever. In certain cases, an increase in the relative abundance of one type of leukocyte may occur in the absence of an increase in the total white blood cell count—for example, lymphocytosis due to pernicious anemia, influenza, infectious hepatitis, rubella ("German measles"), or mumps.

A decrease in the white blood cell count **(leukopenia)** is usually due to either a decrease in the number of neutrophils or a decrease in the number of eosinophils. A decrease in the number of neutrophils occurs in typhoid fever, measles, infectious hepatitis, rubella, and aplastic anemia. *Eosinopenia* is produced by an elevated secretion of the corticosteroids, which occurs under various conditions of stress, such as severe infections and shock, and in adrenal hyperfunction (Cushing's syndrome).

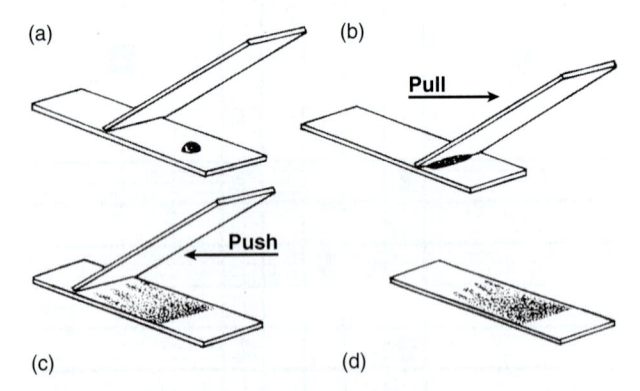

Figure 6.5 A procedure (a–d) for making a blood smear for a differential white blood cell count.

Procedure

Staining a Slide Using Wright's Stain
(Follow the manufacturer's protocol or use the following procedure):

1. Place the slide on a slide rack and flood the surface of the slide with Wright's stain. Rock the slide back and forth gently for 1 to 3 minutes.

Note: *The stain is dissolved in methyl alcohol, which evaporates easily. If any part of the slide should dry during this procedure, the stain will precipitate, ruining the slide.*

2. Drip buffer or distilled water on top of the Wright's stain, being careful not to wash the stain off the slide. Mixing Wright's stain with water is crucial for proper staining; this mixing can be aided by gently blowing on the surface of the stain. Proper staining is indicated by the presence of a metallic sheen on the surface of the stain. The diluted stain should be left on the slide for a full 5 minutes.
3. Wash the stain off the slide with a jet of distilled water from a water bottle, and allow the slide to drain at an angle for a few minutes.
4. Using the oil-immersion objective, count the different types of white blood cells. Start at one point in the feathered-tip area of the blood smear and systematically scan the slide until you have counted a total of 100 leukocytes.
5. Keep a running count of the different leukocytes in the table provided on plate 1 and indicate the total number of each. Calculate the percentage of the total count contributed by each type of leukocyte, and enter these values in your laboratory report.

Procedure

Alternative: Staining a Slide Using One-Step Wright's Stain (Hardy Diagnostics)

1. Make a blood smear as previously described, and allow it to air dry.
2. Place the slide on a staining rack, and use a dropper or pipette to add the One-Step Wright's Stain to the slide so that it covers the smear.
3. Wait 15–30 seconds and add an equal volume of distilled or deionized water to the slide, being careful not to wash off the stain.
4. Gently blow on the slide for 15–45 seconds to mix the water and stain.
5. Pour the water and stain off the slide.
6. Dip the slide in distilled or deionized water and gently swish it for 25 seconds. Alternatively, the slide can be washed with a stream of water from a wash bottle for 25 seconds.
7. Wipe the back of the slide, and allow the slide to air dry in a vertical position before observing the smear in a microscope.
8. Use the oil-immersion objective lens and perform a differential white blood cell count, as described in steps 4 and 5 of the previous procedure.

Laboratory Report 6.2

Name _____
Date _____
Section _____

DATA FROM EXERCISE 6.2

A. Total White Blood Cell Count
1. Enter your white blood cell count: _____ WBC per mm^3.
2. Compare your measured values to the normal range, and write your conclusions in this space.

B. Differential White Blood Cell Count
1. Count each type of white blood cell (leukocyte), and record the number in the following table. Next, determine your total of leukocytes counted (roughly 100). Now, calculate the percentage of each type of white blood cell, and enter these values in this table (see plate 1 for normal values).

Leukocytes	Cells Counted	Total WBCs Counted	Percentage
Neutrophils			
Eosinophils			
Basophils			
Lymphocytes			
Monocytes			

2. Compare your values to the normal range, and write your conclusions in this space.

REVIEW ACTIVITIES FOR EXERCISE 6.2

Test Your Knowledge

Identify the leukocyte by these descriptions:
- ____ 1. polymorphonuclear with poorly staining granules
- ____ 2. agranular with round nucleus, little cytoplasm
- ____ 3. granules with affinity for red stain
- ____ 4. rarest white blood cell
- ____ 5. agranular and phagocytic

(a) eosinophil
(b) neutrophil
(c) monocyte
(d) lymphocyte
(e) basophil

6. Antibodies are produced by _____ lymphocytes; cell-mediated immunity is provided by _____ lymphocytes.

7. White blood cells leave capillaries by a process called _____.

8. The major phagocytic white blood cells are the _____.

9. Molecules that activate the immune system are called _____.

Test Your Understanding

10. Distinguish between humoral and cell-mediated immunity, identifying the cells involved, their origin, and their functions.

11. Describe the clonal selection theory, and explain how it accounts for the ability to defend against subsequent exposure to a particular antigen.

Test Your Analytical Ability

12. *Active immunizations* involve the exposure of a person to a pathogen whose *virulence* (ability to cause disease) has been reduced without altering its *antigenicity* (nature of its antigens). How do you think this might be accomplished? What are the benefits and dangers of this procedure?

13. *Passive immunizations* involve injecting a person exposed to a pathogen with serum containing antibodies, called *antiserum* or *antitoxin*. Antiserum is developed by injecting an animal with a pathogen. What happens in that animal? What are the benefits and shortcomings of passive immunization compared to active immunization?

Test Your Quantitative Ability

14. Suppose 150 white blood cells (wbc) are counted in a hemocytometer. What is the wbc count per cubic millimeter of blood?

15. Suppose, in a differential white blood cell count, a total of 86 white blood cells are counted, including 50 neutrophils and 30 lymphocytes. What is the percentage of neutrophils and lymphocytes in this blood sample?

Clinical Investigation Questions

16. How did the patient's appendicitis relate to the results of his total and differential white blood cell counts?

17. A year after his appendectomy he was being treated with corticosteroids for an inflammation and had an eosinophil differential count of 1%. What clinical word describes this measurement? How do these observations relate to each other?

EXERCISE 6.3

Blood Types

MATERIALS

1. Sterile lancets, 70% alcohol
2. Anti-A, anti-B, and anti-Rh sera (Hardy Diagnostics)
3. Slide warmer, glass slides, and toothpicks
4. Container for the disposal of blood-containing objects
5. Artificial blood kits obtained from Carolina Biological or Ward's) can be used instead of student blood.

Red blood cells (RBCs) have characteristic molecules on the surface of their membranes that can be different in different people. These genetically determined membrane molecules can function as antigens—capable of bonding to specific antibodies when exposed to plasma from a person with a different blood type. The major blood group antigens are the Rh antigen and the antigens of the ABO system.

LEARNING OUTCOMES

You should be able to:

1. Explain the meaning of *blood type,* and identify the major blood types.
2. Explain how agglutination occurs and how agglutination tests can be used to determine a person's blood type.
3. Identify the different genotypes that can produce the different blood group phenotypes, and explain how different blood types can be inherited.
4. Explain how erythroblastosis fetalis is produced.
5. Explain the dangers of mismatched blood types in blood transfusions.

Textbook/Multimedia Correlations

Before performing this exercise, you should study the introductory material presented here. Further information relating to this exercise can be found in these pages of *Human Physiology,* thirteenth edition, by Stuart I. Fox:

- *Red Blood Cell Antigens and Blood Typing.* Chapter 13, p. 412.

Clinical Investigation

A woman with blood type O negative, whose husband is type A positive, gave birth to her first child, who was type O positive. This child was successfully treated for erythroblastosis fetalis in the neonatal ICU unit and discharged.

- Explain the meaning of these blood types and how the child inherited its specific blood type.
- Identify erythroblastosis fetalis and explain how it developed in the baby.
- Describe the precautions the mother must take in subsequent pregnancies to prevent erythroblastosis fetalis in future children she may have.

When blood from one person is mixed with plasma from another person, the red blood cells will sometimes **agglutinate,** or clump together (fig. 6.6). This agglutination reaction, very important in determining the safety of transfusions (agglutinated cells can block small blood vessels), is due to a mismatch of genetically determined blood types.

On the surface of each red blood cell are a number of molecules that have antigenic properties, and in the plasma each antibody molecule has two combining sites for antigens. In a positive agglutination test, the red blood cells clump together because they are combined through antibody bridges (fig. 6.6).

A. The Rh Factor

One of the antigens on the surface of red blood cells is the **Rh factor** (named because it was first discovered in rhesus monkeys). The Rh factor is found on the red blood cell membranes of approximately 85% of the people in the United States. The presence of this antigen on the red blood cells (an **Rh positive** phenotype) is inherited as a dominant trait and is produced by both the *homozygous (RR)* genotype and the *heterozygous (Rr)* genotype. Individuals who have the *homozygous recessive* genotype *(rr)* do not have this antigen on their red blood cells and are said to have the **Rh negative** phenotype.

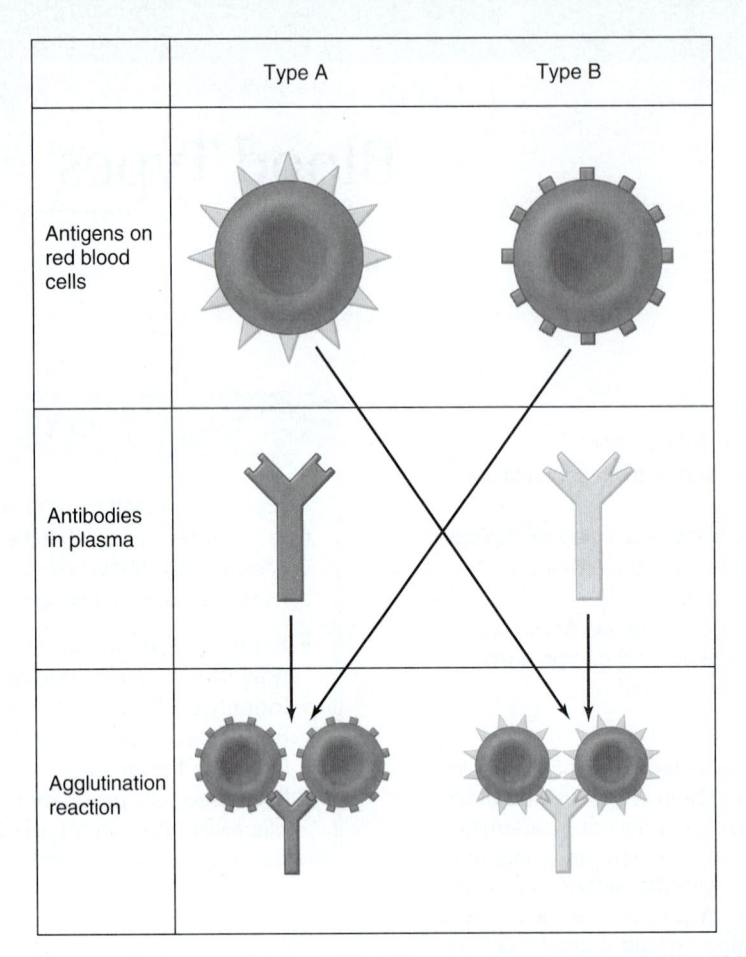

Figure 6.6 Agglutination reaction. A person with type A blood has type A antigens on their red blood cells and antibodies in their plasma against the type B antigen. A person with type B blood has type B antigens on their red blood cells and antibodies in their plasma against the type A antigen. Therefore, if red blood cells from one blood type are mixed with antibodies from the plasma of the other blood type, an agglutination reaction occurs. In this reaction, red blood cells stick together because of antigen-antibody binding.

(For a full-color version of this figure, see fig. 13.5 in *Human Physiology*, thirteenth edition, by Stuart I. Fox.)

Suppose an Rh positive man who is heterozygous *(Rr)* mates with an Rh positive woman who is also heterozygous.

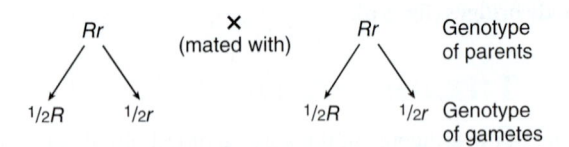

$$Rr \times Rr \quad \text{Genotype of parents}$$
$$\frac{1}{2}R \quad \frac{1}{2}r \quad \frac{1}{2}R \quad \frac{1}{2}r \quad \text{Genotype of gametes}$$

Since the mother is Rh positive, her immune system cannot be stimulated to produce antibodies by the presence of an Rh positive fetus. The development of **immunological competence** does not occur until shortly after birth, so that an Rh negative fetus in an Rh positive mother would not yet have an immune response during normal gestation (pregnancy).

However, when an Rh negative mother is carrying an Rh positive fetus, some of the Rh antigens may enter her circulation when the placenta tears at birth (red blood cells do not normally cross the placenta during pregnancy). Because these red blood cells express an antigen (the Rh factor) foreign to the mother, her immune system will eventually be stimulated to produce antibodies capable of destroying the red blood cells of subsequent Rh positive fetuses, a condition known as *hemolytic disease of the newborn,* or **erythroblastosis fetalis.** However, erythroblastosis fetalis can be prevented by the administration of exogenous Rh antibodies known as **Rho(D) immune globulin** (e.g., *RhoGAM*) to the mother within 72 hours after delivery. These antibodies prevent the Rh positive red blood cells that have entered the maternal circulation from stimulating an immune response in the mother.

Procedure

1. Place one drop of anti-Rh serum on a clean glass slide.
2. Add an equal amount of fingertip or artificial blood, and mix it with the antiserum (use an applicator stick or a toothpick).
3. Place the slide on a slide warmer (45° C to 50° C), and rock it back and forth. This step is optional.
4. Examine the slide for agglutination. If no agglutination is observed after a 2-minute period, examine the slide under the low-power objective of the microscope. The presence of grains of agglutinated red blood cells indicates Rh positive blood.

⚠ **Caution:** Handle only your own blood, and be sure to discard the slide, toothpicks, and lancet in the container provided by the instructor.

5. Enter your Rh factor type (positive or negative) in the laboratory report.

B. THE ABO ANTIGEN SYSTEM

Each individual inherits two genes, one from each parent, that control the synthesis of red blood cell antigens of the ABO classification. Each gene contains the information for one of three possible phenotypes: antigen A, antigen B, or no antigen (written O). Thus, an individual may have one of six possible genotypes: **AA, AO, BB, BO, AB, or OO.**

An individual who has the genotype AO will produce type A antigens just like an individual who has the genotype AA, and therefore both are said to have **type A** blood. Likewise, an individual with the genotype BO and one with the genotype BB will both have **type B** blood. Lack of the antigen is a recessive trait, so an individual with **type O** blood must have the genotype OO.

Unlike many other traits, the heterozygous genotype AB has a phenotype different from either of the homozygous genotypes (AA or BB). There is no dominance between A and B, therefore individuals with the genotype AB produce red blood cells with *both* the A and B antigens (a condition known as *codominance*) and have **type AB** blood. The most common blood types are type O and type A; the rarest is type AB (table 6.1).

Also, unlike the other immune responses considered, antibodies against the A and B antigens are not induced by prior exposure to these blood types. A person with type A blood, for example, has antibodies in the plasma against type B blood even though that person may never have been exposed to this blood. A transfusion with type B blood into the type A person would be extremely dangerous because the anti-B antibodies in the recipient's plasma would agglutinate the red blood cells in the donor's blood. The outcome would be the same if the donor were type A and the recipient type B (see **plate 2,** following p. 234).

Table 6.1 Incidence of Blood Types—Approximate Incidence in the U.S. (%)

Blood Types	Caucasian	Black	Asian
O	45	48	36
A	41	27	28
B	10	21	23
AB	4	4	13

Antigen on RBC Surface	Antibody in Plasma
A (type A)	Anti-B
B (type B)	Anti-A
O (type O)	Anti-A and anti-B
AB (type AB)	No antibody

Procedure

⚠ **Always follow precautions when handling blood.**

1. Draw a line down the center of a clean glass slide with a marking pencil, and label one side A and the other side B.
2. Place a drop of anti-A serum on the side marked A and a drop of anti-B serum on the side marked B.
3. Add a drop of blood to each antiserum, and mix each with a separate applicator stick.
4. Tilt the slide back and forth, and examine for agglutination over a 2-minute period. *Do not heat the slide on the slide warmer.*
5. Enter your ABO blood type in the laboratory report.

Laboratory Report 6.3

Name _____

Date _____

Section _____

DATA FROM EXERCISE 6.3

1. Did your blood agglutinate with the anti-Rh serum? _____
   Are you Rh positive or negative? _____

2. Indicate (with a yes or no) whether your blood agglutinated with the anti-A and anti-B serum.

   Anti-A: _____

   Anti-B: _____

   What is your blood type? _____

REVIEW ACTIVITIES FOR EXERCISE 6.3

Test Your Knowledge

1. Name the antigens present and absent on the surface of a red blood cell if the person is:

   (a) type A negative _____

   (b) type O positive _____

   (c) type AB negative _____

2. If a person has blood type A, the possible genotypes that the person may have are _____ and _____.

3. If a person who is blood type O marries a person who is blood type A, what are the possible blood types their children could have? _____ or _____

4. The universal blood donor is blood type _____.

5. The rarest blood type is blood type _____.

6. The most common Rh type is _____.

7. The person most in danger of having a child who develops erythroblastosis fetalis is a woman who has the blood type _____ when her husband has the blood type _____.

Test Your Understanding

8. What are the dangers of giving a person a transfusion when the blood types don't match?

9. Explain how hemolytic disease of the newborn is produced. How may the disease be prevented?

Test Your Analytical Ability

10. Can blood types be used in paternity cases to prove or disprove possible fatherhood? Give examples to support your answer.

11. Suppose a person who has type A blood receives large amounts of whole blood from a person who has the universal donor blood type. Will that be safe? Explain.

Clinical Investigation Questions

12. What is the meaning of the blood types given for the parents and baby? How did the baby inherit type O positive blood?

13. What is erythroblastosis fetalis, and why did this baby get it? What can be done to prevent this from recurring in the mother's subsequent pregnancies?

Blood Clotting System

EXERCISE 6.4

MATERIALS

1. Pipettes (0.10–0.20 mL) and small test tubes
2. Constant-temperature water-bath set at 37° C
3. 0.02 M calcium chloride, activated thromboplastin, activated cephaloplastin (Dade), fresh plasma
4. Plasma samples with citrate or oxalate used as an anticoagulant (may be obtained from a local hospital)

Two interrelated clotting pathways—the intrinsic system and the extrinsic system—require the successive activation of specific plasma clotting factors. Defects in these factors can be detected by means of two clotting-time tests.

LEARNING OUTCOMES

You should be able to:

1. Describe the intrinsic and extrinsic clotting systems.
2. Describe why bleeding time is prolonged in cases of hemophilia and vitamin K deficiency.
3. Demonstrate the tests for prothrombin time and for activated partial thromboplastin time (APTT). Identify the normal values for each, and explain how these tests are used to diagnose bleeding disorders.

Textbook/Multimedia Correlations

Before performing this exercise, you should study the introductory material presented here. Further information relating to this exercise can be found in these pages of *Human Physiology*, thirteenth edition, by Stuart I. Fox:

- Blood Clotting. Chapter 13, p. 414.

Clinical Investigation

A patient in danger of forming emboli is treated with a coumarin drug and is told that it will take several days to be effective.

- Describe the action of coumarin drugs and explain why it takes time for them to become effective.
- Explain how treatment with a coumarin drug would affect the prothrombin time and activated partial thromboplastin time tests.

Damage to a blood vessel initiates a series of events that, if successful, culminate in *hemostasis* (the arrest of bleeding).

1. The first event is **vasoconstriction**, which decreases the flow of the blood in the damaged vessel.
2. The next event is the formation of a **platelet plug** (fig. 6.7). This response occurs in two steps:
   a. In the first step, platelets adhere to the exposed collagen (connective tissue protein) of the damaged vessel and then release *adenosine diphosphate (ADP)*.
   b. The second step occurs when the ADP, by making the adherent platelets sticky, causes other platelets to cling at this site and form a platelet clump.
3. The third event is the sequential activation of **clotting factors** in the plasma, resulting in the formation of an insoluble fibrous protein, *fibrin,* around the platelet clump (fig. 6.8). This produces a blood clot.

The formation of fibrin from its precursor, *fibrinogen,* requires the presence of the enzyme *thrombin.*

$$\text{fibrinogen} \xrightarrow{\text{thrombin}} \text{fibrin}$$

The insoluble fibrin is formed instantly whenever the enzyme thrombin is present, and thus the formation of thrombin must be a carefully regulated event in the body. The formation of thrombin from its precursor, *prothrombin,* requires the sequential activation of a number of other clotting factors.

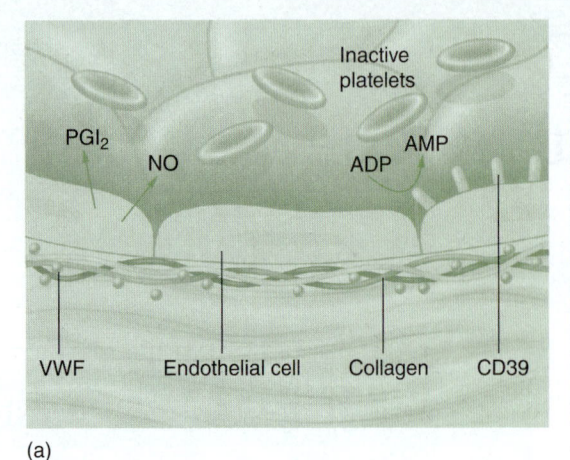

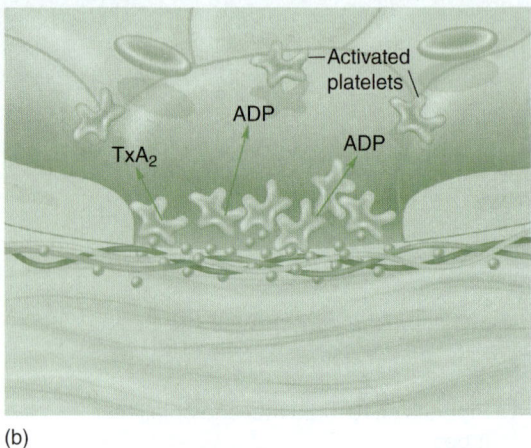

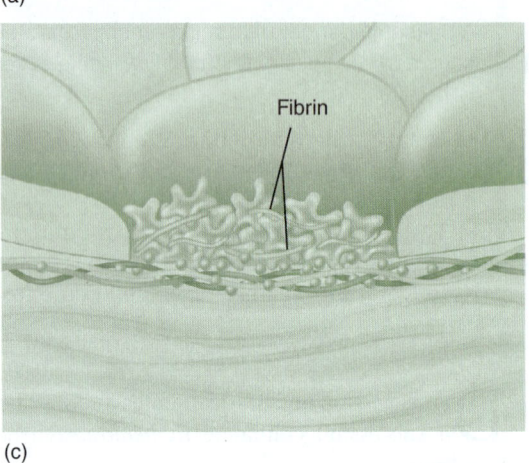

Figure 6.7 Platelet aggregation. (a) Platelet aggregation is prevented in an intact endothelium because it separates the blood from collagen, a potential platelet activator. Also, the endothelium secretes nitric oxide (NO) and prostaglandin I₂ (PGI₂), which inhibit platelet aggregation. An enzyme called CD39 breaks down ADP in the blood, which would otherwise promote platelet aggregation. (b) When the endothelium is broken, platelets adhere to collagen and to von Willebrand's factor (VWF), which helps to anchor the platelets activated by this process and by the secretion of ADP and thromboxane A₂ (TxA₂), a prostaglandin. (c) A platelet plug is formed.

(For a full-color version of this figure, see fig. 13.7 in *Human Physiology*, thirteenth edition, by Stuart I. Fox.)

The conversion of fibrin may occur through two pathways, designated intrinsic and extrinsic. The **intrinsic (contact) pathway** is initiated by exposure of the plasma to a negatively charged surface, such as that provided by collagen in a wound or by the glass of a test tube. This activates a plasma protein called *factor XII,* a proteinase enzyme that activates another clotting factor, which activates yet another (fig. 6.8). The **extrinsic pathway** is initiated by *tissue factor* (also called *tissue thromboplastin*), a membrane protein found in many different tissues. This pathway, shown on the left side of figure 6.8, generates thrombin and fibrin more rapidly than the intrinsic pathway, and is believed to be the more significant pathway for clot formation *in vivo.*

Many people have either an acquired or an inherited inability to form fibrin threads within the normal time interval. This inability may be due to a **vitamin K deficiency.** Vitamin K is needed for the conversion of glutamate, an amino acid found in clotting factor proteins, into a derivative called *gamma-carboxyglutamate.* This derivative is more effective at bonding to Ca^{2+}, and such bonding is needed for proper function of clotting factors II, VII, IX, and X.

Vitamin K deficiency may occur in the newborn who has an inadequate intake of milk, in a person with fat malabsorption and thus inadequate absorption of this vitamin (bile salts facilitate absorption), and as a result of antibiotic therapy destroying intestinal flora (a source of vitamin K). The *coumarin* drugs (*warfarin* and *dicumarol*) block a cellular enzyme needed for the conversion of vitamin K into its active form. Because of this mechanism, and because the effect of activated vitamin K on blood clotting is indirect, the coumarin drugs must be given orally for several days before becoming effective. These drugs are the only oral anticoagulants used clinically.

There are many hereditary conditions in which a clotting factor is either missing or defective. The best known of these conditions, classical **hemophilia,** is due to the genetic inability to synthesize normal factor VIII. This condition, as well as *Christmas disease* (defective factor IX), is inherited as a sex-linked recessive trait. Other genetic defects inherited as autosomal traits include those associated with factors II, VII, X, XI, and XII.

In this exercise, two tests will be performed to screen for defective clotting factors. The formation of thrombin in the plasma samples will be inhibited by an anticoagulant.

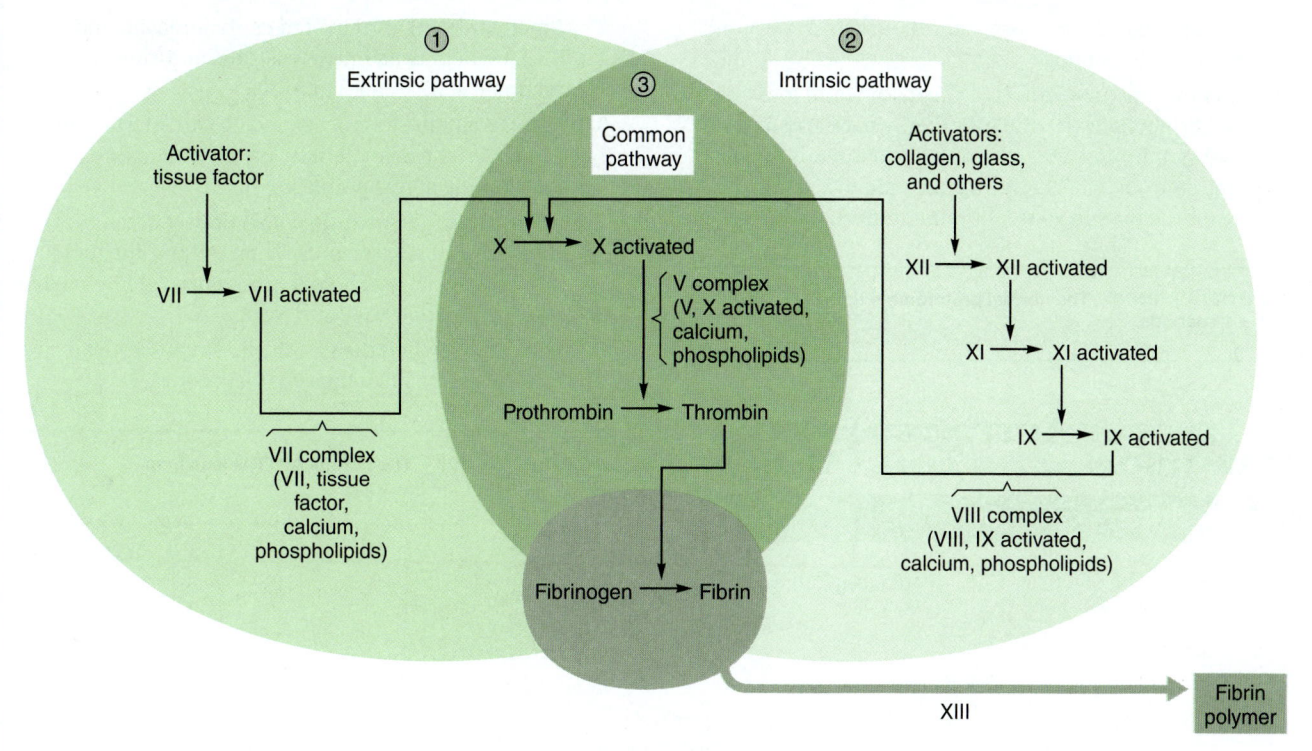

Figure 6.8 The clotting pathways. (1) The extrinsic clotting pathway is initiated by the release of tissue factor. (2) The intrinsic clotting pathway is initiated by the activation of factor XII by contact with collagen or glass. (3) The extrinsic and intrinsic clotting pathways converge when they activate factor X, eventually leading to the formation of fibrin.

(For a full-color version of this figure, see fig. 13.9 in *Human Physiology*, thirteenth edition, by Stuart I. Fox.)

The anticoagulant used, either *citric acid* or *oxalic acid*, removes calcium ion (Ca^{2+}) from the plasma. The removal of Ca^{2+} has an anticoagulant effect because calcium is a necessary cofactor in the activation of a number of the clotting factors. This inhibition can be easily reversed by adding calcium ions during the clotting tests.

Clinical Applications

The test for **prothrombin time** is used to determine deficiencies in the *extrinsic* clotting system and is prolonged when factors V, VII, or X are defective. The test for **activated partial thromboplastin time (APTT)** is used to determine deficiencies in the *intrinsic* clotting system and is sensitive to all defective factors except factor VII. A person with classical hemophilia (defective factor VIII), for example, would have a normal prothrombin time but an abnormal APTT. In this way, these two tests complement each other and can be used, together with other tests, to determine the exact cause of prolonged clotting time (see table 6.2).

Table 6.2 Test Results for Clotting Factors

Defective Factor	Prothrombin Time	APTT
V	Abnormal	Abnormal
VII	Abnormal	Normal
VIII	Normal	Abnormal
IX	Normal	Abnormal
X	Abnormal	Abnormal
XI	Normal	Abnormal
XII	Normal	Abnormal

A. Test for Prothrombin Time

Procedure

1. Pipette 0.10 mL of activated thromboplastin and 0.10 mL of 0.02 M $CaCl_2$ into a test tube. Place the tube in a 37° C water-bath, and allow it to warm for at least 1 minute.
2. Warm a sample of plasma in the water-bath for at least 1 minute. Then use a pipette to forcibly expel 0.10 mL of plasma into the warmed tube containing the thromboplastin–$CaCl_2$ mixture. Start timing at this point.

3. Agitate this tube mixture continuously in the water-bath for 10 seconds.
4. Remove the tube, quickly wipe it, and hold it in front of a bright light. Tilt the tube gently back and forth and stop timing when the first fibrin threads appear (the solution will change from a fluid to a semigel). Enter the time in your laboratory report.

> **Normal Values** The normal prothrombin time is 11 ± 1 seconds.

B. Test for Activated Partial Thromboplastin Time (APTT)

Procedure

1. Warm a tube of 0.02 M $CaCl_2$ by placing it in a 37° C water-bath.
2. Pipette 0.10 mL of activated cephaloplastin and 0.10 mL of plasma into a test tube, and allow it to incubate at 37° C for 3 minutes.
3. Using a pipette, forcibly expel 0.10 mL of warmed $CaCl_2$ into the cephaloplastin-plasma mixture. Start timing at this point.
4. Agitate this tube mixture continuously in the 37° C water-bath for 30 seconds. Then remove the tube, quickly wipe it, and hold it against a bright light while rocking the tube back and forth.
5. Stop timing when the first fibrin threads appear. Enter the time in your laboratory report.

> **Normal Values** The normal APTT is less than 40 seconds.

Laboratory Report 6.4

Name _____
Date _____
Section _____

DATA FROM EXERCISE 6.4

A. Prothrombin Time
1. Enter your prothrombin time measurement: _____ seconds.

B. Activated Partial Thromboplastin Time (APTT)
1. Enter your APTT measurement: _____ seconds.
2. Compare your data to the normal measurements, and write your conclusions here.

REVIEW ACTIVITIES FOR EXERCISE 6.4

Test Your Knowledge

1. The factor that starts the extrinsic clotting pathway: _____.
2. Which clotting pathway is faster? _____.
3. The factor that converts fibrinogen into fibrin is _____.
4. The factor name in question 3 is derived from _____.
5. Which vitamin is needed for the formation of some of the clotting factors? _____.
6. Citric acid (citrate) is an anticoagulant because it _____.
7. The _____ test is used to detect defects in the extrinsic clotting system, whereas the _____ test is used to detect defects in the intrinsic clotting system.
8. In the formation of a platelet plug, platelets release _____, which makes other platelets sticky.
9. The general category of disorders in which a person's clotting time is abnormally long: _____.

Test Your Understanding

10. Which factors are common to both the intrinsic and extrinsic clotting pathway? Which clotting test(s) would be abnormally prolonged if a person had a deficiency in factor VIII? Explain.

11. Which factor(s) would be defective if a person had a prolonged prothrombin time but a normal partial thromboplastin time (APTT)? Explain.

Test Your Analytical Ability

12. Heparin is a mucopolysaccharide extracted from beef lung and liver that inhibits the action of thrombin. What effect would thrombin have on the prothrombin time and APTT? Why was citric acid or oxalic acid used as an anticoagulant instead of heparin in these tests?

13. Why might a person with an abnormally slow clotting time be given vitamin K? Would treatment with vitamin K immediately improve the clotting time? Explain.

14. You might expect that almost all people with hemophilia due to factor VIII or IX deficiency would be males, whereas those with hemophilia due to factor XI or XII deficiency would be equally likely to be males or females. Explain.

Clinical Investigation Questions

15. What is the action of coumarin drugs, and why do these drugs take time to be effective?

16. What would be true of the prothrombin time and APTT tests of a person treated with a coumarin drug? Explain.

Section 7

The Cardiovascular System

The blood transports glucose, amino acids, fatty acids, and other monomers from the digestive tract, liver, and adipose tissue to all the cells of the body. The waste products of cellular metabolism are carried by the blood to the kidneys and lungs for elimination. Hormones, secreted by endocrine glands, are carried by the blood to target organs. Blood is thus the major channel of communication between the different specialized organs of the body.

The interchange of molecules between blood and tissue cells occurs across the walls of **capillaries,** which are composed of only a single layer of epithelial cells (endothelium). Blood is delivered to the capillaries in **arterioles,** microscopic vessels with walls of endothelium, smooth muscle, and connective tissue. The arterioles receive their blood from larger, more muscular **arteries.** Blood is drained from the capillaries into microscopic **venules.** The venules drain their blood into larger **veins** that are less muscular and more distensible than arteries.

Since the tissue cells must be located within 0.10 mm of a capillary for molecules to diffuse adequately, the vascular system within an organ is highly branched. The many narrow, muscular arterioles of this *vascular tree* offer great resistance to blood flow *(peripheral resistance)* through the organs. For organs to receive an adequate blood flow *(perfusion)*, the arterial blood must be under sufficient pressure to overcome this resistance to blood flow. This arterial pressure is routinely measured with a device known as a *sphygmomanometer.*

The blood pressure required to overcome peripheral resistance and maintain adequate tissue perfusion is generated by a muscular pump—the *heart.* The heart has four chambers: two atria and two ventricles. The *right atrium* receives oxygen-depleted blood returning from body cells in the superior and inferior venae cavae; and the *left atrium* receives oxygen-rich blood from the pulmonary veins. The *right ventricle* pumps blood into the pulmonary arteries to the lungs where oxygen enters and carbon dioxide exits the blood. The *left ventricle* pumps blood into the large *aorta,* which by means of its many branches perfuses all the organs in the body. Because the blood pumped out of the heart by each of the two ventricles, the **cardiac output,** is carried by arteries to the body's organs, and blood from the organs is returned by veins to the heart, the cardiovascular system forms a closed circle called the **circulatory system** (fig. 7.1).

The heart's ability to maintain adequate perfusion of the body's organs depends on proper electrical stimulation and muscular contraction, proper functioning of *valves* (directing blood flow within the heart), and the integrity of the blood vessels. These functions can be assessed by various techniques that will be explored in these exercises.

Exercise 7.1	Effects of Drugs on the Frog Heart
Exercise 7.2	Electrocardiogram (ECG)
Exercise 7.3	Effects of Exercise on the Electrocardiogram
Exercise 7.4	Mean Electrical Axis of the Ventricles
Exercise 7.5	Heart Sounds
Exercise 7.6	Measurements of Blood Pressure
Exercise 7.7	Cardiovascular System and Physical Fitness

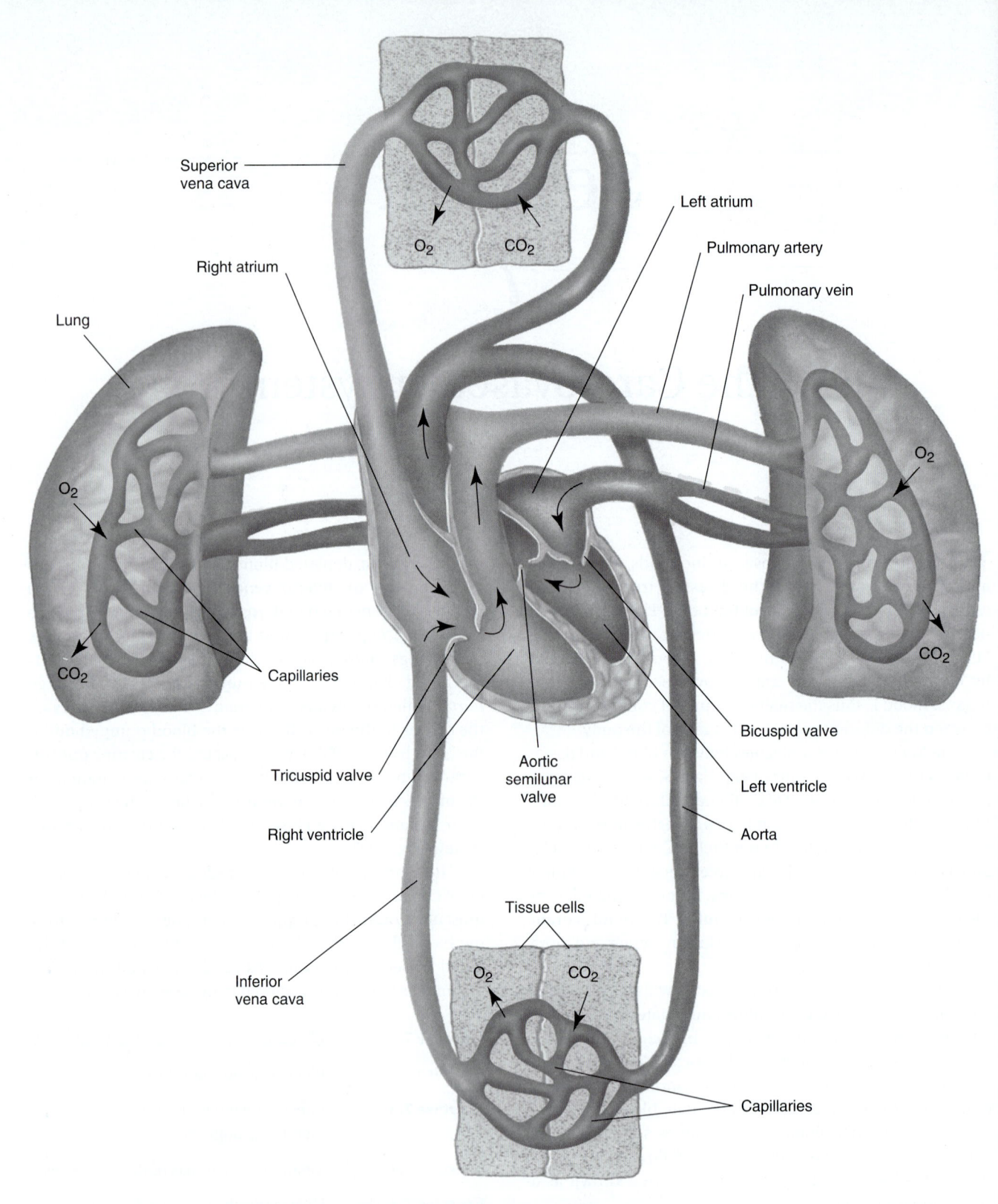

Figure 7.1 A diagram of the circulatory system. The pulmonary arteries and veins compose the pulmonary circulation, whereas other arteries and veins are part of the systemic circulation. The right ventricle pumps blood into the pulmonary circulation, while the left ventricle pumps blood into the systemic circulation.

(For a full-color version of this figure, see fig. 13.10 in *Human Physiology*, thirteenth edition, by Stuart I. Fox.)

EXERCISE 7.1

Effects of Drugs on the Frog Heart

MATERIALS

1. Frogs, dissecting instruments, trays
2. Copper wire or bent pin, thread
3. Recording apparatus: physiograph, transducer coupler, and myograph transducer (Narco); or kymograph, kymograph paper, and kerosene burner
4. As an alternative to pen-and-paper recording equipment, a computerized data acquisition and analysis system, such as those provided by Biopac, iWorx, and others, may be used to perform the exercises in this section
5. Ringer's solution (see exercise 5.1—all drugs to be prepared using Ringer's solution as a solvent); calcium chloride (2.0 g/100 mL); digitoxin (0.2 g/100 mL); pilocarpine (2.5 g/100 mL); atropine (5.0 g/100 mL); potassium chloride (2.0 g/100 mL); epinephrine (0.01 g/100 mL); caffeine (0.2 g/dL); and nicotine (1.0 g/2L; or 6.16 mL liquid/dL)

The heart of a pithed frog may continue to beat after the frog's central nervous system has been destroyed. By this means, the function of the heart and the effects of various drugs on the heart can be studied.

LEARNING OUTCOMES

You should be able to:

1. Describe the pattern of contraction in the frog heart.
2. Describe the effect of various drugs on the heart, and explain their mechanisms of action.

Textbook/Multimedia Correlations

Before performing this exercise, you should study the introductory material presented here. Further information relating to this exercise can be found in these pages of *Human Physiology*, thirteenth edition, by Stuart I. Fox:

- *Cardiac Muscle.* Chapter 12, p. 392.
- *Structure of the Heart.* Chapter 13, p. 418.
- *Electrical Activity of the Heart.* Chapter 13, p. 425.

Clinical Investigation

An anxious patient drank a pot of coffee before an eye exam in which atropine was instilled into the eyes. This patient was also taking atropine (along with other drugs) for gastritis as well as digitalis for atrial fibrillation.

- Explain how anxiety, coffee, atropine, and digitalis affect the heart.
- Explain the mechanisms whereby atropine is used for eye exams and for the treatment of gastritis.
- Explain the mechanisms by which digitalis can help treat atrial fibrillation.

A **drug** is an **exogenous** ("originating outside") substance that affects some aspects of physiology when given to the body. Drugs may be foreign molecules *(xenobiotics)* produced by plants or fungi. However, drugs can also be the same as molecules produced by the human body—**endogenous** ("originating inside") molecules, such as hormones—but that are administered exogenously. Many drugs marketed by pharmaceutical companies are derived from natural products whose chemical structure has been slightly modified to alter the biological activity of the native compounds.

The biological effects of endogenous compounds vary with their concentration. A normal blood potassium concentration, for example, is necessary for good health, but too high a concentration can be fatal. Similarly, the actions exhibited by many hormones at abnormally high concentrations may not occur when the hormones are at normal concentrations. It is important, therefore, to distinguish between the **physiological effects** (normal effects) of these substances and their **pharmacological effects** (those that occur when the substances are administered as drugs). A study of the pharmacology of various substances, however, can reveal much about the normal physiology of the body.

In this exercise, we will test the effects of various pharmacological agents on the heart of a pithed frog. Although the heart, like skeletal muscle, is striated, it differs from skeletal muscles in several respects. The heartbeat is automatic; it does not have to be stimulated by nerves or electrodes

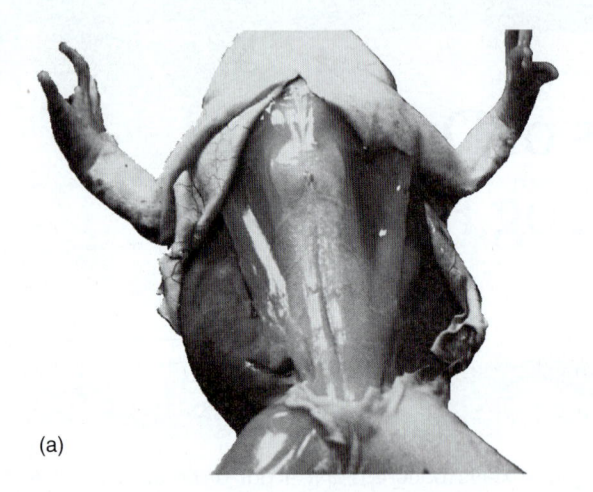

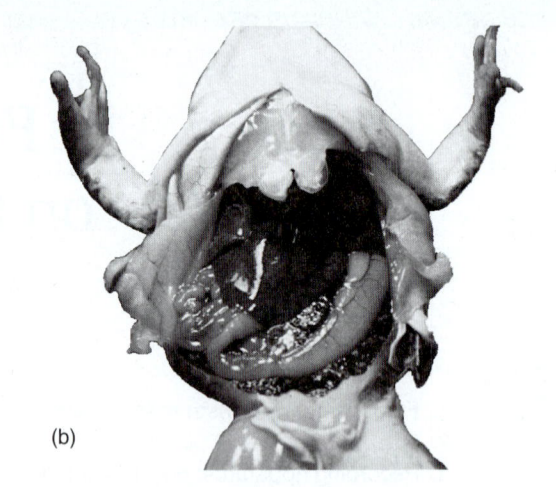

(a)

(b)

Figure 7.2 Procedure for exposing the frog heart. (a) First, the skin is cut. (b) The body cavity is exposed by cutting through the muscles to the sternum, split to expose the heart.

to contract. Action potentials begin spontaneously in the *pacemaker region*—the *sinoatrial* or *SA node* region—of the right atrium and spread through the ventricles in an automatic, rhythmic cycle. As can be seen in the exposed frog heart, this causes the atria to contract before the ventricle. (Unlike mammals, frogs have only one ventricle.)

When the frog heart is connected by a thread to the recording equipment, contractions of the atria and ventricle produce two successive peaks in the recordings. The strength of contraction is related to the amplitude (height) of these peaks, and the rate of beat can be determined by the distance between the ventricular peaks if the chart speed is known. The rate of impulse conduction between the atria and ventricle is related to the distance between the atrial and ventricular peaks in the recording of each cycle. Therefore, the effects of various drugs on the strength of contraction, rate of contraction, and rate of impulse conduction from the atria to the ventricle can be determined.

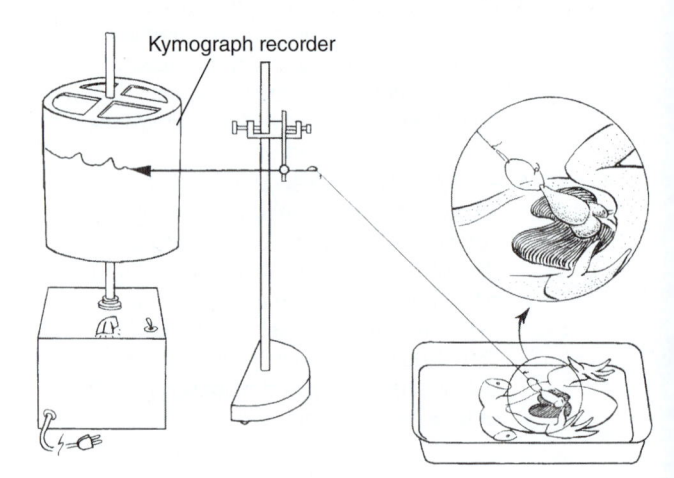

Preparation for Recording

1. Double-pith a frog, and expose its heart (see exercise 5.1 and fig. 7.2). Skewer the apex of the heart muscle with a short length of thin copper wire or a bent pin, being careful not to let the wire enter the chamber of the ventricle. (The frog heart has only one ventricle and two atria.)
2. Bend the copper wire into a loop, and tie one end of cotton thread to this loop or to the head of a bent pin (see the enlarged insert in fig. 7.3).
3. Procedure for **kymograph** recording:
   (a) Tie the other end of the thread to a heart lever. The thread tension should be fairly taut so that contractions of the heart produce movements of the lever.

Kymograph recorder

Figure 7.3 Frog heart setup. The contractions of the heart pull a lever that writes on a moving chart (kymograph). The setup using electronic recording equipment, such as a physiograph recorder, is similar to that shown here, except that the string from the heart will be connected to a force transducer. In the Biopac setup, the force transducer is connected to the MP30 or MP35 amplifier (see fig. 7.5).

   (b) Attach kymograph paper (shiny side out) to the kymograph drum, and rotate the drum slowly over a kerosene burner until the paper is uniformly blackened. Arrange the heart lever so that it lightly drags across the smoked paper. Too much pressure of the writing stylus against the kymograph will prevent movement of the heart lever. (See fig. 7.3 for the proper setup.)
4. Procedure for **physiograph** recording (see physiograph, exercise 5.1):
   (a) Tie the other end of the thread to the hook below the myograph transducer. (Make sure the myograph is plugged into the transducer

(a) (b)

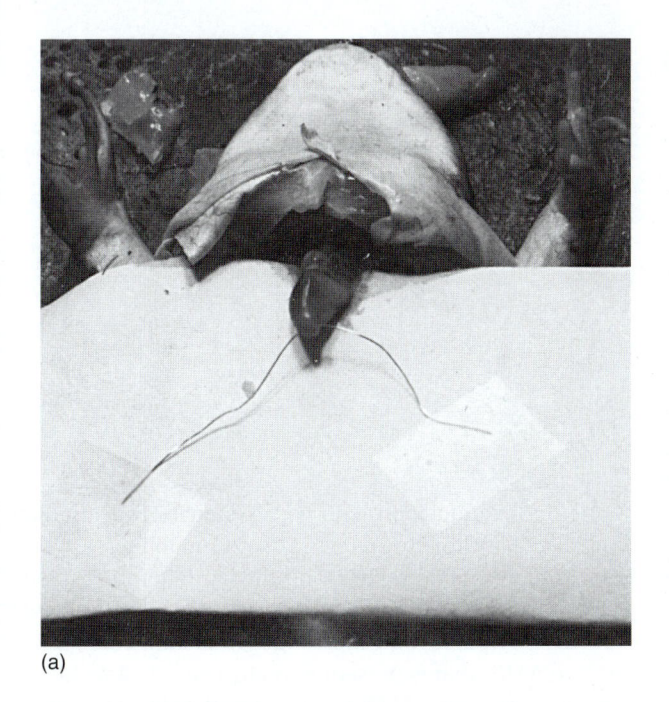

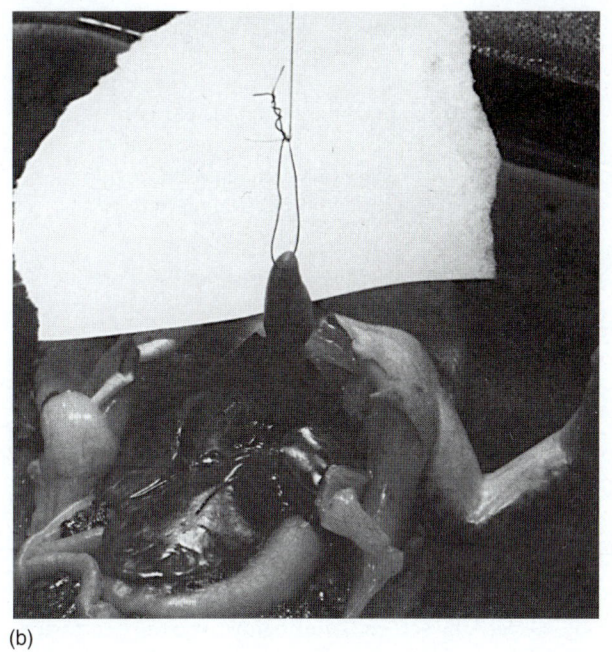

Figure 7.4 **Procedure for setting up the frog heart to record its contractions.** (a) A small length of thin copper wire is passed through the tip of the ventricle. (b) This wire is then twisted together to form a loop, tied by a cotton thread to the hook in the myograph transducer.

coupler on the physiograph.) The heart should be positioned directly below the myograph. Adjust the height of the myograph on its stand so that the heart is pulled out of the chest cavity (fig. 7.4b).

(b) Make sure the physiograph is properly balanced and set the paper speed at *0.5 cm per second.* Depress the record button, and press the gray paper advance button when ready to record.

5. Procedure for **computerized data acquisition and analysis** equipment.
   (a) For the **Biopac system,** the force transducer assembly (SS12LA) and amplifier (MP30 or MP35) are connected to each other and to the computer as illustrated in figure 7.5. The procedures in this exercise or in Biopac *PRO* Lesson A04 can then be followed.
   (b) When using the Biopac system, notice that the force transducer is connected to channel 1 of the MP30 (fig. 7.5), rather than to channel 2 (as it was for the muscle exercises 5.1 and 5.2).
   (c) The Biopac recording of the heart beat will appear much like that shown in figure 7.6 and be displayed in the top half of the computer screen.
   (d) For the **iWorx system,** connect the force transducer to channel 3 of the data acquisition unit. The procedures in this exercise or in iWorx Experiment 10 can then be followed.

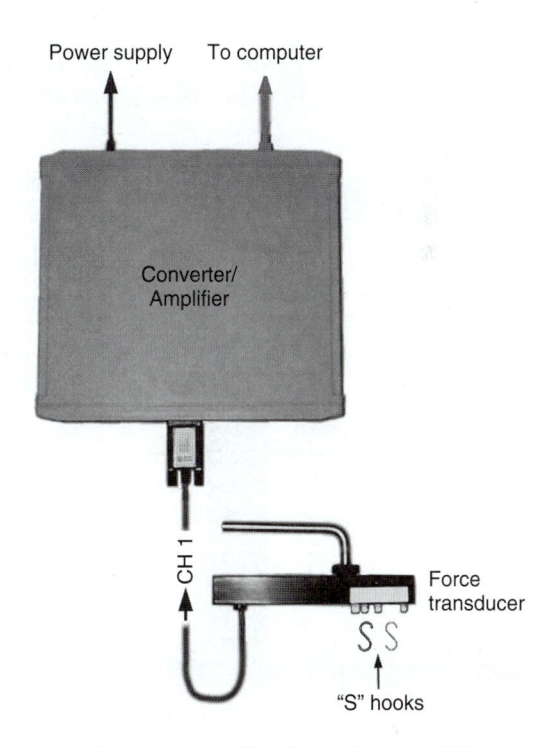

Figure 7.5 **Connection of hardware for use of Biopac system with frog heart exercises.** The details of the setup are described in the Biopac PRO Lesson A04. Once the equipment is set up, it can be used to perform the exercises in this section.

6. Observe the pattern of the heartbeat prior to the addition of drugs; this tracing will serve as the normal, or *control*, record (fig. 7.6). Distinguish the atrial from the ventricular beats. Measure the heart rate (in beats per minute), the strength of contraction (in millimeters deflection above baseline), and the distance (in millimeters) between atrial and ventricular peaks of the heartbeat.
   (a) For the Biopac system, recording begins when the "start" button on the bottom right of the computer screen is clicked. After recording for a couple of minutes, stop the recording and choose "select all" from the "File" drop-down menu. Then, the beats per minute (in the small window labeled "bpm") and the amplitude (strength) of the recording (in the small window labeled "p-p") can be read near the top of the screen.
   (b) Select "clear all" from the drop-down menu to clear the screen. Thoroughly rinse the heart with Ringer's solution and record the heart tracings again to establish new control values before adding the next drug.
   (c) When you add a new drug, wait 2–3 minutes before pressing the "start" button on the bottom right of the computer screen. After recording, you can repeat step (a) to obtain measurements with the new drug. Be sure to rinse with Ringer's and obtain a new control recording prior to each new drug.
7. Record this data in the laboratory report.

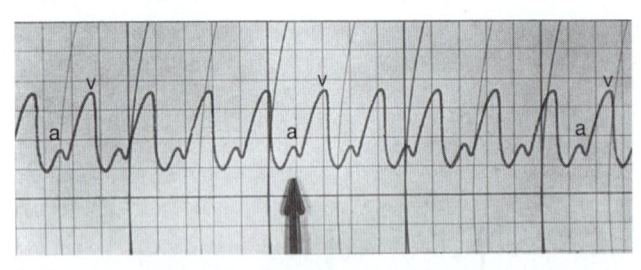

A. Effect of Caffeine on the Heart

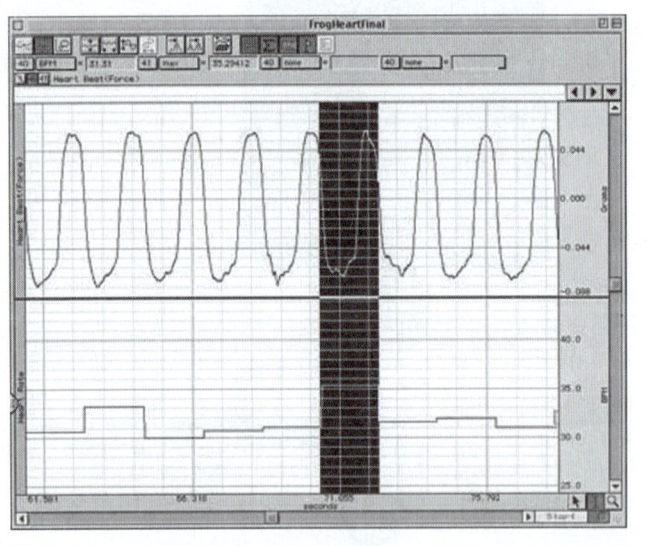

Caffeine is a mild central nervous system (CNS) stimulant that also acts directly on the myocardium to increase both the strength of contraction and the cardiac rate. Caffeine inhibits activity of the enzyme *phosphodiesterase*, which breaks down a second messenger molecule called *cyclic AMP (cAMP)* that is present in many cells. As a result, the concentration of cAMP rises in heart cells. This duplicates the action of the hormone epinephrine, which uses cAMP as a second messenger. Caffeine's usefulness as a central nervous system stimulant is limited because, in high doses, it can promote the formation of ectopic pacemakers (foci), resulting in serious arrhythmias (see exercise 7.2).

Procedure

1. Obtain a record of the control heartbeat. Then, bathe the heart with a saturated solution of caffeine.
2. Record the results; then stop recording and rinse the heart thoroughly with Ringer's solution.
3. In the table of your laboratory report, tape the recording or draw a facsimile of the normal heartbeat and of the changed heartbeat after the caffeine solution was added.

B. Effect of Nicotine on the Heart

Nicotine promotes electrochemical transmission at the autonomic ganglia by stimulating nicotinic receptors for acetylcholine in the postganglionic neurons. When applied directly to the heart, the major effect of nicotine will be stimulation of parasympathetic ganglia located within the epicardium. Activation of postganglionic parasympathetic neurons, in turn, will cause slowing of the heart rate. When nicotine is administered into the blood in pharmacological doses (as a drug), it can stimulate sympathetic ganglia and the adrenal medulla (the sympathoadrenal system), resulting in an increase in the heart rate.

Figure 7.6 Recordings of the frog heart contractions. (a) Physiograph recording. The arrow points to a smaller atrial contraction (a) followed by the larger ventricular contraction (v). (b) The frog heart contraction as recorded with the Biopac system. Courtesy of and © BIOPAC Systems, Inc.

Procedure

1. Obtain a record of the control heartbeat. Then, bathe the heart in a 0.2% solution of nicotine.
2. Record the results; then stop recording and rinse the heart thoroughly with Ringer's solution.
3. In the table of your laboratory report, tape the recording or draw a facsimile of the normal heartbeat and of the changed heartbeat after the nicotine solution was added.
4. Analyze your data and record your results for parts A through H in the Results table of your laboratory report.

C. Effect of Epinephrine on the Heart

Epinephrine is a hormone secreted by the adrenal medulla. Together with norepinephrine, epinephrine is released in response to sympathetic nerve stimulation. Epinephrine acts to increase both the strength of contraction (contractility) of the heart and the cardiac rate. Exogenous epinephrine is a *sympathomimetic drug*, because it mimics the effect of sympathetic nerve stimulation.

Procedure

1. Obtain a record of the control heartbeat. Then, bathe the heart in epinephrine (adrenaline).
2. Record the results; then stop recording and rinse the heart thoroughly with Ringer's solution.
3. In the table of your laboratory report, tape the recording or draw a facsimile of the normal heartbeat and of the changed heartbeat after the epinephrine solution was added.

D. Effect of Atropine on the Heart

Atropine is an alkaloid drug derived from the nightshade plant *Atropa belladonna* (the species name, *belladonna*, is often also used as the drug name). Atropine specifically blocks the ability of ACh to bind to its muscarinic ACh receptors. These receptors for acetylcholine are located in the smooth muscles, cardiac muscles, and glands, which normally respond to ACh released by postganglionic parasympathetic axons. Thus, atropine inhibits the effects of parasympathetic activity on the heart, smooth muscles, and glands. If the cardiac rate is decreased as a result of vagal stimulation (or the presence of pilocarpine), the administration of atropine will increase this rate.

Clinical Applications

Atropine specifically blocks the muscarinic subtype of ACh receptors, which are the receptors stimulated by the ACh released by postganglionic parasympathetic neurons. The ability of atropine to block parasympathetic nerve effects is medically important. Atropine is used clinically to dilate pupils during eye exams (because parasympathetic nerves cause constriction of the pupils), inhibit secretions of the respiratory tract during general anesthesia, inhibit spasmodic contractions of the intestines, and inhibit stomach acid secretion in a person with gastritis. Atropine may also be given to block the muscarinic effects of *nerve gas*, which prevents the breakdown of ACh and thereby overstimulates cholinergic pathways.

Procedure

1. Bathe the heart in a 5.0% solution of atropine *while it is still under the influence of pilocarpine*.
2. Record the results; then stop recording and rinse the heart thoroughly with Ringer's solution.
3. In the table of your laboratory report, tape the recording or draw a facsimile of the normal heartbeat and of the changed heartbeat after the atropine solution was added.

E. Effect of Pilocarpine on the Heart

Pilocarpine is termed a *parasympathomimetic* drug because it mimics the effect of parasympathetic nerve stimulation. Pilocarpine acts to facilitate the release of the neurotransmitter acetylcholine from the vagus nerve, resulting in a marked decrease in the cardiac rate.

Procedure

1. Obtain a record of the control heartbeat. Then, bathe the heart in a 2.5% solution of pilocarpine.

   Caution: *Pilocarpine is very effective; be prepared to add atropine if necessary to counter the effects of pilocarpine.*

2. Rinse the heart thoroughly with Ringer's solution until the heartbeat returns somewhat to normal; this new rhythm will serve as the next control.
3. In the table of your laboratory report, tape the recording or draw a facsimile of the normal heartbeat and of the changed heartbeat after the pilocarpine solution was added.

F. Effect of Digitalis on the Heart

People with congestive heart failure are often treated with the drug digitalis. Digitalis binds to and inactivates the Na^+/K^+ pumps in the plasma membrane of myocardial cells. This reduces the ability of these pumps to extrude Na^+ from the cytoplasm, producing a rise in the cytoplasmic Na^+ concentration. There is now less of a difference in Na^+ concentration between the inside and outside of the cells, so the diffusion of Na^+ into the cells is decreased. The decreased diffusion of Na^+ into the cells then reduces the ability of the Na^+/Ca^{2+} exchangers in the plasma membrane to extrude Ca^{2+}. Thus, reduced Na^+/Ca^{2+} exchange produces a rise in the intracellular Ca^{2+} concentration, which then causes more Ca^{2+} to be taken into the sarcoplasmic reticulum. Finally, it is this increased amount of Ca^{2+} in the sarcoplasmic reticulum that is responsible for the increased strength of myocardial contraction when a person takes digitalis.

Clinical Applications

Digitalis glycosides, such as digoxin, or digitalis, are frequently used to treat congestive heart failure, atrial flutter, and atrial fibrillation. Digitalis relieves these conditions by (1) increasing the force of contraction, (2) decreasing the cardiac rate directly by inhibiting the SA node, and (3) by slowing conduction through the bundle of His.

Procedure

1. Obtain a record of the control heartbeat. Then, bathe the heart in a 2.0% solution of digitalis.
2. Rinse the heart thoroughly with Ringer's solution until the heartbeat returns somewhat to normal; this new rhythm will serve as the next control.
3. In the table of your laboratory report, tape the recording or draw a facsimile of the normal heartbeat and of the changed heartbeat after the digitalis solution was added.

G. Effect of Calcium Ions on the Heart

In addition to the role of calcium in coupling excitation to contraction, the extracellular Ca^{2+}/Mg^{2+} ratio also affects the permeability of the cell membrane. An increase in the extracellular concentration of calcium (above the normal concentration of 4.5–5.5 mEq/L) affects both the electrical properties and the contraction strength of heart muscle.

The heart is affected in a number of ways by an increase in extracellular calcium. These include (1) an increased force of contraction; (2) a decreased cardiac rate; and (3) the appearance of ectopic pacemakers in the ventricles, producing abnormal rhythms (extrasystoles and idioventricular rhythm—see exercise 7.2).

Procedure

1. Obtain a control record of the normal heartbeat. Then, while the paper (or computer recording) continues to run, use a dropper to bathe the heart in a 2.0% solution of calcium chloride ($CaCl_2$). On the moving recording paper, indicate the time at which calcium was added. Observe the effects of the added calcium solution for a few minutes, then stop the recording.
2. Rinse the heart thoroughly with Ringer's solution until the heartbeat returns somewhat to normal; this new rhythm will serve as the next control.
3. In the table of your laboratory report, tape the recording or draw a facsimile of the normal heartbeat and of the changed heartbeat after the calcium solution was added.

H. Effect of Potassium Ions on the Heart

The resting membrane potential of all cells is dependent in large part on the maintenance of a higher concentration of potassium ions (K^+) on the inside of the cell than on the outside, so an increase in the concentration of extracellular K^+ results in a *decrease in the resting membrane potential* (the potential becomes more positive). This, in turn, produces a decrease in the force of contraction and a slower conduction rate of the action potentials. In **hyperkalemia** (high blood potassium), the strength of myocardial contractions is weakened and the cardiac cells become more electrically excitable because the resting potential has risen closer to the threshold required for generating action potentials. In extreme hyperkalemia, the conduction rate may be so depressed that ectopic pacemakers appear in the ventricles and fibrillation may develop. Indeed, *lethal injections* (as in legal executions) use this principle to raise the plasma K^+ concentration to levels that cause cessation of the heartbeat.

Procedure

1. Obtain a record of the control heartbeat. Then, bathe the heart in a 2.0% solution of potassium chloride (KCl).
2. Record the results; then stop recording and rinse the heart thoroughly with Ringer's solution.
3. In the table of your laboratory report, tape the recording or draw a facsimile of the normal heartbeat and of the changed heartbeat after the potassium solution was added.

Laboratory Report 7.1

Name _____
Date _____
Section _____

DATA FROM EXERCISE 7.1

Condition	Effects (Tape the Recording or Draw a Facsimile)
Normal	
Caffeine	
Nicotine	
Epinephrine	
Atropine	
Pilocarpine	
Digitalis	
Calcium (Ca^{2+})	
Potassium (K^+)	

RESULTS FROM EXERCISE 7.1

Condition	Rate (beats/min)	Strength (grams or mm of Amplitude from Baseline)	Separation of Atrial and Ventricular Peaks (seconds or mm Distance)	Conclusion About Drug Effects
Normal				
Caffeine				
Nicotine				
Epinephrine				
Atropine				
Pilocarpine				
Digitalis				
Calcium (Ca^{2+})				
Potassium (K$^+$)				

REVIEW ACTIVITIES FOR EXERCISE 7.1

Test Your Knowledge

Match these items:

___ 1. endogenous substance that makes the heart beat faster and stronger

___ 2. makes the heart beat slower and stronger

___ 3. facilitates the release of ACh from parasympathetic nerve endings

___ 4. mimics the action of epinephrine by inhibiting the action of phosphodiesterase

___ 5. stimulates the acetylcholine receptors of autonomic ganglia

___ 6. blocks the acetylcholine receptors for the target cells of postganglionic neurons

(a) digitalis
(b) nicotine
(c) caffeine
(d) epinephrine
(e) atropine
(f) pilocarpine

7. The drug used in this exercise that helps people with atrial fibrillation: _____

8. The drug used in this exercise used by ophthalmologists to dilate pupils: _____

9. The term *hyperkalemia* means _____

Test Your Understanding

10. What are the effects of hyperkalemia on the heart? How were these effects produced?

11. What is a sympathomimetic drug? What are its effects on the heart?

12. What are the effects of digitalis on the heart? Describe the clinical uses of this drug.

Test Your Analytical Ability

13. What effect did bathing the heart with Ca^{2+} have on the strength of the contraction and the ability of the heart to relax between beats? Provide a physiological explanation of these results.

14. The drug *theophylline* is in the same chemical class as caffeine (they are methylxanthines). Theophylline is sometimes used clinically to dilate the bronchioles in a person suffering from asthma. By analogy with caffeine, provide a physiological explanation for this clinical application.

Clinical Investigation Questions

15. How do anxiety, coffee, atropine, and digitalis affect the heart? Explain the mechanisms involved.

16. What are the effects of atropine on the stomach and eyes? How does it exert these effects?

17. What are the effects of digitalis on the heart and how does this help treat atrial fibrillation?

EXERCISE 7.2

Electrocardiogram (ECG)

MATERIALS

1. Electrocardiograph or other strip chart recorder, such as Physiograph (Narco), or Lafayette Instrument Company, Inc. recorder with EKG module
2. Alternatively, the Biopac system with Student Lessons 5 and 6 or the PowerLab modules may be used to record the electrocardiogram.
3. Electrode plates, rubber straps, electrolyte gel or paste; disposable electrodes and electrode clips
4. Alternatively, simulated electrocardiograph exercises may be performed using the *Physiology Interactive Lab Simulations (Ph.I.L.S.):* Electrocardiogram and Heart Function (exercises 4 and 6)

The regular pattern of electrical impulse production and conduction in the heart results in the mechanical contraction (systole) and relaxation (diastole) of the myocardium—the cardiac cycle. The recording of these electrical events, the electrocardiogram (ECG), may reveal abnormal patterns associated with abnormal cardiac rhythms.

LEARNING OUTCOMES

You should be able to:

1. Describe the normal cyclical pattern of electrical impulse production in the heart and conduction along specialized tissues of the heart.
2. Describe the normal electrocardiogram (ECG, or EKG), and explain how it is produced.
3. Obtain an electrocardiogram using the limb leads, identify the waves, determine the P-R interval, and measure the cardiac rate.
4. Describe the common ECG abnormalities or arrhythmias.

Textbook/Multimedia Correlations

Before performing this exercise, you should study the introductory material presented here. Further information relating to this exercise can be found in these pages of *Human Physiology,* thirteenth edition, by Stuart I. Fox:

- *Electrical Activity of the Heart and the Electrocardiogram.* Chapter 13, p. 425.
- Multimedia correlations are provided in *MediaPhys 3.0:* Topics 8.17 and 8.18; also *Physiology Interactive Lab simulations (Ph.I.L.S.):* Electrocardiogram and Heart Function (exercises 4 and 6).

Clinical Investigation

A patient was brought to the hospital after briefly losing consciousness due to a fall in blood pressure. Her ECG displayed a wavy baseline between QRS complexes, with an estimated atrial rate of approximately 600 beats per minute. Afterward, a normal sinus rhythm was restored with a PR interval of 0.16 sec; however, occasional runs of PVCs were observed.

- Explain the significance of her ECG pattern when she was first brought to the hospital.
- Explain the meaning of a normal sinus rhythm, a PR interval of 0.16 sec, and PVCs.

Electrical stimulation at any point in the heart musculature (myocardium) results in the almost simultaneous contraction of the individual muscle cells (myocardial cells). This allows the heart to function as an effective pump, with its chambers contracting and relaxing as integrated units. The contraction phase of the cardiac cycle is called **systole** and the relaxation phase is called **diastole**.

Unlike skeletal muscles, cardiac muscle is able to stimulate itself electrically in the absence of neural input (a property termed *automaticity*). The sympathetic and parasympathetic nerves that innervate the heart only modulate the ongoing rate of depolarization-contraction and repolarization-relaxation intrinsic to the heart. The intrinsic regulation of systole and diastole, unique to heart muscle, is termed *rhythmicity*.

Although each individual myocardial cell is potentially capable of initiating its own cycle of depolarization-contraction and repolarization-relaxation, a single group of cells usually regulates the cycle of the entire myocardium. This *pacemaker* region establishes its dominance because its cycle is more rapid than other areas, depolarizing the other myocardial cells before they can depolarize themselves. A region of the right atrium, the **sinoatrial node (SA node)**, serves as the normal pacemaker of the heart (fig. 7.7). The wave of depolarization initiated by the SA node quickly spreads across the right and left atria as a result of electrical synapses *(gap junctions)* between myocardial cells. However, the depolarization wave cannot easily spread from the myocardial cells of the atria to the myocardial cells of the ventricles. For this to occur, the depolarization wave must be carried from the atria to the ventricles along the specialized conducting tissue of the heart.

The depolarization wave that spreads over the atria stimulates a node called the **atrioventricular node (AV node)**, located at the base of the interatrial septum. After a brief delay, the depolarization wave in the AV node is quickly transmitted over a bundle of specialized conducting tissue, the **atrioventricular (or AV) bundle,** otherwise known as the **bundle of His.** The bundle of His splits to form right and left *bundle branches,* carrying the depolarization wave to the apex of the ventricles. The impulse is then carried to the innermost cells of the ventricles on a network of branching **Purkinje fibers** (fig. 7.7).

Because of the depolarization stimulus, the atria contract as a single unit (atrial systole), followed quickly by depolarization and contraction of the ventricles (ventricular systole). Contraction of the atria forces blood into the ventricles, and contraction of the ventricles forces blood out of the heart and into the pulmonary arteries *(pulmonary circulation)* and into the aorta *(systemic circulation).* When the atria repolarize and relax, they refill with blood from the veins; when the ventricles repolarize and relax during diastole, they receive blood from the filling atria. The cardiac cycle is now ready to again be initiated by the spontaneous depolarization of the SA node.

Since the body fluids contain a high concentration of electrolytes (ions), the electrical activity generated by the heart travels throughout the body and can easily be monitored by placing a pair of electrodes on different areas of the skin (fig. 7.8). A graphic representation of these electrical activities is called an **electrocardiogram** (**ECG,** or **EKG,** from the German term *kardio*), and the instrument producing this record is called an *electrocardiograph*. A normal ECG is shown in figure 7.9.

The **P wave** represents depolarization of the atria; the **QRS complex** is produced by depolarization of the ventricles; and the **T wave** represents repolarization of the ventricles at the beginning of diastole (fig. 7.10). An incompletely understood **U wave** sometimes follows the T wave.

In the following exercises we will use the *standard limb leads* I, II, and III. These leads record the difference in potential (the voltage) between two electrodes placed on the arms and legs (see fig. 7.8). In clinical electrocardiography, however, *unipolar leads* are also used. These are the *AVR* (right arm), *AVL* (left arm), *AVF* (left leg), and the chest leads labeled V_1 to V_6.

The standard ECG running at a chart speed of 25 mm (2.5 cm) per second, shows thin vertical lines 0.04 sec apart, with the distance between every fifth, heavier line representing an interval of 0.20 sec. This produces a graphic recording in which the length of time between depolarization of the atria (P wave) and depolarization of the ventricles (QRS complex) can easily be measured. This interval is known as the **P-R interval** (although it is actually measured from the beginning of the P wave to the Q wave—fig. 7.9) and is equal to or less than 0.20 sec in the normal ECG.

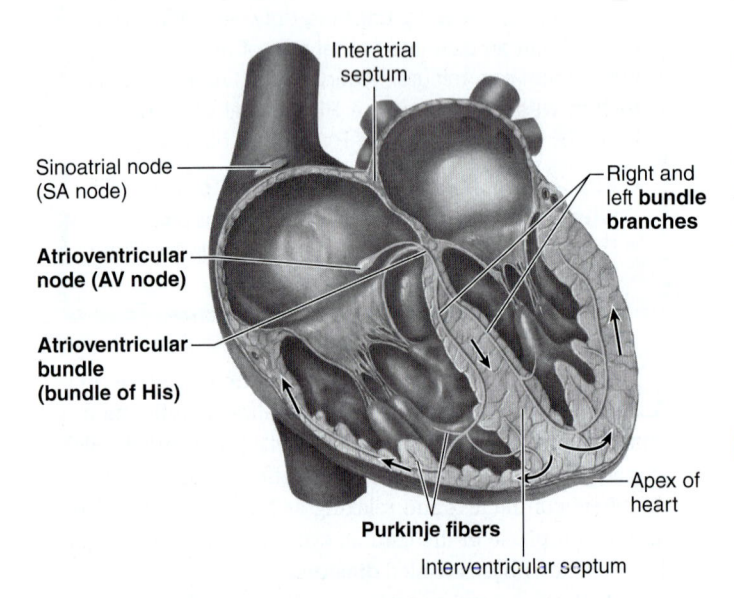

Figure 7.7 The conduction system of the heart. The conduction system consists of specialized myocardial cells that rapidly conduct the impulses from the atria into the ventricles.

(For a full-color version of this figure, see fig. 13.20 in *Human Physiology,* thirteenth edition, by Stuart I. Fox.)

Procedure

1. With the subject comfortably lying down, rub a silver dollar-sized amount of electrolyte gel on the medial surface, about 2 inches above the wrists and ankles. Attach electrode plates to these four spots, using the rubber straps provided (fig. 7.8).
2. Attach the four ECG leads to the appropriate plates.
3. The specific instructions for obtaining ECG tracings vary with the instrument being used. Your instructor will demonstrate the use of the recording equipment

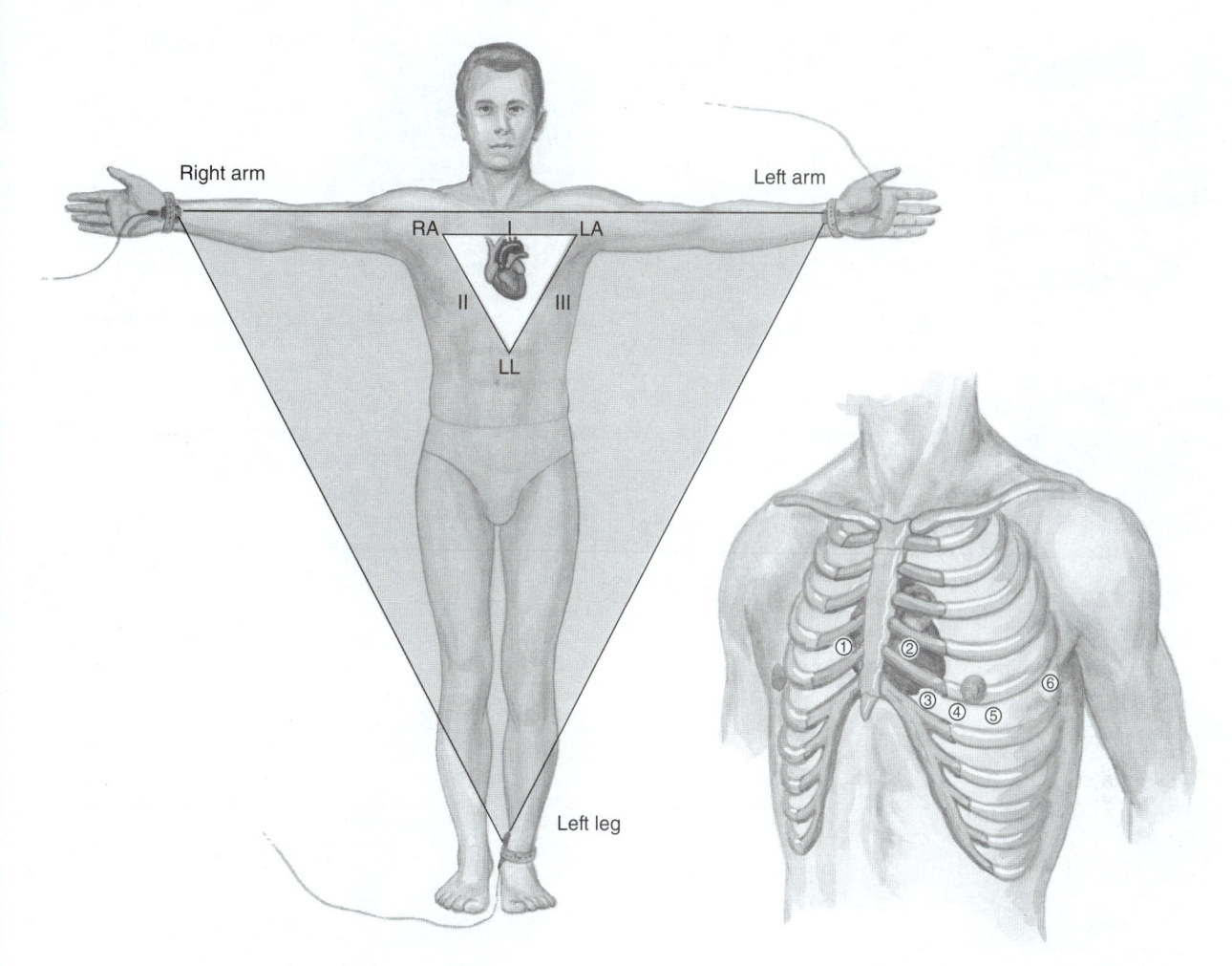

Figure 7.8 Electrocardiograph leads. The placement of the bipolar limb leads (RA, LA, LL) and the positions (1–6) for the unipolar chest leads (V_1–V_6) for recording electrocardiograms. (RA = right arm; LA = left arm; LL = left leg.) *Note:* The right leg is also connected to the electrocardiograph, but is not shown because it serves as the "ground."

(For a full-color version of this figure, see fig. 13.24 in *Human Physiology*, thirteenth edition, by Stuart I. Fox.)

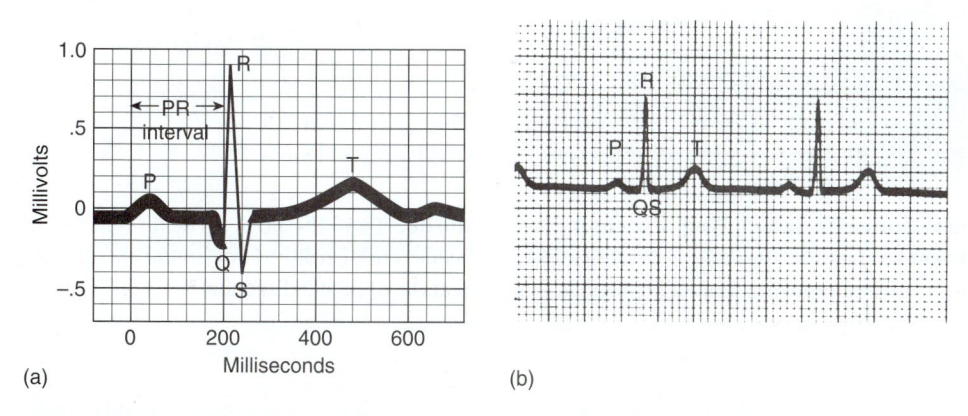

Figure 7.9 The normal electrocardiogram (ECG). (a) A labeled drawing, and (b) an ECG recording.

in your lab. The following instructions are valid only for a single-channel electrocardiograph.
(a) Turn on the power switch.
(b) Set the paper-speed selector switch to 25 mm (2.5 cm) per second.
(c) Set the sensitivity to 1 (most sensitive).
(d) Set the lead selector switch to the first dot to the left of the STD or CAL position.
(e) Turn the control or knob to the run or record position.

249

(a)

(b)

(c) P wave: Atria depolarize and contract

(d)

(e) QRS complex: Ventricles depolarize and contract

(f)

(g) T wave: Ventricles repolarize and relax

Depolarization

Repolarization

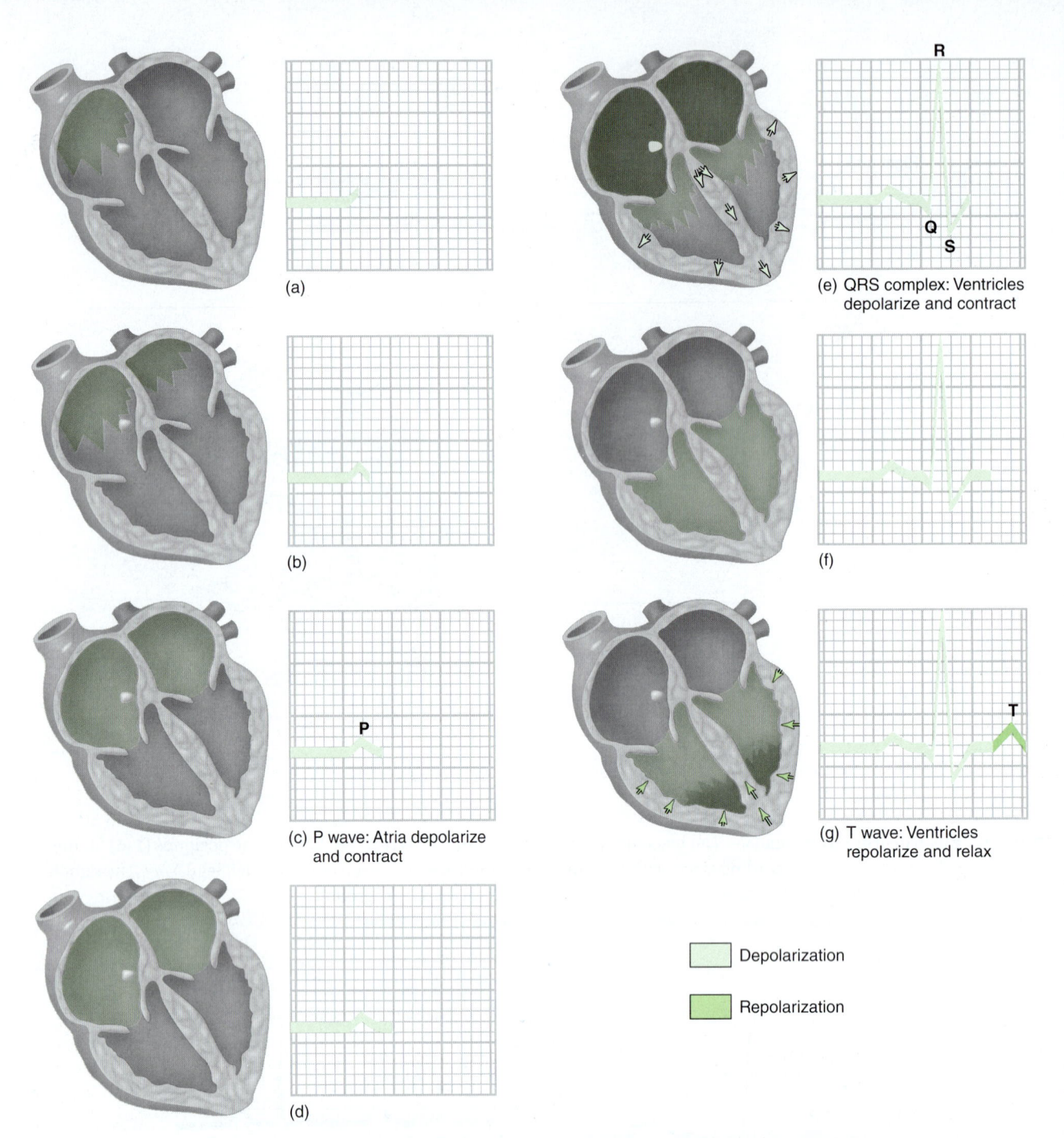

Figure 7.10 **The relationship between impulse conduction in the heart and the ECG.** The direction of the arrows in (e) indicates that depolarization of the ventricles occurs from the inside (endocardium) out (to the epicardium). The arrows in (g), by contrast, indicate that repolarization of the ventricles occurs in the opposite direction.

(For a full-color version of this figure, see fig. 13.23 in *Human Physiology*, thirteenth edition, by Stuart I. Fox.)

4. Turn the position knob until the stylus is centered on the ECG paper.
5. Turn the lead selector switch to the *1* (lead I) position to measure the voltage difference between the right and left arms. As the paper is running, depress the *mark* button once—this makes a single dash at the top of the chart to indicate that the record is from lead I. Continue recording until an adequate sample of the tracing can be provided to each member of the subject's group; then stop the paper drive by turning the lead selector switch to the dot above the *1* position. Each dot is a "rest" position where the movement of the chart will stop between recording from each lead.
6. Turn the lead selector switch to the *2* (lead II) position to measure the voltage difference between

the right arm and left leg. As the chart is running, depress the *mark* button twice—the two dashes produced at the top of the chart will indicate that this is the recording from lead II. Stop the chart by turning the lead selector switch to the dot above the *2* position.

7. Repeat this procedure with lead III to measure the voltage difference between the left arm and the left leg.
8. After recordings from leads I, II, and III have been obtained, turn the lead selector switch to the *STD* or *CAL* position. Run the recording out of the machine, allowing members of the group to cut sample tracings of each lead.
9. Remove the electrode plates from the subject's skin and thoroughly wash the electrolyte gel from both the plate and the skin.
10. Tape samples of the recordings in your laboratory report and label all the waves.
11. Determine the P-R interval of **lead II.** This can be done by counting the number of *small* boxes between the beginning of the P and the Q and multiplying this number by 0.04 sec.

Note: *If you use a multichannel recorder (such as a Physiograph) instead of an electrocardiograph, the recording paper and, consequently, the arithmetic calculation, will be different. If you use a Cardiocomp, the P-R interval will be provided automatically.*

P-R interval _____ sec

> **Normal Values** The normal P-R interval is 0.12–0.20 sec.

12. Determine the cardiac rate by the following methods:
    (a) Count the number of QRS complexes in a 3-sec interval (the distance between two vertical lines at the top of the ECG paper) and multiply by 20.

Note: *If you use a multichannel recorder (such as a Physiograph), the amount of chart paper corresponding to a given time interval must be calculated and will vary with the paper speed. If you use a Cardiocomp, the cardiac rate will be provided automatically.*

Beats per minute = _____

  (b) Count the number of QRS complexes in a 6-sec interval, and multiply by 10.

  Beats per minute = _____

  (c) At a chart speed of 25 mm (2.5 cm) per second, the time interval between one light vertical line and the next is 0.04 sec. The time interval between heavy vertical lines is 0.20 sec. The cardiac rate in beats per minute can be calculated if the time interval between two R waves in two successive QRS complexes is known.

For example, suppose that the time interval from one R wave to the next is exactly 0.60 sec. Therefore,

$$\frac{1 \text{ beat}}{0.60 \text{ sec}} = \frac{x \text{ beats}}{0.60 \text{ sec}}$$

$$x = \frac{1 \text{ beat} \times 60 \text{ sec}}{0.60 \text{ sec}}$$

$$x = 100 \text{ beats per minute}$$

Beats per minute = _____

  (d) The values obtained by *method c* can be approximated by counting the number of heavy vertical lines between one R wave and the next according to the memorized sequence: 300, 150, 100, 75, 60, 50 (at a paper speed of 25 mm per sec):

  The cardiac rate in the previous sample tracing is 75 beats per minute.

  Beats per minute = _____

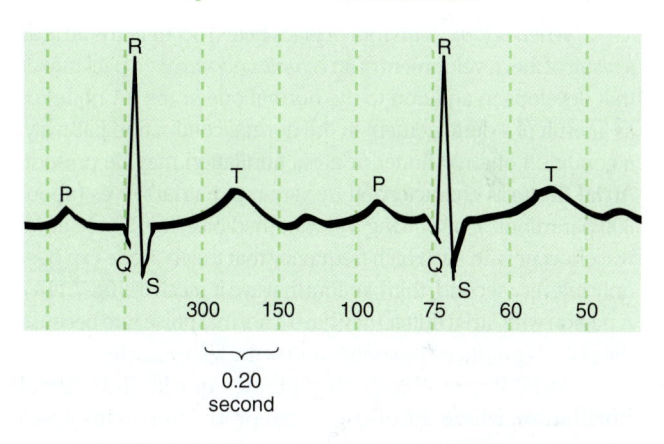

> **Normal Values** The normal cardiac rate is 60–100 beats per minute.

ABNORMAL ECG PATTERNS

Interpretation of the electrocardiogram can provide information about the heart rate and rhythm, as well as possible conditions of *hypertrophy, ischemia* (inadequate blood supply), *necrosis* (death of cells), and other conditions that may produce abnormalities of electrical conduction. According to the standards set by the National Conference on Cardiopulmonary Resuscitation and Emergency Cardiac Care, all health professionals should be able to recognize:

1. bradycardia (a ventricular rate slower than 60 beats per minute)
2. the difference between supraventricular and ventricular rhythms
3. premature ventricular contractions
4. ventricular tachycardia
5. atrioventricular block
6. atrial fibrillation and flutter
7. ventricular fibrillation

(a)

(b)

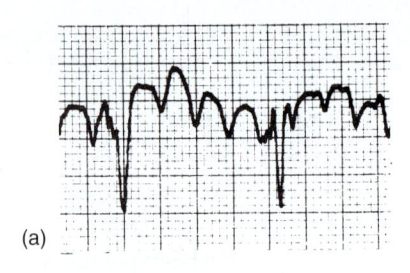

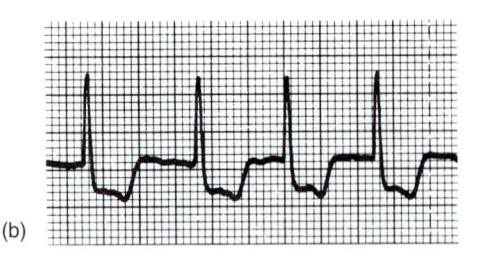

Figure 7.11 **Abnormal atrial rhythms.** (a) Atrial flutter and (b) atrial fibrillation.

When *ectopic beats* (out of place beats) occur in the atria as a result of the development of an *ectopic pacemaker* (a pacemaker that develops in addition to the normal one in the SA node) or as a result of a derangement in the normal conduction pathway, a condition of atrial flutter or atrial fibrillation may be present. **Atrial flutter** is characterized by very rapid atrial waves (about 300 per minute), producing a sawtoothed baseline. These atrial waves occur with such high frequency that the AV node can beat only to every second, third, or fourth wave it receives (fig. 7.11a). A person with atrial flutter may have a normal pulse rate because the pulse is produced by contraction of the left ventricle.

Atrial flutter usually degenerates quickly into **atrial fibrillation,** where the disorganized production of impulses occurs very rapidly (up to 600 times per minute) and contraction of the atria is ineffectual. In an ECG, the P waves are replaced by a wavy baseline (fig. 7.11b). The AV node doesn't respond to all of those impulses, but enough impulses still get through to stimulate the ventricles to beat at a rapid rate (up to 150–180 beats per minute). The ventricles fill to about 80% of their end-diastolic volume before even normal atrial contraction, so atrial fibrillation only reduces the cardiac output by about 15%. People with atrial fibrillation can thus live for many years, although this condition is associated with increased mortality from stroke and heart failure. Approximately 20–25% of all strokes may be the result of thrombi promoted by atrial fibrillation, so patients with atrial fibrillation are often placed on anticoagulant drugs.

Atrial fibrillation is characteristic of atrial enlargement (hypertrophy) as might be produced by *mitral stenosis* (narrowing of the mitral valve), but may occur in all forms of heart disease and occasionally in apparently healthy individuals. *Digitalis* is a drug often used in atrial flutter and fibrillation to decrease the excitability of the AV node, thus maintaining the ventricular rate within the normal range.

A delay in the conduction of the impulse from the atria to the ventricles is known as **atrioventricular (AV) block.** When the ECG pattern is otherwise normal, a P-R

Clinical Applications

Stroke (cerebrovascular accident) is the third leading cause of death in the United States. Most commonly, it is produced when a region of the brain becomes *ischemic* (has inadequate blood flow) due to a thrombus (blood clot), which can be formed as a consequence of atherosclerosis. Thrombolytic drugs help, but must be administered soon after the ischemic injury to be effective. This is because neurons progressively die through a process known as *excitoxicity*. Excitoxicity occurs because the removal of the major excitatory neurotransmitter, glutamate, from synaptic clefts is impaired by the ischemia. Because of the accumulation of glutamate, the postsynaptic neurons become overstimulated. This results in excessive accumulation of intracellular Ca^{2+}, producing death of the neurons. At present there is no clinically effective means of blocking excitotoxicity and its consequences.

interval greater than 0.20 sec represents a *first-degree AV block* (fig. 7.12a). First-degree AV block may be a result of inflammatory states, rheumatic fever, or digitalis treatment.

When the excitability of the AV node is further impaired so that two or more atrial depolarizations are required before the impulse can be transmitted to the ventricles, a *second-degree AV block* is present. This is seen on the ECG as a "dropped" beat (a P wave without an associated QRS complex). In *Wenckebach's phenomenon,* the cycle after the dropped beat is normal, but the P-R interval of successive cycles lengthens until a beat (QRS complex) is again dropped (fig. 7.12b).

Third-degree, or *complete, AV block* occurs when none of the impulses from the atria reach the ventricles (fig. 7.12c). In this case, the myocardial cells of the ventricles are freed from their subservience to the SA node, causing one or more ectopic pacemakers to appear in the ventricles without corresponding P waves. The rhythm produced by these ectopic foci in the ventricles is usually very slow (20–45 beats per minute) compared to the normal pace set by the SA node *(sinus rhythm)*. An artificial pacemaker may be used to compensate for this condition, as shown by the ECG tracing in figure 7.12d.

Premature ventricular contractions (PVCs) are produced by ectopic foci in the ventricles when the sinus rhythm is normal. This results in *extrasystoles* (extra beats or QRS complexes without preceding P waves) in addition to the normal cycle, often subjectively described by patients as "palpitations." The ectopic QRS complexes are broad and deformed and may be abnormally coupled to the preceding normal beats. This coupling is termed *bigeminy* and is often seen in digitalis toxicity (fig. 7.13a).

When an ectopic focus in the ventricles discharges at a rapid rate, a condition of **ventricular tachycardia** (usually

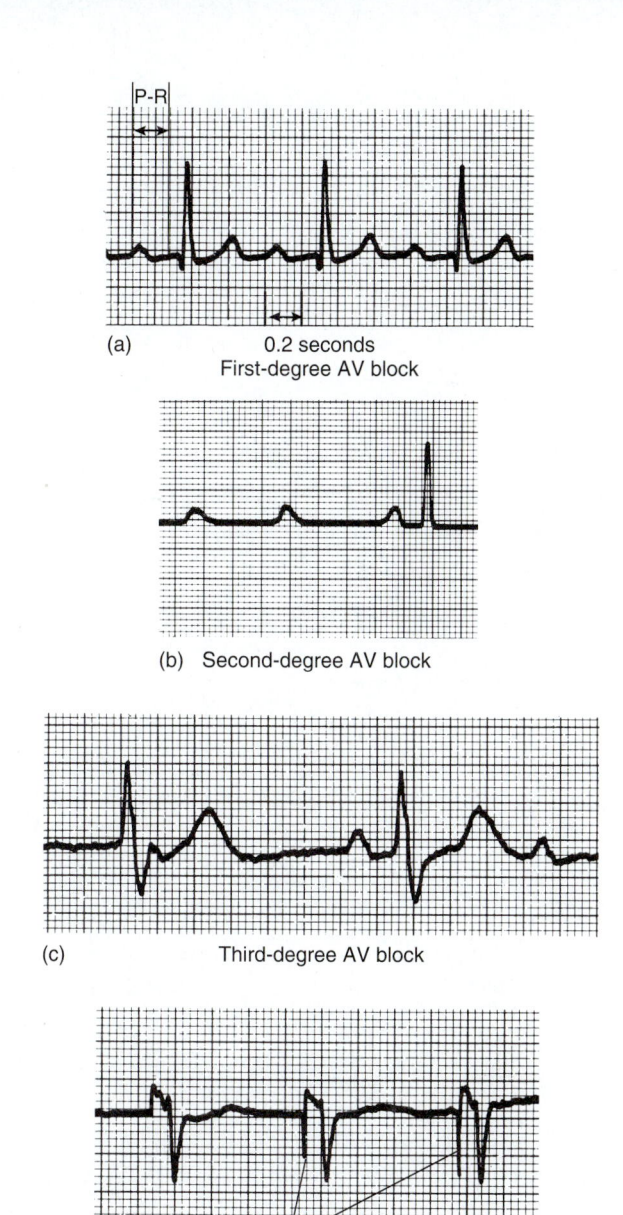

Figure 7.12 Different stages of AV block. (a) First-degree AV block, where the P-R interval is greater than 0.2 seconds (one large square); (b) second-degree AV block, where beats (i.e., QRS complexes) are missed (a 3:1 block is shown); and (c) third-degree, or complete, AV block, where a slower than normal heart rate is set by an ectopic focus in the ventricles. (d) An artificial pacemaker, implanted in a person with complete AV node block, produces the sharp downward spikes that precede the inverted QRS waves.

100–150 beats per minute) develops (fig. 7.13*b*). The ECG shows a widened and distorted QRS complex that often obscures the P wave (although the atria are discharging at their slower, sinus rate). This serious condition should be distinguished from the less serious **supraventricular tachycardia,** in which an ectopic focus above the ventricles results in spontaneous rapid running of the heart (150–250 beats/min) that begins and ends abruptly (is paroxysmal). This condition is appropriately called *paroxysmal atrial tachycardia.*

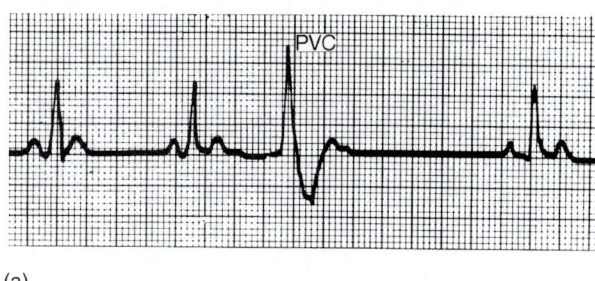

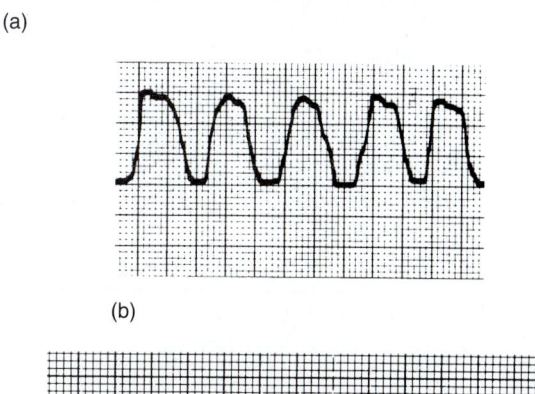

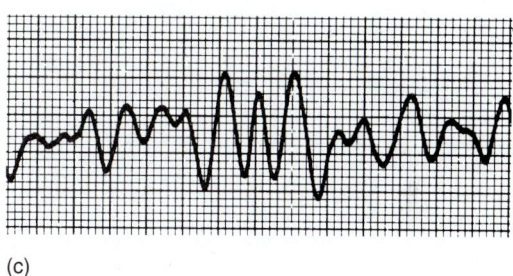

Figure 7.13 Abnormal ventricular rhythms.
(a) Premature ventricular contraction (PVC), or bigeminy; (b) ventricular tachycardia; and (c) ventricular fibrillation.

The most serious of all arrhythmias is **ventricular fibrillation** (fig. 7.13*c*). Ventricular fibrillation may develop as ectopic foci emerge and depolarization waves circle the heart *(circus rhythm).* The recycling of action potentials that occurs in circus rhythms is normally prevented by the entire myocardium entering a refractory period as a single unit. This is due to the rapid transmission of the action potential among the myocardial cells through their gap junction and to the long duration of the myocardial action potential. Circus rhythms can occur when some cells emerge from their refractory periods before others, so that the action potential can be continuously regenerated and produce an impotent tremor rather than a coordinated pumping action.

Sudden death from cardiac arrhythmia usually progresses from ventricular tachycardia through ventricular fibrillation, culminating in *asystole* (the cessation of beating, with a straight line ECG). Sudden death from cardiac arrhythmia is commonly a result of acute myocardial ischemia (insufficient blood flow), most often due to atherosclerosis of the coronary arteries. Fibrillation can sometimes be stopped by a strong electric shock to the chest, a procedure called *electrical defibrillation.* The electric shock depolarizes all of the myocardial cells at the same time, causing them to enter a refractory period. This can stop the conduction of the circus rhythms and allow a normal cycle to begin again.

Laboratory Report 7.2

Name _____
Date _____
Section _____

DATA FROM EXERCISE 7.2

1. Tape your recording in these spaces.
   (a) **Lead I**

   (b) **Lead II**

   (c) **Lead III**

REVIEW ACTIVITIES FOR EXERCISE 7.2

Test Your Knowledge

1. The pacemaker region of the heart is the _____.
2. The conducting tissue of the heart located in the interventricular septum is the _____.
3. Indicate the electrical events that produce each of these waves:
   (a) P wave _____
   (b) QRS wave _____
   (c) T wave _____
4. The electrical synapses between adjacent myocardial cells are called _____.
5. An abnormally fast rate of beat is called _____; an abnormally slow rate is called _____.
6. An abnormally long P-R interval indicates a condition called _____.
7. Leads I, II, and III are collectively called the _____ leads.
8. Which ECG wave must occur before the ventricles can contract? _____
9. Which ECG wave must occur before the ventricles can relax? _____

Test Your Understanding

10. What property makes the normal pacemaker region of the heart function as a pacemaker? Explain.

11. Describe the pathway of conduction from the atria to the ventricles and correlate this conduction with the ECG waves.

12. Compare supraventricular tachycardia with ventricular tachycardia in terms of its nature, ECG pattern, and seriousness.

13. What happens to the beating of the atria and ventricles during third-degree AV node block? Why does this occur?

Test Your Analytical Ability

14. On initial examination, a patient was found to have a P-R interval of 0.24 second. A year later, the patient had a resting pulse of about 40 per minute. A year after that, the patient collapsed with a pulse of 20 per minute and required the insertion of an artificial pacemaker. Explain what caused this sequence of events.

15. Suppose a person had paroxysmal atrial tachycardia, producing "palpitations." Describe this condition, and propose ways that the condition might be treated.

Test Your Quantitative Ability

Refer to figure 7.9b to answer the following questions:

16. What is the PR interval in this ECG?

17. What is the cardiac rate in this ECG?

Clinical Investigation Questions

18. What condition did the ECG indicate when the patient was brought to the hospital? How might this have caused a fall in blood pressure and a momentary loss of consciousness?

19. What does a normal sinus rhythm with a PR interval of 0.16 sec. indicate? What are PVCs and what could a run of them feel like to the patient?

EXERCISE 7.3

Effects of Exercise on the Electrocardiogram

MATERIALS

1. Electrocardiograph or multichannel recorder (e.g., Physiograph) with appropriate ECG module
2. ECG plates, straps, gel
3. Alternatively, the Biopac system may be used with the pulse transducer and electrodes for Biopac Student Lab Lesson 7, or the PowerLab modules may be used
4. Simulated electrocardiograph exercises may be performed using the *Physiology Interactive Lab Simulations (Ph.I.L.S.):* Electrocardiogram and Heart Function (exercises 4 and 6)

During exercise, there is a decrease in the activity of the parasympathetic innervation and an increase in the activity of the sympathetic innervation to the SA node, conductive tissue, and myocardium. More rapid discharge of the SA node, more rapid conduction of impulses, and a faster rate of contraction all result in an increased cardiac rate during exercise.

LEARNING OUTCOMES

You should be able to:

1. Describe the effects of the sympathetic and parasympathetic innervation to the heart.
2. Obtain an ECG before, immediately after, and 2 minutes after exercise and explain the differences observed.
3. Determine cardiac rate, P-R interval, and period of ventricular diastole for the ECG tracings obtained in objective 2.

Textbook/Multimedia Correlations

Before performing this exercise, you should study the introductory material presented here. Further information relating to this exercise can be found in these pages of *Human Physiology,* thirteenth edition, by Stuart I. Fox:

- *Pressure Changes During the Cardiac Cycle.* Chapter 13, p. 423.
- *The Electrocardiogram.* Chapter 13, p. 428.
- *Regulation of Cardiac Rate.* Chapter 14, p. 451.
- Multimedia correlations are provided in *Physiology Interactive Lab simulations (Ph.I.L.S.):* Electrocardiogram and Heart Function (exercises 4 and 6).

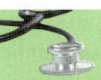

Clinical Investigation

A patient had complained of pain in his left pectoral region several days before seeing the physician, who was concerned that it might be angina pectoris. His resting ECG, however, appeared normal. A stress test was performed using a treadmill that required a high setting to increase the cardiac rate sufficiently because the patient had a history of regular exercise. At the highest cardiac rate achieved, the patient's ECG showed ST segment depression.

- Explain how regular exercise can increase cardiac output and aerobic capacity, and how this influences a stress ECG test.
- Explain why the patient experienced ST segment depression only at the highest cardiac rate achieved, and how that might relate to his possible angina pectoris episode.

During exercise, the cardiac output (blood pumped per minute by each ventricle) can increase from about 5 L per minute to about 25 L per minute for average healthy adults. This is primarily due to an increase in cardiac rate. The cardiac rate, however, can increase only up to a maximum value, determined primarily by the person's age. In well-trained athletes the stroke

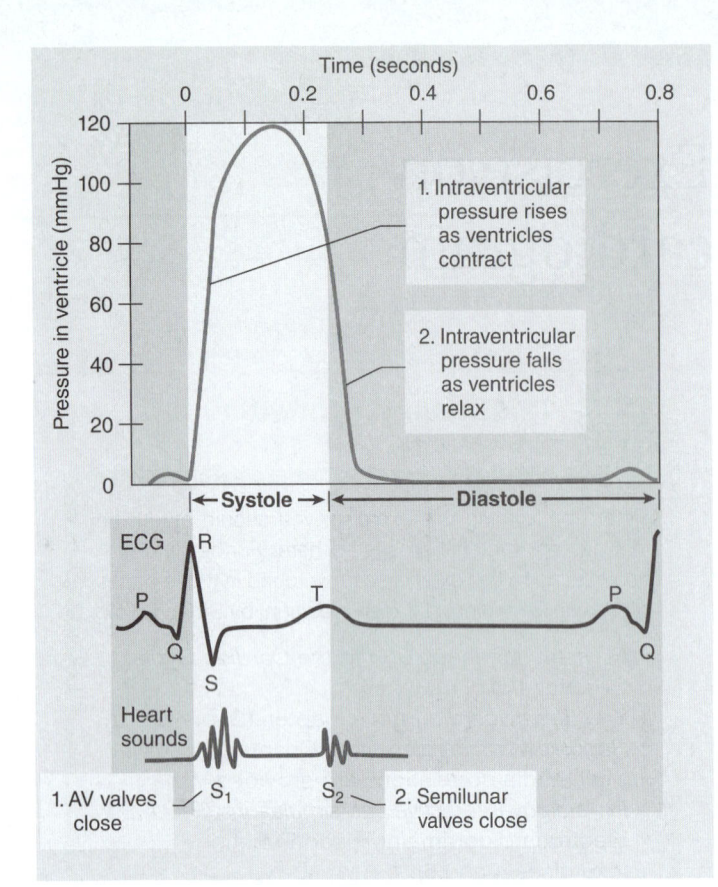

Figure 7.14 **The relationship between changes in intraventricular pressure and the ECG.** The QRS wave (representing depolarization of the ventricles) occurs at the beginning of systole, whereas the T wave (representing repolarization of the ventricles) occurs at the beginning of diastole.

(For a full-color version of this figure, see fig. 13.25 in *Human Physiology,* thirteenth edition, by Stuart I. Fox.)

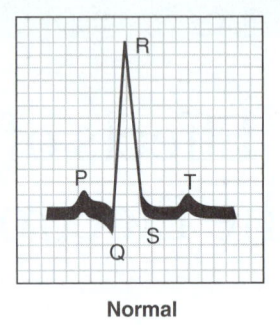

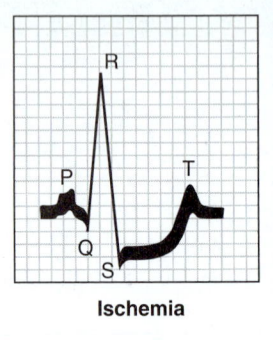

Figure 7.15 **ECG changes during myocardial ischemia.** In myocardial ischemia, the S-T segment of the electrocardiogram may be depressed, as illustrated in this figure.

the velocity of both impulse conduction and ventricular contraction. These effects are most evident at high cardiac rates and contribute only slightly to the increased cardiac rate during exercise. Thus, the increased cardiac rates are mainly due to a shortening of the ventricular diastole (from the peak of the T wave to the beginning of the next QRS complex) and only secondarily due to a shortening of ventricular systole (measured from the QRS peak to the peak of the T wave, fig. 7.14).

Clinical Applications

The rate of blood flowing through the coronary circulation may be adequate to meet the aerobic requirements of the heart at rest, but may be inadequate for the increased metabolic energy demands of the heart during exercise. A portion of the cardiac muscle may receive insufficient blood flow because of a clot in a coronary artery *(coronary thrombosis)* or because of narrowing of the vessel due to *atherosclerosis.* This insufficient blood flow to the heart—**myocardial ischemia**—may, however, be relative to the aerobic demand. Supervised exercise tests such as those on a treadmill with varying levels of aerobic demand help to diagnose this type of ischemia. During these tests, changes in the *S-T segment* of the ECG (fig. 7.15) may indicate ischemia.

volume (blood pumped per minute by each ventricle) can also increase significantly, allowing them to achieve cardiac outputs during strenuous exercise that are six to seven times greater than their resting values. This high cardiac output results in increased oxygen delivery to the exercising muscles, which is the major reason for the much higher than average aerobic capacities of elite athletes.

The heart is innervated by both sympathetic and parasympathetic nerve fibers. At the beginning of exercise, the activity of the parasympathetic fibers that innervate the SA node decreases. Because these fibers have an inhibitory effect on the pacemaker, a decrease in their activity results in an increase in cardiac rate. As exercise becomes more intense the activity of sympathetic fibers that innervate the SA node increases. This has an excitatory effect on the SA node and causes even greater increases in cardiac rate.

Sympathetic fibers also innervate the conducting tissues of the heart and the ventricular muscle fibers. Through these innervations, sympathetic stimulation may increase

Procedure

1. After the resting ECG has been recorded (from exercise 7.2), unplug the electrode leads from the electrocardiograph.
2. Have the subject exercise while holding the lead wires—for example, by walking up and down stairs, using a stationary bicycle, hopping, doing sit-ups or leg lifts.

> **Caution:** *The intensity of exercise should be monitored so that 80% of the maximum cardiac rate is not exceeded. Students not in good health should not serve as subjects.*

3. Immediately after exercise, the subject should lie down and the electrode leads are plugged into the electrocardiograph. Record lead II *only*. Wait 2 minutes and record lead II again.
4. Determine the cardiac rate, period of ventricular diastole, and the duration of the QRS complex for the resting ECG and for the two postexercise ECG recordings. Enter these data in the table in your laboratory report.
5. Alternatively, the ECG may be obtained using the Biopac system (as set up for Biopac Student Lab Lessons 5 and 6). If the pulse transducer is available (fig. 7.16), as used in Biopac Student Lab Lesson 7, the pulse can be followed directly during rest, exercise, and postexercise.

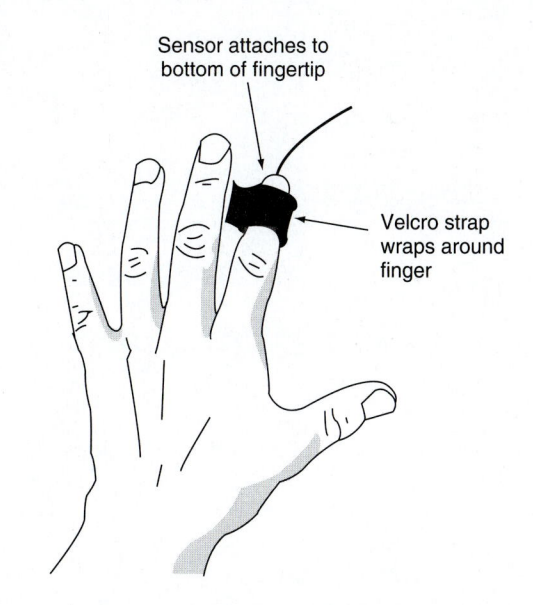

Figure 7.16 **Pulse transducer for Biopac system.**

Laboratory Report 7.3

Name _____
Date _____
Section _____

DATA FROM EXERCISE 7.3

1. Enter your data in this table.

ECG Tracing	Cardiac Rate (beats/min)	P-R Interval (sec) (beginning of P to Q)	Ventricular Diastole (sec) (middle of T to next Q)
Resting ECG			
Immediate postexercise ECG			
Two-minute postexercise ECG			

(a) Which measurement changed the most as you went from resting to exercise, and then to the two-minute postexercise conditions? _____

(b) Which changed the least? What conclusions can you draw regarding the changes in impulse conduction and the cardiac cycle as a result of exercise?

2. Tape your ECG recordings in this space.

REVIEW ACTIVITIES FOR EXERCISE 7.3

Test Your Knowledge

1. The ECG wave that occurs at the beginning of ventricular systole is the _____ wave.
2. The ECG wave that occurs at the end of systole and the beginning of diastole is the _____ wave.
3. The ECG wave completed just before the end of ventricular diastole is the _____ wave.
4. The nerve that increases the rate of discharge of the SA node is a _____ nerve.
5. The specific nerve that, when stimulated, causes a decrease in the cardiac rate is the _____.
6. The scientific term for insufficient blood flow to the heart muscle is _____.

Test Your Understanding

7. Describe the regulatory mechanisms that produce an increase in cardiac rate during exercise. Explain how these changes affect the electrocardiogram (ECG).

8. Describe the pressure changes that occur within the ventricles as each ECG wave is produced, and explain how these pressure changes are related to the ECG waves.

Test Your Analytical Ability

9. Explain how a person could have a normal electrocardiogram at rest, but show evidence of myocardial ischemia during exercise. Do you think all people should have regular treadmill (stress) electrocardiogram tests or only certain people? Explain.

10. Provide a cause-and-effect explanation as to how the pulse rate relates to the ECG rate. How would the relationship change if the pulse was measured by palpation of the carotid artery instead of by a pulse transducer in the finger?

Clinical Investigation Questions

11. What cardiovascular changes allow trained elite athletes to have a high cardiac output? How would a stress ECG test of an elite athlete differ from that of an average person?

12. What might have been responsible for the patient's depressed ST segment? What might account for the observation that the depressed ST segment appeared only at the highest cardiac rate? How does this relate to a possible episode of angina pectoris?

EXERCISE 7.4

Mean Electrical Axis of the Ventricles

MATERIALS

1. Electrocardiograph or multichannel recorder (e.g., Physiograph) with appropriate ECG module
2. ECG plates, straps, gel
3. Alternatively, the Biopac system may be used with Student Lab 6, or PowerLab modules may be used

The voltage changes in the ECG measured by two different leads can be compared and used to determine the mean electrical axis, which corresponds to the average direction of depolarization as the impulses spread into the ventricles. Significant deviations from the normal axis may be produced by specific heart disorders.

LEARNING OUTCOMES

You should be able to:

1. Describe the electrical changes in the heart that produce the ECG waves.
2. Determine the mean electrical axis of the ventricles in a test subject, and explain the clinical significance of this measurement.

Textbook/Multimedia Correlations

Before performing this exercise, you should study the introductory material presented here. Further information relating to this exercise can be found in these pages of *Human Physiology*, thirteenth edition, by Stuart I. Fox:

- *The Electrocardiogram.* Chapter 13, p. 428.
- Multimedia correlations are provided in *Physiology Interactive Lab simulations (Ph.I.L.S.):* Electrocardiogram and Heart Function (exercises 4 and 6).

Clinical Investigation

A patient who had an improperly corrected septal defect when he was a child developed pulmonary hypertension when he was in his fifties. His ECG revealed a mean electrical axis of 120°.

- Describe the mean electrical axis of the ventricles and explain how this measurement can be obtained from ECG recordings.
- Explain the significance of the patient's measurements and how this might relate to his medical history.

The depolarization waves of action potentials spread through the heart in a characteristic pattern. Depolarization waves begin at the SA node and spread from the pacemaker to the entire mass of both atria, producing the *P wave* in an electrocardiogram. After the AV node is excited, the interventricular septum becomes depolarized as the impulses spread through the bundle of His. At this point the septum is depolarized while the lateral walls of the ventricles still have their original polarity, so there is a potential difference (voltage) between the septum and the ventricular walls. This produces the *R wave*. When the entire mass of the ventricles is depolarized, there is no longer a potential difference within the ventricles and the voltage returns to zero (completing the *QRS complex*).

The direction of the depolarization waves depends on the orientation of the heart in the chest and on the particular instant of the cardiac cycle being considered. It is clinically useful, however, to determine the **mean axis** (average direction) **of depolarization** during the cardiac cycle. This can be done by observing the voltages of the QRS complex from two different perspectives using two different leads. Lead I provides a horizontal axis of observation (from left arm to right arm); lead III has an axis of about 120° (from left arm to left leg). Using the recordings of leads I and III, the mean electrical axis can be found (a normal axis of 59° is shown in figure 7.17).

> **Normal Values** The normal mean electrical axis of the ventricles is between 0 and +90°.

(a)

(b)

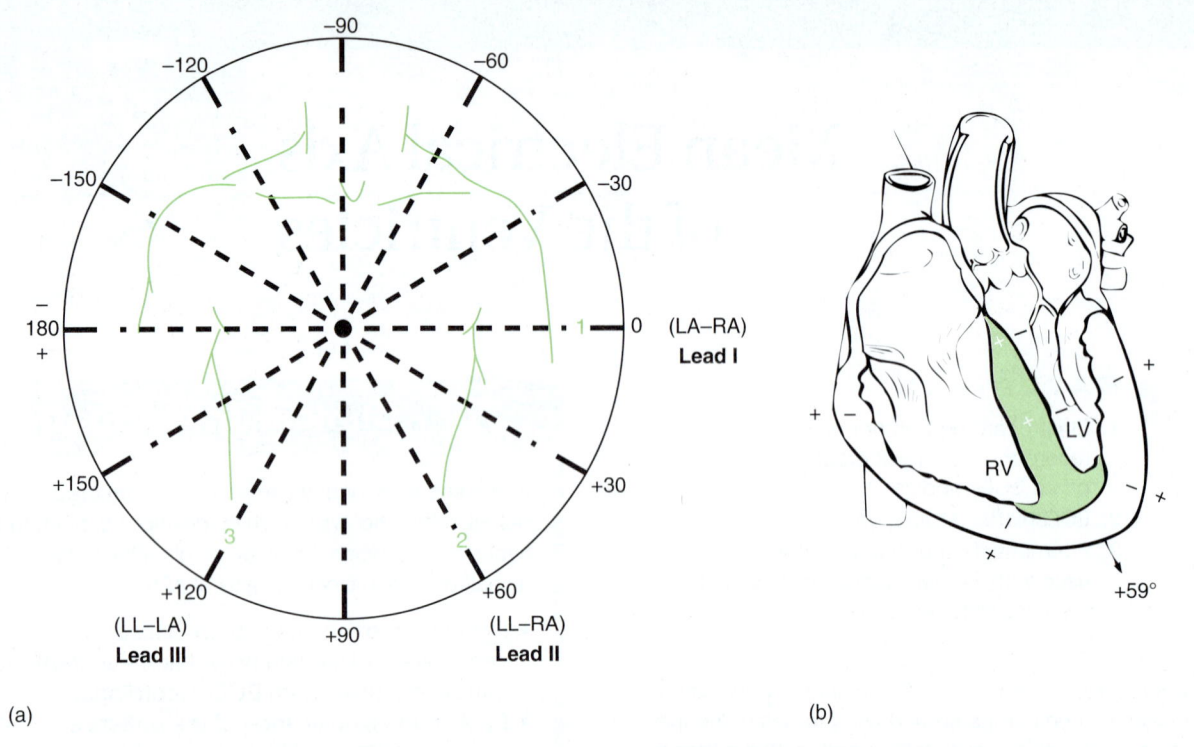

Figure 7.17 Mean electrical axis of the heart.   (a) The convention by which the axis of depolarization is measured. The bottom half of the circle (with the heart at the center) is considered the positive role (LA = left arm, RA = right arm, LL = left leg). (b) When the interventricular septum is depolarized, the surface of the septum is electrically negative compared to the walls of the ventricles, which have not yet become depolarized. The average normal direction of depolarization, or mean electrical axis of the ventricles, is about 59° (RV = right ventricle, LV = left ventricle).

Clinical Applications

Hypertrophy (enlargement) of one ventricle shifts the mean axis of depolarization toward the hypertrophied ventricle because it takes longer to depolarize the larger ventricle. Therefore, a left axis deviation occurs when the left ventricle is hypertrophied (as a result of hypertension or narrowing of the aortic semilunar valve). A right axis deviation occurs when the right ventricle hypertrophies. The latter condition may be secondary to narrowing of the pulmonary semilunar valve, or to such congenital conditions as a septal defect or the tetralogy of Fallot.

The depolarization wave normally spreads through both the right and left ventricles at the same time. However, if there is a conduction block in one of the branches of the bundle of His—a **bundle-branch block**—depolarization will be much slower in the blocked ventricle. In left bundle-branch block, for example, depolarization will occur more slowly in the left ventricle than in the right ventricle, and the mean electrical axis will deviate to the left. In right bundle-branch block, there will be a right axis deviation. Deviations of the electrical axis also occur to varying degrees as a result of myocardial infarction.

Procedure

Note: *If a Cardiocomp-7 or –12 is used, the mean electrical axis of the ventricles can be determined by examining the QRS loop of the vectorgram displayed on the computer screen. Alternatively, this procedure can be used by examining leads I and III previously obtained (using any equipment) in exercise 7.2.*

1. Analysis of lead I:
   (a) Find a QRS complex and count the number of millimeters (small boxes) that it projects above the upper edge of the baseline.
   Enter this value here:
   + _____ mm

   (b) Count the number of millimeters the Q and S waves (or the R wave, if it is inverted) project below the upper edge of the baseline. Add these measurements of downward deflections.
   Enter this sum here:
   _____ mm

   (c) Algebraically, add the two values from steps (a) and (b), keeping the negative sign if the sum is negative.
   Enter this sum in the laboratory report.

2. Analysis of lead III:
   (a) Find a QRS complex and count the number of millimeters (small boxes) it projects above the upper edge of the baseline.

   Enter this value here:

   + _____ mm

   (b) Count the number of millimeters the Q and S waves (or the R wave, if it is inverted) project below the upper edge of the baseline. Add these measurements of downward deflections.

   Enter this sum here:

   _____ mm

   (c) Algebraically, add the two values from steps (a) and (b), keeping the negative sign if the sum is negative.

   Enter this sum in the laboratory report.

3. On the blank grid chart in your laboratory report, use a straight edge to make a line on the axis of *lead I* that corresponds to the sum you obtained in step 1c (see example grid, fig. 7.19).

4. Use a straight edge to make a line on the axis of *lead III* that corresponds to the sum you obtained in step 2c.

5. Use a straight edge to draw an arrow from the center of the grid chart to the intersection of the two lines drawn in steps 3 and 4. Extend this arrow to the edge of the grid chart, and record the mean electrical axis of the ventricles.

Example (fig. 7.18)

Lead I of sample ECG:
Upward deflection:	+7 mm
Downward deflections:	−1 mm
	6 mm

Lead III of sample ECG:
Upward deflection:	+14 mm
Downward deflections:	− 2 mm
	12 mm

A straight line is drawn perpendicular to the horizontal axis of lead I corresponding to position 6 on the scale. Similarly, a straight line is drawn perpendicular to the axis of lead III that corresponds to position 12 on the scale. An arrow is then drawn from the center of the circle through the intersection of the two lines previously drawn (fig. 7.19).

In this example, the mean electrical axis of the ventricles is +71°.

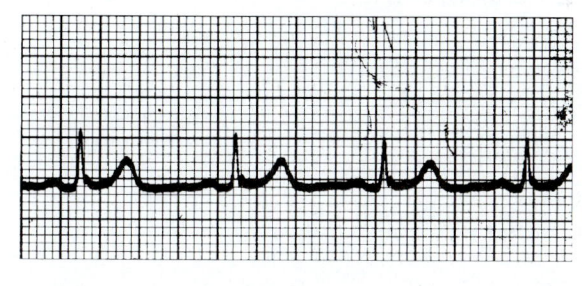

Lead I

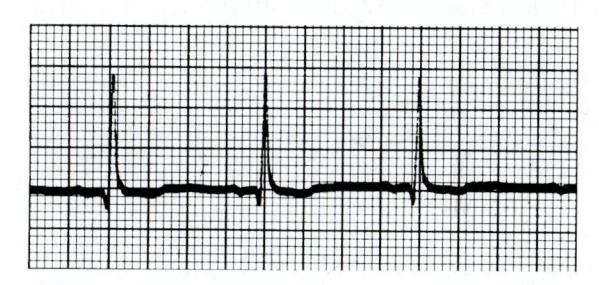

Lead III

Figure 7.18 **Sample electrocardiograms of leads I and III.** These were used in the example for the determination of the mean electrical axis of the heart. (Note: Each small square is 1 mm on a side.)

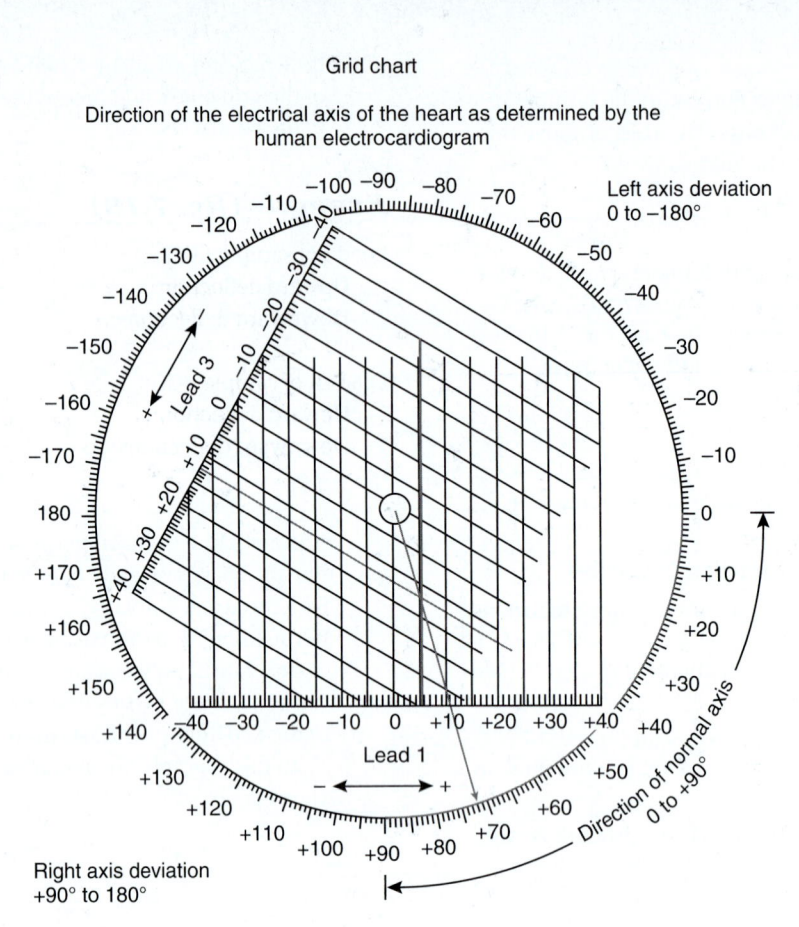

Figure 7.19 Example of the method used to determine the mean electrical axis of the heart. This uses the data from leads I and III in the sample ECG shown in figure 7.18. In this example, the mean electrical axis is 72°.

Laboratory Report 7.4

Name _____
Date _____
Section _____

DATA FROM EXERCISE 7.4

1. Enter the value for the sum of steps 1a and 1b in the space provided here. Be sure to indicate whether it is a positive or negative number. _____

2. Enter the value for the sum of steps 2a and 2b in the space provided here. Be sure to indicate whether it is a positive or negative number. _____

3. Use this grid chart to determine the mean electrical axis of the ventricles from your data, as described in steps 3, 4, and 5 of the procedure.

   Mean electrical axis of the ventricles: _____

Grid chart

Direction of the electrical axis of the heart as determined by the human electrocardiogram

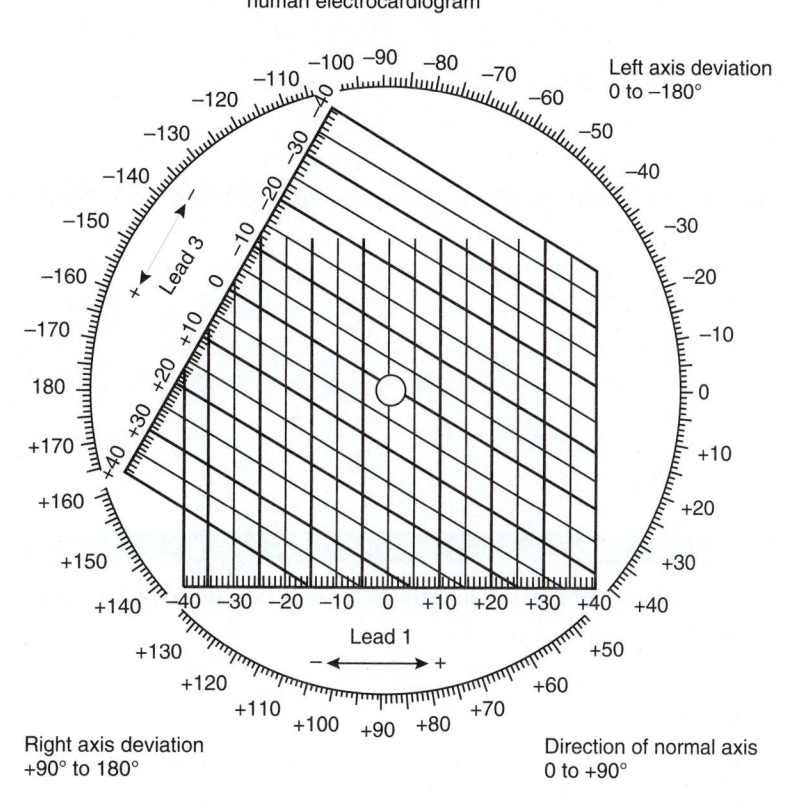

Left axis deviation
0 to −180°

Lead 3

Lead 1

Right axis deviation
+90° to 180°

Direction of normal axis
0 to +90°

269

REVIEW ACTIVITIES FOR EXERCISE 7.4

Test Your Knowledge

1. Depolarization of the atria produces the _____ wave.
2. The difference in polarity between the interventricular septum and the lateral walls of the ventricles produces the _____ wave.
3. The mean electrical axis is determined using two bipolar limb leads, lead _____ and lead _____.
4. Hypertrophy of the left ventricle would shift the mean axis of depolarization to the _____.
5. A blockage in conduction in the right branch of the bundle of _____ would cause the mean axis of depolarization to shift to the _____.

Test Your Understanding

6. Explain how the normal pattern of depolarization is affected by ventricular hypertrophy. What factors may be responsible for hypertrophy of the right or left ventricle?

7. Explain how the normal pattern of depolarization is affected by bundle-branch block. What might cause bundle-branch block?

Test Your Analytical Ability

8. Explain the electrical events in the heart that produce the QRS wave. Why does the tracing go up from Q to R, and then back to baseline from R to S?

9. How do you think obesity or pregnancy might influence the results of this laboratory exercise? Explain.

Clinical Investigation Questions

10. What is a normal measurement of the mean electrical axis and what does this patient's measurement indicate?

11. Identify two possible causes of the patient's mean electrical axis measurement. Which of these two possible causes is the most likely given his medical history? Explain.

EXERCISE 7.5

Heart Sounds

MATERIALS

1. Stethoscopes
2. Physiograph, high-gain couplers, microphone for heart sounds (Narco), ultrasonic flowmeter (such as Doppler)
3. Alternatively, the Biopac system may be used as per Student Lab Lesson 17, employing the Biopac electrode lead set and the amplified stethoscope

Contraction and relaxation of the ventricles is accompanied by pressure changes that cause the one-way heart valves to close. Closing valves produce sounds that aid in the diagnosis of structural abnormalities of the heart.

LEARNING OUTCOMES

You should be able to:

1. Describe the causes of the normal heart sounds.
2. List some of the causes of abnormal heart sounds.
3. Correlate the heart sounds with the waves of the ECG and the events of the cardiac cycle.

Textbook/Multimedia Correlations

Before performing this exercise, you should study the introductory material presented here. Further information relating to this exercise can be found in these pages of *Human Physiology,* thirteenth edition, by Stuart I. Fox:

- *Heart Sounds.* Chapter 13, p. 420.
- *Cardiac Cycle.* Chapter 13, p. 422.
- *The Electrocardiogram.* Chapter 13, p. 428.
- Multimedia correlations are provided in *Physiology Interactive Lab Simulations (Ph.I.L.S.):* Electrocardiogram and Heart Function (exercises 4 and 6).

Clinical Investigation

A patient in her twenties was told that she has a heart murmur. She reported that, as a child, she had a small opening in the septum between her atria. However, further investigation revealed that her murmur was due to a mitral valve prolapse, not to a patent foramen ovale between the atria.

- Explain how blood would flow, in a fetus and in an adult, if there were an opening in the septum between the atria.
- Describe the cause of mitral valve prolapse and how that would affect blood flow and heart sounds.

A. AUSCULTATION OF HEART SOUNDS WITH THE STETHOSCOPE

The cycle of mechanical contraction (**systole**) and relaxation (**diastole**) of the ventricles can be followed by listening to the heart sounds with a **stethoscope.** The contraction of the ventricles produces a rise in intraventricular pressure, resulting in the vibration of the surrounding structures as the atrioventricular valves slam shut. The valves closing produces the *first sound* of the heart, usually verbalized as "**lub.**" As the ventricles relax and the intraventricular pressure falls below the pressure in the arteries, the blood in the aorta and pulmonary arteries pushes the one-way semilunar valves shut. The resulting vibration of these structures produces the *second sound* of the heart, verbalized as "**dub**" (fig. 7.20).

Careful **auscultation** (listening) to the two heart sounds may reveal two further sounds. This *splitting* of the heart sounds into four components is more evident during inhalation than during exhalation. During deep inhalation, the first heart sound may be split into two sounds because the tricuspid and mitral valves close at different times. The second heart sound may also be split into two components because the pulmonary and aortic semilunar valves close at different times.

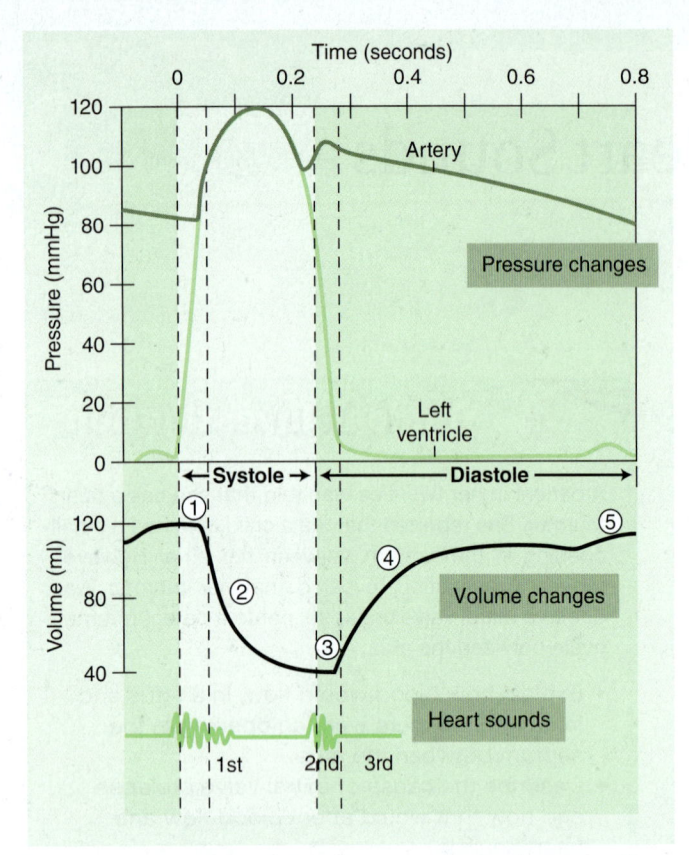

Figure 7.20 **The relationship between heart sounds and the intraventricular pressure and volume.** The numbers refer to the following events within the left ventricle: (1) phase of *isovolumetric contraction,* when all the heart valves are closed; (2) phase of *ejection,* as the semilunar valves open; (3) phase of *isovolumetric relaxation,* as all the heart valves are closed; (4) phase of *rapid filling,* as the AV valves open; and (5) phase of *atrial contraction,* when atria add the final amount of blood to the ventricles.

(For a full-color version of this figure, see fig. 13.17 in *Human Physiology,* thirteenth edition, by Stuart I. Fox.)

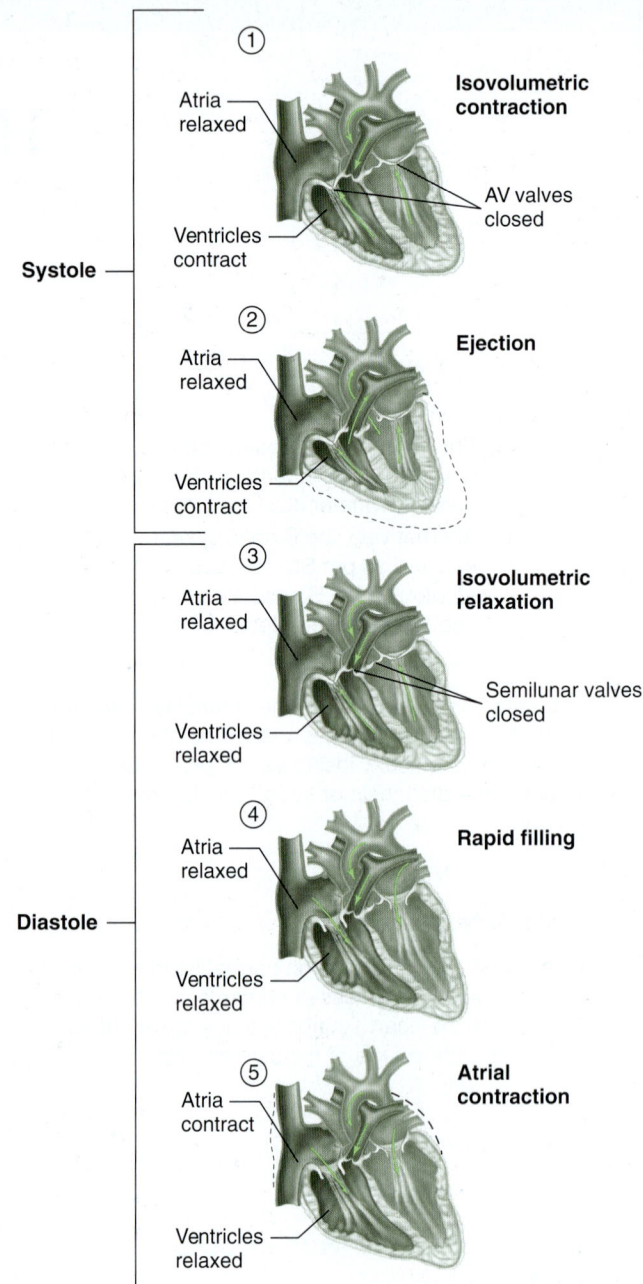

Clinical Applications

Auscultation of the chest aids in the diagnosis of many cardiac conditions, including **heart murmurs.** A murmur may be caused by an irregularity in a valve; a septal defect; or the persistent fetal opening *(foramen ovale)* between the right and left atria after birth, resulting in the audible regurgitation of blood in the reverse direction of normal flow. Abnormal splitting of the first and second heart sounds may be due to heart block, septal defects, aortic stenosis, hypertension, or other abnormalities.

Mitral valve prolapse is the most common cause of chronic *mitral regurgitation,* where blood flows backward into the left atrium. It has both congenital and acquired forms. In younger people, it is usually caused by excess valve leaflet material. Although most people with this condition are without symptoms and have an apparently normal life span, in some people the condition can progress to the point where the mitral valve must be repaired or replaced.

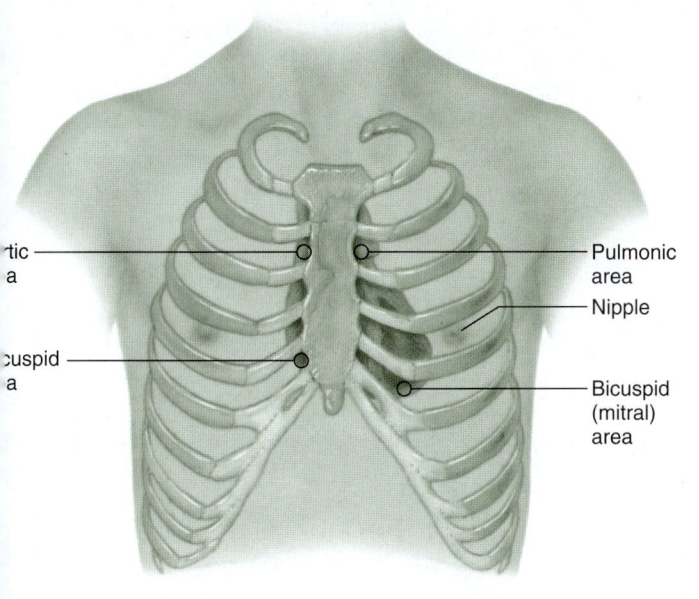

Figure 7.21 Routine stethoscope positions for listening to the heart sounds. The first heart sound is caused by closing of the AV valves; the second by closing of the semilunar valves.

(For a full-color version of this figure, see fig. 13.13 in *Human Physiology*, thirteenth edition, by Stuart I. Fox.)

Procedure

1. Clean the earpieces of the stethoscope with an alcohol swab. To best hear the first heart sound, auscultate the apex beat of the heart by placing the diaphragm of the stethoscope in the *fifth left intercostal space* (the bicuspid area in fig. 7.21).

2. To best hear the second heart sound, place the stethoscope to the right or left of the sternum in the *second intercostal space* (the aortic or pulmonic area in fig. 7.21).

3. Compare the heart sounds in the three stethoscope positions described during quiet breathing, slow and deep inhalation, and slow exhalation.

B. Correlation of the Phonocardiogram with the Electrocardiogram

If the heart sounds are monitored with a device known as a **phonocardiograph** in conjunction with the monitoring of the electrical patterns of the heart with an electrocardiograph, a correlation of the two recordings will show that the first heart sound occurs at the end of the QRS complex of the ECG, and the second heart sound occurs at the end of the T wave (fig. 7.22; also see fig. 7.20). The arterial pulse is palpated (felt) in either the radial artery or carotid artery in the time interval between the two heart sounds.

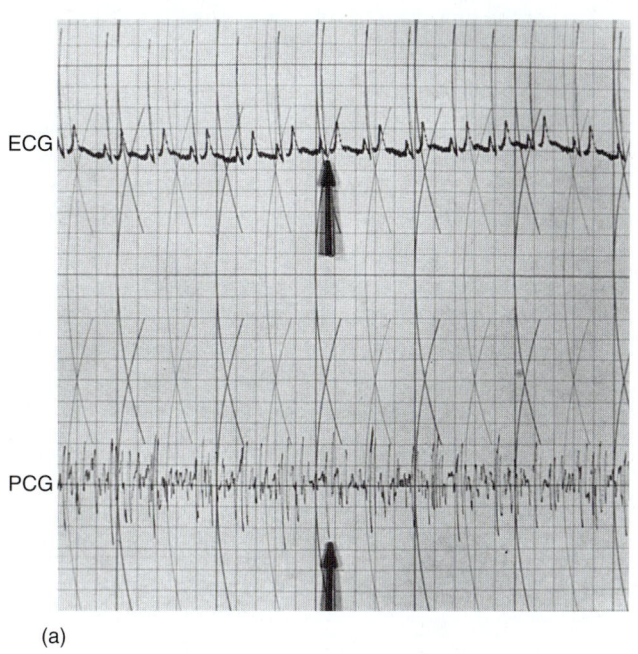

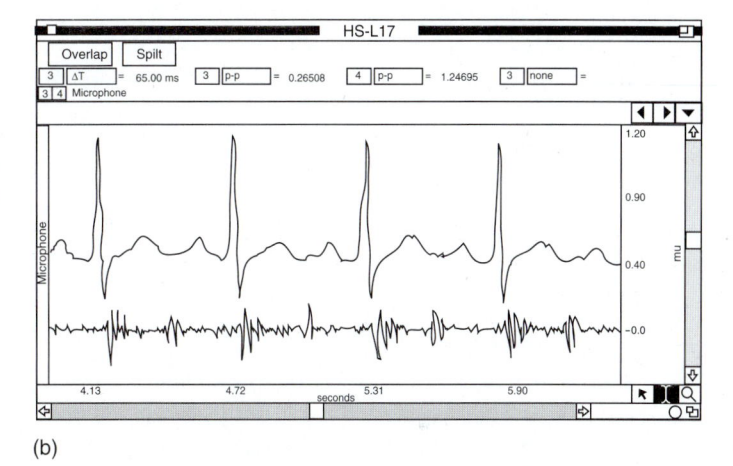

Figure 7.22 Simultaneous recording of an electrocardiogram (ECG) and a phonocardiogram (PCG).
(a) Photograph of recordings taken with a Narco physiograph.
(b) Idealized computer image of recordings taken with the Biopac system ECG electrodes and amplified stethoscope.

Procedure

1. The following steps relate to the procedure for Physiograph recording. A different procedure for steps 2–7 should be followed if different equipment is used to perform this exercise.

2. Insert two *high-gain couplers* into the physiograph. Plug the cable from the ECG lead selector box into the coupler for channel 1 and the cable from the phonocardiograph microphone into the coupler for channel 2.

3. Attach the ECG electrode plates to the subject in the standard limb lead positions, and plug the ECG cable into the lead selector box (see exercise 7.2).

4. For the high-gain coupler in channel 1:
   (a) Turn the time constant knob to the *3.2* position.
   (b) Turn the gain knob to the × *2* position.
   (c) Turn the knob on the ECG lead selector box to the calibrate position. Turn the outer sensitivity knob on the amplifier to the *10* position. The inner knob should be turned all the way to the right until it clicks.
   (d) Lower the pen lift lever, raise the inkwells, and squeeze the rubber bulbs on the inkwells until ink flows freely. Release the paper-advance button, and position the pen for channel 1 so that it writes on the appropriate heavy horizontal line.
5. For the high-gain coupler in channel 2 adjust the time constant and gain knobs as described in step 3, and position the pen to write on the appropriate heavy horizontal line. The sensitivity knob can be adjusted for the individual subject once recording begins.
6. Place the microphone to the fifth left intercostal space (fig. 7.21) and start the recording. Continue recording during normal breathing and deep inhalation. Move the microphone to the second right (or left) intercostal space and continue recording during normal breathing and deep inhalation. Repeat the procedure on the other side of the sternum.
7. Place the ultrasonic flowmeter crystal on either the radial artery or carotid artery with a dab of electrode paste. Plug the flowmeter into another high-gain coupler (in multichannel recorders), and record the arterial pulse simultaneously with the ECG and phonocardiogram.

Tape your recordings or draw facsimiles in the laboratory report.

Laboratory Report 7.5

DATA FROM EXERCISE 7.5

1. In this table, tape your phonocardiogram recordings or draw facsimiles.

Microphone Position	Normal Breathing	Deep Inhalation
Fifth left intercostal space		
Second left intercostal space		
Second right intercostal space		

REVIEW ACTIVITIES FOR EXERCISE 7.5

Test Your Knowledge

1. The scientific term for listening carefully (as with a stethoscope) is _____.
2. The first heart sound (lub) is caused by _____.
3. During which phase of the cardiac cycle (systole or diastole) does the first heart sound occur? _____
4. The second heart sound is caused by _____.
5. The second heart sound occurs at the (beginning or end) _____ of (systole or diastole) _____.
6. Abnormal heart sounds are called _____.
7. The first heart sound (lub) is correlated with which ECG wave? _____
8. The second heart sound (dub) is correlated with which ECG wave? _____

Test Your Understanding

9. Using a complete cause-and-effect sequence, explain the correlation of the heart sounds with the ECG waves.

10. Explain how the heart sounds are normally produced, and describe some of the conditions that can cause heart murmurs.

11. What is meant by the "splitting" of the heart sounds? What can cause this splitting?

Test Your Analytical Ability

12. Can a defective valve be detected by electrocardiography, auscultation, palpation of the radial artery, or all three techniques? Explain your answer.

13. Can the beating of the atria be detected by electrocardiography, auscultation, palpation of the radial artery, or all three techniques? Explain your answer.

Clinical Investigation Questions

14. Which way would blood flow through an opening in the septum between the atria in a fetus and in an adult? How could this cause a murmur?

15. What is the usual cause of mitral valve prolapse? How does blood flow through the prolapsed valve cause a murmur? What would happen if the abnormal blood flow were medically significant?

279

EXERCISE 7.6

Measurements of Blood Pressure

MATERIALS

1. Sphygmomanometer
2. Stethoscope
3. Alternatively, the Biopac system for Student Lab Lesson 16 may be employed, using the Biopac blood pressure cuff and Biopac stethoscope

Blood exerts a hydrostatic pressure against the walls of arteries that can be measured indirectly by means of a sphygmomanometer and a stethoscope. Abnormally high or low arterial blood pressures are a health risk.

LEARNING OUTCOMES

You should be able to:

1. Describe how the sounds of Korotkoff are produced.
2. Demonstrate the ability to take blood pressure measurements, and explain why measurements of systolic and diastolic pressure correspond to the first and last sounds of Korotkoff.
3. Describe how pulse pressure and mean arterial pressure are calculated.

Textbook/Multimedia Correlations

Before performing this exercise, you should study the introductory material presented here. Further information relating to this exercise can be found in these pages of *Human Physiology,* thirteenth edition, by Stuart I. Fox:

- *Blood Pressure.* Chapter 14, p. 475.
- *Hypertension.* Chapter 14, p. 482.
- Multimedia correlations are provided in *MediaPhys 3.0:* Topics 9.15 and 9.16; Topics 9.39–9.42.

Clinical Investigation

A patient has a blood pressure of 140/110 and is prescribed a drug that reduces vasoconstriction and increases urine volume.

- Describe the medical significance of the patient's blood pressure.
- Explain the mechanisms by which the drug lowers the patient's blood pressure.

The delivery of oxygen and nutrients necessary for normal metabolism depends on the adequate flow of blood through the tissues (*perfusion*). Many narrow, muscular arteries and arterioles branch to form *arterial trees*, which offers great resistance to blood flow through peripheral tissues. For normal perfusion of tissue capillaries, therefore, the arterial blood must be under sufficient pressure to overcome this vascular resistance.

This arterial blood pressure is directly dependent on **cardiac output** (the amount of blood pumped by each ventricle per minute) and **peripheral resistance** (the resistance to blood flow through the arterioles). In this way, blood pressure can be raised by constriction and lowered by dilation of the arterioles.

The arterial blood pressure is routinely measured in an indirect manner with a **sphygmomanometer,** or blood pressure cuff. This device consists of an inflatable rubber bag connected by rubber hoses to a hand pump and to a pressure gauge (manometer) graduated in millimeters of mercury (mm Hg). The rubber bag is wrapped around the upper arm at the level of the heart and inflated to a pressure greater than the suspected systolic pressure, thus occluding the brachial artery. The examiner auscultates the brachial artery by placing the bell of a stethoscope in the cubital fossa and slowly opening a screw valve on the hand pump (bulb), allowing the pressure in the rubber bag to fall gradually (fig. 7.23).

Blood normally travels smoothly through the arteries. **Laminar flow** occurs when all parts of a fluid move in the same direction, parallel to the axis of the vessel. The term laminar means "layered"—blood in the central axial stream moves the fastest, and blood closer to the artery wall moves more slowly. If flow is perfectly laminar, there is no

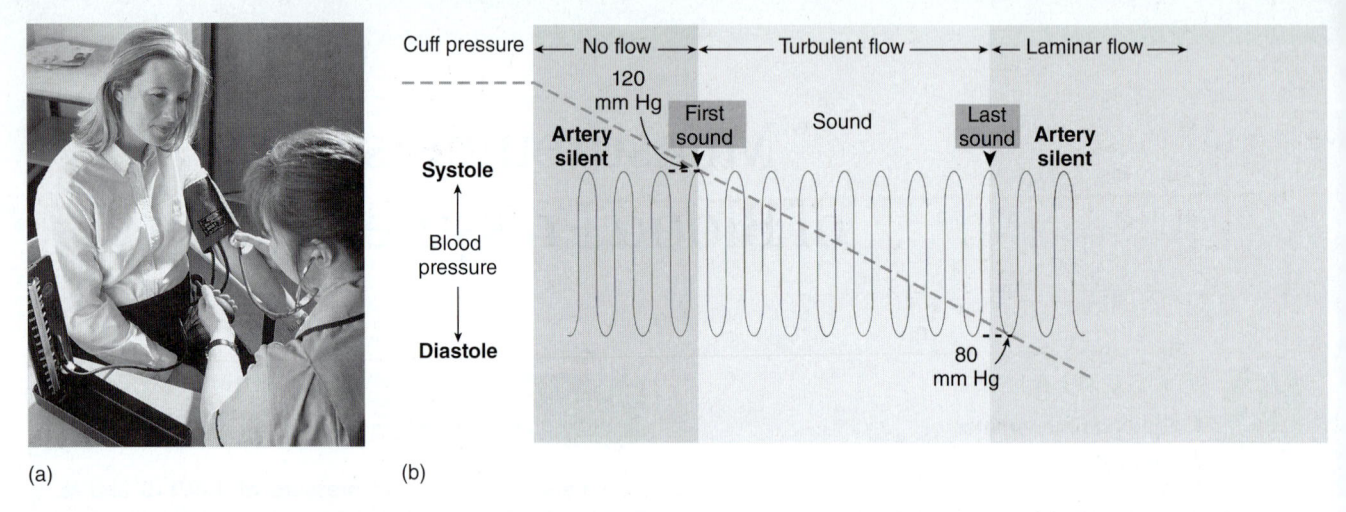

Figure 7.23 **The use of a stethoscope and a sphygmomanometer to measure blood pressure.** (a) The locations where the arm cuff and stethoscope bell are placed. (b) Korotkoff sounds are heard when there is turbulent blood flow through a constriction in the brachial artery. The systolic pressure corresponds to the cuff pressure when the first sound is heard, and the diastolic pressure corresponds to the cuff pressure when the last sound is heard.

(For a full-color version of this figure, see figs. 14.29 and 14.31 in *Human Physiology*, thirteenth edition, by Stuart I. Fox.)

transverse movement between these layers that would produce mixing. By contrast, **turbulent flow** occurs when some parts of the fluid move in radial and circumferential directions, churning and mixing the blood. Turbulent flow causes vibrations of the vessel, which may produce sounds. Blood flow in the brachial artery before the cuff is inflated is mostly laminar and so is smooth and silent.

When the sphygmomanometer bag is inflated to a pressure above the systolic pressure, the flow of blood in the artery is stopped and the artery is again silent. As the pressure in the bag gradually drops to levels between the systolic and diastolic pressures of the artery, the blood is pushed through the partially compressed walls of the artery, creating turbulent flow. Under these conditions the layers of blood are mixed by eddies that flow at right angles to the axial stream, and the turbulence sets up vibrations in the artery heard as sounds in the stethoscope. These sounds are known as the **sounds of Korotkoff,** after Nikolai S. Korotkoff, the Russian physician who first described them.

The sounds of Korotkoff are divided into *five phases* on the basis of the loudness and quality of the sounds, as shown in figure 7.24. Start on the left of the figure with a cuff pressure at approximately 130 millimeters of mercury (mm Hg), and note the Korotkoff sounds as the cuff pressure slowly falls.

- **Phase 1.** A loud, clear *tapping* (or snapping) sound is evident, which increases in intensity as the cuff is deflated. This phase begins at a cuff pressure of 120 mm Hg and ends at a pressure of 106 mm Hg.
- **Phase 2.** A succession of *murmurs* can be heard. Sometimes the sounds seem to disappear during this time (auscultatory gap), perhaps a result of inflating or deflating the cuff too slowly. This phase begins at a cuff pressure of 106 mm Hg and ends at a pressure of 86 mm Hg.
- **Phase 3.** A loud, *thumping* sound, similar to phase 1 but less clear, replaces the murmurs. This phase begins at a cuff pressure of 86 mm Hg and ends at a pressure of 81 mm Hg.
- **Phase 4.** A *muffled* sound abruptly replaces the thumping sounds of phase 3. This phase begins at a cuff pressure of 81 mm Hg and ends at a pressure of 76 mm Hg.
- **Phase 5.** Silence again as all sounds disappear. This phase is absent in some people.

The cuff pressure at which the first sound is heard (the beginning of phase 1) is the **systolic pressure.** The cuff pressure at which the sound becomes muffled (the beginning of phase 4) and the pressure at which the sound disappears (the beginning of phase 5) are taken as measurements of the **diastolic pressure.** Although the phase 5 measurement is closer to the true diastolic pressure than the phase 4 measurement, the beginning of phase 4 is easier to detect and the results are more reproducible. It is often recommended that both measurements of diastolic pressure be recorded. In the example shown in figure 7.24, the pressure would be indicated as 120/81/76 mm Hg. Frequently, however, the blood pressure would be recorded in this example as 120/76 mm Hg, or spoken as "120 over 76." The **pulse pressure** is calculated as the difference in these two pressures; and the **mean arterial pressure** is equal to the diastolic pressure plus one-third of the pulse pressure.

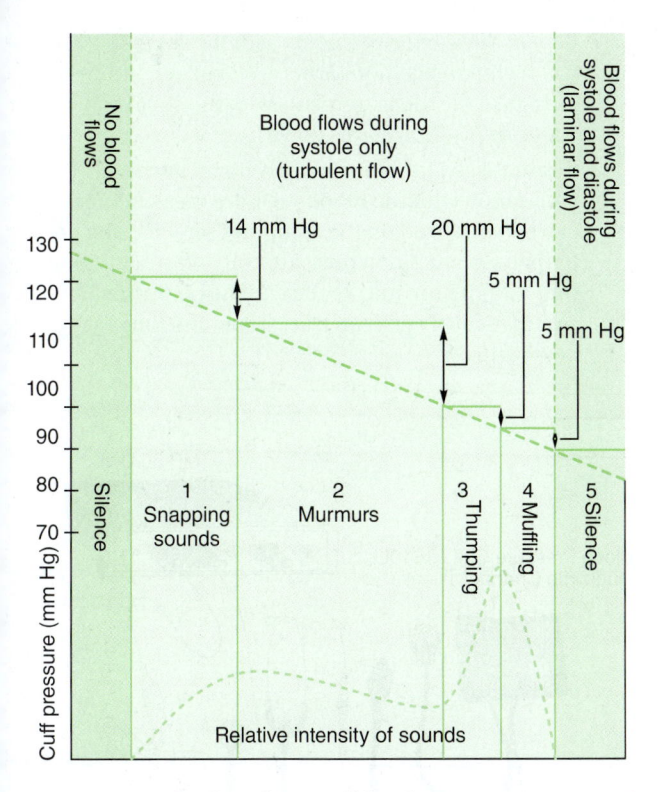

Figure 7.24 The five phases of blood pressure measurement. Not all phases are heard in all people. The cuff pressure is indicated by the falling dashed line.

(For a full-color version of this figure, see fig. 14.32 in *Human Physiology*, thirteenth edition, by Stuart I. Fox.)

Clinical Applications

A normal blood pressure measurement for a given individual depends on the person's age, sex, heredity, and environment. Considering these factors, chronically elevated blood pressure measurements may indicate an unhealthy state called **hypertension**, a major risk factor in heart disease and stroke. Hypertension may be defined as a systolic pressure of 140 mm Hg or higher, or a diastolic pressure of 90 mm Hg or higher (table 7.1).

Hypertension may be divided into two general categories. *Primary hypertension* (95% of all cases) refers to hypertension of unknown etiology. This category is, in turn, divided into *benign hypertension* (also known as *essential hypertension*) and *malignant hypertension*. *Secondary hypertension* refers to hypertension for which the pathological process is known. Newer evidence suggests that cardiovascular risk begins to increase when a person's systolic blood pressure exceeds 115 mm Hg or diastolic pressure exceeds 75 mm Hg. The new medical goal of maintaining a blood pressure that does not exceed 120/80 is expected to reduce the risk of heart attacks, heart failure, stroke, and kidney disease.

As the blood pressure rises from diastolic to systolic values within an artery, the rise in the hydrostatic pressure against the artery wall causes the artery to expand somewhat. This is the "pulse" you feel when you press your fingers against the outside of the artery. In this way, the pulse pressure causes the pulse. Importantly, this rise in blood pressure (the pulse pressure), a result of the contraction of the ventricle, serves as the pressure head that drives the blood through the arterial tree. The mean arterial pressure represents the combined systolic and diastolic pressure head that drives the blood from the arterial tree into the blood capillaries.

Different examiners can record different values for systolic and diastolic pressures, even on the same subject. Blood pressure measurements can vary with the instruments used, the anxiety of the subject, and even body position. Although the systolic pressure (caused by the left ventricle peak ejection pressure) tends to remain fairly constant with changes in body position, the diastolic pressure is affected by gravity. The arm positioned below the heart level may cause a rise in diastolic pressure, whereas positioning the arm above the heart lowers diastolic pressure.

Table 7.1 Blood Pressure Classifications for Adults in a Resting State

Blood Pressure Classification	Systolic Blood Pressure		Diastolic Blood Pressure	Drug Therapy
Normal	Under 120 mm Hg	and	Under 80 mm Hg	No drug therapy
Prehypertension	120–139 mm Hg	or	80–89 mm Hg	Lifestyle modification;* no antihypertensive drug indicated
Stage 1 Hypertension	140–159 mm Hg	or	90–99 mm Hg	Lifestyle modification; antihypertensive drugs
Stage 2 Hypertension	160 mm Hg or greater	or	100 mm Hg or greater	Lifestyle modification; antihypertensive drugs

From the Seventh Report of the Joint National committee on Prevention, Detection, Evaluation, and Treatment of High Blood Pressure: The JNC 7 Report. *Journal of the American Medical Association;* 289 (2003): 2560–2572.
*Lifestyle modifications include weight reduction; reduction in dietary fat and increased consumption of vegetables and fruit; reduction in dietary sodium (salt); engaging in regular aerobic exercise, such as brisk walking for at least 30 minutes a day, most days of the week; and moderation of alcohol consumption.

Procedure

1. Have the subject sit with his or her right or left arm resting on a table at the level of the heart. Wrap the cuff of the sphygmomanometer around the arm about 2.5 cm above the elbow.

2. Palpate the brachial artery in the cubital fossa and place the bell of the stethoscope where the arterial pulse is felt. Gently close the screw valve and pump the pressure in the cuff up to about 20 mm Hg above the point where sounds disappear, or to about 20 mm Hg above the point where the radial pulse can no longer be felt.

3. Open the screw valve to allow the pressure in the cuff to fall slowly, at a rate of about 2 or 3 mm Hg per second. The most common errors in blood pressure measurements occur because the cuff pressure is allowed to fall too rapidly.

4. Record the systolic pressure (beginning of phase 1) and the two measurements of diastolic pressure (beginning of phases 4 and 5). Enter these values in your laboratory Report, and compare your pressures with the range of normal values listed in table 7.1.

5. Calculate the subject's **pulse pressure** (systolic minus diastolic pressure). Enter this value in your laboratory report.

6. Calculate the subject's **mean arterial pressure** (equal to the diastolic pressure plus one-third of the pulse pressure). Enter this value in the laboratory report.

7. Repeat these measurements with the same subject in a reclining position (arms at sides). A few minutes later, take the measurements with the subject in a standing position (arms down).

8. Alternatively, the sounds of Korotkoff and their correlation with the blood pressure measurements can be visualized on the computer screen using the Biopac system blood pressure cuff and stethoscope (fig. 7.25). These may or may not be performed simultaneously with an ECG, as per the Biopac Lesson 16.

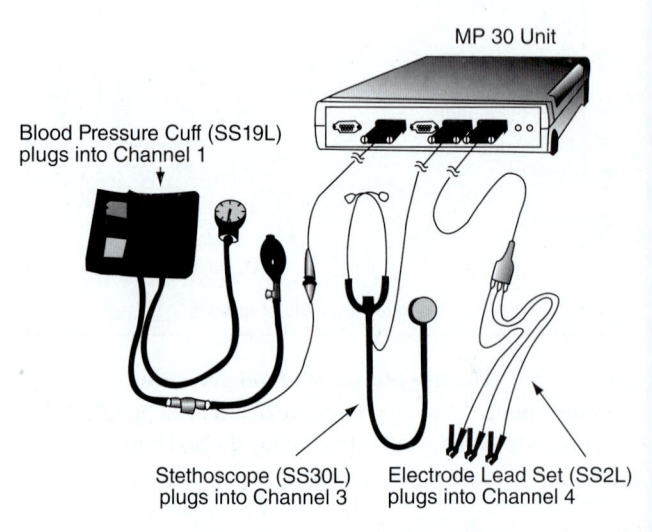

Figure 7.25 **Biopac equipment for visualizing Korotkoff sounds and blood pressure.**

Laboratory Report 7.6

Name _____
Date _____
Section _____

DATA FROM EXERCISE 7.6

Enter your data in this table.

Pressure	Sitting Down	Reclining	Standing Up
Systolic pressure			
Diastolic pressure, phases 4 and 5			
Pulse pressure			
Mean arterial pressure			

REVIEW ACTIVITIES FOR EXERCISE 7.6

Test Your Knowledge

1. When blood pressure measurements are taken, the first sound of Korotkoff occurs when the cuff pressure equals the _____ pressure.

2. The last Korotkoff sound occurs when the cuff pressure equals the _____ pressure.

3. The sounds of Korotkoff are produced by _____.

4. Suppose a person's blood pressure is 168/112.
   (a) What is the systolic pressure? _____
   (b) What is the diastolic pressure? _____

5. What condition does the person described in question 4 have? _____

6. The arterial blood pressure is directly proportional to two factors: the _____ and the _____.

7. The scientific name of the device used to take a blood pressure reading (hint: one word) is the _____.

Test Your Understanding

8. Describe what is meant by "laminar flow" and "turbulent flow." Before you inflate the cuff, which term more closely describes the blood flow in the brachial artery? Explain.

9. How are the Korotkoff sounds produced? When do you hear the first Korotkoff sound? When do you hear the last Korotkoff sound? Explain why this is true of the first and last sounds.

10. How is the pulse pressure calculated, and how does its value relate to the pulse? Also, describe how the mean arterial pressure is calculated, and explain its significance.

Test Your Analytical Ability

11. Were the blood pressure measurements different when they were taken in the standard sitting position, reclining, and standing up? Explain what might have caused any differences observed.

12. How do you think the arterial blood pressure, pulse pressure, and mean arterial pressure would change during exercise? Explain.

Test Your Quantitative Ability

13. If a person has a blood pressure of 142/92, what is the pulse pressure?

   _____

14. If a person has a blood pressure of 142/92, what is the mean arterial pressure?

   _____

Clinical Investigation Questions

15. What is the medical significance of a blood pressure of 140/110?

16. How does the patient's drug affect cardiac output, total peripheral resistance, and blood pressure? Explain.

EXERCISE 7.7
Cardiovascular System and Physical Fitness

MATERIALS
1. Sphygmomanometer
2. Stethoscope, watch with second hand
3. Chair, stair, or platform 18 inches high
4. Optional: pulse and blood pressure may be monitored with the Biopac system using the pulse transducer (see fig. 7.16 in exercise 7.3) and blood pressure cuff (see fig. 7.25 in exercise 7.6).

Physical fitness is dependent on adaptations of the cardiovascular system that include increased stroke volume, decreased resting cardiac rate, and improved cardiovascular responses to exercise. Controlled exercise can be used to assess physical fitness and also to detect heart disease.

LEARNING OUTCOMES

You should be able to:

1. Describe the relationship between age, maximum cardiac rate, and training cardiac rate.
2. Describe the cardiovascular changes that occur when a person becomes physically fit.
3. Explain how controlled exercise protocols may be used to detect heart disease.

Textbook/Multimedia Correlations

Before performing this exercise, you should study the introductory material presented here. Further information relating to this exercise can be found in these pages of *Human Physiology*, thirteenth edition, by Stuart I. Fox:

- *Adaptations of Muscles to Exercise Training.* Chapter 12, p. 383.
- *Circulatory Changes During Exercise.* Chapter 14, p. 470.

Clinical Investigation

A 30-year-old professional athlete with a resting cardiac rate of 50 beats per minute has a routine treadmill stress test, and an ST segment elevation is observed only when his cardiac rate reaches 170 beats per minute.

- Explain the significance of his resting cardiac rate and calculate what percentage 170 represents of his probable maximum cardiac rate.
- Explain how the ST segment change could appear only when his cardiac rate reached 170 beats per minute and the medical significance of this observation.

Both breathing and pulse rate increase within one second of exercise, suggesting that the motor cortex responsible for originating the exercise also influences the cardiovascular adjustments to exercise. However, cardiovascular changes during exercise are also affected by sensory feedback from the contracting muscles and from baroreceptors (blood pressure receptors) in the arteries. These mechanisms increase the activity of the sympathoadrenal system and reduce parasympathetic nerve activity during exercise.

Although the **maximum cardiac rate** (beats per minute) can be estimated for adults in a given age group, those physically fit have a higher *stroke volume* (milliliters per beat) and, therefore, a greater *cardiac output* (milliliters per minute) than those more sedentary. Cardiac output is equal to heart rate multiplied by stroke volume, so at any given cardiac output the athlete with a higher stroke volume usually will have a lower heart rate than the nonathlete. A person in poor physical condition reaches his or her maximum cardiac rate at a lower work level than a person of comparable age in better shape. Maximum cardiac rates for adults can be estimated by subtracting your age from 220, and training rates are usually 60–80% of this maximum, as shown in table 7.2.

Table 7.2 Maximum and Training Cardiac Rates, in Beats per Minute (bpm)

Age	Maximum Cardiac Rate	Training Cardiac Rate (% of maximum)
20	200 bpm	120 bpm (60%) / 160 bpm (80%)
25	195 bpm	117 bpm (60%) / 156 bpm (80%)
30	190 bpm	114 bpm (60%) / 152 bpm (80%)
35	185 bpm	111 bpm (60%) / 148 bpm (80%)
40	180 bpm	108 bpm (60%) / 144 bpm (80%)
45	175 bpm	105 bpm (60%) / 140 bpm (80%)
50	170 bpm	102 bpm (60%) / 136 bpm (80%)
55	165 bpm	99 bpm (60%) / 132 bpm (80%)
60	160 bpm	96 bpm (60%) / 128 bpm (80%)

The **maximal oxygen uptake,** or **aerobic capacity,** measures the maximum rate of oxygen consumption by the body. The intensity of exercise can be rated by the percent of the aerobic capacity attained. In most healthy people, blood levels of lactic acid (lactate) rise significantly when exercise is performed at about 50–70% of their aerobic capacity. This is called the **lactate** (or **anaerobic**) **threshold.** Endurance trained athletes have a higher aerobic capacity and may not reach their lactate threshold until they exercise at about 80% of their maximal oxygen uptake.

The primary cause of the higher aerobic capacity in endurance trained athletes is their higher maximum cardiac outputs, and thus their higher rates of oxygen delivery to the muscles. Endurance training increases the cardiac output through a stronger contraction of the ventricles (thus ejecting more blood per beat) and an increase in blood volume. The proportion of the end-diastolic volume ejected per stroke can increase from 60% at rest to as much as 90% during heavy exercise. This increased **ejection fraction** is produced by the increased contractility that results from sympathoadrenal stimulation. Endurance-trained athletes not only have a higher aerobic capacity, their lactate threshold is a higher percentage of their aerobic capacity (about 80%) than in other people (50–70%). This higher percentage is a result of changes in respiratory enzymes within the trained muscles.

The higher stroke volume of an athlete allows the same cardiac output to be achieved at a slower heart rate, as previously mentioned. The slower resting heart rate of endurance trained athletes—a condition called *athlete's bradycardia*—results from higher levels of inhibitory activity by the vagus nerve innervation to the SA node.

Clinical Applications

Exercise testing using clinical protocols have proved extremely useful in the diagnosis of heart disease, particularly coronary artery disease and **myocardial ischemia** (inadequate blood flow to the heart). People who appear to be healthy and record normal electrocardiograms at rest may develop *angina pectoris* and abnormal ECG patterns during or after exercise. During these tests and under the watchful eye of a physician, the patient follows a standardized exercise protocol while the ECG is continuously recorded and blood pressures are taken automatically at regular intervals. Most cardiologists prefer a treadmill test, during which the speed and incline are automatically varied from easy to difficult, with the patient reaching target heart rates as high as 95% of maximum for the patient's age. During the test, irregularities in the ECG or blood pressure, such as depressed or elevated S-T segments, can be observed, which may indicate myocardial ischemia.

Procedure

1. Measure your reclining (lying down) pulse by placing your fingertips (not the thumb) on the radial artery in the ventrolateral region of the wrist.* Count the number of pulses in 30 seconds and multiply by two. Score points as indicated.

Reclining Pulse

Rate	Points
50–60	3
61–70	3
71–80	2
81–90	1
91–100	0
101–110	−1

Score: _____

2. After the reclining pulse has been measured, stand up and measure the pulse rate *immediately* upon standing.

---

*The procedure for the exercise is from "A Cardiovascular Rating as a Measure of Physical Fatigue and Efficiency" by E. C. Schneider. *JAMA* 74(1920):1507. Copyright 1920 American Medical Association.

Rate	Points
60–70	3
71–80	3
81–90	2
91–100	1
101–110	1
111–120	0
121–130	0
131–140	−1

Score: _____

Measure the pulse as described for 30, 60, 90, and 120 sec after completion of the exercise. Record the time it takes for the pulse to return to normal rate while standing (step 2). Score points as indicated.

Time for Return of Pulse to Normal, Standing Postexercise

Seconds	Points
0–30	4
31–60	3
61–90	2
91–120	1
After 120	0

Score: _____

3. Subtract the pulse rate of step 1 from the pulse rate of step 2 to obtain the pulse rate increase on standing.

Pulse Rate Increase on Standing

Reclining Pulse	0–10 Beats	11–18 Beats	19–26 Beats	27–34 Beats	35–43 Beats
50–60	3	3	2	1	0
61–70	3	2	1	0	−1
71–80	3	2	0	−1	−2
81–90	2	1	−1	−2	−3
91–100	1	0	−2	−3	−3
101–110	0	−1	−3	−3	−3

Score: _____

5. Subtract your normal standing pulse rate (step 2) from your pulse rate immediately after exercise (step 4).

Pulse Rate Increase Immediately Postexercise

Standing Pulse	0–10 Beats	11–20 Beats	21–30 Beats	31–40 Beats	Over 41 Beats
60–70	3	3	2	1	0
71–80	3	2	1	0	−1
81–90	3	2	1	−1	−2
91–100	2	1	0	−2	−3
101–110	1	0	−1	−3	−3
111–120	1	−1	−2	−3	−3
121–130	0	−2	−3	−3	−3
131–140	0	−3	−3	−3	−3

Score: _____

4. Place your right foot on a chair or stair 18 inches high. Raise your body so that your left foot comes to rest by your right foot. Return your left foot to the original position, followed by the right foot. Repeat this exercise five times, allowing 3 sec for each step up. Immediately upon completion of this exercise, measure the pulse for 15 sec and multiply by four. Record this pulse rate.

Pulse: _____

6. Calculate the change in systolic blood pressure as you go from a reclining (lying down position) to a standing position (refer to data in exercise 7.6 or take new measurements).

Change in Systolic Pressure from Reclining to Standing

Change (mm Hg)	Points
Rise of 8 or More	3
Rise of 2–7	2
No rise	1
Fall of 2–5	0
Fall of 6 or more	−1

Score: _____

7. Determine your total score for all the tests and evaluate the score on this basis:

Excellent   18–17
Good   16–14
Fair   13–8
Poor   7 or less

Enter your score and rating in the laboratory report.

Laboratory Report 7.7

Name _____
Date _____
Section _____

DATA FOR EXERCISE 7.7

1. Write your total score in the space provided here, and indicate whether this score is excellent, good, fair, or poor according to the rating scale in the procedure.
   Total Score: _____
   Overall Rating: _____

REVIEW ACTIVITIES FOR EXERCISE 7.7

Test Your Knowledge

1. As a person gets older, the maximum cardiac rate _____.
2. If a person has athlete's bradycardia, the resting heart rate is _____ than the average.
3. The condition described in question 2 is caused by _____.
4. Define *aerobic capacity*. _____
5. Define *lactate threshold*. _____
6. The primary cause of the higher aerobic capacity of endurance-trained athletes is _____
   _____.

Test Your Understanding

7. What cardiovascular adaptations are associated with endurance training? How do these changes help to improve performance?

8. How does the increase in blood pressure and pulse rate after exercise, and the return of these values to baseline following exercise, compare in people who are and who are not physically fit?

9. Athletes not only have a higher maximal oxygen uptake, they can exercise at a higher percentage of this maximum before reaching their lactate threshold. Explain what these terms mean and how they relate to the adaptations within the athlete's cardiovascular and muscular systems.

Test Your Analytical Ability

10. How does a person's aerobic capacity and lactate threshold change with age? What causes these changes? How might these changes be minimized?

11. What is myocardial ischemia? How might exercise provoke this condition? How might a regular exercise program in a healthy person help reduce the chances of developing myocardial ischemia? What other measures should be taken to minimize the risk of developing myocardial ischemia?

Clinical Investigation Questions

12. What is the probable reason for the value of the resting cardiac rate in this person?

13. What percentage of his maximum cardiac rate did he achieve before the ECG abnormality appeared? What is a likely explanation for this?

Section 8

Respiration and Metabolism

All of the chemical reactions that occur in the body are collectively termed **metabolism.** Those reactions that build larger molecules out of smaller ones are dehydration synthesis reactions and require input of energy (are *endergonic*); they are collectively termed **anabolism.** Reactions that break down larger molecules into smaller ones (by hydrolysis) release energy (are *exergonic*) and are collectively termed **catabolism.** The energy required for anabolic reactions is provided by the hydrolysis of *adenosine triphosphate (ATP)*. This catabolic reaction produces *adenosine diphosphate (ADP)* and inorganic *phosphate* (PO_4^{3-}).

$$ATP + H_2O \rightarrow ADP + PO_4^{3-} + energy$$

The energy released by the hydrolysis of ATP is used to power all of the anabolic reactions in the cell. ATP is frequently referred to as the *universal energy carrier* of the cell. Because ATP is continuously hydrolyzed to release energy, energy must also be continuously supplied to resynthesize ATP by the reverse reaction, condensation of ADP with phosphate.

$$ADP + PO_4^{3-} + energy \rightarrow ATP + H_2O$$

The energy required for this anabolic synthesis of ATP is provided by a sequence of catabolic reactions known as **cellular respiration.** In this complex process, monosaccharides, amino acids, and fatty acids serving as fuels are broken down to smaller molecules in a series of hydrolytic steps, with the accompanying release of energy.

Cell respiration can be simplified by considering only one fuel molecule and its final products, skipping the intermediate steps of the metabolic pathway. Plasma glucose is the major energy source of the CNS and an important fuel for many other organs. In the presence of oxygen, glucose is eventually converted through the metabolic pathway of *aerobic respiration* into CO_2 and H_2O. This provides the energy for the formation of a theoretical maximum of 36 to 38 ATP per glucose. However, because not all of the ATP formed inside mitochondria enters the cytoplasm, the cell obtains an average of 30 to 32 ATP through the aerobic respiration of a molecule of glucose.

$$Glucose + O_2 + 30\ ADP + 30\ PO_4^{3-} \rightarrow CO_2 + H_2O + 30\ ATP$$

The continuous hydrolysis of ATP by body cells for energy requires the continuous delivery of glucose and other fuel molecules, plus oxygen for aerobic respiration. In the absence of oxygen, each molecule of glucose is broken down to lactic acid in a process called *anaerobic metabolism,* or *fermentation.* This provides a net gain of two molecules of ATP.

$$Glucose + 2\ ADP + 2\ PO_4^{3-} \rightarrow lactic\ acid + 2\ ATP$$

With the production of thirty molecules of ATP per molecule of glucose, aerobic respiration is obviously more economical than anaerobic metabolism. Moreover, the aerobic pathway is also favored because the accumulation of lactic acid can be toxic to the body.

Oxygen is delivered to the cells of the body by *hemoglobin* molecules located within *red blood cells (erythrocytes).* Overall, the adequate delivery of oxygen to the tissue cells depends on two factors: **ventilation** and **oxygen transport.**

Ventilation (breathing) results in the exchange of "old," oxygen-poor air with "new," oxygen-rich air within the alveoli of the lungs, so that hemoglobin within the red blood cells can load with oxygen (O_2) and transport it in the blood.

Ventilation is also needed for the elimination of the carbon dioxide (CO_2) produced by aerobic respiration within the body cells. The process of ventilation thus helps to maintain homeostasis of CO_2 in the blood and thereby helps to maintain homeostasis of the blood pH.

Oxygen transport includes:

1. the diffusion of oxygen from the alveoli of the lungs into the red blood cells within pulmonary capillaries,
2. the *loading* of oxygen onto hemoglobin molecules located within these pulmonary capillary red blood cells,
3. the *unloading* of oxygen from the hemoglobin in red blood passing through systemic capillaries,
4. the diffusion of oxygen from the red blood cells and tissue capillaries into the body cells.

Exercise 8.1	Measurements of Pulmonary Function
Exercise 8.2	Effect of Exercise on the Respiratory System
Exercise 8.3	Oxyhemoglobin Saturation
Exercise 8.4	Respiration and Acid-Base Balance

EXERCISE 8.1

Measurements of Pulmonary Function

MATERIALS

1. Collins, Inc. 9-L respirometer or Spirocomp program and equipment
2. Disposable mouthpieces and nose clamp
3. Alternatively, the Biopac system may be used with the set up for Biopac Lesson 12 (for part A of this exercise) and Lesson 13 (for part B of this exercise)
4. Alternatively, the iWorx system may be used with the setup as described in iWorx Exercise 10. Once set up, part A of this exercise can be performed
5. Another alternative is the PowerLab system with their spirometry extension

Spirometry is used to measure lung volumes and capacities and to measure ventilation as a function of time. Such measurements are clinically useful in the diagnosis of restrictive and obstructive pulmonary disorders.

LEARNING OUTCOMES

You should be able to:

1. Identify the major muscles involved in inspiration and expiration, and explain the mechanics of breathing.
2. Define the different lung volumes and capacities.
3. Perform a normal spirogram, and determine measurements of the different lung volumes and capacities.
4. Describe and perform the forced expiratory volume (FEV) test; and determine FEV from the spirogram.
5. Explain how pulmonary function tests are used in the diagnosis of restrictive and obstructive pulmonary disorders.

Textbook/Multimedia Correlations

Before performing this exercise, you should study the introductory material presented here. Further information relating to this exercise can be found in these pages of *Human Physiology,* thirteenth edition, by Stuart I. Fox:

- *Physical Aspects of Ventilation.* Chapter 16, p. 536.
- *Mechanics of Breathing.* Chapter 16, p. 541.
- Multimedia correlations are provided in *MediaPhys 3.0:* Topics 10.10–10.13, and 10.22–10.26; also *Physiology Interactive Lab Simulations (Ph.I.L.S):* Respiration (exercises 7, 8, 9, and 10).

Clinical Investigation

A 32 year-old woman smoker complained of difficulty breathing and was given pulmonary function tests. Her vital capacity measured 2,630 mL, and her FEV_1 was 70%. However, her FEV_1 increased to 78% after using albuterol.

- Describe the tests for vital capacity and forced expiratory volume (FEV).
- Explain the clinical significance of these tests and what may account for the patient's test results.
- Explain the action of albuterol and the significance of the patient's response to this drug.

Spirometry is a technique for measuring lung volumes and capacities. A Collins respirometer can be used for this purpose (fig. 8.1). As the subject exhales into a mouthpiece, the oxygen bell rises, causing a pen to move downward on a moving chart (kymograph). This is a closed system, so soda lime is provided in the system to remove CO_2 from the exhaled air. As the subject inhales, the oxygen bell falls, causing the pen to move upward on the moving chart. The *y axis* of the chart is graduated in milliliters, and the *x axis* is graduated in millimeters. Because the speed with which the chart moves is known, a graph of

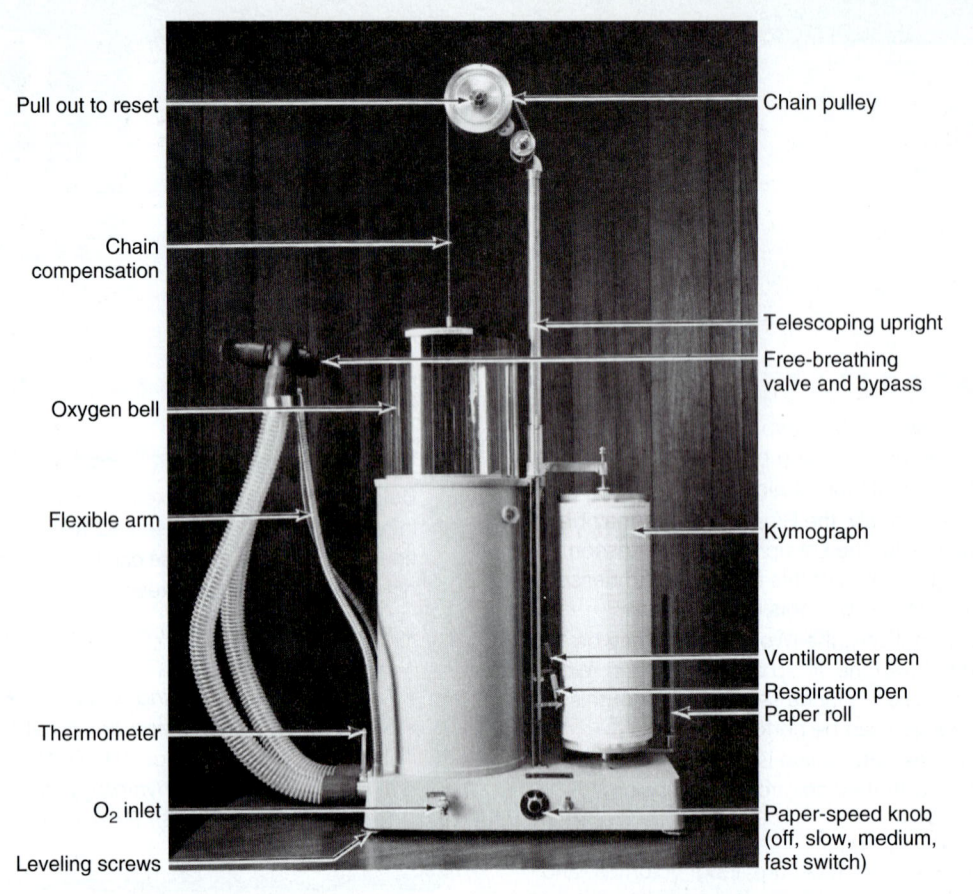

Figure 8.1 The Collins 9-liter respirometer.

air volume (mL) moved into and out of the lungs in a given time interval can be obtained.

Alternatively, a computerized setup using a Phipps and Bird spirometer and a program for analyzing the data (Spirocomp) may be used (fig. 8.2). In this case, the treated data are displayed on a computer screen. Or, if the Biopac system is employed, a wet spirometer is avoided because the flow of air is transduced and fed into the Biopac MP30 unit (fig. 8.3), which then displays the data on a computer screen. By means of spirometry, many important aspects of pulmonary function can be visualized and measured, as described here and shown in figure 8.4.

> The **total lung capacity (TLC)** is the total volume of gas in the lungs after a maximum (forced) inhalation.
> The **vital capacity (VC)** is the maximum volume of gas that can be exhaled after a maximum inhalation.
> The **tidal volume (TV)** is the volume of gas inspired or expired during each normal (unforced) ventilation cycle.
> The **inspiratory capacity (IC)** is the maximum volume of gas that can be inhaled after a normal (unforced) exhalation.
> The **inspiratory reserve volume (IRV)** is the maximum volume of gas that can be forcefully inhaled after a normal (tidal) inhalation.
> The **expiratory reserve volume (ERV)** is the maximum volume of gas that can be forcefully exhaled after a normal (tidal) exhalation.

Figure 8.2 Equipment involved in using the Spirocomp program.

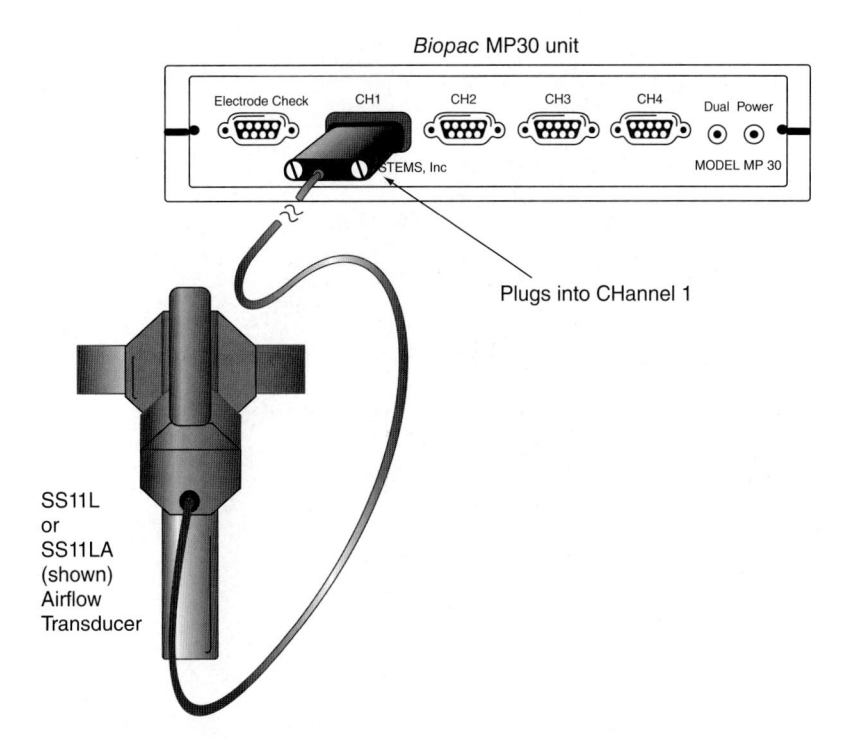

Figure 8.3 Biopac system equipment for spirometry.

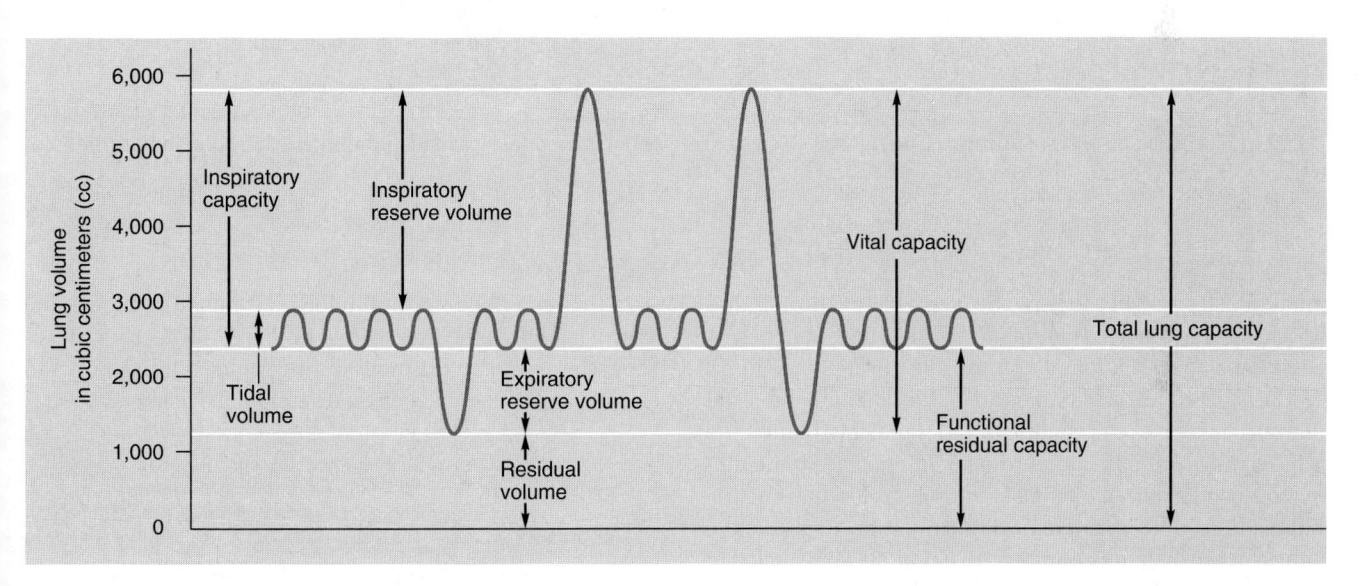

Figure 8.4 Spirogram recording of lung volumes and capacities.

The **functional residual capacity (FRC)** is the volume of gas remaining in the lungs after a normal (unforced) exhalation.

The **residual volume (RV)** is the volume of gas remaining in the lungs after a maximum (forced) exhalation.

The movement of air into and out of the lungs (ventilation) results from a pressure difference between the pulmonary air and the atmosphere. This pressure difference is created by a change in the volume of the thoracic cavity. According to **Boyle's law,** the *pressure of a gas is inversely proportional to its volume,* so an increase in thoracic volume results in a decrease in intrapulmonary pressure. During inhalation, therefore, air is pushed into the lungs by the greater pressure of the atmosphere. When the thoracic volume decreases during exhalation, the intrapulmonary pressure rises above the atmospheric pressure and air is pushed out of the lungs.

In normal (unforced) ventilation, the thoracic volume is regulated by action of the *diaphragm* and the *external intercostal muscles* (fig. 8.5). At rest, the diaphragm forms the convex floor to the thoracic cavity. During inhalation, the diaphragm contracts and pulls itself into a more

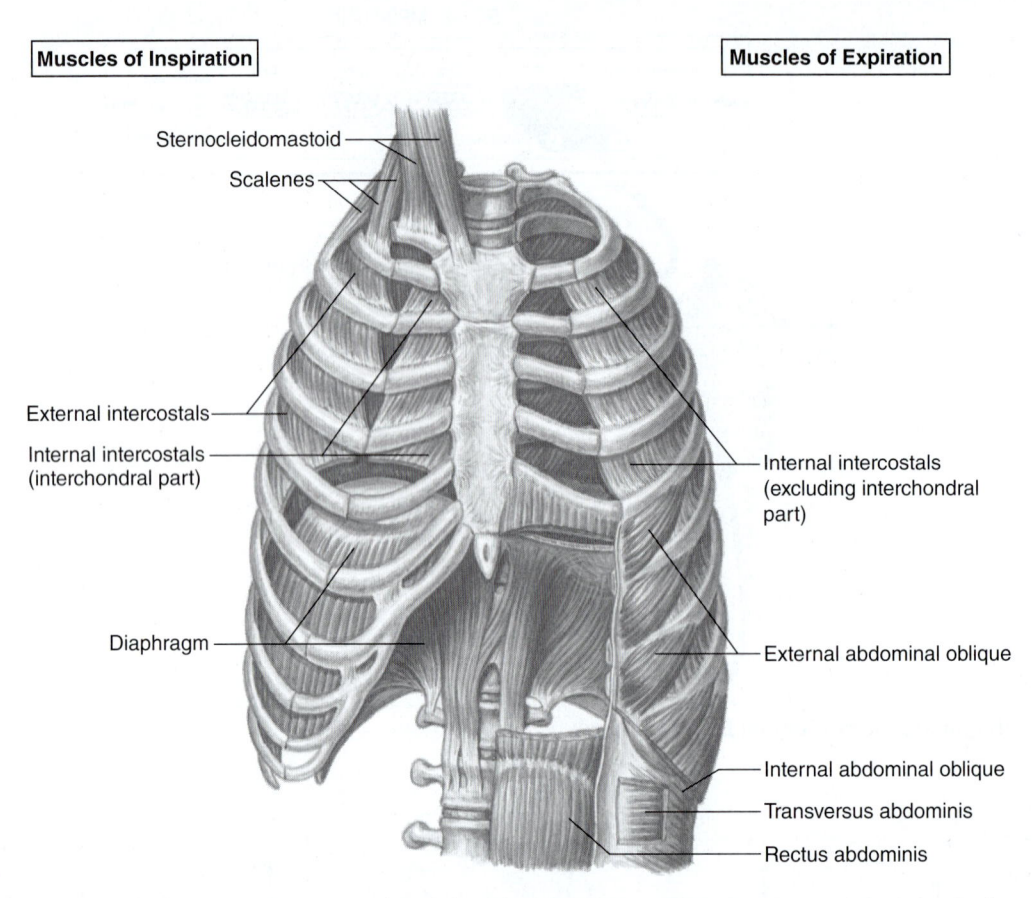

Figure 8.5 Muscles of respiration. The right side (of the trunk) shows the muscles of inspiration; the left side depicts the muscles involved in forced expiration.

(For a full-color version of this figure, see fig. 16.13 in *Human Physiology*, thirteenth edition, by Stuart I. Fox.)

flattened form. This lowers the floor of the thorax and increases the thoracic volume (and pushes down on the viscera, causing the abdomen to protrude during inhalation). At the same time, the contraction of the external intercostal muscles increases the volume of the thorax by rotating the ribs upward and outward. At the end of inhalation the diaphragm and external intercostal muscles relax, causing the thorax to resume its original volume and the air inside the lungs to be exhaled. The amount of air inhaled or exhaled in this manner is the tidal volume.

Inhalation can become difficult if the air passages are obstructed. In these cases, the affected person relies increasingly on muscles not usually used in normal tidal ventilation: the *scalenus, sternocleidomastoid,* and *pectoralis major muscles.* These muscles are also used in healthy people during forced inhalation to obtain the inspiratory reserve volume (IRV).

During forced exhalation, the *internal intercostal muscles* contract, depressing the rib cage, and the *abdominal muscles* contract, pushing the viscera up against the diaphragm. The push of the viscera increases the convexity of the diaphragm and decreases the thoracic volume to a greater extent than in normal exhalation. The amount of air forcefully exhaled by contraction of both groups of muscles is the expiratory reserve volume (ERV).

Even after a maximum forced exhalation there is still some air left in the lungs. This residual volume of air makes it easier to inflate the lungs during the next inhalation and oxygenates the blood between ventilation cycles.

Normal Values Measurements of vital capacity consistently below 80% of the predicted value on repeated tests suggest the presence of a restrictive lung disease, such as emphysema.

A. Measurement of Simple Lung Volumes and Capacities

Procedure

(for Biopac System)

1. With the Biopac MP30 unit off, set up the equipment as shown in fig. 8.3.
2. Turn on the MP30 unit, and place a clean filter on the end of the calibration syringe (fig. 8.6).
3. On the computer, select L12-Lung-1 from the menu.

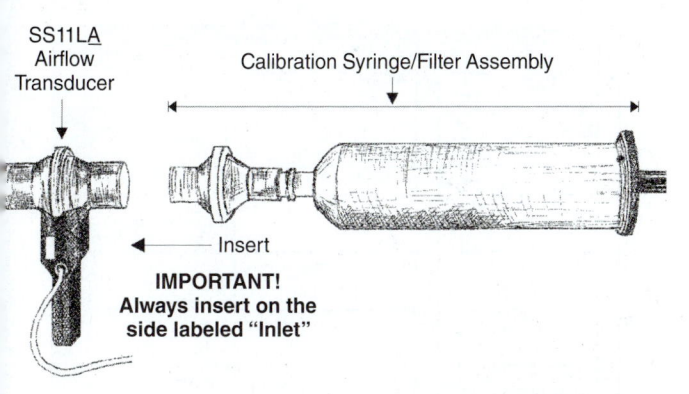

Figure 8.6 Equipment needed to calibrate the Biopac spirometry system. The air in the syringe will simulate a specific volume of inhaled and exhaled air when the plunger is moved.

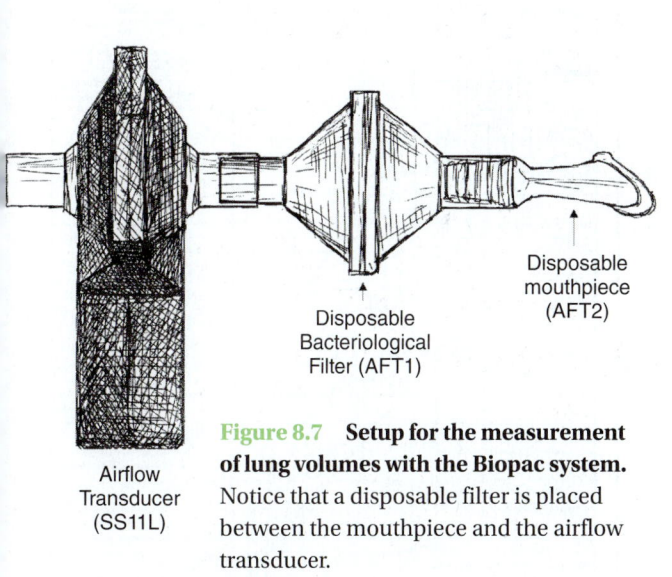

Figure 8.7 Setup for the measurement of lung volumes with the Biopac system. Notice that a disposable filter is placed between the mouthpiece and the airflow transducer.

4. Be sure to hold the calibration syringe parallel to the ground, pull the plunger all the way out, and click "calibrate" on the computer screen.
5. You will simulate 5 breathing cycles with the syringe. To do this, smoothly push the plunger in for one second, wait two seconds, and pull the plunger out for one second. Repeat this four more times. Click "end calibration" on the computer screen.
6. Place a clean mouthpiece into the air filter, already attached to the airflow transducer (fig. 8.7).
7. With the nose pinched closed and the lips tightly sealed around the mouthpiece, perform the following procedure:
   (a) Click "Record" on the computer.
   (b) Take three normal (unforced) breaths for the tidal volume measurements.
   (c) Inhale to your maximum (see fig. 8.4), and then exhale back to your normal level (the tidal volume exhalation).
   (d) Take three normal (unforced) breaths again.
   (e) Exhale as much as you can (to the end of your expiratory reserve volume; see fig. 8.4), and then inhale back to your normal breathing.
   (f) Click "Stop" on the computer. If the data seems all right, click "Done."
8. Select "Analyze current data file" on the computer. At the top of the screen, select "CH 2" for the channel, and select "p-p" for the measurement.
9. At the bottom right of the screen, select the I-beam cursor. Use this to highlight the region of the screen showing the first three breaths.
10. The computer will automatically calculate the tidal volume of the first three breaths. Enter this value in your laboratory report.
11. Use this I-beam cursor to highlight the parts of your recording that will provide the inspiratory reserve volume and the expiratory reserve volume (see fig. 8.4), and enter these values in the laboratory report.
12. Highlight the vital capacity (see fig. 8.4) on your computer screen and enter the p-p value provided by the computer into the laboratory report. The Biopac system does not provide a way to measure the FEV_1.

Procedure

(for Spirocomp)

1. Press the "T" key on the computer and the words "Breathe Normal Cycles" will appear on the computer screen. After three normal tidal volume cycles, the data will appear on the screen.
2. Press the "E" key on the computer and the words "Breathe Normal Cycles" will appear on the screen. At the third breathing cycle, the words "Stop After Normal Exhale" will appear.
3. After the pause in breathing, the words "Exhale Forcefully" will appear on the screen. Forcefully exhale all you can at this point.
4. Press the "V" key and the words "Inhale Max Then Press V Exhale Fully" will appear on the screen. After a maximal inhalation, press "V" and forcefully exhale all the air you can as fast as possible.
5. Record the data displayed on the screen, and use it to complete your laboratory report.

Procedure

(for Collins Respirometer)

1. Raise and lower the oxygen bell (see fig. 8.1) several times to get fresh air into the spirometer. As the bell moves up and down, one of the pens moves a corresponding distance down and up on a shaft. By adjusting the height of the oxygen bell, position this pen so that it will begin writing in the middle of the

chart paper. This pen (the *ventilometer* pen) usually has black ink; the other *(respiration)* pen usually has red ink and will not be used for this exercise. (It should be covered and locked away from the paper.)

2. With the free-breathing valve set to the *open position,* place the mouthpiece in the buccal cavity (as in breathing through a snorkel), and go through several ventilation cycles to become accustomed to the apparatus. (When the free-breathing valve is open, you will breathe room air.) If a disposable cardboard mouthpiece is used, be particularly careful to prevent air leakage from the corners of the mouth. Breathing through the nostrils can be prevented by means of a nose clamp or by pinching the nose tightly with the thumb and forefinger.

3. Turn the respirometer to the *slow* position (32 mm/min) and close the free-breathing valve so that the oxygen bell and the pen go up and down with each ventilation cycle.

4. Breathe in a normal, relaxed manner for 1–2 min. The breaths should appear relatively uniform, and the slope should go upward (fig. 8.8).

   A downward slope indicates that there is an air leak; in this event, tighten the grip of the mouth on the mouthpiece and the nose clamp on the nose, and begin again.

Note: *At this speed (32 mm/min), the distance between heavy vertical lines on the chart is traversed in 1 min.*

   This procedure measures tidal volume—the amount of air inhaled or exhaled during each resting ventilation cycle.

5. With the drum still turning, perform the test for vital capacity (the maximum amount of air that can be exhaled after a maximum inhalation). At the end of a normal exhalation, inhale as much as possible, and then exhale completely to the fullest possible extent, and stop the recording. At the completion of this exercise, the chart should resemble the one shown in figure 8.8.

6. Remove the chart from the kymograph drum. The chart is marked horizontally in milliliters.

Note: *The temperature and pressure of the respirometer are different from those existing in the body, so the volume the air occupies in the respirometer will be subject to changes in ambient (room) conditions. To standardize the volumes measured in spirometry, multiply these measured volumes by a correction factor known as the **BTPS factor** (body temperature, atmospheric pressure, saturated with water vapor). The BTPS factor is very close to **1.1** at normal room temperatures, so we will use this figure in the calculations.*

Calculations

1. Obtain the measured tidal volume from the chart by subtracting the milliliters corresponding to the trough from the milliliters corresponding to the peak of a typical resting ventilation cycle.

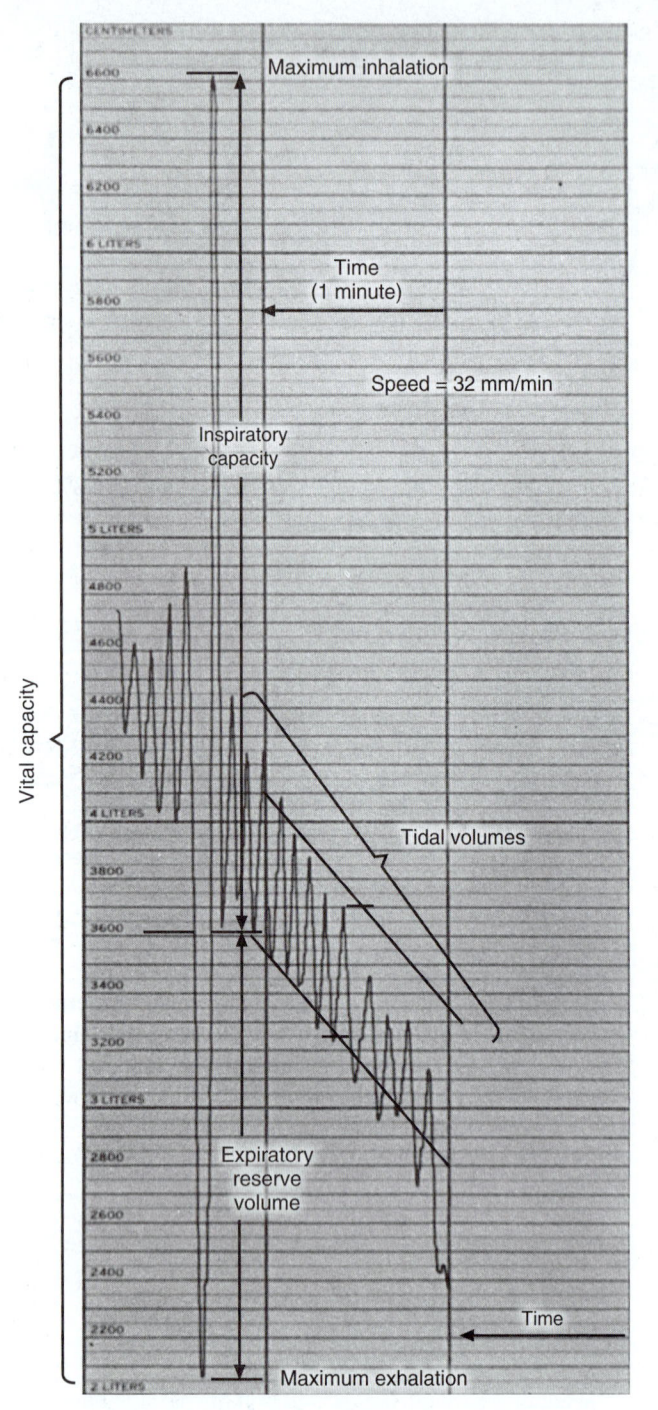

Figure 8.8 A spirometry recording. This chart shows tidal volume, inspiratory capacity, expiratory reserve volume, and vital capacity.

Example (from fig. 8.9)

Step 1   3,700 mL (inhalation)
       − 3,250 mL (exhalation)
          450 mL

Step 2   450 mL (measured tidal volume)
       ×   1.1   (BTPS factor)
          495 mL

300

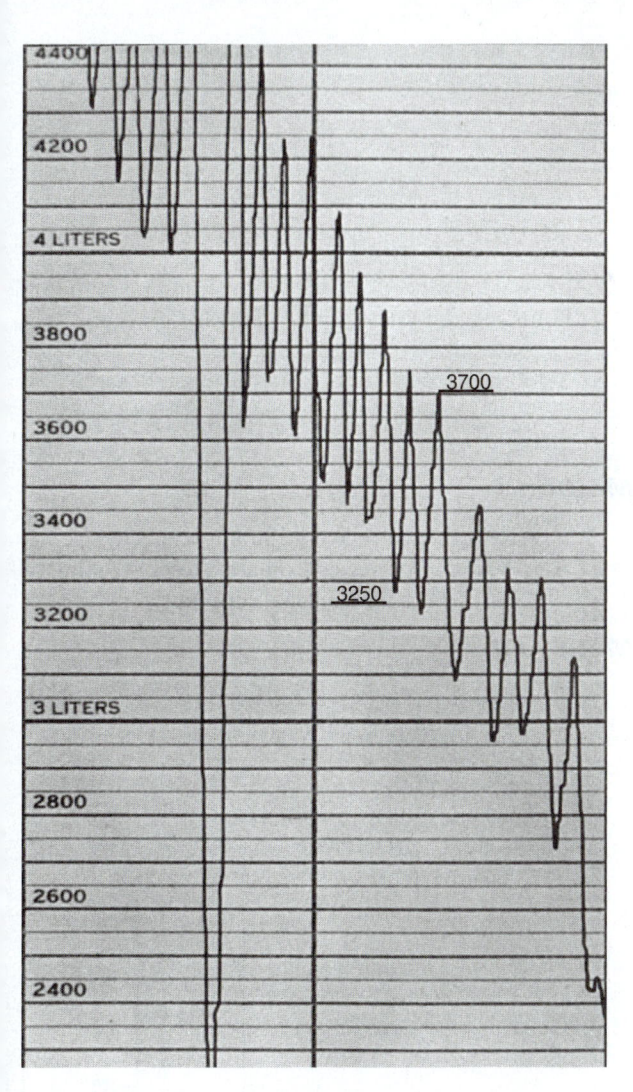

Figure 8.9 A close-up of tidal volume measurements from figure 8.8.

Enter the corrected measured tidal volume (TV) in the *Measured* column of the table in your laboratory report.

2. Obtain the measured inspiratory capacity from the chart. To do this, subtract the milliliters corresponding to the last normal exhalation before performing the vital capacity maneuver from the milliliters corresponding to the maximum inhalation peak.

Example (from fig. 8.10)

Step 1   6,650 mL   (maximum inhalation)
         − 3,650 mL   (normal exhalation)
            3,000 mL

Step 2   3,000 mL   (measured inspiratory capacity)
         ×   1.1   (BTPS factor)
            3,300 mL

Enter the corrected measured inspiratory capacity (IC) in the *Measured* column of the table in your laboratory report.

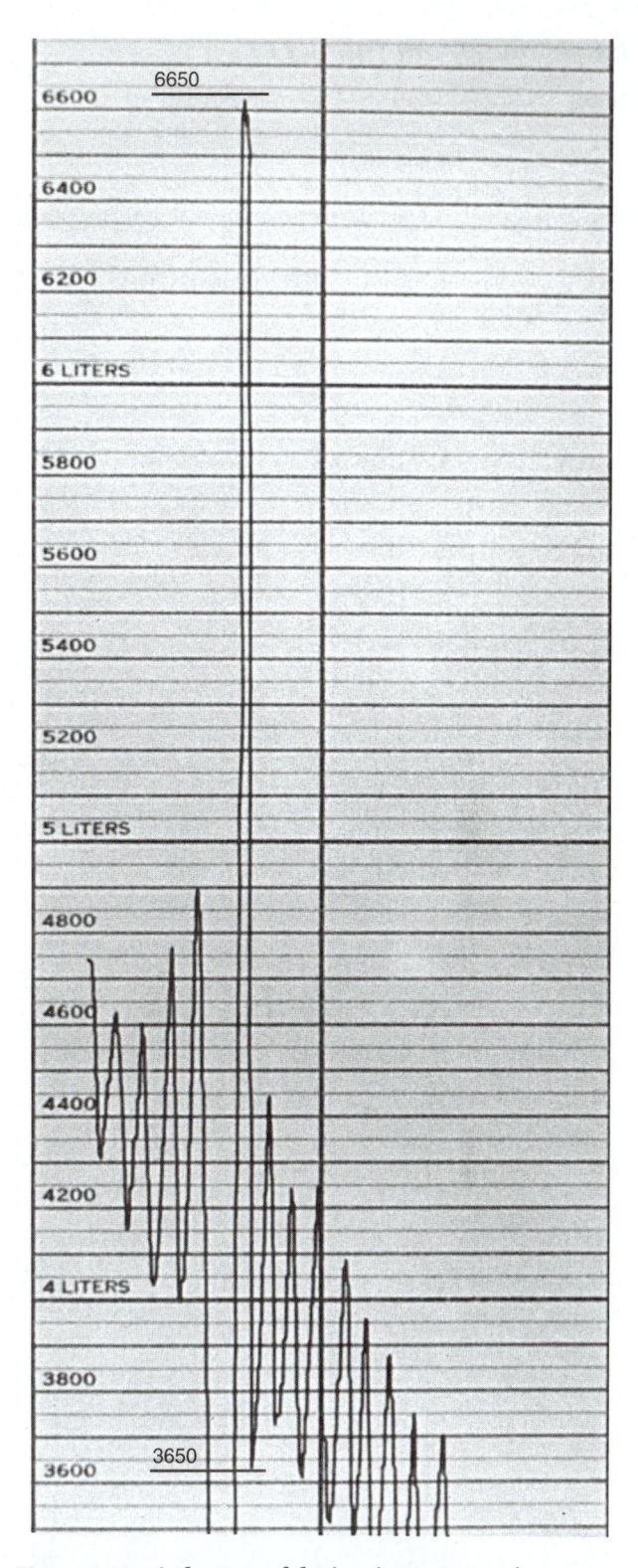

Figure 8.10 A close-up of the inspiratory capacity measurement from figure 8.8.

3. Obtain the measured expiratory reserve volume by subtracting the milliliters corresponding to the trough for maximum exhalation from the milliliters corresponding to the last normal exhalation before the vital capacity maneuver (same value used for step 2).

301

Example (from fig. 8.11)

Step 1
$$\begin{array}{rl} 3{,}650 \text{ mL} & \text{(normal exhalation)} \\ -\ 2{,}050 \text{ mL} & \text{(maximum exhalation)} \\ \hline 1{,}600 \text{ mL} & \end{array}$$

Step 2
$$\begin{array}{rl} 1{,}600 \text{ mL} & \text{(measured expiratory reserve)} \\ \times\ \ 1.1 & \text{1,600 mL (volume)} \\ \hline 1{,}760 \text{ mL} & \text{(BTPS factor)} \end{array}$$

Enter the corrected expiratory reserve volume (ERV) in the *Measured* column of the table in your laboratory report.

4. Obtain the measured vital capacity by either (1) adding the corrected inspiratory capacity (from step 2) and the corrected expiratory reserve volume (step 3) (these values have already been BTPS standardized, so an additional correction step is unnecessary); or (2) subtracting the milliliters corresponding to maximum exhalation from the milliliters corresponding to maximum inhalation. This value must then be multiplied by the BTPS factor.

Method 1

$$\begin{array}{rl} 3{,}300 \text{ mL} & \text{(corrected inspiratory capacity)} \\ +\ 1{,}760 \text{ mL} & \text{(corrected expiratory reserve volume)} \\ \hline 5{,}060 \text{ mL} & \text{(corrected vital capacity)} \end{array}$$

Method 2

$$\begin{array}{rl} 6{,}650 \text{ mL} & \text{(maximum inhalation)} \\ -\ 2{,}050 \text{ mL} & \text{(maximum exhalation)} \\ \hline 4{,}600 \text{ mL} & \end{array}$$

$$\begin{array}{rl} 4{,}600 \text{ mL} & \text{(measured vital capacity)} \\ \times\ \ 1.1 & \text{(BTPS factor)} \\ \hline 5{,}060 \text{ mL} & \text{(corrected capacity)} \end{array}$$

Enter the corrected vital capacity (VC) in the *Measured* column of the table in your laboratory report.

5. Obtain the *predicted vital capacity* for the subject's sex, age, and height from tables 8.1 and 8.2. The height in centimeters can be most conveniently obtained by referring to the height conversion scale in figure 8.12.

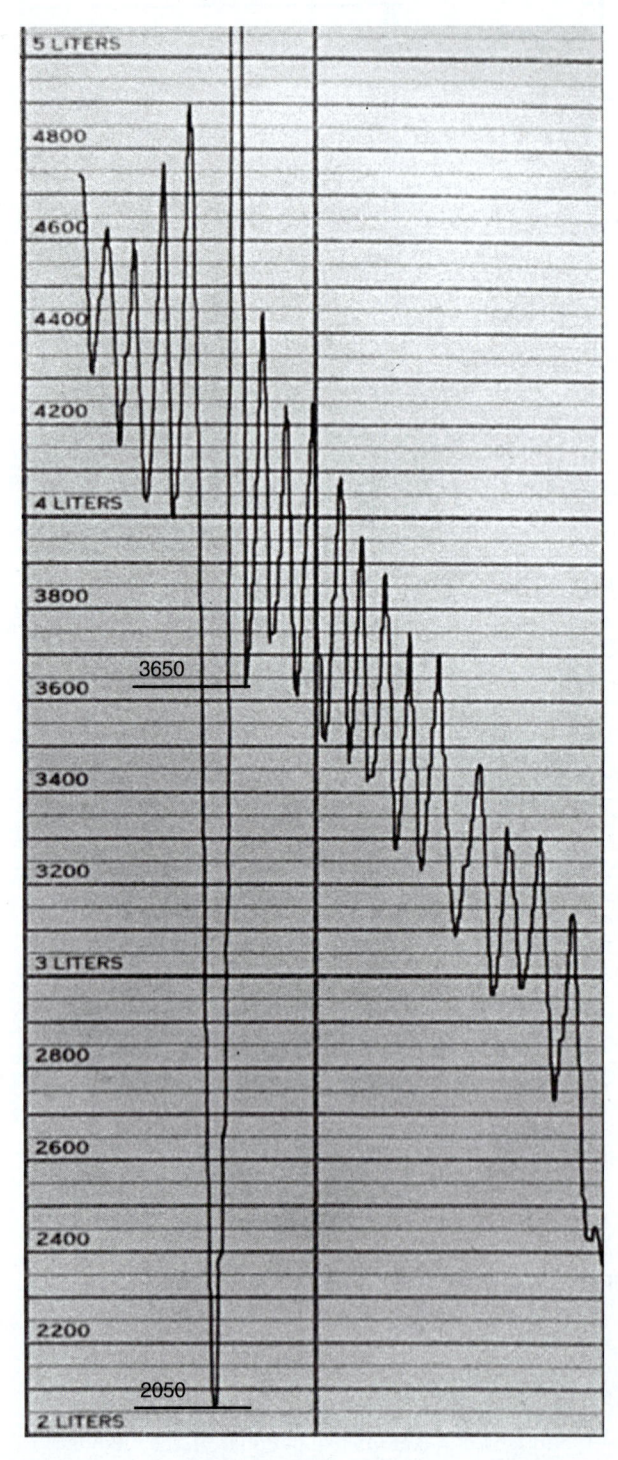

Figure 8.11 A close-up of the expiratory reserve volume measurement from figure 8.8.

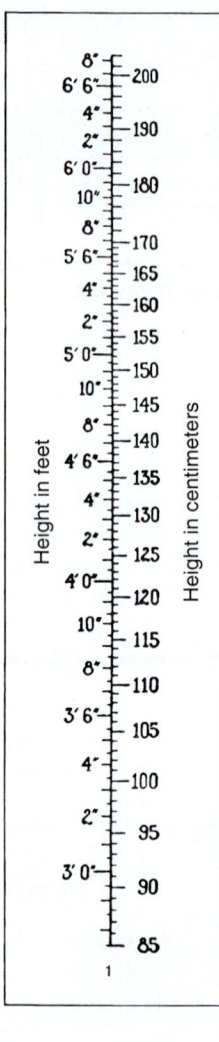

Figure 8.12 A scale for converting height in feet and inches to height in centimeters.

Table 8.1 Predicted Vital Capacities, Females (mL)

Height in Centimeters

Age	146	148	150	152	154	156	158	160	162	164	166	168	170	172	174	176	178	180	182	184	186	188	190	192	194
16	2950	2990	3030	3070	3110	3150	3190	3230	3270	3310	3350	3390	3430	3470	3510	3550	3690	3630	3670	3715	3755	3800	3840	3880	3920
17	2935	2975	3015	3055	3095	3135	3175	3215	3255	3295	3335	3375	3415	3455	3495	3535	3575	3615	3655	3695	3740	3780	3820	3860	3900
18	2920	2960	3000	3040	3080	3120	3160	3200	3240	3280	3320	3360	3400	3440	3480	3520	3560	3600	3640	3680	3720	3760	3800	3840	3880
20	2890	2930	2970	3010	3050	3090	3130	3170	3210	3250	3290	3330	3370	3410	3450	3490	3525	3565	3605	3645	3695	3720	3760	3800	3840
22	2860	2900	2940	2980	3020	3060	3095	3135	3175	3215	3255	3290	3330	3370	3410	3450	3490	3530	3570	3610	3650	3685	3725	3765	3800
24	2830	2870	2910	2950	2985	3025	3065	3100	3140	3180	3220	3260	3300	3335	3375	3415	3455	3490	3530	3570	3610	3650	3685	3725	3765
26	2800	2840	2880	2920	2960	3000	3035	3070	3110	3150	3190	3230	3265	3300	3340	3380	3420	3455	3495	3530	3570	3610	3650	3685	3725
28	2775	2810	2850	2890	2930	2965	3000	3040	3070	3115	3155	3190	3230	3270	3305	3345	3380	3420	3460	3495	3535	3570	3610	3650	3685
30	2745	2780	2820	2860	2895	2935	2970	3010	3045	3085	3120	3160	3195	3235	3270	3310	3345	3385	3420	3460	3495	3535	3570	3610	3645
32	2715	2750	2790	2825	2865	2900	2940	2975	3015	3050	3090	3125	3160	3200	3235	3275	3310	3350	3385	3425	3460	3495	3535	3570	3610
34	2685	2725	2760	2795	2835	2870	2910	2945	2980	3020	3055	3090	3130	3165	3200	3240	3275	3310	3350	3385	3425	3460	3495	3535	3570
36	2655	2695	2730	2765	2805	2840	2875	2910	2950	2985	3020	3060	3095	3130	3165	3205	3240	3275	3310	3350	3385	3420	3460	3495	3530
38	2630	2665	2700	2735	2770	2810	2845	2880	2915	2950	2990	3025	3060	3095	3130	3170	3205	3240	3275	3310	3350	3385	3420	3455	3490
40	2600	2635	2670	2705	2740	2775	2810	2850	2885	2920	2955	2990	3025	3060	3095	3135	3170	3205	3240	3275	3310	3345	3380	3420	3455
42	2570	2605	2640	2675	2710	2745	2780	2815	2850	2885	2920	2955	2990	3025	3060	3100	3135	3170	3205	3240	3275	3310	3345	3380	3415
44	2540	2575	2610	2645	2680	2715	2750	2785	2820	2855	2890	2925	2960	2995	3030	3060	3095	3130	3165	3200	3235	3270	3305	3340	3375
46	2510	2545	2580	2615	2650	2685	2715	2750	2785	2820	2855	2890	2925	2960	2995	3030	3060	3095	3130	3165	3200	3235	3270	3305	3340
48	2480	2515	2550	2585	2620	2650	2685	2715	2750	2785	2820	2855	2890	2925	2960	2995	3030	3060	3095	3130	3160	3195	3230	3265	3300
50	2455	2485	2520	2555	2590	2625	2655	2690	2720	2755	2785	2820	2855	2890	2925	2955	2990	3025	3060	3090	3125	3155	3190	3225	3260
52	2425	2455	2490	2525	2555	2590	2625	2655	2690	2720	2755	2790	2820	2855	2890	2925	2955	2990	3020	3055	3090	3125	3155	3190	3220
54	2395	2425	2460	2495	2530	2560	2590	2625	2655	2690	2720	2755	2790	2820	2855	2885	2920	2950	2985	3020	3050	3085	3115	3150	3180
56	2365	2400	2430	2460	2495	2525	2560	2590	2625	2655	2690	2720	2755	2790	2820	2855	2885	2920	2950	2980	3015	3045	3080	3110	3145
58	2335	2370	2400	2430	2460	2495	2525	2560	2590	2625	2655	2690	2720	2750	2785	2815	2850	2880	2920	2945	2975	3010	3040	3075	3105
60	2305	2340	2370	2400	2430	2460	2495	2525	2560	2590	2625	2655	2685	2720	2750	2780	2810	2845	2875	2915	2940	2970	3000	3035	3065
62	2280	2310	2340	2370	2405	2435	2465	2495	2525	2560	2590	2620	2655	2685	2715	2745	2775	2810	2840	2870	2900	2935	2965	2995	3025
64	2250	2280	2310	2340	2370	2400	2430	2465	2495	2525	2555	2585	2620	2650	2680	2710	2740	2770	2805	2835	2865	2895	2925	2955	2990
66	2220	2250	2280	2310	2340	2370	2400	2430	2460	2495	2525	2555	2585	2615	2645	2675	2705	2735	2765	2800	2825	2860	2890	2920	2950
68	2190	2220	2250	2280	2310	2340	2370	2400	2430	2460	2490	2520	2550	2580	2610	2640	2670	2700	2730	2760	2795	2820	2850	2880	2910
70	2160	2190	2220	2250	2280	2310	2340	2370	2400	2425	2455	2485	2515	2545	2575	2605	2635	2665	2695	2725	2755	2780	2810	2840	2870
72	2130	2160	2190	2220	2250	2280	2310	2335	2365	2395	2425	2455	2480	2510	2540	2570	2600	2630	2660	2685	2715	2745	2775	2805	2830
74	2100	2130	2160	2190	2220	2245	2275	2305	2335	2360	2390	2420	2450	2475	2505	2535	2565	2590	2620	2650	2680	2710	2740	2765	2795

Courtesy of Warren E. Collins, Inc., Braintree, MA.

Table 8.2 Predicted Vital Capacities, Males (mL)

Height in Centimeters

Age	146	148	150	152	154	156	158	160	162	164	166	168	170	172	174	176	178	180	182	184	186	188	190	192	194
16	3765	3820	3870	3920	3975	4025	4075	4130	4180	4230	4285	4335	4385	4440	4490	4540	4590	4645	4695	4745	4800	4850	4900	4955	5005
18	3740	3790	3840	3890	3940	3995	4045	4095	4145	4200	4250	4300	4350	4405	4455	4505	4555	4610	4660	4710	4760	4815	4865	4915	4965
20	3710	3760	3810	3860	3910	3960	4015	4065	4115	4165	4215	4265	4320	4370	4420	4470	4520	4570	4625	4675	4725	4775	4825	4875	4930
22	3680	3730	3780	3830	3880	3930	3980	4030	4080	4135	4185	4235	4285	4335	4385	4435	4485	4535	4585	4635	4685	4735	4790	4840	4890
24	3635	3685	3735	3785	3835	3885	3935	3985	4035	4085	4135	4185	4235	4285	4330	4380	4430	4480	4530	4580	4630	4680	4730	4780	4830
26	3605	3655	3705	3755	3805	3855	3905	3955	4000	4050	4100	4150	4200	4250	4300	4350	4395	4445	4495	4545	4595	4645	4695	4740	4790
28	3575	3625	3675	3725	3775	3820	3870	3920	3970	4020	4070	4115	4165	4215	4265	4310	4360	4410	4460	4510	4555	4605	4655	4705	4755
30	3550	3595	3645	3695	3740	3790	3840	3890	3935	3985	4035	4080	4130	4180	4230	4275	4325	4375	4425	4470	4520	4570	4615	4665	4715
32	3520	3565	3615	3665	3710	3760	3810	3855	3905	3950	4000	4050	4095	4145	4195	4240	4290	4340	4385	4435	4485	4530	4580	4625	4675
34	3475	3525	3570	3620	3665	3715	3760	3810	3855	3905	3950	4000	4045	4095	4140	4190	4225	4285	4330	4380	4425	4475	4520	4570	4615
36	3445	3495	3540	3585	3635	3680	3730	3775	3825	3870	3920	3965	4010	4060	4105	4155	4200	4250	4295	4340	4390	4435	4485	4530	4580
38	3415	3465	3510	3555	3605	3650	3695	3745	3790	3840	3885	3930	3980	4025	4070	4120	4165	4210	4260	4305	4350	4400	4445	4495	4540
40	3385	3435	3480	3525	3575	3620	3665	3710	3760	3805	3850	3900	3945	3990	4035	4085	4130	4175	4220	4270	4315	4360	4410	4455	4500
42	3360	3405	3450	3495	3540	3590	3635	3680	3725	3770	3820	3865	3910	3955	4000	4050	4095	4140	4185	4230	4280	4325	4370	4415	4460
44	3315	3360	3405	3450	3495	3540	3585	3630	3675	3725	3770	3815	3860	3905	3950	3995	4040	4085	4130	4175	4220	4270	4315	4360	4405
46	3285	3330	3375	3420	3465	3510	3555	3600	3645	3690	3735	3780	3825	3870	3915	3960	4005	4050	4095	4140	4185	4230	4275	4320	4365
48	3255	3300	3345	3390	3435	3480	3525	3570	3615	3655	3700	3745	3790	3835	3880	3925	3970	4015	4060	4105	4150	4190	4235	4280	4325
50	3210	3255	3300	3345	3390	3430	3475	3520	3565	3610	3650	3695	3740	3785	3830	3870	3915	3960	4005	4050	4090	4135	4180	4225	4270
52	3185	3225	3270	3315	3355	3400	3445	3490	3530	3575	3620	3660	3705	3750	3795	3835	3880	3925	3970	4010	4055	4100	4140	4185	4230
54	3155	3195	3240	3285	3325	3370	3415	3455	3500	3540	3585	3630	3670	3715	3760	3800	3845	3890	3930	3975	4020	4060	4105	4145	4190
56	3125	3165	3210	3255	3295	3340	3380	3425	3465	3510	3550	3595	3640	3680	3725	3765	3810	3850	3895	3940	3980	4025	4065	4110	4150
58	3080	3125	3165	3210	3250	3290	3335	3375	3420	3460	3500	3545	3585	3630	3670	3715	3755	3800	3840	3880	3925	3965	4010	4050	4095
60	3050	3095	3135	3175	3220	3260	3300	3345	3385	3430	3470	3500	3555	3595	3635	3680	3720	3760	3805	3845	3885	3930	3970	4015	4055
62	3020	3060	3110	3150	3190	3230	3270	3310	3350	3390	3440	3480	3520	3560	3600	3640	3680	3730	3770	3810	3850	3890	3930	3970	4020
64	2990	3030	3080	3120	3160	3200	3240	3280	3320	3360	3400	3440	3490	3530	3570	3610	3650	3690	3730	3770	3810	3850	3900	3940	3980
66	2950	2990	3030	3070	3110	3150	3190	3230	3270	3310	3350	3390	3430	3470	3510	3550	3600	3640	3680	3720	3760	3800	3840	3880	3920
68	2920	2960	3000	3040	3080	3120	3160	3200	3240	3280	3320	3360	3400	3440	3480	3520	3560	3600	3640	3680	3720	3760	3800	3840	3880
70	2890	2930	2970	3010	3050	3090	3130	3170	3210	3250	3290	3330	3370	3410	3450	3480	3520	3560	3600	3640	3680	3720	3760	3800	3840
72	2860	2900	2940	2980	3020	3060	3100	3140	3180	3210	3250	3290	3330	3370	3410	3450	3490	3530	3570	3610	3650	3680	3720	3760	3800
74	2820	2860	2900	2930	2970	3010	3050	3090	3130	3170	3200	3240	3280	3320	3360	3400	3440	3470	3510	3550	3590	3630	3670	3710	3740

Courtesy of Warren E. Collins, Inc., Braintree, MA.

Table 8.3 Factors for Obtaining the Predicted Residual Volume and Total Lung Capacity

Age	Residual Volume: Vital Capacity × Factor	Total Lung Capacity: Vital Capacity × Factor
16–34	0.250	1.250
35–49	0.305	1.305
50–69	0.445	1.445

Example
Sex: male
Age: 34
Height: 174 cm
Predicted vital capacity: 4,140 mL

Enter the subject's predicted vital capacity from the tables of normal values in the *Predicted* column in your laboratory report.

6. To obtain an estimate of the predicted residual volume and the predicted total lung capacity, refer to table 8.3.

Note: *These values cannot be measured by spirometry because residual volume cannot be exhaled; total lung capacity equals the vital capacity plus residual volume.*

7. Obtain the percent predicted value for all the measurements in this way:

$$\text{Percent predicted} = \frac{\text{corrected measured value}}{\text{predicted value}} \times 100\%$$

Example
Measured vital capacity 5,060 mL (corrected to BTPS)
Predicted vital capacity 4,140 mL (from table 8.1 or 8.2)

$$\% \text{ predicted} = \frac{5,060 \text{ mL}}{4,140 \text{ mL}} \times 100\%$$

$$= 122\%$$

Enter the percent predicted values in the appropriate places in the table in your laboratory report.

B. MEASUREMENT OF FORCED EXPIRATORY VOLUME

The ability to ventilate the lungs in a given amount of time is often of greater diagnostic value than measurements of simple lung volumes and capacities. One measurement that considers time intervals is the **forced expiratory volume (FEV),** otherwise known as the *timed vital capacity.*

In the forced expiratory volume test, the subject performs a vital capacity maneuver by inhaling maximally, holding, and then exhaling maximally as rapidly as possible. While holding at the peak of inhalation, the respirometer kymograph drum speed is set to its fastest setting (1,920 mm/min); then the subject is instructed to exhale forcefully and maximally. This fast speed of the kymograph stretches out the exhalation tracing and the distance between heavy vertical lines is now traversed in 1 sec. From the recording, the percentage of the total vital capacity exhaled in the *first* second (FEV_1), the *second* second (FEV_2), and the *third* second (FEV_3) can be determined. A sample record of the forced expiratory volume is shown in figure 8.13.

PULMONARY DISORDERS

Chronic (long-term) pulmonary dysfunctions can be divided into two general categories: obstructive disorders and restrictive disorders. These two categories can be distinguished, in part, by the use of the spirometry tests performed in this exercise.

The flow of air through a tube is proportional to the fourth power of its radius (r^4); thus, a small obstruction in the pulmonary airways results in a greatly magnified resistance to airflow. **Obstructive disorders** of the bronchioles are characteristic of *emphysema, bronchitis,* and *asthma.* This obstruction can result from bronchiolar secretions, inflammation and edema, or contraction of bronchiolar smooth muscle. These conditions make it difficult for sufferers to move air rapidly into and out of the lungs. The bronchioles may be weakened to such a degree that they collapse during exhalation before all the air has been emptied from the lungs. This condition, known as *air trapping,* is revealed by an increase in the functional residual capacity.

In **restrictive disorders,** actual damage to the lung tissue results in an abnormal vital capacity test. However, if the disease is purely restrictive (as in *pulmonary fibrosis*), the airways may be clear, resulting in a normal forced expiratory volume (FEV) test. The vital capacity is reduced, but it can be quickly exhaled. This is not true of emphysema, in which the loss of elastic lung tissue decreases the FEV due to the collapse of small airways during exhalation.

Caused primarily by cigarette smoking and aggravated by air pollution, emphysema reduces the number of alveoli in the lungs. This results in an abnormally low vital capacity due to decreased expiratory reserve volume. However, because the elastic alveolar tissue normally helps to keep the thin-walled bronchioles open during exhalation, the loss of alveoli in emphysema also results in a reduction in the elastic support of the bronchioles. During exhalation, such bronchioles may narrow (increasing the resistance to airflow) and even collapse. This adds an obstructive component to the disease, resulting in an abnormal forced expiratory volume test.

Figure 8.13 A recording of the forced expiratory volume (FEV) measurement.

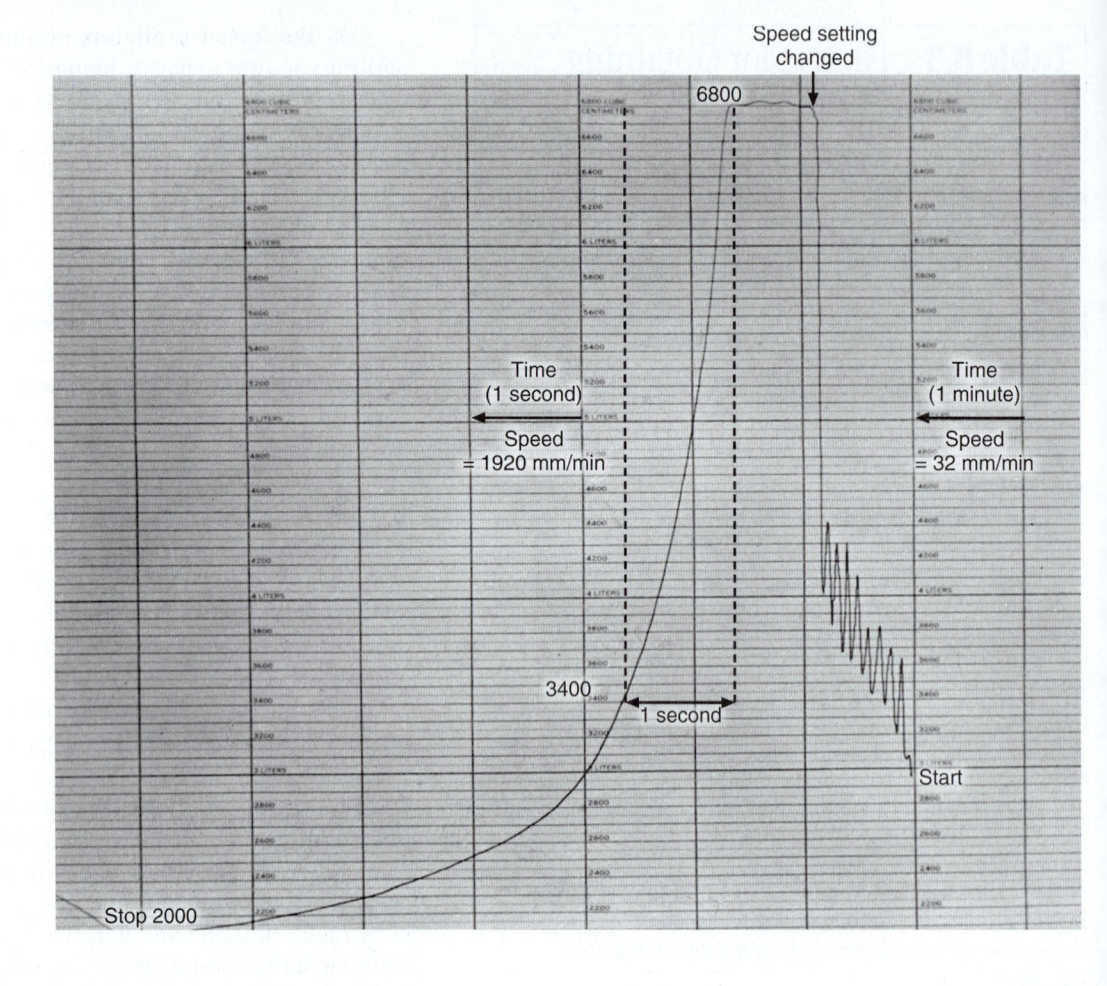

Clinical Applications

Chronic obstructive pulmonary disease (COPD) is characterized by chronic inflammation with narrowing of the airways and destruction of alveolar walls. Included in the COPD category is *chronic obstructive bronchiolitis,* with fibrosis and obstruction of the bronchioles, and *emphysema.* This condition results in an accelerated age-related decline in the FEV_1. Although asthma is also classified as a chronic inflammatory disorder, it is distinguished from COPD in that the obstruction in asthma is largely reversible with inhalation of a bronchodilator (Albuterol). Also, asthma (but not COPD) is characterized by airway hyper-responsiveness—an abnormal bronchoconstrictor response to a stimulus. About 90% of people with COPD are (or have been) smokers. Although not all smokers get COPD, a substantial proportion do—estimated at 10% to 20%. Unfortunately, quitting smoking after COPD has begun does not seem to stop its progression. In chronic pulmonary disorders, alveolar hypoxia may induce constriction of pulmonary arterioles throughout the lungs. This can produce pulmonary hypertension leading to right heart (ventricle) failure, a condition called *cor pulmonale.* Inhaled corticosteroids, useful in treating the inflammation of asthma, are of limited value in the treatment of COPD. COPD is the fifth leading cause of death, and it has been estimated that by 2020 it will become the third leading cause of death worldwide.

The **FEV_1 test** detects increased airway resistance, as occurs in emphysema, bronchitis, and asthma. Also used preoperatively, this test is used extensively to predict a patient's response to general anesthesia and to estimate the length of time the patient must be kept on a respirator post-operatively. The FEV_1 test is also valuable in research on the effects of air pollutants—such as cigarette smoke and ozone—on pulmonary function.

Procedure

Note: *If the Spirocomp was used, the FEV_1 has already been computed and was displayed on the computer screen when the "V" test in part A was performed. The following procedure is for the Collins respirometer.*

1. Start the kymograph slowly rotating (32 mm/min), and have the subject breathe normally into the respirometer for a few breaths, until comfortable.
2. After a normal (unforced) exhalation, instruct the subject to take a deep, forceful inhalation and to hold this inhalation momentarily.
3. Switch the kymograph to the fast speed (1,920 mm/min), and instruct the subject to exhale as rapidly and as forcefully as possible.

Note: At a speed of 1,920 mm/min, the time interval between two heavy vertical lines on the chart is 1 sec.

Calculations: Forced Expiratory Volume (FEV$_1$)

1. Measure the vital capacity from the chart by subtracting the exhalation trough (flat, because all air has been expelled) from the inhalation peak (also flat, because the subject's breath is held). Do not multiply this value by the BTPS factor.

Example (from fig. 8.13)

    6,800 mL   (maximum inhalation)
  − 2,000 mL   (maximum exhalation)
    4,800 mL   (vital capacity)

Enter the *uncorrected* vital capacity here: _____ mL

2. Measure the amount of air exhaled in the first second by subtracting the milliliters corresponding to the exhalation line after 1 sec (3,400 mL, in fig. 8.13) from the milliliters of the inhalation peak at the moment of exhale (6,800 mL). Remember that here the distance between heavy vertical lines is passed in 1 sec. If the subject does not begin to exhale exactly on a vertical line, use a ruler to measure 3.2 cm horizontally from the start of exhalation. (This is the distance between vertical lines and is equivalent to 1 sec at this chart speed.)

Example (from fig. 8.13)

    6,800 mL   (maximum inhalation)
  − 3,400 mL   (exhalation line after 1 sec)
    3,400 mL   (amount exhaled in first second)

Enter the amount exhaled in the first second here: _____ mL

3. Calculate the percentage of the vital capacity exhaled in the first second (the FEV$_1$).

Example

$$FEV_1 = \frac{3,400 \text{ mL (step 2)}}{4,800 \text{ mL (step 1)}} \times 100\%$$
$$= 70.8\%$$

4. If the Biopac system is used, a recording such as the one shown in figure 8.14 will be obtained. The same calculations can be performed as described for the Collins respirometer to obtain the FEV$_1$.
5. Enter the FEV$_1$ in the laboratory report. Refer to table 8.4, and enter the predicted percentage for the FEV$_1$ in the laboratory report.

Table 8.4 Predicted Percentage of the Vital Capacity (VC) Exhaled during the First Second (FEV$_1$)

Age	Predicted Percent VC (FEV$_1$)
18–29	82–80%
30–39	78–77%
40–44	75.5%
45–49	74.5%
50–54	73.5%
55–64	72–70%

From *Archives of Environmental Health*, Volume 12, p. 146, 1966. Reprinted with permission of the Helen Dwight Reid Educational Foundation. Published by Heidref Publications, 1319 Eighteenth St., N.W., Washington, D.C. 20036-1802. Copyright © 1966.

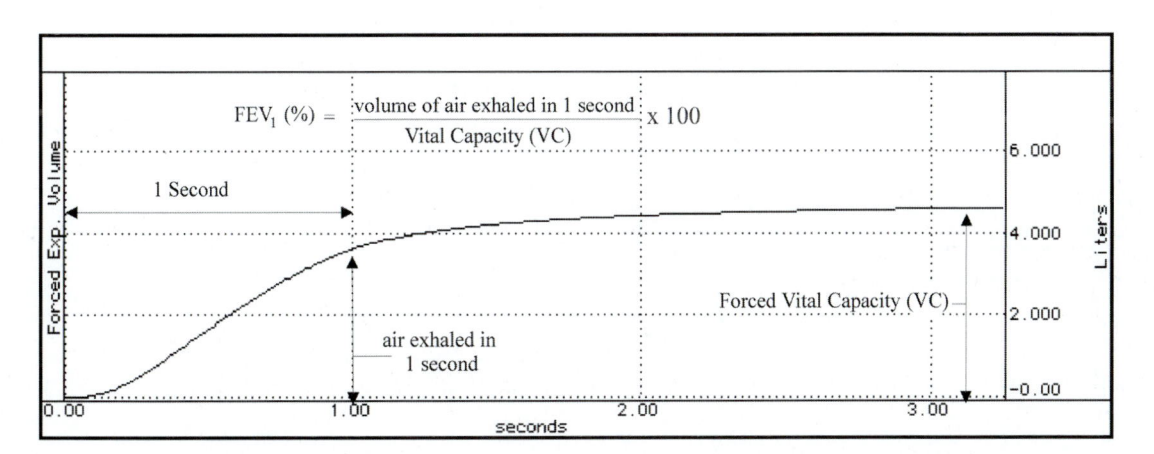

Figure 8.14 Recording of the FEV$_1$ using the Biopac system. Courtesy of and © BIOPAC Systems, Inc.

Laboratory Report 8.1

Name _____

Date _____

Section _____

DATA FROM EXERCISE 8.1

A. Measurement of Simple Lung Volumes and Capacities

1. Enter your data (corrected to BTPS) under the *Measured* column, and enter your calculated *Percent Predicted,* in this table.

Volume/Capacity	Measured	Predicted	Percent Predicted
Tidal volume		500 mL (avg. normal)	
Inspiratory capacity		2,800 mL (avg. normal)	
Expiratory reserve volume		1,200 mL (avg. normal)	
Vital capacity		(from tables)	
Residual volume	not measured	(vital capacity × factor)	cannot calculate
Total lung capacity	not measured	(vital capacity × factor)	cannot calculate

B. Measurement of Forced Expiratory Volume

1. Enter your measured and predicted FEV_1: measured = _____%; predicted range = _____%.

2. Compare your values to the normal range, and enter your conclusions here.

REVIEW ACTIVITIES FOR EXERCISE 8.1

Test Your Knowledge

1. Identify these lung volumes and capacities:
   a. Maximum amount of air that can be expired after a maximum inspiration: _____.
   b. Maximum amount of air that can be expired after a normal expiration: _____.
   c. Maximum amount of air that can be inspired after a normal expiration: _____.
   d. Amount of air left in the lungs after a maximum expiration: _____.

2. Category of pulmonary disorders in which the alveoli are normal but there is an abnormally high resistance to air flow: _____.

3. An example of a disorder in the category described in question 5 is _____.

4. A pulmonary function test for the category of disorders named in question 5 is the _____ test.

Test Your Understanding

5. Describe Boyle's law and how inhalation and exhalation follow from this law. Identify the muscles involved in quiet inhalation and those additionally required for forced inhalation. Which spirometry measurements indicate the extent of quiet and forced inhalation?

6. Explain how quiet exhalation is accomplished, and compare this to forced exhalation. Which spirometry measurements indicate quiet and forced exhalation?

7. Can you measure the residual volume and total lung volume by spirometry? Explain. Also, explain how the residual volume, vital capacity, and total lung capacity change with age.

8. Distinguish between obstructive and restrictive pulmonary disorders. Explain how spirometry aids in their diagnosis.

Test Your Analytical Ability

9. Does your chest expand because your lungs inflate, or do your lungs inflate because your chest expands? Explain.

10. When a person's lungs are ventilated by a tank of gas, the technique is known as *intermittent positive pressure breathing (IPPB)*. Analyzing this term, explain how this technique inflates the lungs for inspiration and deflates the lungs for expiration.

Test Your Quantitative Ability

11. Calculate these values for the spirogram shown here. Be sure to correct your values to BTPS (use a BTPS value of 1.1).

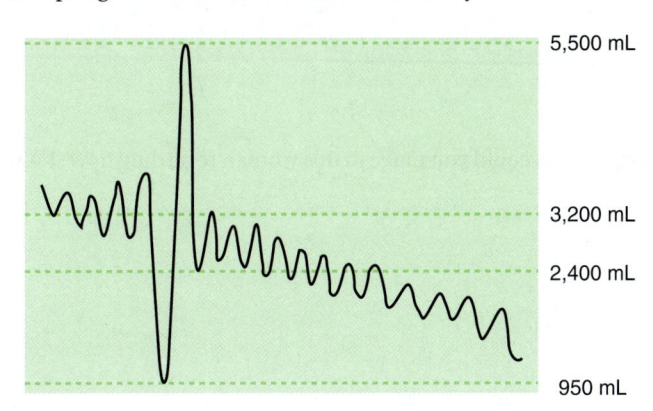

5,500 mL
3,200 mL
2,400 mL
950 mL

(a) tidal volume _____ mL
(b) inspiratory capacity _____ mL
(c) expiratory reserve volume _____ mL
(d) vital capacity _____ mL

12. Calculate the FEV_1 value for the spirogram shown here.

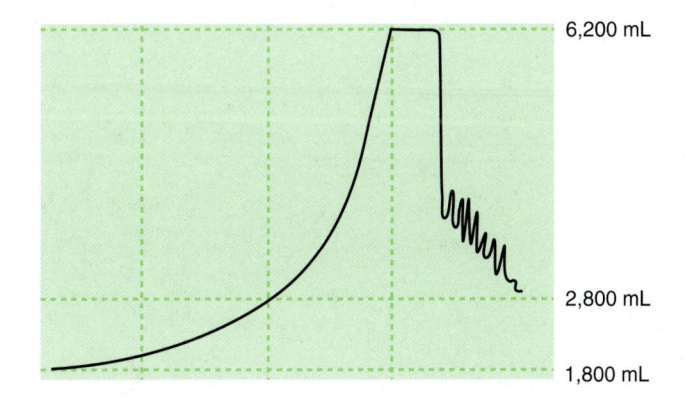

6,200 mL
2,800 mL
1,800 mL

$FEV_1 =$ _____ %

311

Clinical Investigation Questions

13. What is this woman's predicted vital capacity? What percentage of this predicted value was actually measured? What does this result indicate about the health of her lungs?

14. What does the measured FEV$_1$ indicate about this patient's lungs? What does albuterol do, and what does this patient's response to albuterol suggest about the underlying pulmonary problem?

15. Given these results, what suggestions could you make to this woman regarding how she could improve her health?

EXERCISE 8.2

Effect of Exercise on the Respiratory System

MATERIALS

1. Collins 9-L respirometer or Spirocomp program and equipment (Intelitool)
2. Nose clamps and disposable mouthpieces
3. Alternatively, the Biopac system may be used with the setup for Biopac Lesson 12
4. Or, the iWorx system may be used with the setup as described in iWorx Exercise 10; once set up, the measurements for this exercise can be obtained
5. Another alternative is the PowerLab system with their spirometry extension

Total minute volume is the product of the rate of breathing per minute and the depth of each breath. Oxygen consumption per minute is a measure of the metabolic rate. Total minute volume is adjusted by physiological mechanisms to compensate for changes in the metabolic rate.

LEARNING OUTCOMES

You should be able to:

1. Define *total minute volume,* and explain how this measurement is obtained.
2. Describe how the rate of oxygen consumption is measured, and explain how it is used as a measure of the metabolic rate.
3. Describe the relationship between the total minute volume and the rate of oxygen consumption; and explain how, and why, these measurements are changed during exercise.
4. Explain why oxygen consumption and total minute volume remain elevated after exercise has ceased.

Textbook/Multimedia Correlations

Before performing this exercise, you should study the introductory material presented here. Further information relating to this exercise can be found in these pages of *Human Physiology,* thirteenth edition, by Stuart I. Fox:

- *Gas Exchange in the Lungs.* Chapter 16, p. 547.
- *Regulation of Breathing.* Chapter 16, p. 553.
- *Ventilation During Exercise.* Chapter 16, p. 570.
- Multimedia correlations are provided in *Physiology Interactive Lab Simulations (Ph.I.L.S):* Respiration (exercises 7, 8, 9, and 10).

Clinical Investigation

A patient expressed great anxiety over the loss of his job and complained of dizziness. His breathing frequency at rest was 28 breaths/minute and his total minute volume was 13,000 mL/minute. Blood gas measurements revealed that this patient had a resting arterial P_{CO_2} of 20 mm Hg.

- Define hyperpnea, hyperventilation, and hypoventilation.
- Explain the significance of abnormal arterial P_{CO_2} measurements and how this could relate to the patient's dizziness.

The volume of air exhaled in a minute of unforced breathing is known as the **total minute volume** and is equal to the tidal volume (milliliters per breath) multiplied by the frequency of breathing (breaths per minute). Only about two-thirds of this volume reaches the alveoli (this is known as the *alveolar minute volume*). The remaining one-third stays within the dead space of the lungs and is not involved in gas exchange.

Gas exchange occurs in the alveoli of the lungs. Oxygen diffuses from the alveolar air into the blood, while carbon dioxide diffuses from the blood into the alveolar air. Thus, the blood leaving the lungs is rich in oxygen and reduced in carbon dioxide. When the blood reaches the tissue capillaries, oxygen diffuses from the blood into the tissues, where it can be used by the cells in aerobic respiration. Meanwhile carbon dioxide, formed as a waste product of aerobic respiration, diffuses from the tissues into the capillary blood. The concentration of oxygen and carbon dioxide in the blood plasma is indicated by their partial pressures (measured in millimeters of mercury), abbreviated P_{O_2} and P_{CO_2}, respectively. As a result of gas exchange in the pulmonary capillaries, the P_{O_2} of the blood is increased and its P_{CO_2} is decreased. Conversely, as a result of gas exchange in the systemic capillaries, the blood P_{O_2} is decreased while its P_{CO_2} is raised (fig. 8.15).

Oxygen is consumed by the body's cells in aerobic respiration, so the initial volume of air within the oxygen bell of the respirometer decreases as the subject breathes through the mouthpiece. (The exhaled carbon dioxide is removed by

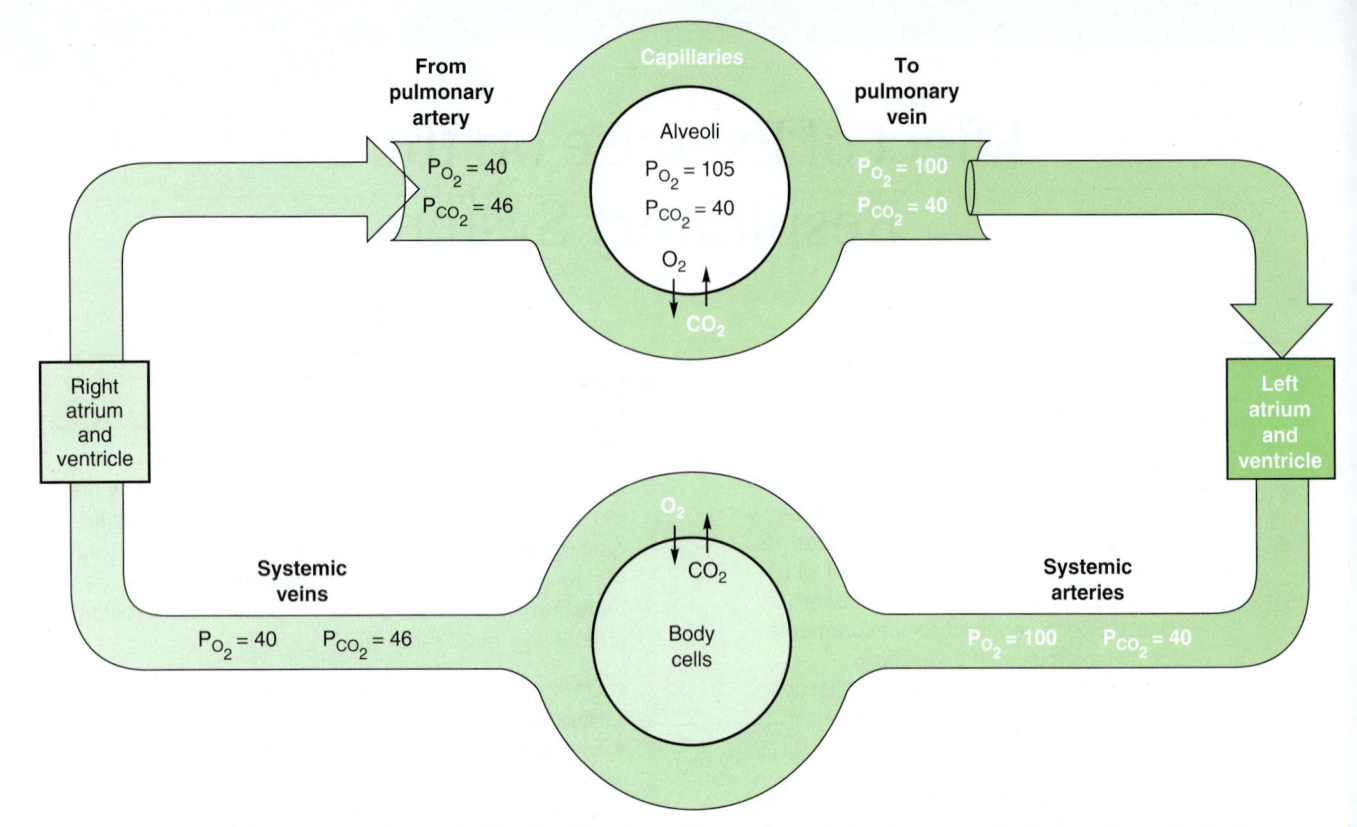

Figure 8.15 **Partial pressures of gases in blood.** The P_{O_2} and P_{CO_2} values of blood are a result of gas exchange in the lung alveoli and gas exchange between systemic capillaries and body cells.
(For a full-color version of this figure, see fig. 16.22 in *Human Physiology*, thirteenth edition, by Stuart I. Fox.)

soda lime within the respirometer.) Thus, oxygen consumption results in the removal of air from the bell. On a spirogram, this is seen as an upward slope of the tidal volume tracing. The amount of oxygen consumed per minute can be calculated as the difference between milliliter levels before and after 1 minute of resting ventilation (fig. 8.16).

The rate at which oxygen is consumed and carbon dioxide is produced by the body cells during aerobic respiration is related to the *metabolic rate* of the person. When the individual is relaxed and comfortable and has not eaten for 12 to 15 hours, the metabolic rate is lowest. This rate is referred to as the **basal metabolic rate (BMR).** Under these conditions the metabolic rate is set primarily by the activity of the thyroid gland; in the past, measurements of BMR were used to assess thyroid function.

When a person exercises, however, the metabolic rate increases greatly (the metabolism of muscles can increase as much as sixtyfold during strenuous exercise). As a result, oxygen is consumed and carbon dioxide is produced at much more rapid rates than during resting conditions. The respiratory system keeps pace with this increased demand by increasing the total minute volume.

The **respiratory control center** in the medulla oblongata contains inspiratory and expiratory neurons that regulate breathing largely via axons to the **phrenic motor nuclei** in cervical regions C3 through C6 of the spinal cord. Lower motor neurons here send axons to the phrenic nerves that control the diaphragm. This is why many people with spinal cord injuries above C4 cannot breathe independently. The respiratory control center's activity is influenced by higher brain centers and by chemical changes in the blood and brain. These chemical changes are detected by chemoreceptors.

Clinical Applications

Hypoventilation occurs when the alveoli are inadequately ventilated (alveolar minute volume is reduced). Due to a reduction in the total minute volume where either the tidal volume or the frequency of breathing is depressed, hypoventilation can also result from a pathological increase in dead air space in the lungs. The latter may be caused by any condition that affects lung tissue (such as emphysema) or by inadequate blood flow to well-ventilated alveoli. Such ventilated alveoli lacking the appropriate perfusion of capillary blood (producing an abnormal *ventilation/perfusion ratio*) are incapable of fully oxygenating the blood. As a result of hypoventilation, there is also inadequate elimination of carbon dioxide from the blood. Because the plasma carbon dioxide levels are directly affected by ventilation, hypoventilation may be operationally defined as an *abnormally increased plasma carbon dioxide level* (P_{CO_2} above 40 mm Hg).

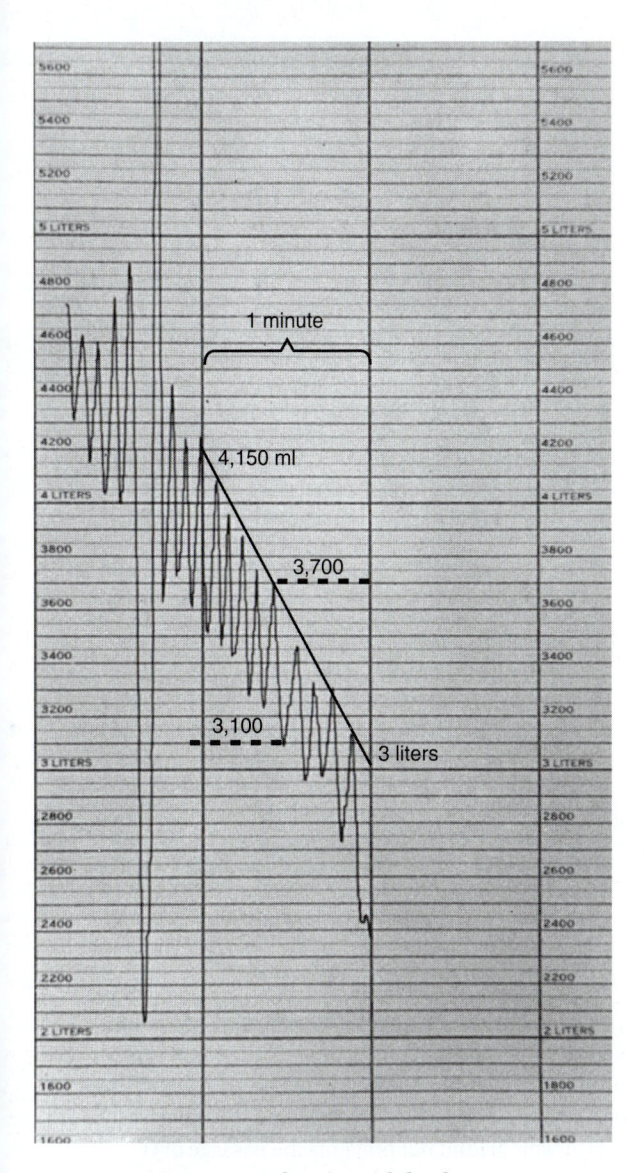

Figure 8.16 Spirogram showing tidal volume measurements over a 1-minute interval at rest. The rising slope of the tidal volume measurements indicates oxygen consumption.

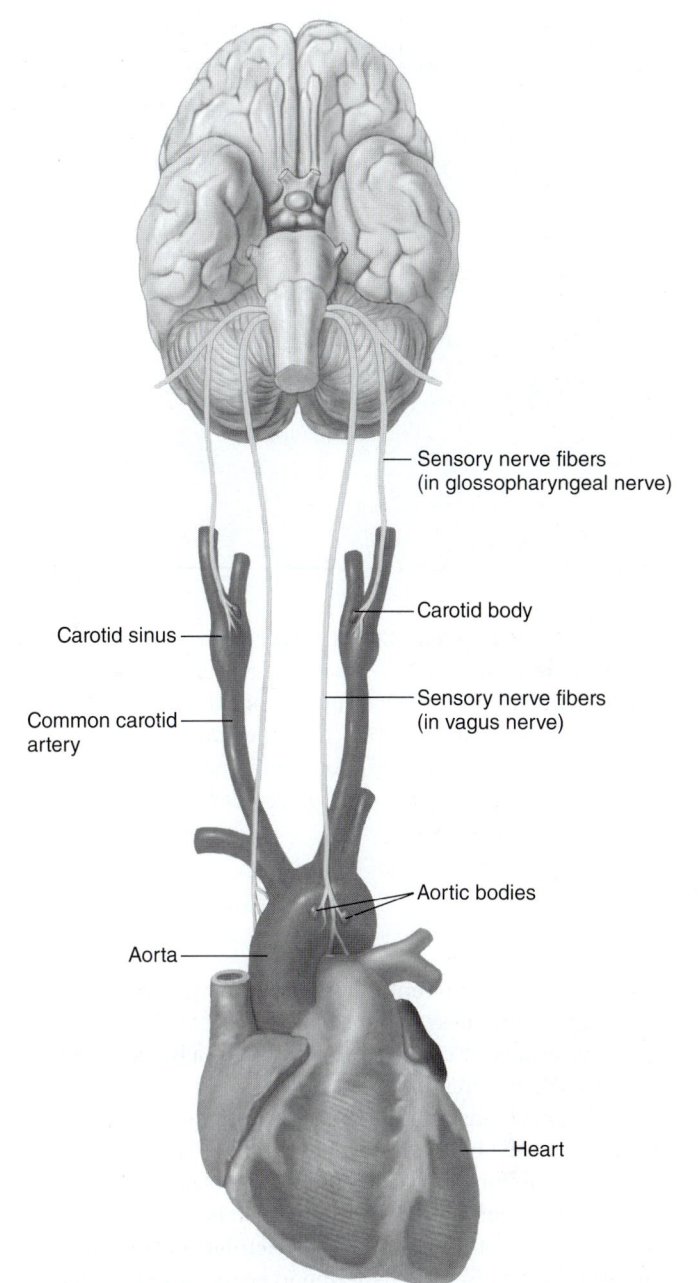

Figure 8.17 Sensory input from the aortic and carotid bodies. The peripheral chemoreceptors (aortic and carotid bodies) regulate the brain stem respiratory centers by means of sensory nerve stimulation.

(For a full-color version of this figure, see fig. 16.25 in *Human Physiology*, thirteenth edition, by Stuart I. Fox.)

Normal, resting breathing at sea level is regulated primarily by the blood CO_2 levels. A rise in the plasma CO_2 concentration (as measured by the P_{CO_2}) during hypoventilation causes a fall in the plasma pH. This stimulates the *peripheral chemoreceptors* in the **aortic** and **carotid bodies** (fig. 8.17). After a delay, the plasma CO_2 reaches the brain and lowers the pH of CSF and brain interstitial fluid, stimulating *central chemoreceptors* in the medulla oblongata (fig. 8.18). The peripheral and central chemoreceptors stimulate the respiratory control center in the medulla, which directs an increase in breathing. Thus, a negative feedback loop is completed to help restore homeostasis (fig. 8.19).

The increase in total minute volume that occurs during exercise may be due in part to an increase in CO_2 production, although concentrations of arterial CO_2 during exercise are not usually increased. Anticipation and excitement coming from conscious brain areas and sensory feedback from the exercising muscles may also contribute to the **hyperpnea,** or increased total minute volume, of exercise. This increased breathing during exercise usually matches the increased metabolic rate, so that the arterial blood P_{CO_2} remains relatively constant. However, if the total minute volume increases more than the metabolic rate, the arterial P_{CO_2} will fall because CO_2 is "blown off" faster than it is produced. Whenever a high total minute volume causes a fall in the arterial P_{CO_2} the breathing

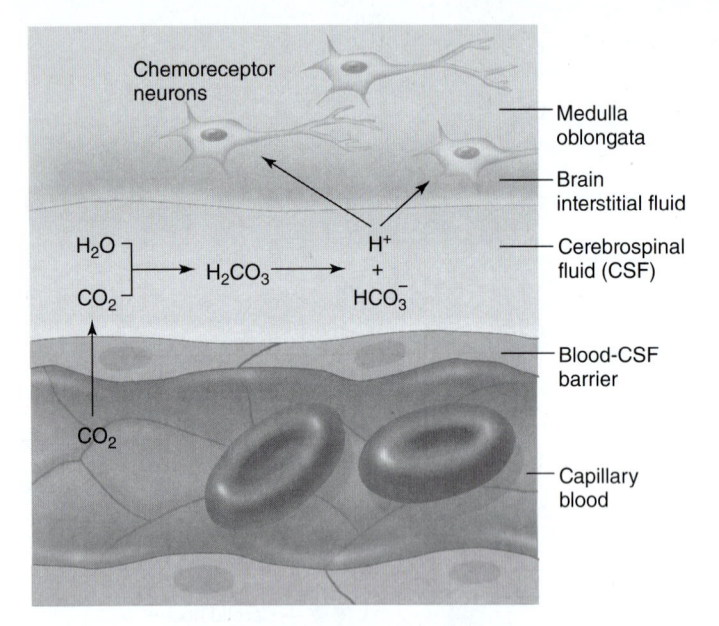

Figure 8.18 How blood CO₂ affects chemoreceptors in the medulla oblongata. An increase in blood CO_2 stimulates breathing indirectly by lowering the pH of blood and brain interstitial fluid. This figure illustrates how a rise in blood CO_2 increases the H^+ concentration (lowers the pH) and thereby stimulates chemoreceptor neurons in the medulla oblongata.

(For a full-color version of this figure, see fig. 16.29 in *Human Physiology*, thirteenth edition, by Stuart I. Fox.)

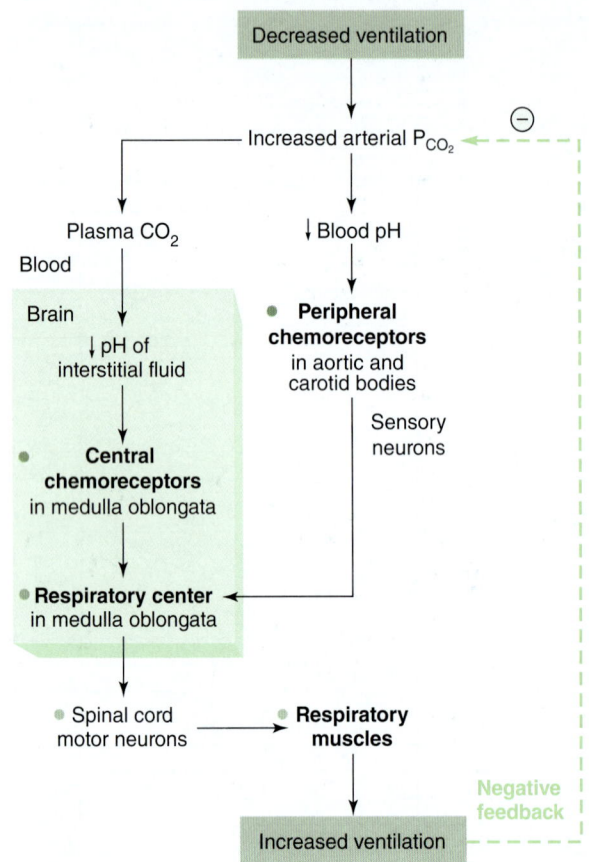

Figure 8.19 Chemoreceptor control of breathing. This figure depicts the negative feedback control of ventilation through changes in blood P_{CO_2} and pH. The blood-brain barrier, represented by the box, allows CO_2 to pass into the cerebrospinal fluid but prevents the passage of H^+.

(For a full-color version of this figure, see fig. 16.28 in *Human Physiology*, thirteenth edition, by Stuart I. Fox.)

is described as **hyperventilation.** Because hyperventilation can produce dizziness (because the low arterial P_{CO_2} causes cerebral vasoconstriction), people do not usually hyperventilate when they exercise.

Oxygen consumption and the total minute volume remain elevated immediately after exercise. This extra oxygen consumption (over resting levels) following exercise is called the **oxygen debt.** The extra oxygen is used to oxidize lactic acid produced as a result of anaerobic metabolism in the exercising muscles and to support an increased metabolism within the warmed muscles.

Procedure

(BIOPAC SYSTEM)

1. Set up the Biopac system and calibrate it with the air syringe, as described in exercise 8.1.
2. Assemble the mouthpiece setup (see fig. 8.7 in exercise 8.1) and record your resting tidal volume measurements for one minute. Click "Record" to begin recording and "Stop" when you are done.
3. Select "Analyze current data file" on the computer. At the top of the screen, select "CH 2" for the channel, and select "p-p" for the measurement.
4. At the bottom right of the screen, select the I-beam cursor. Use this to highlight the region of the screen showing your tidal volume breaths for one minute.
5. The computer will automatically calculate the tidal volume. Enter this value in your laboratory report. Count the number of breaths in the minute so that you can multiply this by the tidal volume to obtain the total minute volume. Enter the tidal volume and total minute volume at rest in your laboratory report.
6. Perform light exercise, such as several jumping jacks or pedaling for 15–30 seconds on a bicycle ergometer.
7. When you are done, analyze the data as described in steps 2 and 3. Record your immediate postexercise tidal volumes and total minute volume in the laboratory report.
8. Note that you cannot obtain an oxygen consumption measurement using the Biopac system.

Procedure

(COLLINS RESPIROMETER)

1. Set the Collins respirometer to the slow speed of 32 mm/min. (At this speed the distance between two heavy vertical lines is traversed in 1 minute.) To avoid air leakage, position the mouthpiece securely in the mouth with the lips tightly sealed around it, and clamp the nostrils closed.
2. Under resting conditions, breathe normally into the respirometer for 1 minute (that is, perform the procedure for measuring tidal volume—see exercise 8.1).

Note: *The tidal volume measurements must have an upward slope, indicating that the oxygen being consumed is from the air trapped in the oxygen bell. If the slope is downward, air is leaking into the system, usually through the corners of the mouth or the nose. In this event, reposition the mouthpiece, check the nose clamp, and begin the measurements again.*

3. After 1 minute, stop the respirometer, remove the chart from the kymograph drum, and determine the frequency of ventilation (number of breaths per minute) and the tidal volume (corrected to BTPS).

Example (from fig. 8.16)

```
   3,700 mL   (inhalation peak of tidal volume)
 - 3,100 mL   (exhalation trough of tidal volume)
     600 mL   (uncorrected tidal volume)

     600 mL
 ×     1.1    (BTPS factor)
     660 mL   (corrected tidal volume)
```

From figure 8.16, frequency = 10 breaths/min

Enter the subject's corrected tidal volume here: _____ mL

Enter the subject's frequency of ventilation here: _____ breaths/min

> **Normal Values** The average ventilation frequency is 14 breaths/min.

4. Determine the subject's total minute volume at rest by multiplying the frequency of ventilation by the tidal volume.
   Enter this value in the data table in your laboratory report.

> **Normal Values** The average total minute volume is 6,750 mL/min.

Example (from fig. 8.16)

```
660 mL/breath × 10 breaths/min
   = 6,600 mL/min total minute volume
```

5. Use a straight edge to draw a line that touches either the peaks or the troughs of the tidal volume measurements. Determine the oxygen consumption per minute by subtracting the milliliters where this straight line intersects the heavy vertical chart line at the beginning of one minute from the milliliters where the straight line intersects the next heavy vertical line at the end of one minute.

Example (from fig. 8.16)

Using a line that averages the peaks:

```
   4,150 mL   (at end of 1 minute)
 - 3,000 mL   (at begining of 1 minute)
   1,150 mL   (oxygen consumption)
```

Enter the resting oxygen consumption in the data table in your laboratory report.

6. Now, have the subject perform light exercise, such as 5 to 10 jumping jacks, and then repeat the respirometer measurements and data calculation described in steps 1–5. Alternatively, if a bicycle ergometer is available, the total minute volume may be determined while the student pedals lightly for one minute.

> ⚠ **Caution:** *If breathing into the respirometer becomes difficult after exercise, the subject should stop the procedure. Results obtained in less than a minute can then be extrapolated to 1 minute. Alternatively, the bell can be filled with 100% oxygen to prevent the possible occurrence of hypoxia.*

Enter the corrected tidal volume after exercise in this space:
_____ mL

Enter the frequency of ventilation after exercise in this space:
_____ breaths/min

Enter the total minute volume and the oxygen consumption per minute after exercise in the data table of the laboratory report.

7. Calculate the percent increase after exercise for total minute volume and for oxygen consumption. This is the difference between the exercise and resting measurements, divided by the resting measurement and multiplied by 100%. Enter these values in the data table in your laboratory report.

317

Laboratory Report 8.2

Name _____

Date _____

Section _____

DATA FROM EXERCISE 8.2

1. Enter the total minute volume and rate of oxygen consumption during rest and after exercise in this table.
2. Calculate and enter the percent increases in your measurements after exercise, and enter these values in this table.

Measurement	Resting	During or After Exercise	Percent Increase
Tidal volume			
Oxygen consumption			

REVIEW ACTIVITIES FOR EXERCISE 8.2

Test Your Knowledge

1. The net diffusion of oxygen and carbon dioxide in opposite directions across the wall of alveoli is known as _____.
2. Define *tidal volume*. _____
3. The total minute volume is obtained by multiplying the _____ times the _____.
4. The peripheral chemoreceptors include the _____ and the _____.
5. The stimulus that directly activates the peripheral chemoreceptors is _____.
6. Define *hyperpnea*. _____
7. *Hypoventilation* may be operationally defined as _____.

Test Your Understanding

8. Distinguish between the peripheral and central chemoreceptors in terms of their locations and the mechanisms by which they are stimulated.

9. Define the term *hypoventilation*, explaining its relationship to the carbon dioxide concentration of the blood plasma. Explain why a person could have a much lower total minute volume at rest than another person, without hypoventilating.

10. What happened to the total minute volume during exercise? What happened to the total minute volume right after exercise was completed? What physiological mechanisms might account for these changes?

319

11. Define *oxygen debt,* and explain the functions of the oxygen debt.

Test Your Analytical Ability

12. High total minute volume during mild to moderate exercise is more accurately termed hyperpnea than hyperventilation. Distinguish between these two terms, and explain why this statement is true.

13. How might the plasma CO_2 levels differ during the performance of a particular exercise in a trained versus an untrained person? Explain your answer.

Test Your Quantitative Ability

14. Given a tidal volume spirogram with a peak inspiration at 4,200 mL and expiration at 3,400 mL; a BTPS value of 1.1; and a respiratory rate of 15 breaths per minute, calculate the total minute volume.

Clinical Investigation Questions

15. What terms would you use to describe the measurements of this patient's breathing? Explain.

16. What is the most likely explanation for the patient's dizziness, and what can be done about it? Explain.

Oxyhemoglobin Saturation

EXERCISE 8.3

MATERIALS

1. Graduated cylinder and a 1-cc syringe
2. Test tube and distilled water
3. Colorimeter and cuvettes
4. Sodium dithionite (hydrosulfite), 1.0 g per 100 mL
5. Alcohol swabs and lancets for preparing fingertip blood. Alternatively, dog or cat blood (obtained from a veterinarian) may be used

The iron atoms within the heme groups of hemoglobin may be free (unbound), or they may be bonded to oxygen or carbon monoxide gas. Each of these different forms of hemoglobin has a slightly different color, which allows the percentage of each type in a mixture to be measured using the absorption spectrum of each hemoglobin type. Measurement of the percent oxyhemoglobin is used clinically to assess lung function and the capacity of blood to transport oxygen.

LEARNING OUTCOMES

You should be able to:

1. Define the term *percent saturation*.
2. Explain the clinical significance of the percent oxyhemoglobin measurement.
3. Explain the clinical significance of the percent carboxyhemoglobin measurement.
4. Describe how an absorption spectrum is obtained, and explain how the absorption spectra of the different forms of hemoglobin are used to determine the percent saturation.

Textbook/Multimedia Correlations

Before performing this exercise, you should study the introductory material presented here. Further information relating to this exercise can be found in these pages of *Human Physiology*, thirteenth edition, by Stuart I. Fox:

- *Partial Pressure of Gases in Blood.* Chapter 16, p. 548.
- *Hemoglobin and Oxygen Transport.* Chapter 16, p. 558.
- Multimedia correlations are provided in *MediaPhys 3.0:* Topics 10.45–10.49; also *Physiology Interactive Lab Simulations (Ph.I.L.S):* Respiration (exercises 7, 8, 9, and 10).

Clinical Investigation

A person visiting a mountain ski area at an altitude of 9,000 feet felt weak and went to the emergency room at the local hospital. His finger was placed in a pulse oximeter and a percent oxyhemoglobin saturation of 88% was measured. When questioned, he admitted that he was a heavy smoker.

- Explain the significance of the percent oxyhemoglobin saturation measurement.
- Explain the significance of this person's measurement and the probable reasons for any abnormality.

The ability of the blood to carry oxygen depends on (1) ventilation; (2) gas exchange across the alveoli of the lungs; (3) the red blood cell count and hemoglobin concentration; and (4) the chemical form of the hemoglobin.

There are two chemical forms of normal hemoglobin. Normal hemoglobin without oxygen is called **deoxyhemoglobin;** after it binds to oxygen, it is called **oxyhemoglobin.** If a person suffers from *carbon monoxide poisoning,* however, the abnormal hemoglobin called **carboxyhemoglobin** (hemoglobin bound to carbon monoxide) causes the blood to carry a lower amount of oxygen. This is because the carbon monoxide displaces oxygen and binds to hemoglobin with a higher affinity than does oxygen. Therefore, when health professionals need to determine the oxygen carrying capacity of the blood, they need to learn the relative proportion of each hemoglobin type as well as the total hemoglobin concentration.

The relative proportion of each type of hemoglobin is given as its **percent saturation.** The percent oxyhemoglobin saturation, for example, is the proportion of hemoglobin bound to oxygen. Normally, this value is approximately 97% in arterial blood and 75% in venous blood.

$$\% \text{ oxyhemoglobin saturation} = \frac{\text{oxyhemoglobin}}{\text{total blood hemoglobin}} \times 100$$

The determination of percent oxyhemoglobin saturation is a very sensitive means of assessing the effectiveness of pulmonary function. When pulmonary function and blood hemoglobin are normal, the arterial blood has a percent oxyhemoglobin saturation at sea level of about 97%. Even when pulmonary function is normal, the percent oxyhemoglobin saturation can be decreased by carbon monoxide poisoning. When a person's blood has a carboxyhemoglobin saturation of 6.9%, for example (obtained in one study from cigarette-smoking New York taxicab drivers), the percent oxyhemoglobin saturation is decreased accordingly. Another abnormal form of hemoglobin is **methemoglobin.** This is oxidized hemoglobin that lacks the electron needed to bond with oxygen and therefore cannot participate in oxygen transport. These impairments in oxygen transport cannot be detected by measurement of red blood cell count, hematocrit, or total blood hemoglobin.

The percent saturation of the different types of hemoglobin is measured by comparing the absorption spectrum of an unknown sample of blood with the absorption spectra of pure oxyhemoglobin, pure reduced hemoglobin, and pure carboxyhemoglobin. These hemoglobins have different colors, so they absorb different amounts of light at each wavelength. A graph of absorbance versus wavelength (where the concentration is constant) is called an **absorption spectrum** (fig. 8.20).

The absorption spectrum of an unknown sample of blood will display some combination of these three absorption spectra because the blood contains all three types of hemoglobin. The relative contribution of each hemoglobin type to the absorption spectrum is proportional to the relative amount of each type in the blood. This complex analysis is usually performed by a laboratory instrument specifically manufactured for this purpose.

In this exercise, you will construct an absorption spectrum for 100% oxyhemoglobin and 100% reduced hemoglobin. Bubbling air into a flask containing blood until the blood is in equilibrium with the air can produce 100% oxyhemoglobin. This process essentially duplicates the process that occurs in the capillaries surrounding the lung alveoli. A simpler (but less accurate) method is to obtain a sample of blood from the fingertip, which has a high percent oxyhemoglobin saturation. The sample of 100% reduced hemoglobin may be obtained by adding sodium hydrosulfite ($Na_2S_2O_4$) to a second sample of blood. The sodium hydrosulfite (also called sodium dithionite) removes oxygen from oxyhemoglobin.

Clinical Applications

A **pulse oximeter** is commonly used in hospitals for measuring oxyhemoglobin saturation noninvasively. This device typically clips on the finger or pinna and gives readings of oxygen saturation and pulse rate within a short time, making it useful in many clinical contexts such as in emergency medicine and during anesthesia. The pulse oximeter has two light-emitting diodes (LEDs), one that emits red light (wavelengths of 600–750 nm) and the other that emits light in the infrared range (850–1000 nm), which pass through the tissues to a sensor. Oxyhemoglobin and deoxyhemoglobin absorb light differently in these ranges, allowing microprocessors to determine the percent oxyhemoglobin saturation fairly accurately under most conditions.

Procedure

1. Add 8.0 mL of distilled water to a test tube. Obtain a large drop of blood by wiping the fingertip with 70% alcohol. Let it dry, and then puncture it with a sterile lancet. Mix this blood with the distilled water by inverting the test tube over the punctured finger.
2. Transfer half the contents of the test tube (4.0 mL) to a second tube.
3. Add 0.20 mL of 1.0% sodium dithionite solution to the second test tube and mix thoroughly.

Note: The dithionite solution should be freshly prepared just prior to use, and the absorbance values of the two tubes should be determined within 5 minutes of the time the dithionite is added to the second tube.

4. Transfer the two solutions to two cuvettes. Fill a third cuvette with distilled water, and use it as a blank to standardize the spectrophotometer at 500 nm. (See exercise 2.1 for a description of the spectrophotometer, standardizing procedures, and Beer's law.)
5. Record the absorbances of solutions 1 and 2. Continue using the blank to standardize the spectrophotometer at each of the successive wavelengths from 510 to 600 nm, and then record the absorbances of the two solutions in the laboratory report.
6. Graph the absorption spectra of oxyhemoglobin and reduced hemoglobin on the graph provided in the laboratory report.

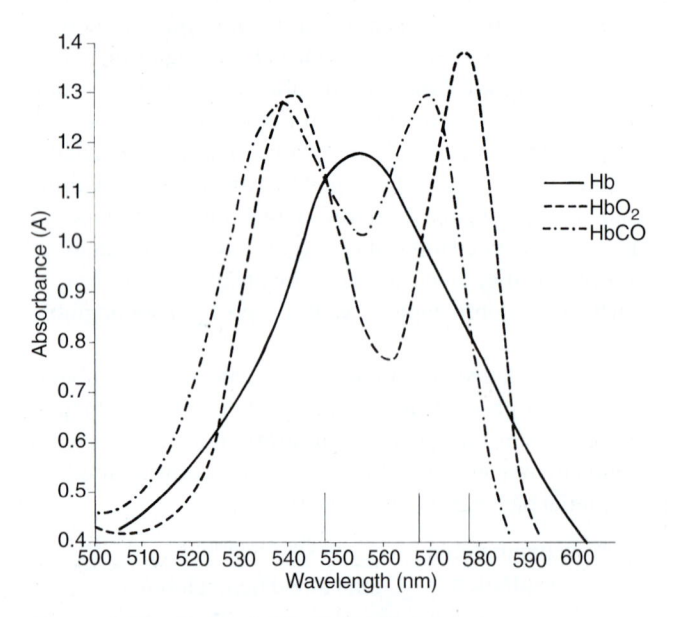

Figure 8.20 Hemoglobin absorption spectra.
The absorption spectra for reduced hemoglobin (Hb), oxyhemoglobin (HbO_2), and carboxyhemoglobin (HbCO) are shown. Source: *Instrumentation Lab, Inc.*

Laboratory Report 8.3

Name _____
Date _____
Section _____

DATA FOR EXERCISE 8.3

Wavelength	Oxyhemoglobin Absorbance	Reduced Hemoglobin Absorbance	Wavelength	Oxyhemoglobin Absorbance	Reduced Hemoglobin Absorbance
500 nm			560 nm		
510 nm			570 nm		
520 nm			580 nm		
530 nm			590 nm		
540 nm			600 nm		
550 nm					

323

REVIEW ACTIVITIES FOR EXERCISE 8.3

Test Your Knowledge

1. In the lungs, normal hemoglobin without oxygen, or _____, binds to oxygen to become _____.

2. The type of hemoglobin bound to carbon monoxide: _____.

3. The type of hemoglobin where the heme iron is in the oxidized Fe^{2+} state: _____.

4. What is the normal percent oxyhemoglobin saturation of arterial blood? _____ %

5. A graph of the absorbance of light as a function of the wavelength of light is called a(n) _____.

6. The different forms of hemoglobin can be distinguished visually because they have different _____.

Test Your Understanding

7. What blood measurement would be abnormally increased in a person with carbon monoxide poisoning? What are the dangers of carbon monoxide poisoning?

8. In what way are carbon monoxide poisoning and anemia different? In what way are they similar? Explain.

Test Your Analytical Ability

9. Suppose a person's percent oxyhemoglobin saturation in the venous blood was 75% at rest, but decreased to 35% during moderate exercise. (a) Explain why the venous percent oxyhemoglobin saturation decreased during the exercise; and (b) predict how the arterial percent oxyhemoglobin saturation might change during exercise. Explain your answer.

10. The percent oxyhemoglobin saturation is measured in babies under treatment for *respiratory distress syndrome (RDS)* and in patients under general anesthesia. What information would these measurements provide under these conditions? Explain.

Clinical Investigation Questions

11. What does a pulse oximeter measure? What is a normal measurement of percent oxyhemoglobin at sea level, and why would this measurement be different at high altitude?

12. Given that most healthy people at that altitude measure a percent oxyhemoglobin saturation of at least 93%, what could account for this person's lower measurement? Explain.

EXERCISE 8.4

Respiration and Acid-Base Balance

MATERIALS

1. pH meter, droppers, beakers, straws
2. Buffer, pH = 7 (made from purchased concentrate); concentrated HCl
3. Concentrated NaOH
4. Phenolphthalein solution (saturated)

Carbon dioxide in plasma can combine with water to produce carbonic acid, which in turn dissociates to produce protons (H$^+$) and bicarbonate ions (HCO$_3^-$). Ventilation regulates the carbon dioxide concentration of the plasma and has an important role in acid-base balance.

LEARNING OUTCOMES

You should be able to:

1. Describe the pH scale, and define the terms *acid* and *base*.
2. Explain how carbonic acid and bicarbonate are formed in the blood, and describe their functions.
3. Define the terms *acidosis* and *alkalosis*, and explain how these conditions relate to hypoventilation and hyperventilation.
4. Explain how ventilation is adjusted to help maintain acid-base balance.

Textbook/Multimedia Correlations

Before performing this exercise, you should study the introductory material presented here. Further information relating to this exercise can be found in these pages of *Human Physiology*, thirteenth edition, by Stuart I. Fox:

- *Regulation of Breathing.* Chapter 16, p. 553.
- *Carbon Dioxide Transport.* Chapter 16, p. 565.
- *Acid-Base Balance of the Blood.* Chapter 16, p. 567.
- *Ventilation During Exercise.* Chapter 16, p. 570.
- Multimedia correlations are provided in *MediaPhys 3.0:* Topics 10.50–10.56 and 12.12–12.18; also *Physiology Interactive Lab Simulations (Ph.I.L.S):* Respiration (exercises 7, 8, 9, and 10).

Clinical Investigation

Blood gas measurements are taken on a patient with a history of chronic obstructive pulmonary disease (COPD). Measured values include an arterial P$_{CO_2}$ of 50 mm Hg; a bicarbonate concentration of 24 mEq/L; and a blood pH of 7.30.

- Explain the significance of each of these measurements.
- Explain how the blood gas measurements relate to this patient's medical history.

Ventilation has two different but related functions: (1) oxygenation of the blood, accomplished by bringing new air into the alveoli during the inhalation phase, and (2) elimination of carbon dioxide from the blood, accomplished by the diffusion of CO$_2$ from the blood into the alveoli and the extrusion of this CO$_2$ by exhalation. The first function serves to maintain aerobic cell respiration; the second serves to maintain the normal pH of the blood.

The pH (see appendix 1) indicates the concentration of H$^+$ (hydrogen ion) in a solution and is defined by the following formula:

$$pH = \log \frac{1}{[H^+]}$$

where H$^+$ is the concentration of H$^+$ in moles (atomic weight in grams) per liter. Some water molecules ionize to produce equal amounts of H$^+$ and OH$^-$ (hydroxyl ion). In pure water, the H$^+$ concentration is 10^{-7} moles/L. (Because hydrogen has an atomic weight of 1, this is the same as 10^{-7} g/L.) This is equal to a pH of 7.0 and is called a *neutral solution*. An *acidic solution* has a higher H$^+$ concentration and a lower pH; a *basic solution* has a lower H$^+$ concentration and a higher pH (table 8.5).

An **acid** is a molecule that can donate free H$^+$ to a solution and lower its pH. *Carbonic acid* (H$_2$CO$_3$) is formed from the combination of CO$_2$ and water within the red blood cells. This reaction is catalyzed by an enzyme called *carbonic anhydrase* (fig. 8.21).

$$CO_2 + H_2O \xrightarrow{\text{carbonic anhydrase}} H_2CO_3$$

Table 8.5   The pH Scale

	H+ Concentration (Molar)	pH	OH− Concentration (Molar)
	1.0	0	10^{-14}
	0.1	1	10^{-13}
	0.01	2	10^{-12}
Acids	0.001	3	10^{-11}
	0.0001	4	10^{-10}
	10^{-5}	5	10^{-9}
	10^{-6}	6	10^{-8}
Neutral	10^{-7}	7	10^{-7}
	10^{-8}	8	10^{-6}
	10^{-9}	9	10^{-5}
	10^{-10}	10	0.0001
Bases	10^{-11}	11	0.001
	10^{-12}	12	0.01
	10^{-13}	13	0.1
	10^{-14}	14	1.0

Some of the carbonic acid formed can immediately dissociate to yield H+ and *bicarbonate ion* (HCO3−). The H+ derived from carbonic acid and other acids in the blood gives normal arterial blood a pH of 7.40 ± 0.05 (table 8.6 and fig. 8.21).

$$H_2CO_3 \longrightarrow HCO_3^- + H^+$$

A. Ability of Buffers to Stabilize the pH of Solutions

Plasma has a particular concentration of bicarbonate as a result of the dissociation of carbonic acid. Bicarbonate serves as the major *buffer* of the blood, helping to stabilize the pH of plasma despite the continuous influx of H+ from molecules of lactic acid, fatty acids, ketone bodies, and other metabolic products. The H+ released by these acids is prevented from lowering the blood pH because it is combined with bicarbonate. Although a new acid molecule (carbonic acid) is formed, this reaction prevents a rise in the free H+ concentration (fig. 8.21).

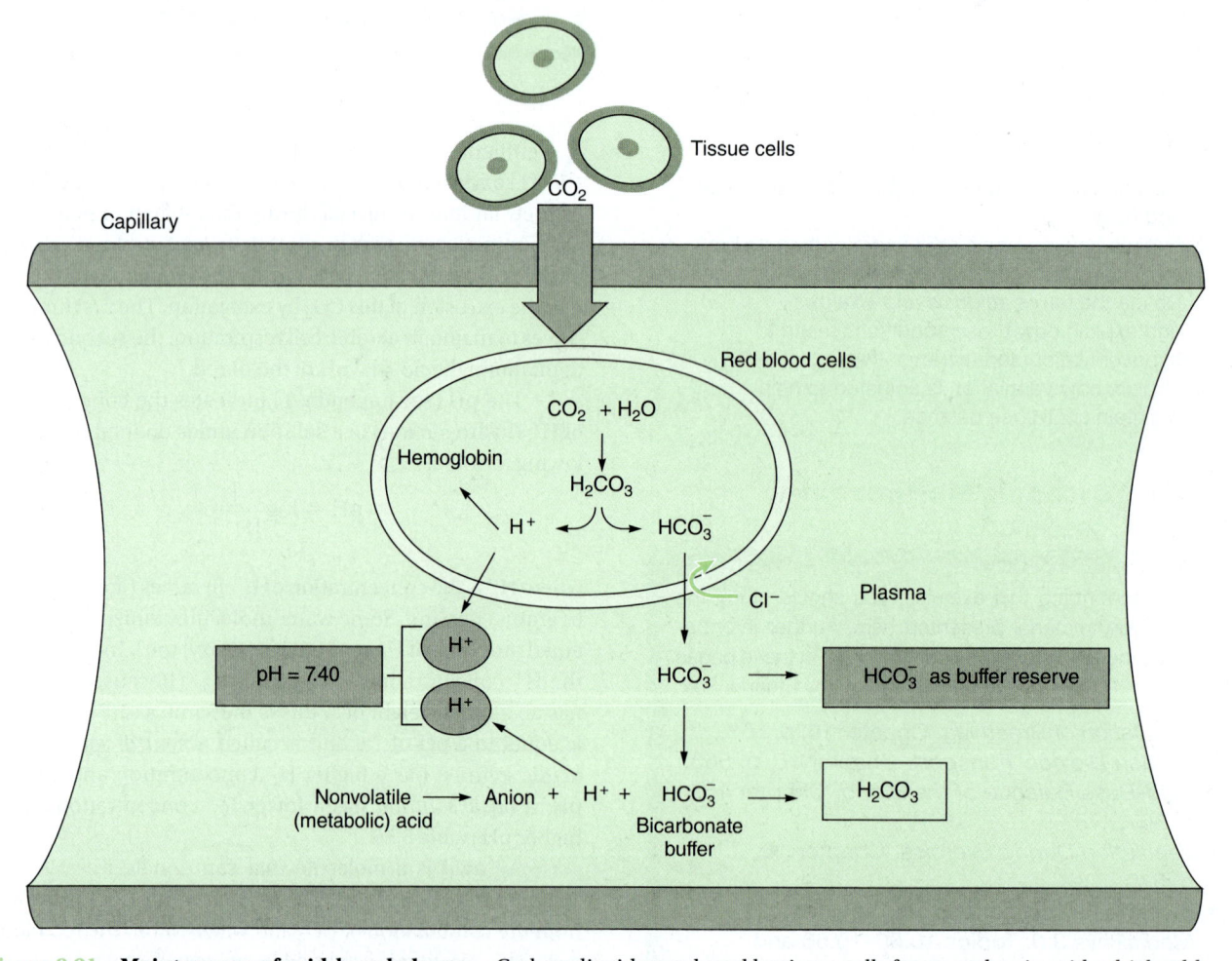

Figure 8.21   Maintenance of acid-base balance. Carbon dioxide produced by tissue cells forms carbonic acid, which adds both H+ and bicarbonate (HCO3−) to the plasma. The bicarbonate released from red blood cells buffers the H+ produced by ionization of metabolic (nonvolatile) acids, such as lactic acid and ketone bodies.

(For a full-color version of this figure, see fig. 16.40 in *Human Physiology*, thirteenth edition, by Stuart I. Fox.)

Table 8.6 The Effect of Respiration on Blood pH

P_{CO_2} (mm Hg)	H_2CO_3 (mEq/L)[1]	HCO_3^- (mEq/L)[1]	HCO_3^-/H_2CO_3 Ratio	Blood pH	Condition
20	0.6	24	40/1	7.70	Respiratory alkalosis
30	0.9	24	26.7/1	7.53	Respiratory alkalosis
40	1.2	24	20/1	7.40	Normal
50	1.5	24	16/1	7.30	Respiratory acidosis
60	1.8	24	13.3/1	7.22	Respiratory acidosis

1. Ion concentrations are commonly measured in milliequivalents (mEq) per liter. This measurement is equal to the millimolar concentration of the ion multiplied by its number of charges.

$$H^+ + HCO_3^- \longrightarrow H_2CO_3$$

Carbonic acid formed in this way can provide a source of new H^+ if the blood pH should begin to rise (from a loss of blood H^+) beyond normal levels. The carbonic acid/bicarbonate buffer system helps to stabilize the blood pH under normal conditions. Disease states, however, may cause the blood pH to fall below 7.35 or to rise above 7.45. These conditions are called *acidosis* and *alkalosis*, respectively.

Normally, the rate of ventilation is matched to the rate of CO_2 production by the tissues, so that the carbonic acid, bicarbonate, and H^+ concentrations in the blood remain within the normal range. If **hypoventilation** occurs, however, the carbonic acid levels will rise above normal and the pH will fall below 7.35. This condition is called **respiratory acidosis** (table 8.6). **Hyperventilation,** conversely, causes an abnormal decrease in carbonic acid and a corresponding rise in blood pH. This condition is called **respiratory alkalosis.** Thus, respiratory acidosis or alkalosis occurs when the blood CO_2 level (as measured by its partial pressure or P_{CO_2} in millimeters of mercury) is different from the normal value (40 mm Hg) as a result of abnormal breathing patterns (fig. 8.22).

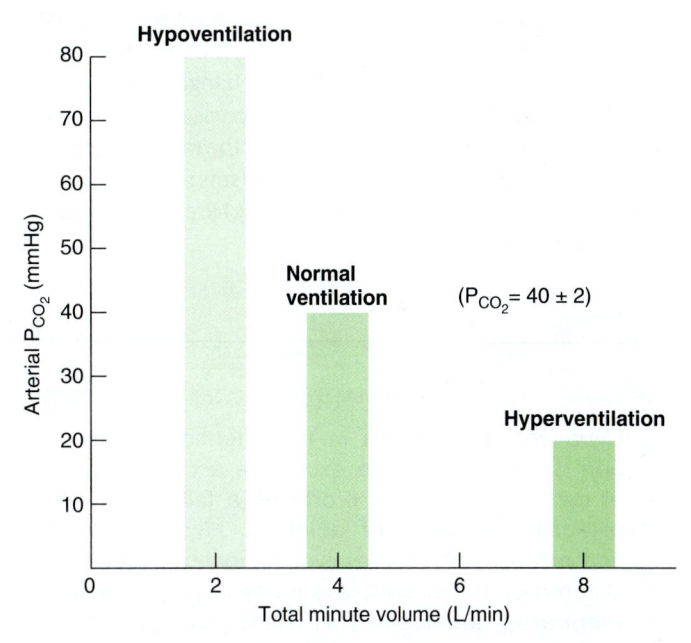

Figure 8.22 **The relationship between total minute volume and arterial P_{CO_2}.** These are inversely related. When the total minute volume increases by a factor of 2, the arterial P_{CO_2} decreases by half. The total minute volume measures breathing and is equal to the amount of air in each breath (the tidal volume) multiplied by the number of breaths per minute. The P_{CO_2} measures the CO_2 concentration of arterial blood plasma.

(For a full-color version of this figure, see fig. 16.27 in *Human Physiology,* thirteenth edition, by Stuart I. Fox.)

Procedure

1. Allow the pH meter to warm up by setting the selector switch to the *standby* position. Be sure that the pH electrodes are immersed in buffer and are not allowed to dry. Verify that the temperature selector switch is set at the current room temperature.
2. Turn the selector switch to *pH* and take a reading of the buffer. Use the calibration knob to set the pH meter to the correct pH of the buffer (7.000). Now, turn the selector switch back to the *standby* position and transfer the pH electrodes to a beaker containing 50 mL of distilled water. Turn the selector switch to *pH* and record the pH of distilled water. Return the selector switch to the *standby* position and the electrodes back to the buffer.
3. Add one drop of concentrated hydrochloric acid (HCl) to the beaker of distilled water and mix thoroughly. Transfer the electrodes to this solution, turn the selector switch to the *pH* position, and record the pH of the solution in your laboratory report.

Note: *After recording the pH of a solution, always turn the selector switch to standby and use a squeeze bottle of distilled water to rinse the electrodes thoroughly. Wipe the electrodes with lint-free paper and return them to the buffer solution. Check the pH of the buffer solution after the cleaning procedure to be sure you have adequately cleaned the electrodes.*

4. Add one drop of concentrated NaOH to a fresh beaker containing 50 mL of distilled water, and record the pH of the water before and after adding the NaOH.

5. Add one drop of concentrated HCl to a beaker containing standard buffer solution (pH = 7.000).
6. Add one drop of concentrated NaOH to a fresh beaker of standard buffer solution (pH = 7.000).
7. Add three drops of concentrated HCl to a beaker of fresh standard buffer solution (pH = 7.000).
8. Add three drops of concentrated NaOH to a beaker of fresh standard buffer solution (pH = 7.000).

B. Effect of Exercise on the Rate of CO_2 Production

Increased muscle metabolism during exercise results in an increase in CO_2 production. Despite this, the CO_2 levels and pH of arterial blood do not normally change significantly during exercise. This is because the increased rate of CO_2 production is matched by an increase in the rate of its elimination through ventilation. The mechanisms responsible for exercise *hyperpnea* (increased breathing) are complex and incompletely understood.

Clinical Applications

Hypoventilation results in the retention of carbon dioxide and in the excessive accumulation of carbonic acid; this produces a fall in blood pH called **respiratory acidosis**. *Hyperventilation* results in the excessive elimination of CO_2, lowered carbonic acid, and a rise in pH causing **respiratory alkalosis**. This differs from the normal hyperpnea (increased total minute volume) that occurs during exercise, where increased respiration matches increased CO_2 production so that the arterial CO_2 levels and pH remain in the normal range.

Procedure

1. Fill a beaker with 200 mL of distilled water and add 5.0 mL of 0.10N NaOH and a few drops of phenolphthalein indicator. This indicator is pink in alkaline solutions and clear in neutral or acidic solutions. Divide this solution into two beakers.
2. While sitting quietly, exhale through a glass tube or straw (or double straws) into the solution in the first beaker. Carefully record the time required to turn the solution from pink to clear in your laboratory report.
3. Exercise vigorously for 2 to 5 minutes by running up and down stairs or by doing jumping jacks. Exhale through a glass tube or straw (or double straws) into the second beaker, and again record the time it takes to clear the pink solution.

C. Role of Carbon Dioxide in the Regulation of Ventilation

The carbon dioxide concentration of the blood reflects a balance between the rate of its production (by aerobic cell respiration) and the rate of its elimination through the lungs. When a person consciously holds his or her breath for a sufficiently long time, the carbon dioxide level rises (and the pH falls) to such an extent that reflex breathing occurs. Remember that the chemoreceptor control of breathing is usually stimulated by a rise in blood CO_2 and consequent fall in pH, rather than by a decrease in blood O_2. It is the fall in blood pH that stimulates the peripheral chemoreceptors (aortic and carotid bodies) and the fall in CSF and brain interstitial fluid pH that stimulates the chemoreceptors in the medulla oblongata.

During hyperventilation, conversely, the arterial CO_2 falls and blood pH rises. As a result, the chemoreceptor drive to breathe is inhibited. When breathing is reduced, the blood CO_2 is allowed to rise until it reaches the level where it will again stimulate breathing.

Procedure

1. Count the number of breaths you take in 1 minute of relaxed, unforced breathing. Enter this number in your laboratory report.
2. Force yourself to hyperventilate for about 10 seconds; stop if you begin to feel dizzy.
3. Immediately after hyperventilation, count the number of breaths you take in 1 minute of relaxed, unforced breathing.

Laboratory Report 8.4

Name _____

Date _____

Section _____

DATA FROM EXERCISE 8.4

A. Ability of Buffers to Stabilize the pH of Solutions

1. Enter your data in these spaces:

    pH of distilled water:

    pH of water + 1 drop HCl: _____

    pH of water + 1 drop NaOH: _____

    pH of buffer: 7.000

    pH of buffer + 1 drop HCl: _____

    pH of buffer + 1 drop NaOH: _____

    pH of buffer + 3 drops HCl: _____

    pH of buffer + 3 drops NaOH: _____

2. Does your data support the statement that "buffers help to stabilize the pH of solutions?" Explain your answer.

B. Effect of Exercise on the Rate of CO_2 Production

1. Enter your data in these spaces:

    Time for color change at rest: _____

    Time for color change after exercise: _____

2. Explain your results.

C. Role of Carbon Dioxide in the Regulation of Ventilation

1. Enter your data in these spaces:

    Rate of breathing at rest: _____ breaths/min

    Rate of breathing after hyperventilation: _____ breaths/min

2. Explain your results.

REVIEW ACTIVITIES FOR EXERCISE 8.4

Test Your Knowledge

1. A solution with a H^+ concentration of 10^{-9} molar has a pH of _____.
2. Hypoventilation produces a condition called respiratory _____; hyperventilation produces a condition called respiratory _____.
3. Define:
   (a) *acid* _____
   (b) *base* _____
   (c) *acidosis* _____
   (d) *alkalosis* _____
4. What is the normal measurement of arterial carbon dioxide levels? _____ mm Hg
5. The free bicarbonate in the plasma serves as the major _____ of the blood.
6. The enzyme in red blood cells that catalyzes the formation of carbonic acid is _____.

Test Your Understanding

7. Draw equations to show how carbon dioxide affects the blood concentration of H^+ (and thus the pH) and the blood concentration of HCO_3^-. Indicate the directions of change in these values if blood carbon dioxide levels were to rise.

8. Use the equations shown in question 7 to explain how hyperventilation and hypoventilation affect the blood pH.

9. Describe how your breathing rate changed following 10 seconds of hyperventilation. Explain the physiological mechanisms responsible for the changed breathing pattern.

Test Your Analytical Ability

10. People who hyperventilate may get dizzy (due to cerebral vasoconstriction), causing anxiety and further hyperventilation. Such people are sometimes urged to breathe into a paper bag. What good would this do? Explain the physiological mechanisms involved.

11. Intravenous infusions of sodium bicarbonate are often given to acidotic patients to correct the acidosis and relieve the strain of rapid breathing. Explain why bicarbonate is helpful in this situation. What would happen if too much bicarbonate were given? Explain.

Clinical Investigation Questions

12. How do the arterial P_{CO_2}, plasma bicarbonate concentration, and blood pH relate to each other? What diagnosis do the patient's measurements indicate?

13. How might the patient's history of COPD relate to the measured values of an arterial P_{CO_2} of 50 mm Hg; plasma bicarbonate of 24 mEq/L; and a blood pH of 7.30?

Section 9

Renal Function and Homeostasis

The kidneys are responsible for the elimination of most of the waste products of metabolism. These wastes include urea and creatinine (derived from protein catabolism) and ketone bodies (derived from fat catabolism). The kidney must also retain (or *reabsorb*) molecules essential for normal body function, such as glucose, amino acids, and bicarbonate.

Through these actions, the kidneys are involved in maintaining a constant internal environment (homeostasis), including the regulation of *electrolyte concentrations, fluid balance,* and *acid-base balance.* The fluid volume of the blood is maintained by the reabsorption of 98% to 99% of the water that leaves the blood in the initial step of urine formation, while the electrolyte and pH balance of the blood is maintained by the selective reabsorption of such ions as Na^+, K^+, and HCO_3^-.

Each kidney contains approximately one million functional units called **nephrons** (fig. 9.1c). Each nephron is composed of two parts: (1) the *glomerulus,* a tightly woven, highly permeable capillary bed at the end of an arteriole; and (2) the *renal tubule,* a bent and convoluted tubule composed of epithelial cells. Each glomerulus is enveloped by the mouth of a tubule, termed the **glomerular capsule (Bowman's capsule).** The last part of the renal tubule (the collecting duct) empties its contents into the renal pelvis as urine, which is then funneled into the ureters (fig. 9.1).

The formation of urine in the nephron occurs in two stages. (1) The hydrostatic pressure of the blood forces fluid out of the capillary wall of the glomerulus, producing a *filtrate* (or *ultrafiltrate*) of blood. Except for proteins, usually too large to enter the glomerular capsule, the glomerular filtrate contains the same solute molecules as plasma and is isotonic to plasma. (2) As the glomerular filtrate passes through the renal tubules, the cells of the tubules selectively reabsorb and secrete solute molecules and ions. The solution that emerges from the collecting duct is urine, which is very different in composition and concentration from the glomerular filtrate that enters the tubule.

Exercise 9.1	Renal Regulation of Fluid and Electrolyte Balance
Exercise 9.2	Renal Plasma Clearance of Urea
Exercise 9.3	Clinical Examination of Urine

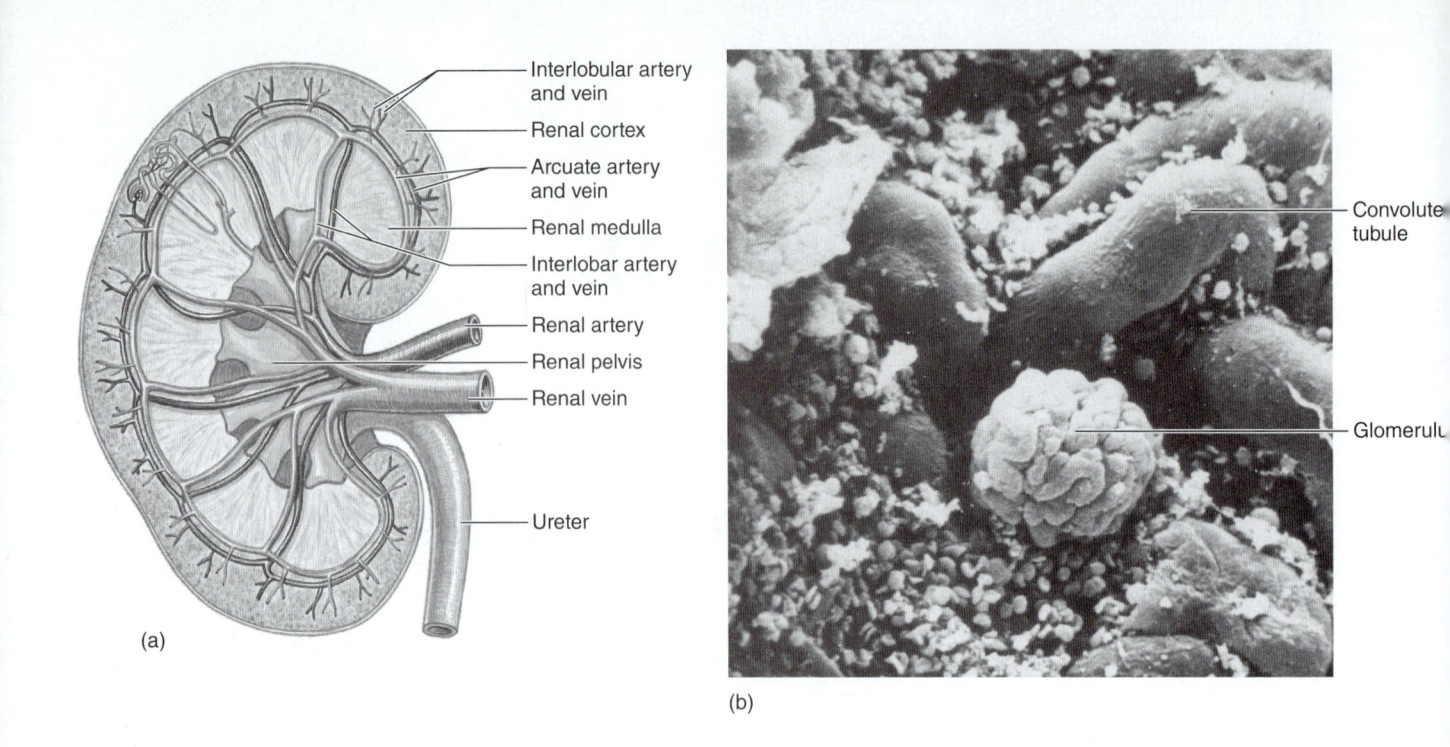

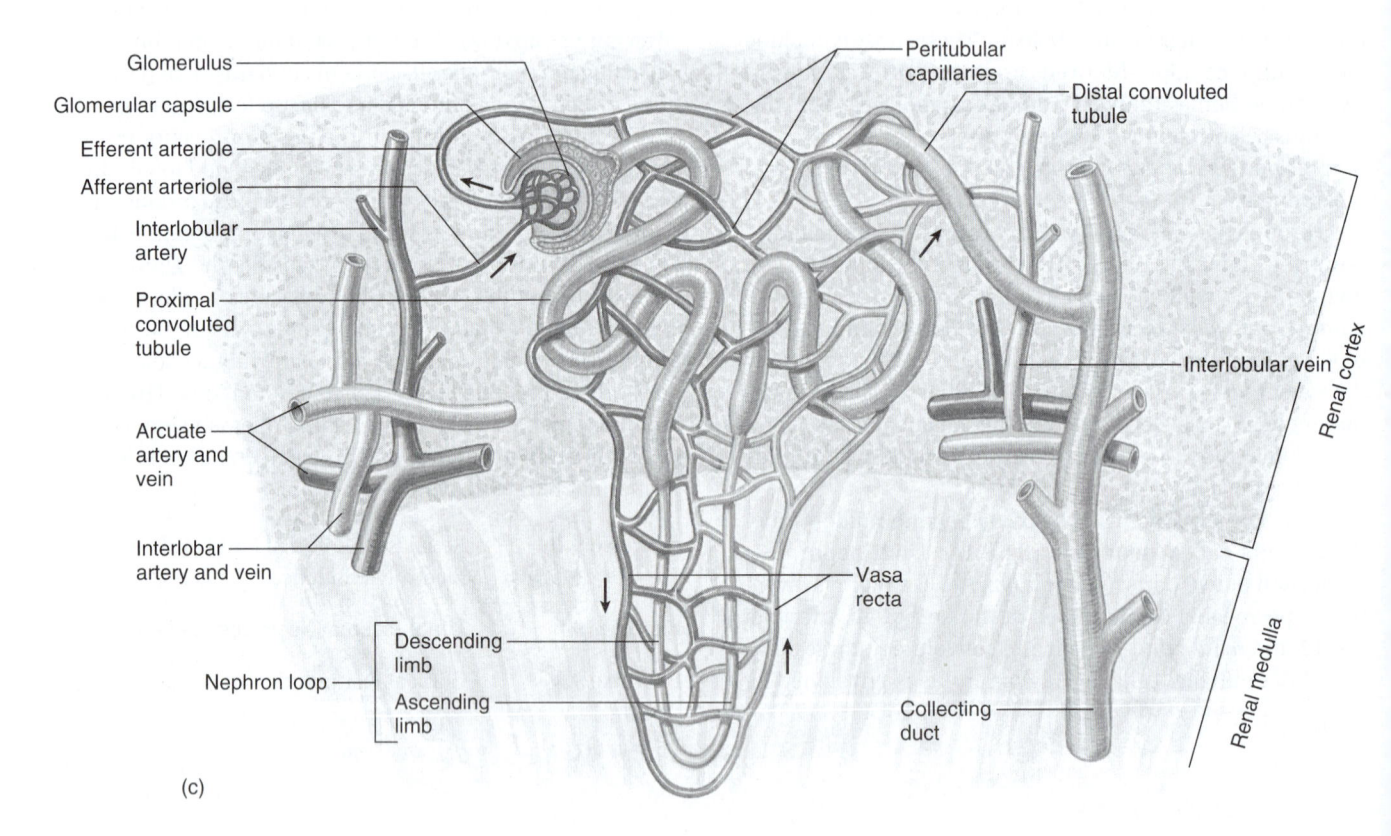

Figure 9.1 Structure of the kidney.   (a) A diagram of a sectioned kidney, illustrating the arrangement of blood vessels. (b) A scanning electron micrograph of glomeruli and tubules. (c) The tubules and associated blood vessels that compose a nephron.

(For full-color versions of these figures, see figs. 17.4 and 17.5 in *Human Physiology*, thirteenth edition, by Stuart I. Fox.)

EXERCISE 9.1

Renal Regulation of Fluid and Electrolyte Balance

MATERIALS

1. Urine collection cups
2. Urinometers and droppers

Note: *Specific gravity can also be measured visually using disposable urine dip strips (Miles Inc., Curtin Matheson Scientific Inc.) or Multistix (Ames Laboratories).*

3. pH paper (pH range 3–9), potassium chromate (20 g per 100 mL), silver nitrate (2.9 g per 100 mL)
4. NaCl crystals or salt tablets
5. Alternatively, normal and abnormal artificial urine is available (Wards Biology). However, modifications must be made by the instructor to simulate the conditions in this exercise.

Urine volume, solute concentration, and electrolyte content are adjusted by the kidneys to maintain homeostasis of the blood. Drinking excess water or eating salty foods results in a rising blood volume, followed by compensatory increases in the urinary excretion of the salt and water.

LEARNING OUTCOMES

You should be able to:

1. Describe the roles of ADH and aldosterone in the regulation of fluid and electrolyte balance.
2. Calculate the concentration of ions in solution (in milliequivalents per liter).
3. Demonstrate and explain how the kidneys respond to water and salt loading by changes in urinary volume, specific gravity, pH, and electrolyte composition.

Textbook/Multimedia Correlations

Before performing this exercise, you should study the introductory material presented here. Further information relating to this exercise can be found in these pages of *Human Physiology,* thirteenth edition, by Stuart I. Fox:

- *Reabsorption of Salt and Water.* Chapter 17, p. 590.
- *Renal Control of Electrolyte and Acid-Base Balance.* Chapter 17, p. 604.
- Multimedia correlations are provided in *MediaPhys 3.0:* Topics 11.46–11.56.

Clinical Investigation

A patient's hypertension was usually well controlled with only an ACE inhibitor, a drug that blocks the action of angiotensin converting enzyme. However, for a period of time she had to combine this with a diuretic drug to bring her blood pressure back down to the normal range. When she took the diuretic, her plasma K^+ concentration was measured at 2.6 mEq/L.

- Explain the physiological significance of angiotensin converting enzyme, how an ACE inhibitor would affect plasma and urine Na^+, and how an ACE inhibitor could help control hypertension.
- Explain the significance of this patient's plasma K^+ concentration measurement.

The reabsorption of fluid and electrolytes (ions) into the blood from the renal tubular filtrate is adjusted to meet the needs of the body by the action of hormones. The major hormones involved in this process are **antidiuretic hormone (ADH),** released by the posterior pituitary gland, and **aldosterone,** secreted by the cortex of the adrenal gland.

Figure 9.2 Homeostasis of plasma concentration is maintained by ADH. In dehydration (left side of figure), a rise in ADH secretion results in a reduction in the excretion of water in the urine. In overhydration (right side of figure), the excess water is eliminated through a decrease in ADH secretion. These changes provide negative feedback correction, maintaining homeostasis of plasma osmolality and, indirectly, blood volume.

(For a full-color version of this figure, see fig. 17.20 in *Human Physiology*, thirteenth edition, by Stuart I. Fox.)

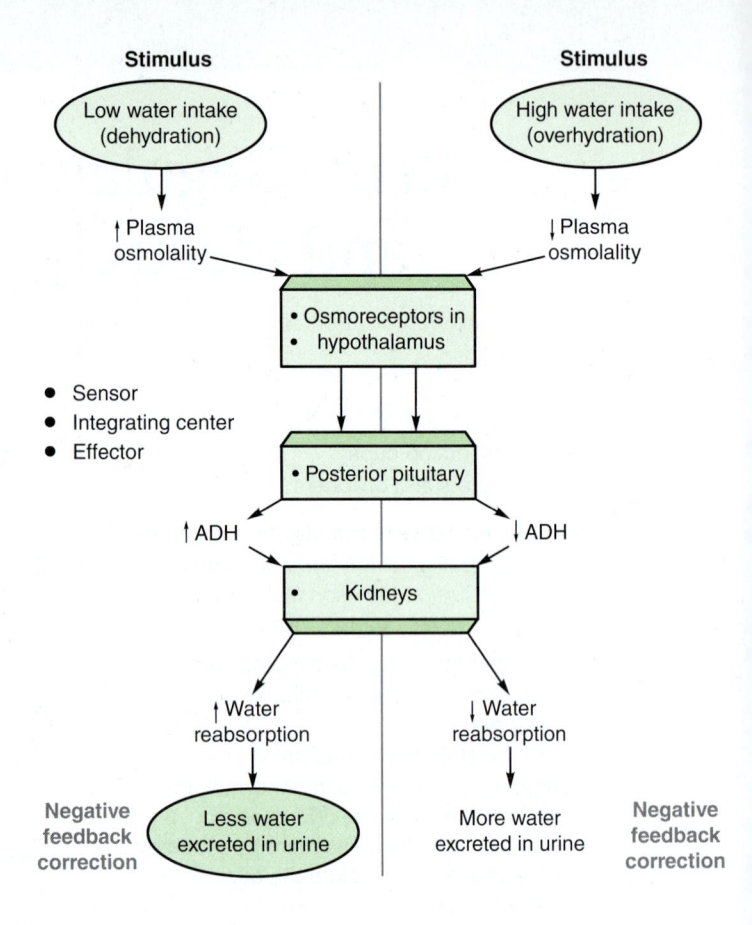

- Sensor
- Integrating center
- Effector

The release of antidiuretic hormone by the posterior pituitary is regulated by osmoreceptors in the hypothalamus (fig. 9.2). These receptors are stimulated by an increase in the osmotic pressure of the blood, as might occur in dehydration. The ADH released in response to this stimulus promotes the reabsorption of water from the renal tubules, resulting in (1) the retention of water and therefore a decrease in the osmotic pressure of the blood back to the normal level, and (2) the excretion of a small volume of highly concentrated (hypertonic) urine.

Clinical Applications

Diabetes insipidus may be caused by (1) drinking too much water *(polydipsia)*; (2) inadequate secretion of ADH; or (3) inadequate ADH action due to a genetic defect in either the ADH receptors or the aquaporin channels. Without adequate ADH secretion or action, the collecting ducts are not very permeable to water. This results in the excretion of a large volume of dilute urine—greater than 3 L per day, and often 5–10 L per day. Excretion of so much water can cause dehydration, but symptoms are usually absent if the person drinks a sufficient amount of water.

The secretion of aldosterone by the adrenal cortex may be stimulated by an increase in blood K^+ or by a decrease in blood Na^+ or blood volume. An increase in blood K^+ directly stimulates the adrenal cortex to secrete aldosterone. A decrease in blood Na^+ or blood volume indirectly affects aldosterone secretion by stimulating the kidneys to secrete the enzyme **renin** into the blood (fig. 9.3). Renin catalyzes the formation of *angiotensin I* from *angiotensinogen*, a plasma protein produced by the liver. Angiotensin I is changed to **angiotensin II** by *angiotensin converting enzyme (ACE)*. Angiotensin II: (1) stimulates vasoconstriction, thus increasing the blood pressure, and (2) stimulates the secretion of aldosterone from the adrenal cortex. Aldosterone then promotes the reabsorption of Na^+ from the glomerular filtrate into the blood, in exchange for K^+ secreted from the blood into the renal tubules. When Na^+ ions are reabsorbed, water follows passively owing to the osmotic gradient created. In this way, aldosterone prompts a rise in blood Na^+ and volume, correcting the original deviation and maintaining homeostasis.

The secretion of K^+ into the nephron tubule (specifically into the late distal tubule and cortical collecting duct) matches the amount of K^+ ingested in the diet. As a result, the blood K^+ concentration remains in the normal range. When a person eats a K^+-rich meal, the rise in blood K^+ stimulates the adrenal cortex to secrete aldosterone. Aldosterone then stimulates an increase in the secretion of K^+ into the filtrate, as it reabsorbs Na^+ from the filtrate. In addition to this aldosterone-dependent K^+ secretion, there is also an aldosterone-independent K^+ secretion. In this process, the rise in blood K^+ directly causes

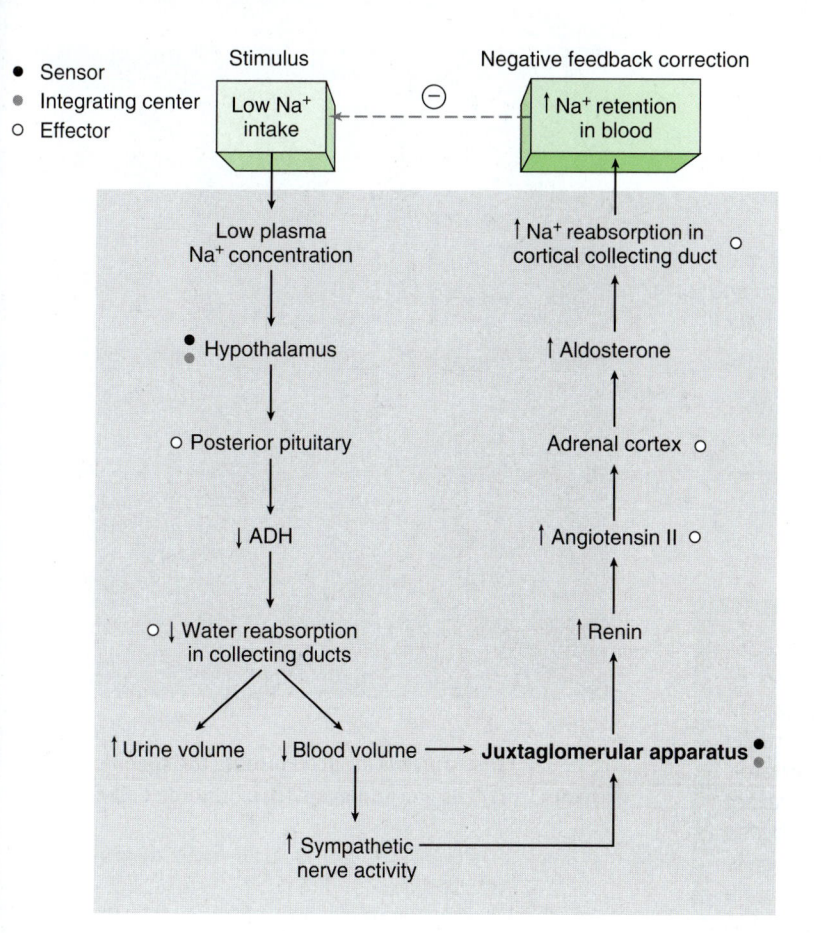

Figure 9.3 Homeostasis of plasma Na⁺.
This is the sequence of events by which a low sodium (salt) intake leads to increased sodium reabsorption by the kidneys. The dashed arrow and negative sign indicate the completion of the negative feedback loop.

(For a full-color version of this figure, see fig. 17.27 in *Human Physiology*, thirteenth edition, by Stuart I. Fox.)

additional K⁺ channels to become inserted into the membrane of the cortical collecting duct. When the blood K⁺ falls, those K⁺ channels are removed from the membrane by endocytosis and K⁺ secretion is thereby reduced. We can thus safely eat bananas and other potassium-rich foods without worry about developing hyperkalemia (high blood potassium—see the next clinical applications box).

MILLIEQUIVALENTS

The concentrations of ions (electrolytes) in body fluids are usually given in terms of milliequivalents (mEq) per liter. As an example of the meaning and significance of this unit of measurement, we will consider the chloride (Cl⁻) concentration of the urine.

Suppose a urine sample had a chloride concentration of 610 mg per 100 mL. How does this number of ions and this number of charges compare with the number of other ions and charges present in the urine? To determine this, we must first convert the chloride concentration from milligrams per 100 mL to millimoles per liter.

Example
The atomic weight of chloride is 35.5. Therefore,

$$\frac{610 \text{ mg of Cl}^-}{100 \text{ mL}} \times \frac{1 \text{g}}{1{,}000 \text{ mg}} \times \frac{1{,}000 \text{ mL}}{1 \text{ L}} \times \frac{1 \text{ mole}}{35.5 \text{ g}}$$

$$= 0.171 \text{ M} \times \frac{1{,}000 \text{ mM}}{1 \text{ M}} = 171 \text{ mM}$$

One mole of chloride has the same number of ions as 1 mole of Na⁺ or 1 mole of Ca²⁺ or 1 mole of anything else. One mole of Ca²⁺, however, has twice the number of charges (*valence*) as 1 mole of Cl⁻. Therefore, 2 moles of Cl⁻ are required to neutralize 1 mole of Ca²⁺. If charges are taken into account by multiplying the moles by the valence, the product is termed the **equivalent weight** of an ion. One-thousandth of the equivalent weight dissolved in 1 liter of solution gives a concentration of *milliequivalents per liter (mEq/L)*.

Example
$$171 \text{ mM} \times 1 \text{ (the valence of Cl}^-)$$
$$= 171 \text{ mEq/L of Cl}^-$$

The major advantage of expressing the concentrations of ions in milliequivalents per liter is that the total concentration of anions can be easily compared with the total concentration of cations. In an average sample of venous plasma, for example, the total anions and the total cations are each equal to 156 mEq/L. Chloride, the major anion, has a plasma concentration of 103 mEq/L, whereas the chloride concentration in the urine is highly variable, ranging from 61 to 310 mEq/L.

Procedure

1. The students void their urine into collection cups at the beginning of the laboratory session. In the analyses done in step 4, this sample will serve as the control (time zero).

2. The students drink 500 mL of water. One group just drinks the water; another group ingests NaCl (salt tablets are easiest to take) in addition to drinking the water. Most people can tolerate up to 4.5 g of salt, but the amount ingested should not be so great as to cause nausea.

> **Note:** *Students with hypertension or on sodium-restricted diets should not perform the salt-loading exercise.* Interestingly, 4.5 g/500 mL of NaCl is a 0.9 g/100 mL saline solution isotonic to plasma (normal saline).

3. After drinking the solutions described in step 2, the students void their urine every 30 minutes for 2 hours. The urine samples are analyzed as described in step 4.

Clinical Applications

Due in large part to the effects of ADH and aldosterone, the kidneys can vary their excretion of water and electrolytes to maintain homeostasis of the blood volume and composition. Abnormally low blood volume can produce **hypotension** (low blood pressure) and may result in circulatory shock; abnormally high blood volume contributes to **hypertension.** Renal regulation of Na^+ balance is also critical for health. Changes in blood Na^+ cause secondary changes in blood volume, as water follows sodium by osmosis. Changes in blood K^+ affect the bioelectrical properties of all cells, but the effects on the heart are particularly serious. **Hyperkalemia** (high blood K^+) is usually fatal when the K^+ concentration rises from normal (3.5 to 5.0 mEq/L) to over 10 mEq/L. This may be caused by a variety of conditions, including inadequate aldosterone secretion (in Addison's disease), or by an excessive intake of potassium. **Hypokalemia** is a plasma K^+ concentration less than 3.5 mEq/L. Moderate hypokalemia (a K^+ concentration between 2.5 and 3.5 mEq/L) is not uncommon and often results from people taking diuretic drugs.

4. Each of the five urine samples collected are analyzed for pH, specific gravity, and chloride content as follows:
   (a) **Volume (mL).** Measure the approximate volume of urine obtained, and enter the data in the table of the laboratory report.
   (b) **pH.** Determine the pH of the urine samples by dipping a strip of pH paper into the urine and matching the color developed with a color chart. The urine normally has a pH between 5.0 and 7.5.
   (c) **Specific gravity.** Determine the specific gravity of the urine samples by floating a urinometer in a cylinder (fig. 9.4) nearly filled with the specimen. Read the specific gravity at the meniscus on the urinometer scale, making sure that the urinometer float is not touching the bottom or the sides of the cylinder. The specific gravity is directly related to the amount of solutes in the urine and ranges from 1.005 to 1.035. (Pure water should have a specific gravity of 1.000.)
   (d) **Chloride concentration.** When Na^+ is reabsorbed by the renal tubules, Cl^- follows passively by electrostatic attraction. Follow these steps to determine the chloride concentration of the urine samples:
   (1) Measure 10 drops of urine into a test tube (1 drop is approximately 0.05 mL).
   (2) To this tube add 1 drop of 20% potassium chromate solution with a second dropper.
   (3) Add 2.9% silver nitrate solution one drop at a time using a third dropper, while shaking the test tube continuously. Count the number of full drops required to cause a permanent change in the color of the solution from yellow to brown.
   (4) Determine the chloride concentration of the urine sample. Because each drop of 2.9% silver nitrate added in step 3 is equivalent to 61 mg of Cl^- per 100 mL of urine, multiply the number of drops by 61 to obtain the chloride concentration of the urine in milligrams per 100 mL.

Figure 9.4 Instruments for determining the specific gravity of urine. (a) A glass cylinder and (b) a urinometer float.

Example
If 10 drops of 2.9% silver nitrate were required,

$$10 \times 61 \text{ mg of } Cl^-/100 \text{ mL} = 610 \text{ mg } Cl^-/100 \text{ mL}$$

5. Convert the chloride concentration to mEq/L, and enter your data in the appropriate table in the laboratory report.

Laboratory Report 9.1

Name _____
Date _____
Section _____

DATA FROM EXERCISE 9.1

1. Enter your data in the appropriate table.
   (a) Ingestion of water only

Time	Volume (mL)	pH	Specific Gravity	Chloride (mEq/L)
0				
30				
60				
90				
120				

   (b) Ingestion of water and NaCl

Time	Volume (mL)	pH	Specific Gravity	Chloride (mEq/L)
0				
30				
60				
90				
120				

REVIEW ACTIVITIES FOR EXERCISE 9.1

Test Your Knowledge

1. Osmoreceptors in the hypothalamus of the brain are stimulated by a(n) _____ (increase/decrease) in the plasma osmotic pressure.

2. As a result of stimulation, the osmoreceptors stimulate the secretion of _____ from the _____ gland.

3. The hormone named in question 2 specifically stimulates the kidneys to _____.

4. The hormone that stimulates the reabsorption of Na^+ from the nephron tubules, and also stimulates the secretion of K^+ into the tubules, is _____.

5. The substance that stimulates vasoconstriction and also stimulates the secretion of the hormone named in question 4: _____.

6. The measurement that is 1.000 for pure water and that increases in proportion to the general solute concentration of a solution: _____.

Test Your Understanding

7. Describe how the specific gravity of the urine changed after drinking water, and explain the physiological mechanisms responsible for this change.

8. Describe how ADH secretion and action in the kidneys help to maintain homeostasis of blood volume and concentration.

9. Identify the sites of action of aldosterone in the renal nephrons, and relate these to the actions of aldosterone on the kidney. Explain how these actions help to maintain the homeostasis of (a) blood sodium concentration and volume and (b) blood potassium concentration.

Test Your Analytical Ability

10. Imagine a dehydrated desert prospector and a champagne-quaffing partygoer, each of whom drinks a liter of water at time zero and voids urine over a period of 3 hours. Using their urine samples, compare the probable differences in volume and composition. (*Hint:* Alcohol inhibits ADH secretion.) Use relative terms, such as *increased* or *decreased*.

Observation	Prospector	Partygoer
Urine volume		
Specific gravity		
Na⁺ and Cl⁻ Content		

(a) Explain the answers you provided in this table.

11. Many diuretic drugs used clinically inhibit Na⁺ reabsorption in the nephron loop. Predict the effect of these drugs on the urinary excretion of Cl⁻ and K⁺, and explain your answer.

12. Endurance athletes such as marathon runners often ingest salt along with water and/or sports drinks containing salt. Explain the physiological mechanisms by which this would improve hydration.

Test Your Quantitative Ability

13. Calcium is normally present at a concentration of about 0.1 g/L. Calculate the mEq/L of Ca^{2+} in the plasma (the atomic weight of calcium is 40).

14. If it required 7 drops of silver nitrate in procedure step *4d* to turn the urine sample color from yellow to amber, what is the chloride concentration, expressed in mEq/L units, of the urine?

Clinical Investigation Questions

15. What is the action of ACE, and what effect would an ACE inhibitor have on urine Na^+? What effect would it have on plasma Na^+, plasma volume, and blood pressure? Explain.

16. What condition does this patient's plasma K^+ measurement indicate? What could she do to treat that condition?

EXERCISE 9.2

Renal Plasma Clearance of Urea

MATERIALS

1. Pipettes with mechanical pipettors (or Repipettes), mechanical microliter pipettes (capacity µL), disposable tips
2. Colorimeter and cuvettes
3. BUN reagents and standard (Stanbio, through Curtin Matheson Scientific, Inc.)
4. Sterile lancets and 70% alcohol
5. File
6. Microhematocrit centrifuge and heparinized capillary tubes
7. Container for the disposal of blood-containing objects
8. As an alternative, normal and abnormal artificial urine is available (Wards Biology). However, modifications must be made by the instructor to simulate the conditions in this exercise.

Urea and other waste products in the plasma are filtered by the kidneys and excreted in the urine. The efficiency of the kidneys in performing these processes for each solute excreted is measured by its renal plasma clearance.

LEARNING OUTCOMES

You should be able to:

1. Describe the chemical nature of urea, and explain its physiological significance.
2. Define renal plasma clearance, and explain how this value is calculated.
3. Perform a renal plasma clearance measurement for urea, and explain the physiological significance of this measurement.
4. Explain how the renal plasma clearance for a solute is affected by filtration, reabsorption, and secretion.

Textbook/Multimedia Correlations

Before performing this exercise, you should study the introductory material presented here. Further information relating to this exercise can be found in these pages of *Human Physiology*, thirteenth edition, by Stuart I. Fox:

- *Renal Plasma Clearance.* Chapter 17, p. 598.

Clinical Investigation

A woman on a high-protein diet had routine blood tests that suggested normal liver function. Her BUN was 27 mg/dL, and her BUN:creatine ratio was normal.

- Explain the significance of the BUN and plasma creatine concentration measurements.
- Identify the most likely explanation for her BUN measurement.

When amino acids are broken down in the process of cellular respiration, or when they are converted to glucose (a process known as gluconeogenesis), the amine ($-NH_2$) groups are removed and secreted into the blood in the form of **urea.** This function is performed by the liver.

$$NH_2-\underset{\underset{O}{\|}}{C}-NH_2$$
Urea

The urea is filtered by the glomeruli and enters the renal tubules. Although urea is a waste product of amino acid metabolism, some of it is transported (through facilitative diffusion) out of the nephron tubules by specific carriers. This increases the urea concentration in the interstitial fluid of the renal medulla, thereby contributing to the hypertonicity of the renal medulla. Because urea is reabsorbed after filtration, only 60% of the blood filtered by the glomeruli is cleared of urea. The average glomerular filtration rate (GFR) is 125 mL/min (for both kidneys), so this amounts to an average of 75 mL of plasma cleared of urea per minute. This value is termed the **renal plasma clearance** of urea (fig. 9.5).

The renal plasma clearance of a substance depends on the size of the kidneys and the rate of urine production, as well as the other factors discussed. In this exercise, it will be assumed that the kidneys are of average size and that urine production is equal to or greater than 2.0 mL/min. Under these conditions, the clearance of urea can be calculated using the formula

$$\text{Clearance} = \frac{U \times V}{P}$$

345

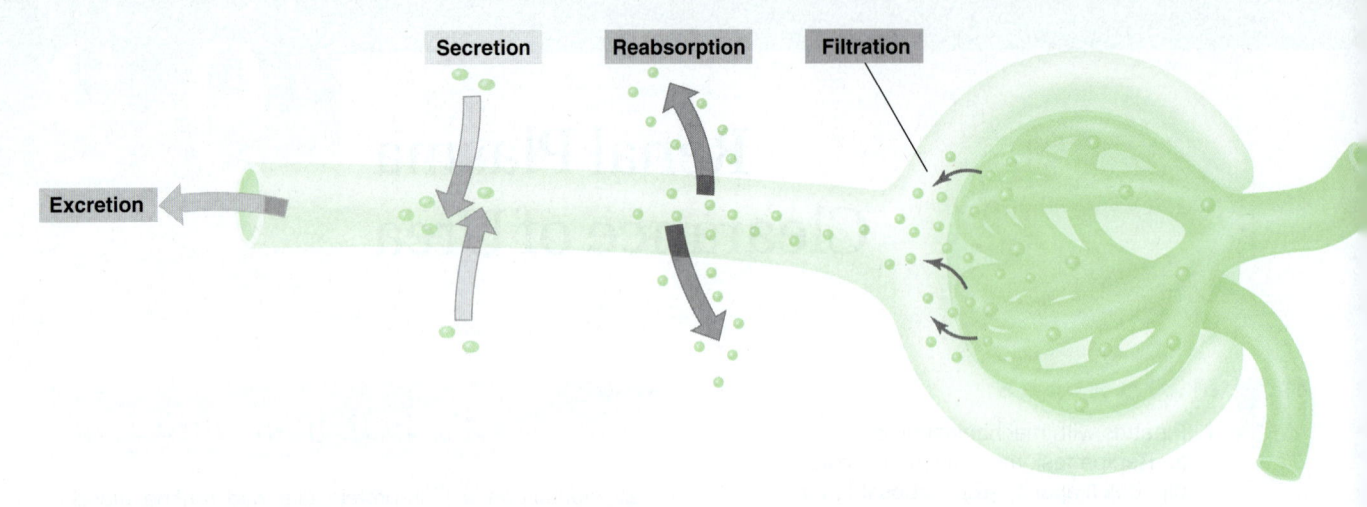

Figure 9.5   Filtration, reabsorption, and secretion by the nephron.   Urea is filtered and about 40% of the amount filtered is passively reabsorbed (there is no secretion of urea). Therefore, approximately 60% of the amount filtered is excreted in the urine. (For a full-color version of this figure, see fig. 17.21 in *Human Physiology,* thirteenth edition, by Stuart I. Fox.)

where:

U is the concentration of urea in the urine, in milligrams per 100 mL

P is the concentration of urea in the plasma, in milligrams per 100 mL

V is the urine excreted in milliliters per minute

The concentration of a substance in urine, *U*, multiplied by the volume, *V*, of urine produced per minute gives the milligrams of the substance excreted in the urine per minute. When this figure is divided by the concentration, *P*, of that substance in the plasma, the result indicates the volume of plasma that contained the amount of the substance excreted per minute. This is the amount of plasma "cleared" by passage through the kidneys per minute.

If the substance is filtered from the glomeruli but is not reabsorbed or secreted (as with the polysaccharide *inulin*), the renal clearance equals the glomerular filtration rate (GFR) (fig. 9.6). If a substance is filtered but then reabsorbed into the blood (as with glucose, amino acids, urea, and many other substances), the renal plasma clearance must be less than the glomerular filtration rate. (How much less depends on the degree of reabsorption.)

If a substance enters the renal tubules both by filtration and by active transport from the capillaries into the nephron tubules (a process called **secretion**), the renal plasma clearance is greater than the glomerular filtration rate. This is the case with the substance *para-aminohippuric acid (PAH),* almost entirely removed or "cleared" from the blood in a single passage through the kidneys. The antibiotic *penicillin* is likewise secreted into the filtrate and eliminated in the urine, and thus is rapidly cleared from the blood by the kidneys.

This information can be summarized (see figure 9.5) as follows: the rate at which a substance in the plasma is excreted in the urine is equal to the rate at which it enters the filtrate (by filtration and secretion) minus the rate at which it is reabsorbed from the filtrate. This is shown in the following equation:

Excretion rate = (filtration rate + secretion rate) − reabsorption rate

Clinical Applications

The plasma concentration of **blood urea nitrogen (BUN)** reflects both the rate of urea formation from protein in the liver and the rate of urea excretion by filtration through the glomeruli of the kidneys. An abnormally high plasma urea level can be due to kidney diseases and kidney stones, congestive heart failure, gastrointestinal bleeding, shock, excessive dietary protein intake, and other causes.

The renal plasma clearance for urea is substantially less than the glomerular filtration rate (GFR) due to its reabsorption, so the urea clearance is not a particularly good indicator of kidney function. More clinically useful are the renal plasma clearances for exogenously administered *inulin* (a large polysaccharide) and endogenous *creatinine* (a byproduct of creatine, a molecule found primarily in muscle). A change in the BUN:creatinine ratio in the plasma can help physicians narrow the possible causes of an abnormal BUN measurement.

It follows that if a substance in the plasma is filtered (enters the filtrate in Bowman's capsule), but is neither reabsorbed nor secreted, its excretion rate must be equal to its filtration rate. This is used to measure the volume of blood plasma filtered per minute by the kidneys, called the

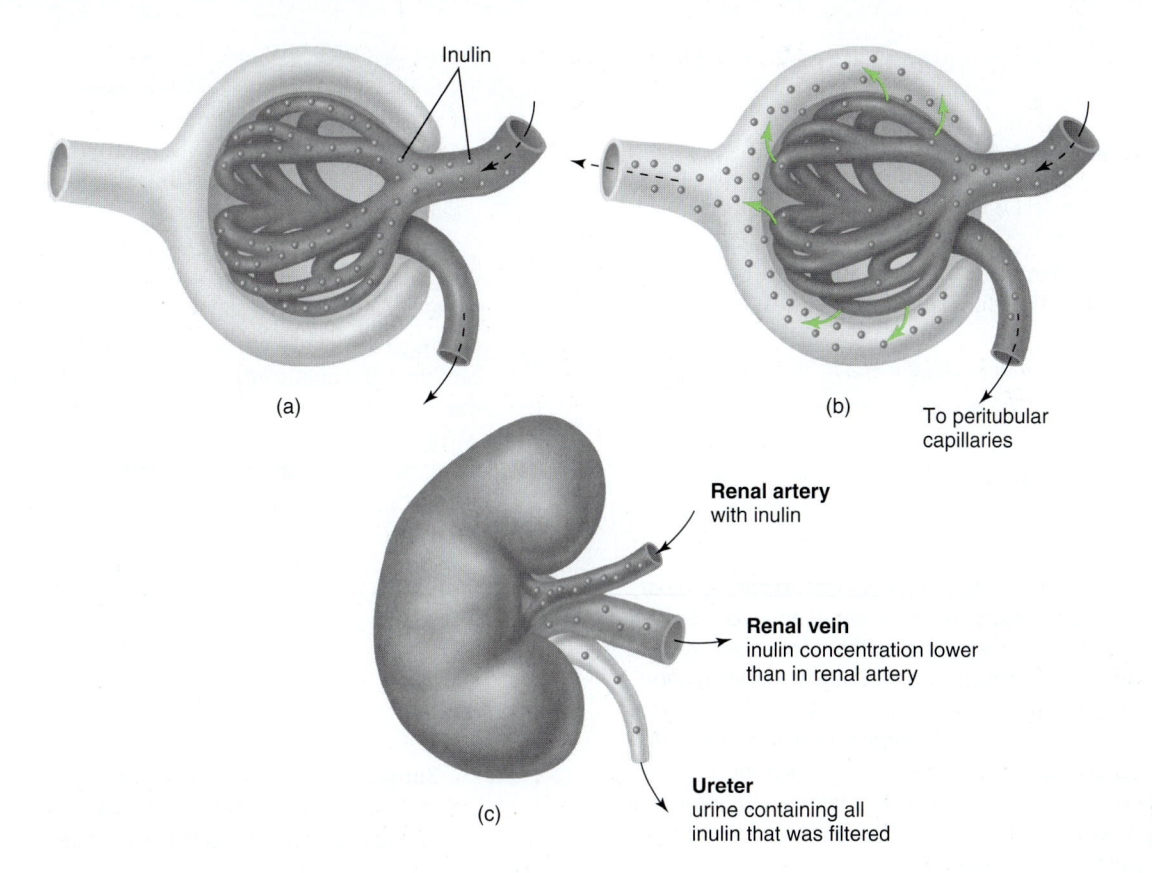

Figure 9.6 **Renal clearance of inulin.** (a) Inulin is present in the blood entering the glomeruli, and (b) some of this blood, together with its dissolved inulin, is filtered. All of this filtered inulin enters the urine, whereas most of the filtered water is returned to the vascular system (is reabsorbed). (c) The blood leaving the kidneys in the renal vein, therefore, contains less inulin than the blood that entered the kidneys in the renal artery. Because inulin is filtered but neither reabsorbed nor secreted, the inulin clearance equals the glomerular filtration rate (GFR).

(For a full-color version of this figure, see fig. 17.22 in Human Physiology, *thirteenth edition, by Stuart I. Fox.)*

glomerular filtration rate (GFR). Measurement of the GFR is important in assessing the health of the kidneys and can be performed easily by measuring the renal plasma clearance of exogenously administered inulin (which is neither reabsorbed nor secreted), or of *creatinine,* an endogenous molecule that is only slightly secreted. A measurement of the blood creatinine concentration is often a routine test, because its concentration in the blood reflects the GFR. An abnormally elevated blood creatinine concentration could indicate an abnormally low GFR, which could be confirmed by performing a creatinine clearance test.

Normal Values The normal plasma concentration of urea is 5–25 mg/dL. The normal urea plasma clearance is 64–99 mL/min.

Procedure

Collection of Plasma and Urine Samples

1. Empty the urinary bladder. Then, have the subject drink 500 mL of water as quickly as is comfortable and record the time (time zero).

2. About 20 minutes later, cleanse the fingertip with 70% alcohol and, using a sterile lancet, obtain a drop of blood from the subject's fingertip. Fill a heparinized capillary tube at least halfway with blood, plug one end, and centrifuge in a microhematocrit centrifuge for 3 minutes (as in performing a hematocrit measurement—see exercise 6.1).

   Caution: *Always handle only your blood and place all objects that have been in contact with blood in the container indicated by the instructor.*

3. Score the capillary tube lightly with a file at the plasma-cell junction and break the tube at the scored mark. Expel the plasma into a labeled small beaker or test tube.

4. Collect a complete urine sample 30 minutes after drinking the 500 mL of water. Measure the milliliters of water produced in 30 minutes, divide by 30, and enter the volume of urine produced per minute in the laboratory report.

5. Dilute the urine 1:20 with water. This can be done by adding 19 mL of water to 1 mL of urine or by adding

1.9 mL of water to 0.10 mL (100 μL) of urine. Mix the diluted urine solution.

Measurement of Urea

6. Label four test tubes or cuvettes: *B* (blank), *S* (standard), *P* (plasma), and *U* (urine).
7. Pipette 2.5 mL of *Reagent A* into each tube; then pipette 20 μL of the following into the indicated tubes:

    Tube *B*: 20 μL of distilled water
    Tube *S*: 20 μL of urea standard (28 mg/dL)
    Tube *P*: 20 μL of plasma from the capillary tube
    Tube *U*: 20 μL of the 30-minute urine sample that was previously diluted to 1/20 of its original concentration

Note: *Since 20 μL is a very small volume, pipetting must be carefully performed to be accurate. If an automatic microliter pipette is used, depress the plunger several times to wet the disposable tip thoroughly in the solution before withdrawing the 20 μL sample. When delivering the sample, slowly depress the plunger to the first stop with the disposable tip against the wall of the test tube. Then pause several seconds before depressing the plunger to the second stop. Be sure to mix the test tube well to wash the sample from the test tube wall.*

8. Mix and incubate the tubes at room temperature for 10 minutes.
9. Pipette 2.5 mL of *Reagent B* into each tube and incubate for an additional 10 minutes.
10. Set the colorimeter at 600 nm, standardize with the reagent blank, and record the absorbance values of the standard, plasma, and urine samples in the data table in your laboratory report.
11. Calculate the urea concentration of the plasma (using Beer's law, as described in exercise 2.1) and enter this value *(P)* in your laboratory report.
12. Calculate the urea concentration of the diluted urine sample in the manner described in step 11. Multiply this answer by the dilution factor *(20)*, and enter this corrected value for the urea concentration of the urine *(U)* in your laboratory report.
13. Using your values for the volume of urine produced per minute *(V)*, plasma urea concentration *(P)*, and urine urea concentration *(U)*, calculate your renal plasma clearance for urea. Enter this value in your laboratory report.

Example

Suppose, V (from step 4) = 2.0 mL/min
P (from step 11) = 10 mg/dL
U (from step 12) = 375 mg/dL, then

$$\text{Renal plasma clearance} = \frac{375 \text{ mg/dL} \times 2.0 \text{ mL/min}}{10 \text{ mg/dL}}$$

Laboratory Report 9.2

Name _____
Date _____
Section _____

DATA FROM EXERCISE 9.2

1. Enter the volume of urine produced per minute: V = _____ mL/min.

2. Enter your absorbance values in this table.

Tube	Contents	Absorbance
P	Plasma	
U	Diluted urine	
S	Standard (28 mg/dL)	

3. Enter the urea concentration of the plasma *(P):* _____ mg/dL and urine *(U):* _____ mg/dL.

4. Calculate your renal plasma clearance rate for urea and enter this value: clearance = _____ mL/min.

REVIEW ACTIVITIES FOR EXERCISE 9.2

Test Your Knowledge

Identify a molecule in the plasma that is:

1. filtered, but neither reabsorbed nor secreted _____.
2. filtered and partially reabsorbed _____.
3. filtered and completely secreted _____.
4. not filtered to a significant degree _____.

Identify the substance from the description of the renal plasma clearance. The clearance is:

5. greater than zero but less than the GFR for _____.
6. equal to the GFR for _____.
7. only slightly greater than the GFR for _____.
8. equal to the total plasma flow rate to the kidneys for
   _____.

Test Your Understanding

9. Explain what is meant by the *renal plasma clearance,* and describe how it is measured.

10. What is the significance of the glomerular filtration rate (GFR) measurement? Describe two ways that this measurement may be obtained.

Test Your Analytical Ability

11. Suppose substance "A" is reabsorbed about 30%, whereas substance "B" is secreted about 30%. Explain how their renal plasma clearance values relate to each other and to the glomerular filtration rate (GFR).

12. Explain the mechanism by which each of the following conditions might produce an increase in the plasma concentration of urea: (a) increased protein catabolism; (b) decreased blood pressure, as in circulatory shock; and (c) kidney failure.

Test Your Quantitative Ability

Suppose a substance has the following values for the determination of renal plasma clearance: $V = 3.5$ mL/min; $P = 0.80$ mg/dL; $U = 125$ mg/dL.

13. Calculate the renal plasma clearance of this substance, and state how the substance is handled (in terms of filtration, reabsorption, and/or secretion) by the kidneys.

Clinical Investigation Questions

14. What can raise the BUN measurement? What can raise the plasma creatinine measurement? Which of these measurements is most indicative of the ability of the kidneys to filter the blood? Explain.

15. Given that the person's BUN:creatinine ratio was normal, what is the most likely explanation for her BUN measurement? Explain.

EXERCISE 9.3

Clinical Examination of Urine

MATERIALS

1. Microscopes
2. Urine collection cups, test tubes, microscope slides, and coverslips
3. Albustix, Clinitest tablets, Ketostix, Hemastix, Ictotest tablets, or Multistix (Ames Laboratories; Ictotest can be obtained from Hardy Diagnostics).
4. Sediment stain (such as Sternheimer-Mablin stain)
5. Centrifuge and centrifuge tubes
6. Transfer pipettes (droppers)
7. Alternatively, normal and abnormal artificial urine is available with a few test strips (Wards Biology), or cat or dog urine may be obtained from a veterinarian.

The presence of abnormally large amounts of proteins and casts in the urine can indicate damage to the glomeruli of the kidney. The presence of bacteria and a large number of white blood cells in the urine sediment indicates urinary tract infection. Abnormal concentrations of glucose, ketone bodies, bilirubin, and other plasma solutes in the urine may reflect abnormally high concentrations in the plasma.

LEARNING OUTCOMES

You should be able to:

1. Describe the physiological processes responsible for normal urinary concentrations of protein, glucose, ketone bodies, and bilirubin in the urine.
2. Describe the pathological processes that may produce abnormal solute concentrations, and explain the clinical significance of this information.
3. Describe the normal constituents of urine sediment, and explain how the microscopic examination of urine sediment can be clinically useful.

Textbook/Multimedia Correlations

Before performing this exercise, you should study the introductory material presented here. Further information relating to this exercise can be found in these pages of *Human Physiology*, thirteenth edition, by Stuart I. Fox:

- *Glomerular Filtration.* Chapter 17, p. 587.
- *Reabsorption of Glucose.* Chapter 17, p. 603.

Clinical Investigation

A student performed a clinical examination of her own urine and found trace amounts of glucose and proteins. There was also mild hematuria. She found no casts when she examined her urine sediment, but did see rod-shaped bacteria and white blood cells.

- Explain the significance of having glucose or proteins in the urine at greater than trace amounts, and possible reasons for mild hematuria.
- Describe the appearance and significance of urinary casts, and the probable reasons for seeing bacteria and white blood cells in the urine sediment.

A clinical examination of urine may provide evidence of urinary tract infection (UTI) or kidney disease. Additionally, because urine is derived from plasma, an examination of the urine provides a convenient, nonintrusive means of assessing the composition of plasma and of detecting a variety of systemic diseases. A clinical examination of the urine includes an observation of its appearance (table 9.1), tests of its chemical composition, and a microscopic examination of urine sediment.

Table 9.1 Appearance of Urine and Cause

Color	Cause
Yellow-orange to brownish green	Bilirubin from obstructive jaundice
Red to red-brown	Hemoglobinuria
Smoky red	Unhemolyzed RBCs from urinary tract
Dark wine color	Hemolytic jaundice
Brown-black	Melanin pigment from melanoma
Dark brown	Liver infections, pernicious anemia, malaria
Green	Bacterial infection (*Pseudomonas aeruginosa*)

Note: Certain foods (e.g., beets, rhubarb) and some commonly prescribed drugs and vitamin supplements may also alter the color of urine.

351

> **Clinical Applications**
>
> When the kidneys are inflamed, there is increased permeability of the filtration barriers into the glomerular capillaries. This may result in **proteinuria** (the leakage of proteins into the urine) and the appearance of **casts** in the urine sediment (see figs. 9.7 and 9.9). The leakage of plasma proteins into the urine causes a loss of the solutes responsible for the colloid osmotic pressure of the plasma. An abnormally low plasma colloid osmotic pressure, in turn, can result in *edema* (the accumulation of interstitial fluid in the tissues).
>
> **Glycosuria** (the appearance of glucose in the urine) due to *hyperglycemia* (high plasma glucose concentration) in a fasting person is the hallmark of **diabetes mellitus.** However, a person can have hyperglycemia without glycosuria, as long as the carrier proteins for glucose in the plasma membrane of the cells in the proximal tubule are not *saturated* (working at 100% capacity). The transport rate at saturation is the *transport maximum,* and the plasma concentration of glucose required to reach this rate is the *renal plasma threshold* for glucose. When the plasma glucose concentration is below the renal plasma threshold, all of the filtered glucose is reabsorbed; when the glucose concentration is at or above the renal plasma threshold, glucose "spills over" into the urine (glycosuria occurs). If diabetes mellitus is suspected, an *oral glucose tolerance* test may be performed.

A. Test for Proteinuria

Because proteins are very large molecules (macromolecules), they are not normally present in measurable amounts in the glomerular ultrafiltrate or the urine. The presence of proteins in the urine may therefore indicate an abnormal increase in the permeability of the filtration barriers into the glomerular capsules. Such permeability changes may be caused by renal infections (glomerulonephritis) or by other diseases that have secondarily affected the kidneys, such as diabetes mellitus, jaundice, or hyperthyroidism.

Procedure

Dip the yellow end of a disposable Albustix strip into a urine sample and compare the color developed with the chart provided. Record the albumin (protein) concentration in the data table in your laboratory report. Alternatively, albumin can be read on a Multistix strip.

B. Test for Glycosuria

Although glucose is easily filtered by the glomerulus, it is not normally present in the urine. This is because all of the filtered glucose is normally reabsorbed from the renal tubules into the blood. This reabsorption process is carrier mediated—that is, the filtered glucose is transported across the wall of the renal tubule by a protein carrier.

When the glucose concentration in the plasma and glomerular ultrafiltrate is within the normal limits (70–110 mg per 100 mL), there is a sufficient number of carrier molecules along the renal tubules to transport all the glucose back into the blood. However, if the blood glucose level exceeds a certain limit, called the **renal plasma threshold** for glucose, (about 180 mg per 100 mL), the concentration of glucose molecules in the glomerular ultrafiltrate will be greater than the capacity of the carrier proteins, and the nontransported glucose will "spill over" into the urine.

People with *type 2 diabetes mellitus* have reduced tissue sensitivity to insulin, often described as an *insulin resistance.* Before they became diabetic, they may have passed through a preliminary stage when they had *impaired glucose tolerance.* This is when they had hyperglycemia in the 140–200 mg/dL range at two hours after ingesting glucose in an oral glucose tolerance test. This blood glucose level, although abnormal, may still be below the renal plasma threshold for glucose. A urine test for glucose would thus be negative. However, once the person becomes diabetic and has a blood glucose concentration above 200 mg/dL, the renal plasma threshold for glucose is exceeded and glycosuria occurs.

Procedure

1. Place 10 drops of water and 5 drops of urine in a test tube.
2. Add a Clinitest tablet.
3. Wait 15 seconds and compare the color developed with the color chart provided.

    Or: Dip the end of a disposable Clinistix strip into a urine sample, wait the required amount of time, then compare the color developed with the chart provided on the bottle. Record the glucose concentration in the data table of the laboratory report.
4. Alternatively, glucose can be read on a Multistix strip.

C. Test for Ketonuria

When there is carbohydrate deprivation, such as in starvation or high-protein diets, the body relies increasingly on the metabolism of fats for energy. This pattern is also seen in people with diabetes mellitus, where lack of the hormone insulin prevents the body cells from using the large amounts of glucose available in the blood. This occurs because insulin is necessary for the transport of glucose from the blood into the body cells.

The metabolism of fat proceeds in a stepwise manner:

1. triglycerides are hydrolyzed to fatty acids and glycerol;
2. fatty acids are converted into smaller intermediate compounds—*acetoacetic acid, β-hydroxybutyric acid,* and *acetone* (collectively known as **ketone bodies**);
3. the ketone bodies are broken down in aerobic cellular respiration, releasing energy. When the production of ketone bodies from fatty acid metabolism exceeds the ability of the body to metabolize these compounds, they accumulate in the blood *(ketonemia)* and spill over into the urine *(ketonuria).*

Procedure

Dip a disposable Ketostix strip into a urine sample; 15 seconds later, compare the color developed with the color chart. Record the ketone concentration in the data table of the laboratory report. Alternatively, ketone bodies can be read on a Multistix strip.

D. Test for Hemoglobinuria

Hemoglobin may appear in the urine in the event of hemolysis in the systemic blood vessels (e.g., in transfusion reactions), of rupture in the capillaries of the glomerulus, or of hemorrhage in the urinary system. In the latter condition, whole red blood cells may be found in the urine *(hematuria),* although the low osmotic pressure of the urine may cause hemolysis and the release of hemoglobin *(hemoglobinuria)* from these red cells. Hemoglobinuria is normally found in the samples from menstruating women.

Procedure

Dip the test end of a disposable Hemastix strip into the urine sample, wait 30 seconds, and compare the color developed with the color chart. Enter the hemoglobin content in the data table of the laboratory report. Alternatively, hemoglobin can be read on a Multistix strip.

E. Test for Bilirubinuria

The fixed phagocytic cells of the spleen and bone marrow *(reticuloendothelial system)* destroy old red blood cells and convert the *heme* groups of hemoglobin into the pigment **bilirubin.** The bilirubin is secreted into the blood and carried to the liver, where it is bonded to *(conjugated* with) glucuronic acid, a derivative of glucose. Most of this conjugated bilirubin is secreted into the bile as bile pigment; the rest is released into the blood.

The blood normally contains a small amount of free and conjugated bilirubin. An abnormally high level of blood bilirubin *(hyperbilirubinemia)* may result from (1) an increased rate of red blood cell destruction, seen in hemolytic anemia, for example; (2) liver damage, as in hepatitis and cirrhosis; or (3) obstruction of the common bile duct, as might occur because of a gallstone. The increase in blood bilirubin results in **jaundice,** a condition characterized by a brownish-yellow pigmentation of the skin, sclera of the eye, and mucous membranes.

Normally, the kidneys can excrete only the more water-soluble bilirubin conjugated with glucuronic acid. An increase in the urine bilirubin *(bilirubinuria),* therefore, may be associated with jaundice due to liver disease or bile duct obstruction, but is not normally observed in jaundice due to hemolytic anemia. In the latter case the excess bilirubin is in a free, nonpolar state, where it binds to plasma proteins and thus is not filtered into the nephron tubules.

Procedure

Bilirubin can be quickly tested using the Multistix strip. If a more accurate determination is desired, the Ictotest procedure may be employed as described in these steps.

1. Place 5 drops of urine on a square of the test mat.
2. Place an Ictotest tablet in the center of the mat.
3. Place 2 drops of water on the tablet.
4. Interpret the test as follows:
   (a) Negative: Mat has no color or a slight pink-to-red color.
   (b) Positive: Mat turns blue to purple. The speed and intensity of color development is proportional to the amount of bilirubin present.
5. Record your observations in the data table in your laboratory report.

F. Microscopic Examination of Urine Sediment

Microscopic examination of the urine sediment may reveal the presence of various cells, crystals, bacteria, and casts (figs. 9.7, 9.8, and 9.9). **Casts** are cylindrical structures formed by the precipitation of protein molecules within the renal tubules. Although a small number of casts are found in normal urine, a large number indicate renal disease such as *glomerulonephritis* or *nephrosis.* The casts may lack cells or may contain cells such as leukocytes, erythrocytes, or epithelial cells (fig. 9.7). The presence of large numbers of erythrocytes, leukocytes, or certain epithelial cells in the urine is indicative of renal disease.

Like casts, a small number of crystals are present in normal urine. Their presence in large numbers, however may suggest a tendency to form kidney stones. Additionally, large numbers of *uric acid* crystals occur in *gout,* a form of arthritis (fig. 9.8).

Red blood cells (erythrocytes)

White blood cells (leukocytes)

Epithelial cells of the renal tubules

Epithelial cells of the bladder

Epithelial cells of the urethra

Bacteria

Hyaline cast

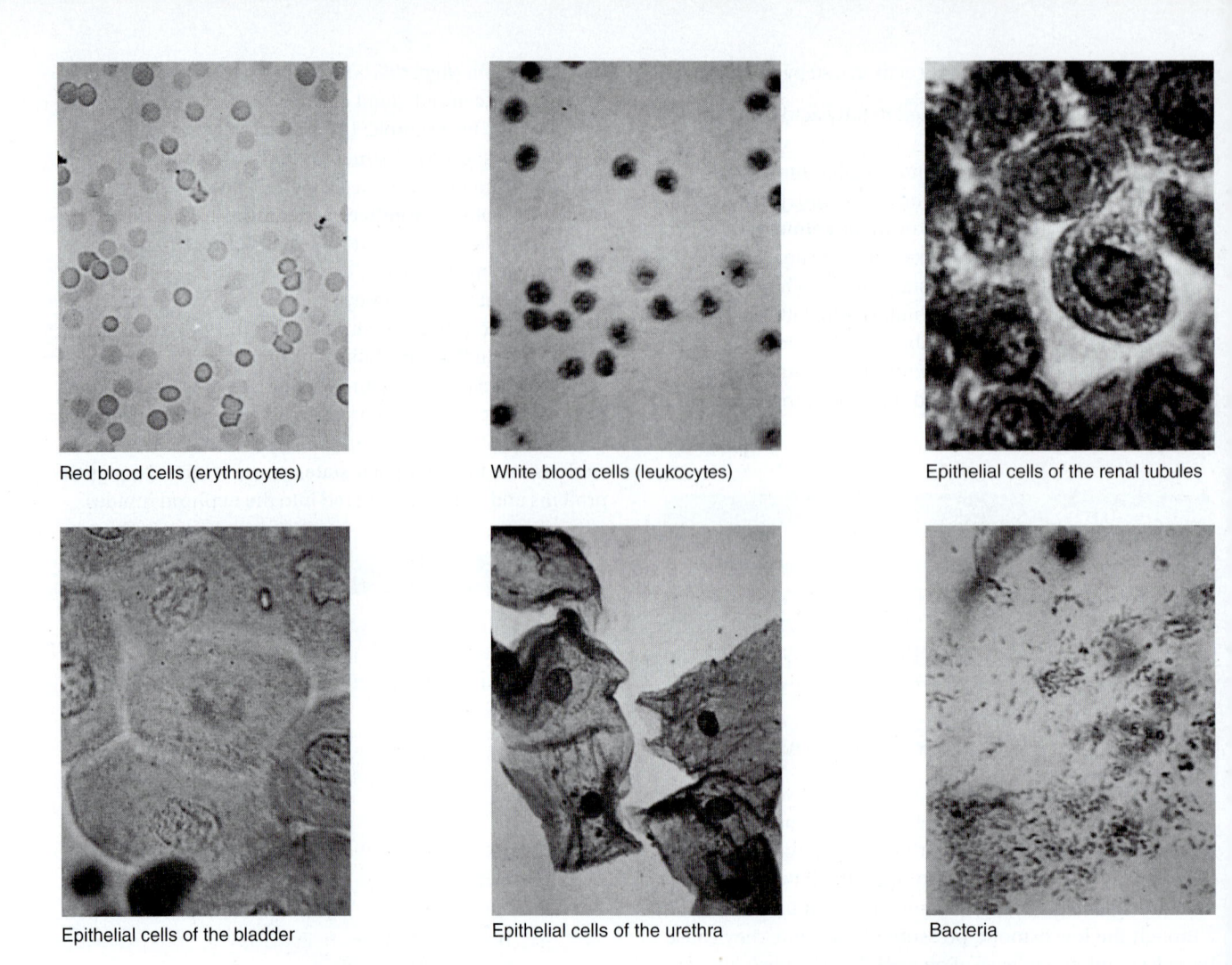

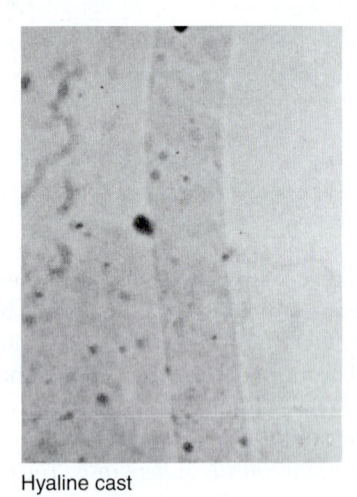

Waxy cast

Figure 9.7 **Components of urine sediment.**

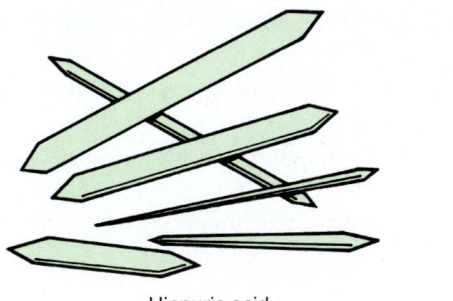

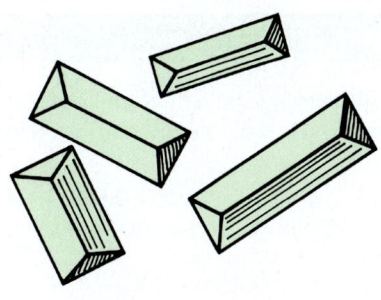

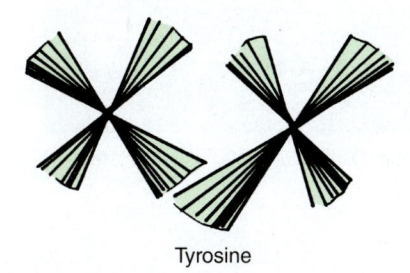

Figure 9.8 **Crystals found in urine sediment.**

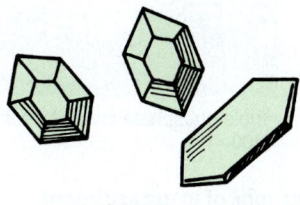

Procedure

1. Fill a conical centrifuge tube three-quarters full of urine, and centrifuge at a moderate speed for 5 minutes.
2. Discard the supernatant and place one drop of Sternheimer-Mablin stain (or similar sediment stain) on the sediment. Mix by aspiration with a transfer pipette.
3. Place 1 drop of the stained sediment on a clean slide, and cover with a coverslip.
4. Scan the slide with the low-power objective under reduced illumination (close the diaphragm), and identify the components of the sediment.

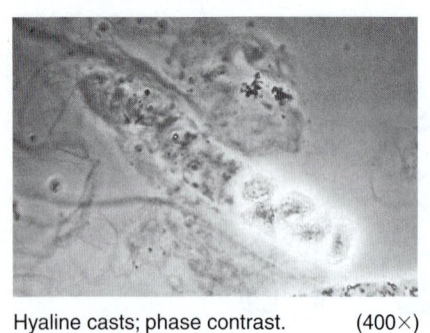

Hyaline casts; phase contrast. (400×)

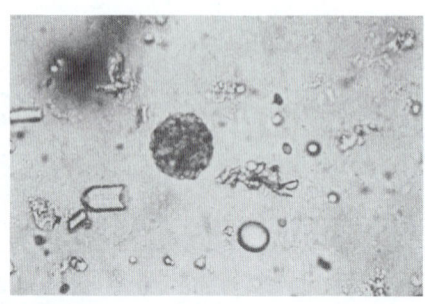

Oval fat bodies; "mulberry cell." Urine from a diabetic rat. (400×)

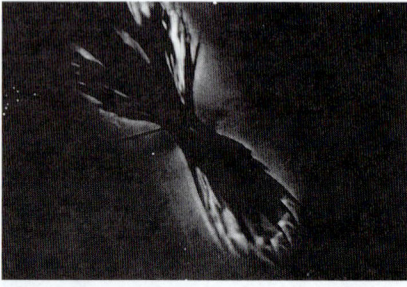

"Drug" crystal (sulfa); polarized light. Note sharp needlelike structure that may cause damage to the tubules. (400×)

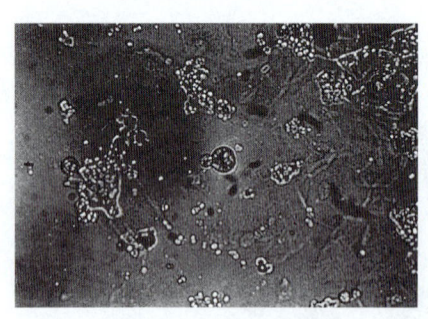

Leucine crystals; note size variation. (400×)

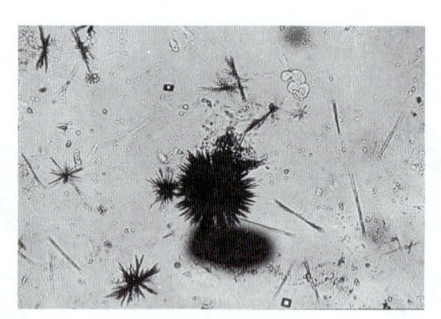

Tyrosine crystals (rosette needles). (100×)

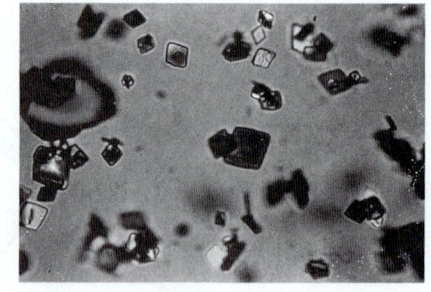

Uric acid crystals (plates); polarized light. (400×)

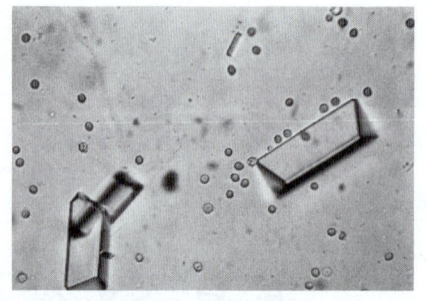

Triple phosphate crystals; "coffin-lid" form. (400×)

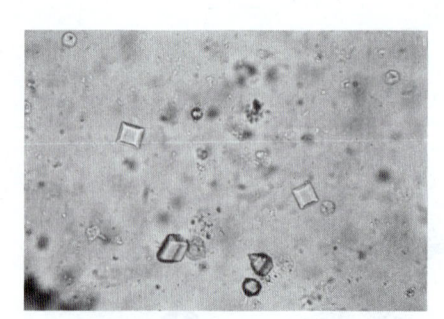

Calcium Oxalate crystals. (400×)

Figure 9.9 **Photographs of urine sediment.**

Laboratory Report 9.3

Name _____
Date _____
Section _____

DATA FROM EXERCISE 9.3

1. Record your data and the interpretations of your data in this table.

Urine Test	Result of Exercise (Positive or Negative)	Physiological Reason for Negative Result	Clinical Significance of Positive Result
Proteinuria			
Glycosuria			
Ketonuria			
Hemoglobinuria			
Bilirubinuria			

(a) Describe the appearance (such as color, intensity, turbidity) of the urine sample, and explain the possible causes of these observations.

(b) List the casts, cells, or crystals of the urine sediment identified under low power. Suggest the possible clinical significance if these components were found in abnormally high concentrations.

REVIEW ACTIVITIES FOR EXERCISE 9.3

Test Your Knowledge

Match these molecules with their descriptions:
- ____ 1. normally not filtered
- ____ 2. filtered, then normally completely reabsorbed
- ____ 3. derived from fat breakdown
- ____ 4. derived from heme groups of hemoglobin

(a) ketone bodies
(b) bilirubin
(c) glucose
(d) protein

5. Common cells found in normal urine sediment are _____.

6. An abnormal component of urine sediment, formed from protein is _____.

7. Indicate a possible cause for each of these conditions:
   (a) proteinuria _____
   (b) glycosuria _____
   (c) ketonuria _____
   (d) bilirubinuria _____

Test Your Understanding

8. Explain why proteins are not normally found (above trace amounts) in urine. What could cause proteinuria? Also, explain why glucose is not normally found (above trace amounts) in urine. What must happen to produce glycosuria?

9. Describe the composition of urinary casts, and explain how they can get into the urine.

10. Which component of the urine would be raised if a person were on a very low-carbohydrate, weight-reducing diet? Explain the process involved.

Test Your Analytical Ability

11. Is it possible for someone to have an abnormally high plasma glucose concentration and yet not have glycosuria? Explain your answer. Relate this answer to a person who eats a couple of sugar-coated doughnuts compared to a person with uncontrolled diabetes mellitus.

12. Proteinuria and the presence of numerous casts in the urine sediment are often accompanied by edema. What is the relationship between these symptoms? (*Hint:* Review exercise 2.1—particularly the Clinical Application box regarding plasma protein concentration.)

Clinical Investigation Questions

13. What might glycosuria and proteinuria indicate? Explain how these could be produced.

14. What did examination of this student's urine sediment indicate? How might this and the mild hematuria relate to the fact that the student is female?

Section 10

Digestion and Nutrition

The digestive tract is a continuous tube open to the external environment at both ends, by way of the mouth and the anus (fig. 10.1). Material inside this digestive tract is outside the body, in the sense that it can contact only the epithelial cells that line the tract. For this material to reach the inner cells of the body, it must pass through the epithelial cells of the digestive tract (a process known as **absorption**) into the blood. Before nutritive material can be absorbed, however, it must first be broken down by physical processes such as chewing (mastication), and by enzymatic hydrolysis into its monomers. The process of hydrolyzing larger food molecules (polymers) into absorbable monomers is known as **digestion.**

The embryonic digestive system consists of a hollow tube only one cell layer thick. As the embryo develops, different regions of the digestive tract become specialized for different functions. Some epithelial cells that line the tract become secretory, forming exocrine glands that secrete mucus, HCl, or particular hydrolytic enzymes characteristic of a certain region of the digestive tract. The liver, which produces bile, and the pancreas, which produces the many digestive enzymes found in *pancreatic juice,* develop as outpouchings *(diverticula)* of the embryonic small intestine and maintain their connections with the intestine by means of the hepatic and pancreatic ducts. Other regions of the small intestine become specialized for absorption through an increase in surface area. This increased surface area is produced by epithelial folds known as **villi** and minute foldings of the epithelial cell membranes called **microvilli.**

The digestive tract may be visualized as a "disassembly" line, where the food is conveyed, by means of muscular movements of the tract *(peristalsis)* and the opening and closing of sphincter muscles, from one stage of processing to the next. Coordination of these processes is achieved by neural reflexes and by hormones secreted by the gastrointestinal tract (table 10.1).

Diet maintains the consistent supply of nutrients to the body cells as energy sources for fuel, for growth, and for the replacement of cellular parts. *Carbohydrates, fats, proteins, vitamins, minerals,* and *water* are six nutrient classes recommended for daily consumption. Only carbohydrates, fats, and proteins can provide energy, measured in **kilocalories** *(kcals).* The **basal metabolic rate (BMR)** is the minimum amount of energy required by the body at rest.

Exercise 10.1	Histology of the Gastrointestinal Tract, Liver, and Pancreas
Exercise 10.2	Digestion of Carbohydrate, Protein, and Fat
Exercise 10.3	Nutrient Assessment, BMR, and Body Composition

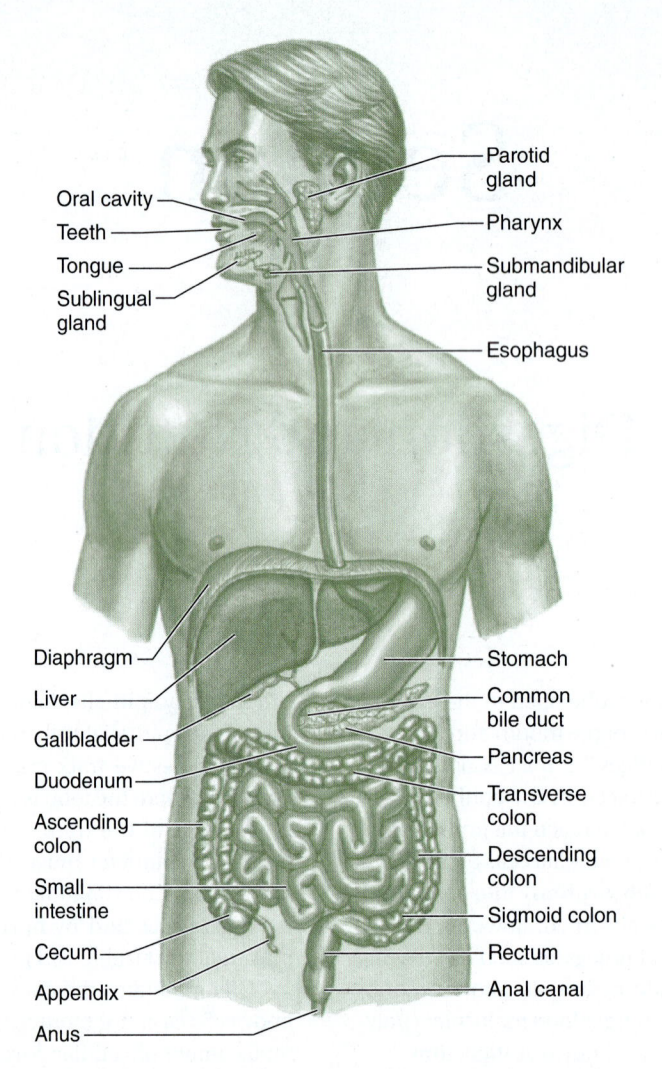

Figure 10.1 **Organs of the digestive system.**
(For a full-color version of this figure, see fig. 18.2 in *Human Physiology*, thirteenth edition, by Stuart I. Fox.)

Table 10.1 Physiological Effects of Gastrointestinal Hormones

Secreted by	Hormone	Effects
Stomach	Gastrin	Stimulates parietal cells to secrete HCl Stimulates chief cells to secrete pepsinogen Maintains structure of gastric mucosa
Small intestine	Secretin	Stimulates water and bicarbonate secretion in pancreatic juice Potentiates actions of cholecystokinin on pancreas
Small intestine	Cholecystokinin (CCK)	Stimulates contraction of gallbladder Stimulates secretion of pancreatic juice enzymes Inhibits gastric motility and secretion Maintains structure of exocrine pancreas (acini)
Small intestine	Gastric inhibitory peptide (GIP)	Inhibits gastric motility and secretion Stimulates secretion of insulin from pancreatic islets
Ileum and colon	Glucagon-like peptide-1 (GLP-1)	Inhibits gastric motility and secretion Stimulates secretion of insulin from pancreatic islets
Ileum and colon	Guanylin	Stimulates intestinal secretion of Cl^-, causing excretion of NaCl and water in the feces

EXERCISE 10.1

Histology of the Gastrointestinal Tract, Liver, and Pancreas

MATERIALS
1. Microscopes
2. Prepared tissue slides of the digestive system

All regions of the gastrointestinal tract have a mucosa, submucosa, muscularis, and serosa, but these layers of the wall display different specializations in different regions of the tract. The histology of the liver and pancreas provides insights into the functions of these organs.

LEARNING OUTCOMES

You should be able to:

1. Identify the mucosa, submucosa, muscularis, and serosa layers of different regions of the gastrointestinal tract.
2. Describe the structure and function of the layers of the esophagus, stomach, small intestine, and large intestine.
3. Describe the microscopic anatomy of the liver, and explain its functional significance.
4. Describe the microscopic anatomy of the pancreas, and distinguish the parts involved in the endocrine and exocrine functions of the pancreas.

Textbook/Multimedia Correlations

Before performing this exercise, you should study the introductory material presented here. Further information relating to this exercise can be found in these pages of *Human Physiology*, thirteenth edition, by Stuart I. Fox:

- *From Mouth to Stomach.* Chapter 18, p. 623.
- *Small Intestine.* Chapter 18, p. 628.
- *Large Intestine.* Chapter 18, p. 632.
- *Liver, Gallbladder, and Pancreas.* Chapter 18, p. 635.
- Multimedia correlations are provided in *MediaPhys 3.0:* Topics 15.9–15.55.

Clinical Investigation

An intestinal biopsy from a child with edema and suspected intestinal problems revealed no ulceration of the mucosa. The lamina propria was filled with a much larger number of eosinophils than normal, suggesting an allergic reaction. The child suffered from low plasma albumin, but levels of urea and of liver enzymes were normal.

- Describe the normal structure of the intestinal mucosa, identifying the lamina propria.
- Describe the functions of the liver that relate to the measurements described, and propose a possible relationship between these observations.

The tubular digestive tract, including the esophagus, stomach, small intestine, and large intestine, consists of four major layers, or *tunics* (fig. 10.2). From the innermost layer outward, they are:

1. The **mucosa,** or mucous membrane, consists of an inner epithelium spread over a thin layer of connective tissue, the *lamina propria,* and bordered by a ribbon of smooth muscle, the *muscularis mucosa.* The epithelium is stratified squamous in the esophagus and anal canal and simple columnar in the stomach, small intestine, and large intestine.
2. The **submucosa** is connective tissue and therefore has abundant extracellular space for blood vessels, nerves, and mucous-secreting glands. Parasympathetic fibers and ganglia can be seen as the *submucosal (Meissner's) plexus* in the submucosa.
3. The **muscularis externa** consists of smooth muscle, arranged in an inner circular and outer longitudinal layer throughout most of the digestive tract. Parasympathetic fibers and ganglia can be seen as the *myenteric (Auerbach's) plexus* in this layer.
4. The **serosa** consists of a *simple squamous* epithelium and connective tissue and is the outermost covering of the digestive tract.

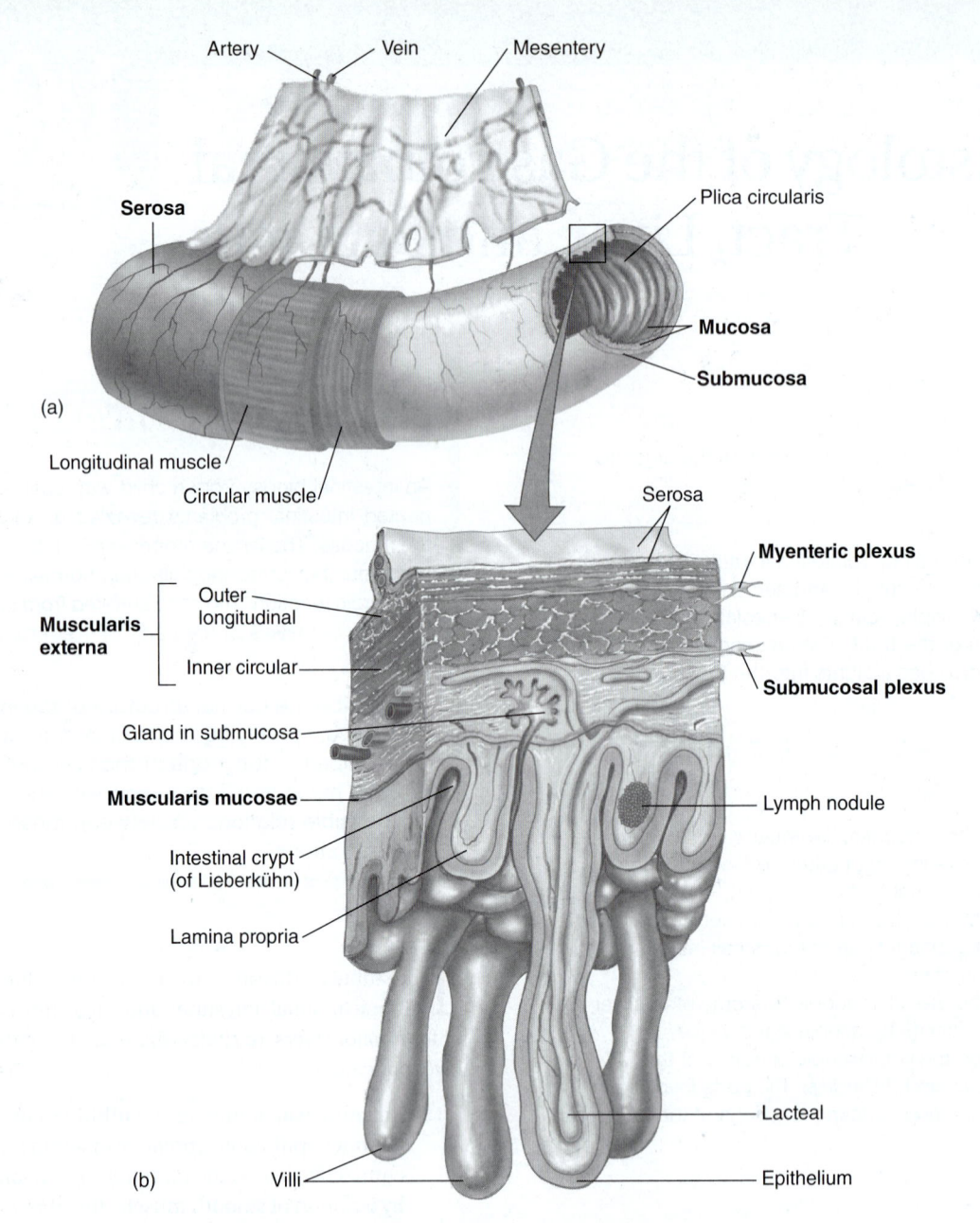

Figure 10.2 The layers of the digestive tract. (a) An illustration of the major tunics, or layers, of the small intestine. The insert shows how folds of mucosa form projections called villi in the small intestine. (b) An illustration of a cross section of the small intestine showing layers and glands.

(For a full-color version of this figure, see fig. 18.3 in *Human Physiology*, thirteenth edition, by Stuart I. Fox.)

A. Esophagus and Stomach

The mucosa layer of the esophagus is lined with a stratified squamous epithelium (fig. 10.3). The muscles of the first third of the esophagus, like those of the pharynx and mouth, are striated to provide voluntary control of swallowing. The middle third contains a mixture of striated and smooth muscle, and the last third of the esophagus contains only involuntary smooth muscle.

The submucosa of the stomach is thrown into large folds, or *rugae,* which can be seen with the unaided eye. Microscopic examination of the mucosa shows that it, too, is folded. The openings of these folds into the stomach lumen are called

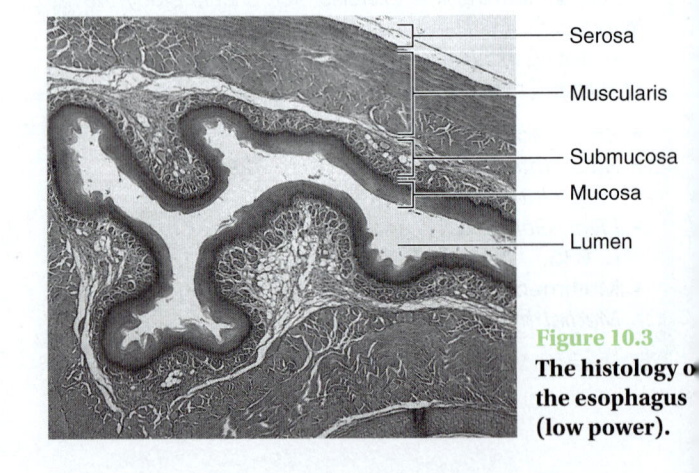

Figure 10.3
The histology of the esophagus (low power).

364

gastric pits. The cells that line the folds of mucosa are secretory and form the **gastric glands** (figs. 10.4 and 10.5).

The gastric glands include: (1) *mucous neck cells,* which secrete mucus (these supplement the surface mucous cells, which are in the epithelium of the stomach lumen and gastric pits); (2) *parietal cells,* which secrete hydrochloric acid (HCl); (3) *chief cells,* which secrete pepsinogen (the inactive precursor of pepsin, a protein-digesting enzyme); (4) *enterochromaffin-like cells (ECLs),* which secrete histamine; (5) *G cells,* which secrete the hormone gastrin into the blood; and (6) *D cells,* which secrete the hormone somatostatin. The gastric mucosa also secretes a polypeptide called **intrinsic factor,** which aids in the absorption of vitamin B_{12} in the intestine. Vitamin B_{12} is needed for the production of red blood cells in the bone marrow. The stomach also secretes a hormone called **ghrelin,** which rises before meals and serves as a signal from the stomach to the brain that helps regulate hunger.

Procedure

1. Observe a cross section of the esophagus under 100× (using the 10× objective lens) and note the four major layers, or tunics.
2. Hold a slide of a stomach section up to a light source and observe a fold, or ruga. Now, place the slide on a microscope and, under 100×, observe the gastric pits and glands in the mucosa, the submucosa, and the muscularis externa.
3. Using the high-dry objective lens (45×), observe the gastric glands in the mucosa under a total magnification of 450×. Identify goblet cells near the surface of the gastric pits. These mucous-secreting cells are numerous and clear in appearance. Near the base of the glands, identify parietal cells (with red-staining cytoplasm) and chief cells (with blue-staining cytoplasm). ECL cells and G cells cannot be easily identified without specially stained slides.

B. SMALL INTESTINE AND LARGE INTESTINE

The small intestine is approximately 21 feet long (in a cadaver) and divided into three regions. The first region, approximately 12 inches long, is called the **duodenum.** The next region, the **jejunum,** is about 8 feet long and constitutes two-fifths of the entire length of the intestine. The **ileum,** about 12 feet long (constituting three-fifths of the intestine), is the terminal region.

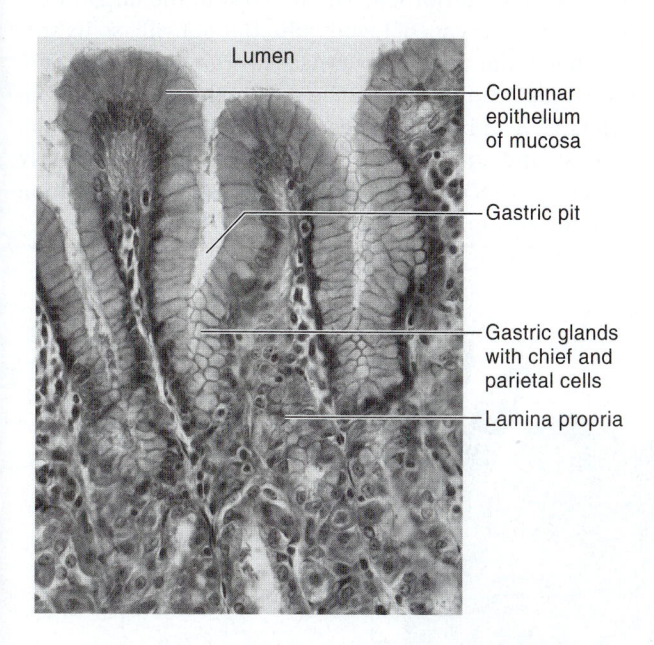

Figure 10.4 The microscopic structure of the stomach.

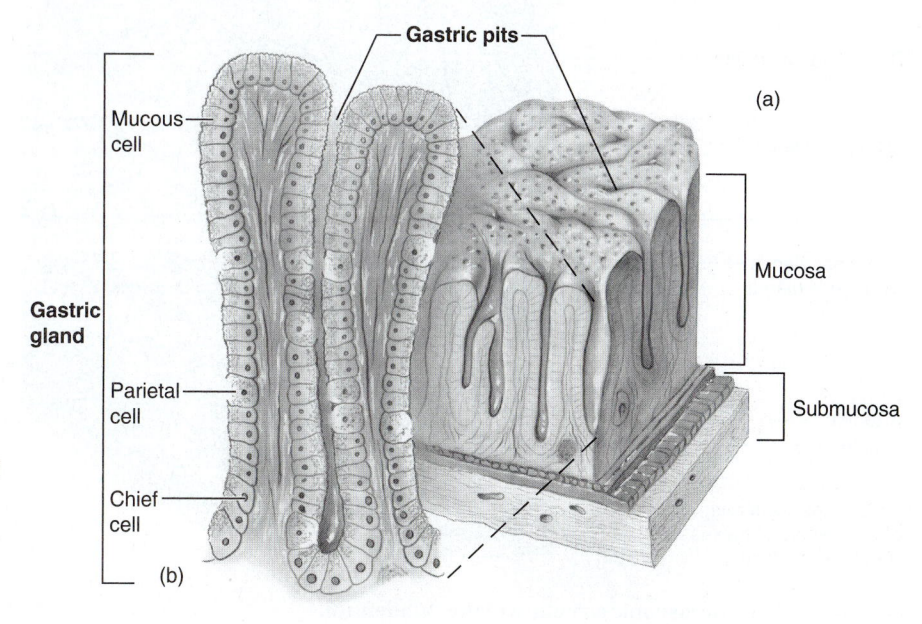

Figure 10.5 Gastric pits and gastric glands of the mucosa.
(a) Gastric pits are the openings of the gastric glands. (b) Gastric glands consist of mucous cells, chief cells, and parietal cells, each of which produces a specific secretion.

(For a full-color version of this figure, see fig. 18.6 in *Human Physiology,* thirteenth edition, by Stuart I. Fox.)

The mucosa and submucosa of the small intestine form large folds called the *plicae circulares*. The surface area of the mucosa is further increased by microscopic folds that form fingerlike projections called *villi* (fig. 10.6). Each villus has a core of connective tissue (the lamina propria) covered with a simple columnar epithelium. The apical surface (facing the lumen) of each epithelial cell has a slightly blurred, "brush border" appearance because of numerous projections of its cell membrane in the form of *microvilli*. Microvilli can be clearly seen only with an electron microscope.

The microvilli, villi, and plicae circulares increase the surface area of the small intestine tremendously, thus maximizing the rate at which the products of digestion can be absorbed by transport through the epithelium into the blood. Various digestive enzymes—called **brush border enzymes**—are fixed to the cell membranes of the microvilli and act together with enzymes from pancreatic juice to catalyze hydrolysis reactions of food molecules (table 10.2). The small intestine epithelium is also the source of important gastrointestinal hormones (see table 10.1).

The epithelium at the base of the villi invaginates to form pouches called *crypts of Lieberkühn*, or **intestinal crypts**. At the bottom of the intestinal crypts in the small (but not the large) intestine are **Paneth cells,** which secrete antibacterial *lysozyme* and bactericidal peptides. The bottoms of the crypts also contain **intestinal stem cells,** which divide by mitosis to replenish themselves and to produce the specialized cells of the intestinal mucosa. Mitosis occurs twice a day in the crypts. At the top of the crypts mitosis stops and the cells differentiate into secretory cells and **enterocytes** (intestinal epithelial cells). The enterocytes migrate upward toward the tips of the villi to replace the epithelial cells continuously shed into the lumen. Within the submucosa of the duodenum are alkaline mucous-secreting *Brunner's glands* (fig. 10.6). The jejunum and ileum lack these glands but are otherwise similar in structure (fig. 10.7).

Waste products from the small intestine pass into the **colon** of the large intestine where water, sodium, and potassium are absorbed. The mucosa of the large intestine contains crypts of Lieberkühn but not villi, so its surface has a flat appearance. As with the small intestine, numerous lymphocytes can be seen in the lamina propria, and large lymphatic nodules appear at the junction of the mucosa and submucosa. Lymphatic nodules are clearly evident in a section of the *appendix,* a short outpouching from the cecum.

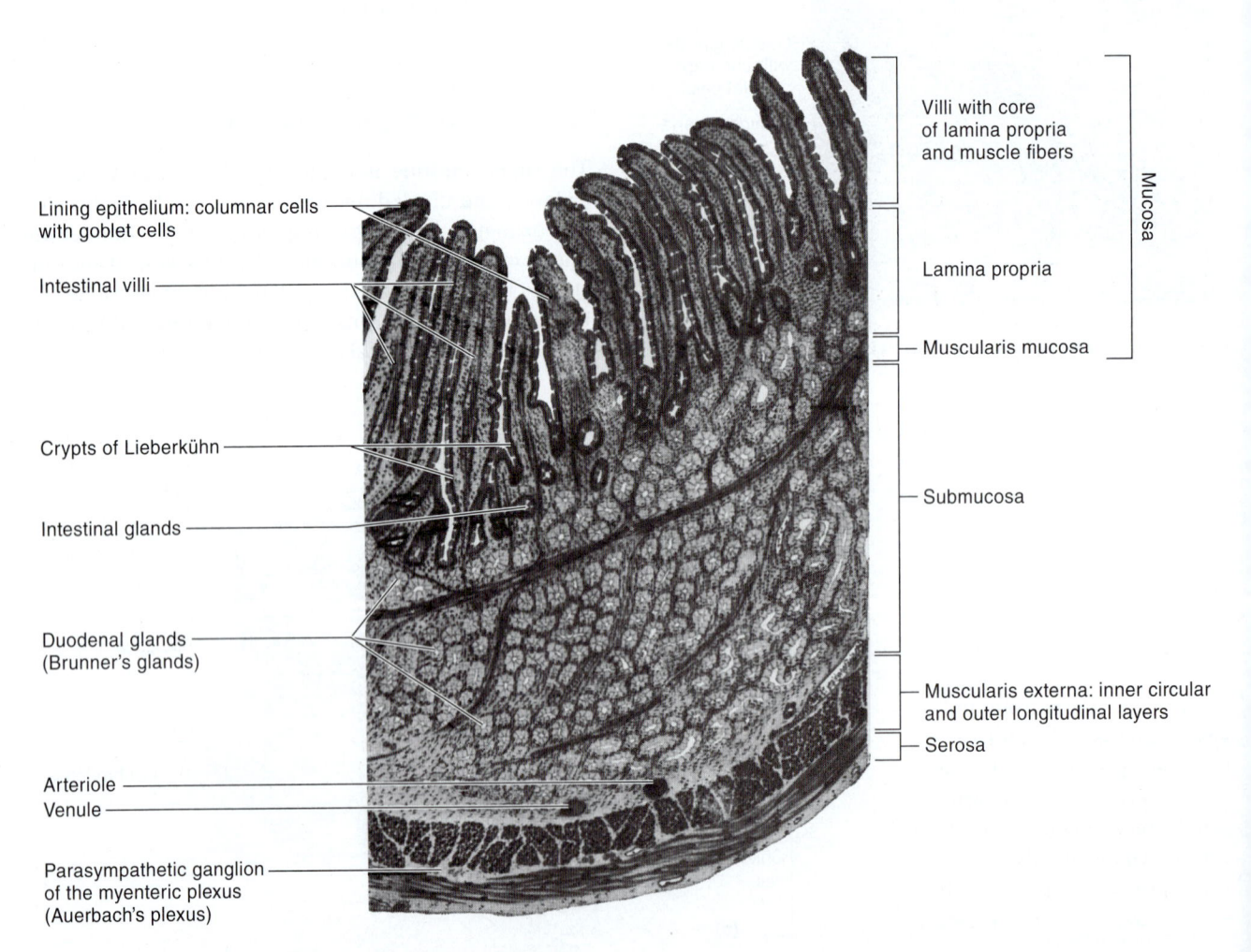

Figure 10.6 **The microscopic structure of the duodenum.**

Table 10.2   Characteristics of the Major Digestive Enzymes

Enzyme	Site of Action	Source	Substrate	Optimum pH	Product(s)
Salivary amylase	Mouth	Saliva	Starch	6.7	Maltose
Pepsin	Stomach	Gastric glands	Protein	1.6–2.4	Shorter polypeptides
Pancreatic amylase	Duodenum	Pancreatic juice	Starch	6.7–7.0	Maltose, maltriose, and oligosaccharides
Trypsin, chymotrypsin, carboxypeptidase	Duodenum	Pancreatic juice	Polypeptides	8.0	Amino acids, dipeptides, and tripeptides
Pancreatic lipase	Duodenum	Pancreatic juice	Triglycerides	8.0	Fatty acids and monoglycerides
Maltase	Brush border	Epithelial membranes	Maltose	5.0–7.0	Glucose
Sucrase	Brush border	Epithelial membranes	Sucrose	5.0–7.0	Glucose + fructose
Lactase	Brush border	Epithelial membranes	Lactose	5.8–6.2	Glucose + galactose
Aminopeptidase	Brush border	Epithelial membranes	Polypeptides	8.0	Amino acids, dipeptides, tripeptides

Clinical Applications

The epithelial layer of the intestine is only about 20 μm across; this presents a thin barrier of high surface area to separate deeper tissues from toxins in the lumen that could provoke inflammation. There is a population of commensal bacteria that live in the intestine, often called the **intestinal microbiota,** which provide a number of benefits, including protection against inflammatory damage from pathogenic bacteria. Other intestinal defenses against bacteria include (1) mucus from goblet cells, which prevent most bacteria from reaching the epithelial cells, (2) secretion of antimicrobial peptides by Paneth cells in the crypts; and (3) secretion of IgA antibodies by plasma cells. When the diversity or numbers of commensal bacteria become reduced (as from the excessive use of antibiotics), protection against inflammation is compromised and **inflammatory bowel disease (IBD)**—including *Crohn's disease* and *ulcerative colitis*—may develop. Bacterial antigens normally present in the intestine may drive the inflammation, and there is increased exposure to these cells because of epithelial injury and other reduced defenses. Many genes have been implicated in the genetic susceptibility to IBD, particularly to Crohn's disease.

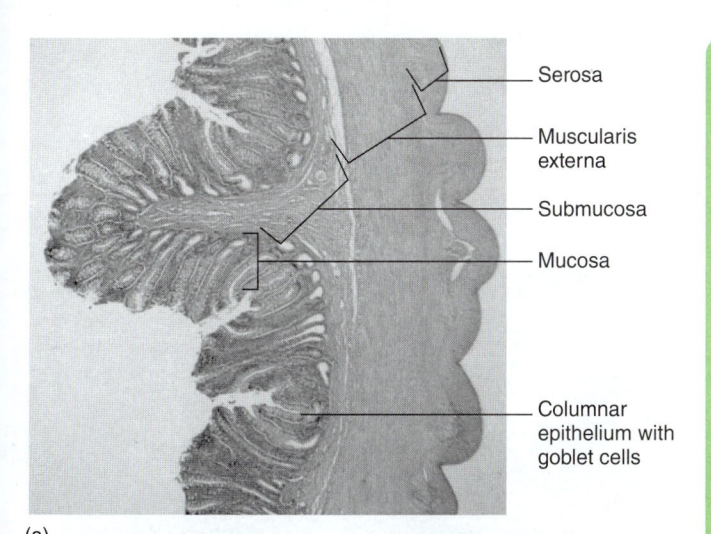

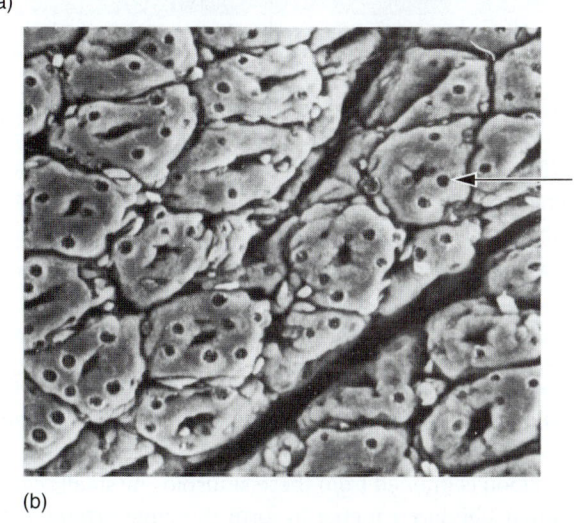

Figure 10.7 The microscopic structure of the ileum and large intestine. (a) A photomicrograph of a cross section of the ileum. (b) A scanning electron micrograph of the luminal surface of the large intestine (the arrow points to the opening of a goblet cell). (b is from R. G. Kessel and R. H. Kardon, *Tissues and Organs: A Text-Atlas of Scanning Electron Microscopy* © 1979 W. H. Freeman and Company.)

Procedure

1. Use the lowest power available on the microscope to observe the layers of a section of small intestine. Identify the villi, submucosa, and muscularis externa.
2. Observe the villi using the 45× objective lens. Identify the goblet cells in the epithelium and the

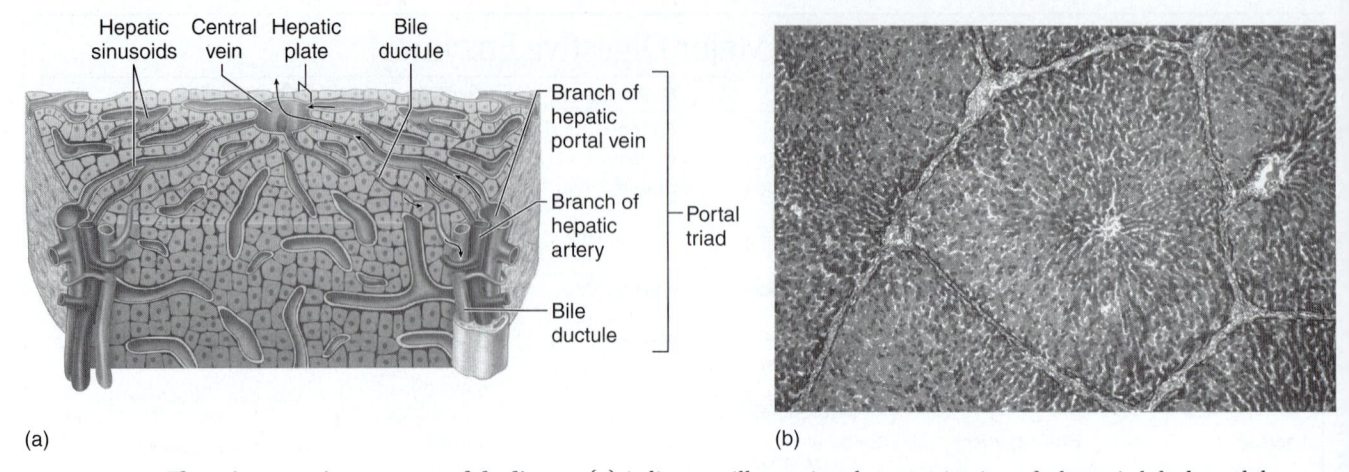

Figure 10.8 **The microscopic structure of the liver.** (a) A diagram illustrating the organization of a hepatic lobule and the relationship between sinusoids carrying blood and bile ductules carrying bile. (b) A photomicrograph of a liver lobule in cross section.

(For a full-color version of this figure, see fig. 18.19 in *Human Physiology*, thirteenth edition, by Stuart I. Fox.)

numerous lymphocytes (small, blue-staining cells) in the lamina propria within each villus. Also within the lamina propria, note the *central lacteal*—a lymphatic vessel that transports absorbed fat from the intestine.

3. Observe a slide of the large intestine using the 10× objective lens (fig. 10.7). Note the four layers, absence of villi, and goblet cells in the columnar epithelium of the mucosa.

C. Liver

Although the liver is the largest internal organ, in a sense it is only one to two cells thick. This is because the liver cells, or **hepatocytes,** form hepatic plates that are one to two cells thick. The plates are separated from each other by large capillary spaces called **sinusoids.** The hepatic plates are arranged into functional units called **liver lobules** (figs. 10.8 and 10.9).

The liver aids digestion by producing and secreting **bile,** which emulsifies fat. Bile leaves the liver in the *common hepatic duct,* which branches to form the *cystic duct* and the *common bile duct.* The cystic duct channels bile to the gallbladder where it is stored and concentrated. The common bile duct joins the pancreatic duct, and together they empty into the duodenum.

The liver also serves to modify the composition of the blood that drains from capillaries of the intestine through the *hepatic portal vein.* Before this rich venous blood can return to the heart and circulation, it must pass through sinusoids in the liver tissue (figs. 10.8 and 10.9). Liver sinusoids, however, are wider than most types of capillaries and are lined with phagocytic cells called *Kupffer cells.* The sinusoids have very large pores (called *fenestrae*) and, unlike other capillaries, lack a basement membrane. The fenestrae,

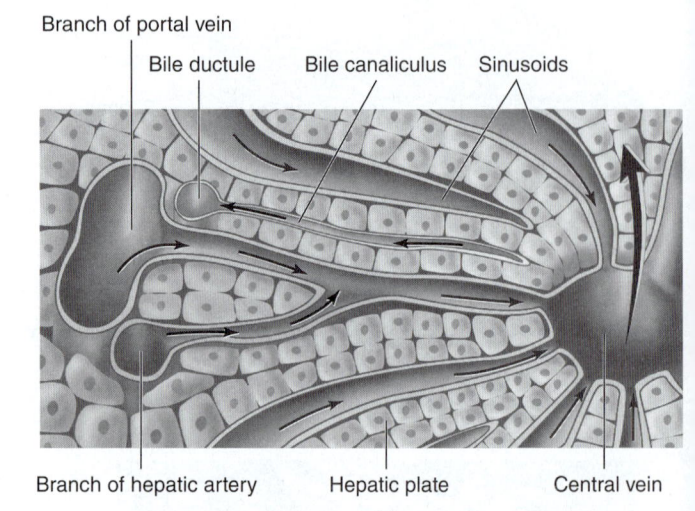

Figure 10.9 **The flow of blood and bile in a liver lobule.** Blood flows within sinusoids from a portal vein to the central vein (from the periphery to the center of a lobule). Bile flows within hepatic plates from the center to bile ductules at the periphery of a lobule.

(For a full-color version of this figure, see fig. 18.20 in *Human Physiology*, thirteenth edition, by Stuart I. Fox.)

lack of a basement membrane, and plate structure of the liver allows more intimate contact between hepatocytes and the contents of the blood.

Blood is drained from these sinusoids by small *central veins* that ultimately merge to form the large *hepatic vein,* which carries blood away from the liver. Liver sinusoids also receive arterial blood from branches of the *hepatic artery.* Arterial blood mixes with blood from the hepatic portal vein, and together they pass through the sinusoids to the central vein (fig. 10.9).

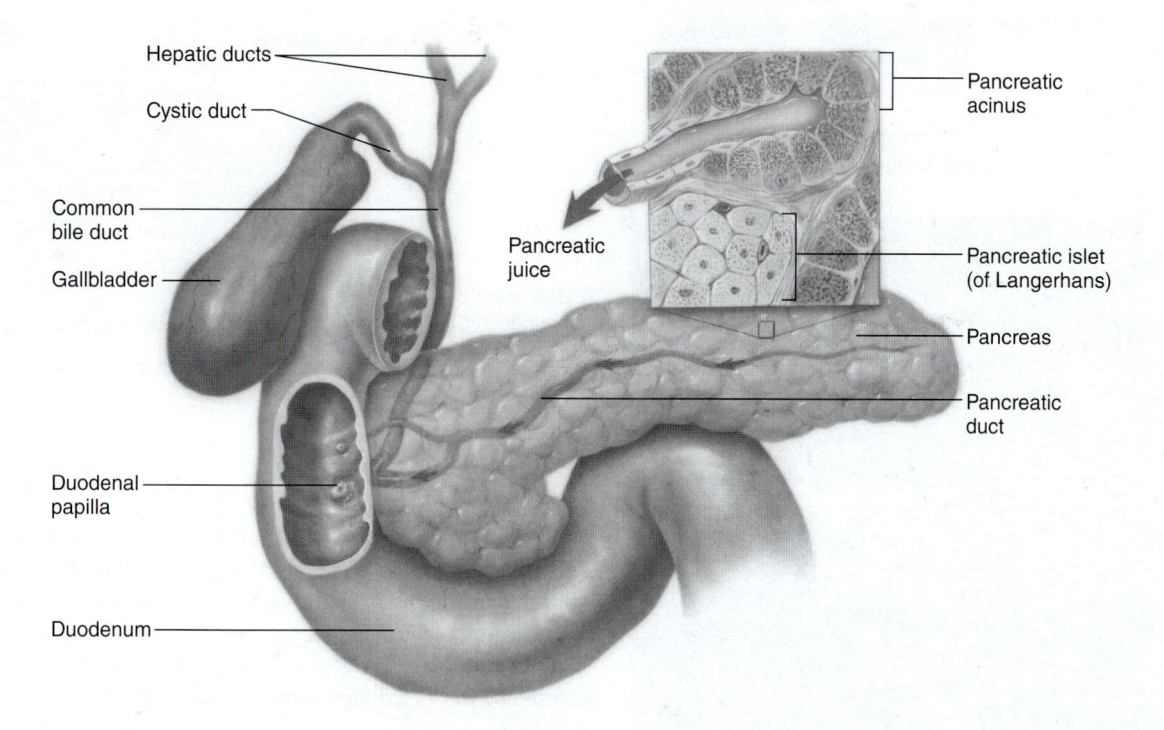

Figure 10.10 **Exocrine and endocrine structures of the pancreas.** The pancreatic acini secrete pancreatic juice into the pancreatic duct, which carries this exocrine secretion into the duodenum. The pancreatic islets (of Langerhans) secrete hormones (insulin and glucagon) into the blood. These are also shown in fig. 10.11.

(For a full-color version of this figure, see fig. 18.25 in *Human Physiology*, thirteenth edition, by Stuart I. Fox.)

Bile is produced and secreted by the liver cells (*hepatocytes*), but does not mix with blood because bile never enters the sinusoids. Instead, the hepatocytes secrete bile into *bile canaliculi,* located between adjacent hepatocytes (fig. 10.9). Bile is drained from the canaliculi into *bile ducts,* located near the entry of the portal vein and hepatic artery into the sinusoid. The grouping of the portal vein, hepatic artery, and bile duct that one sees in a microscopic view of the liver is called a *portal,* or *hepatic, triad* (see fig. 10.8).

Procedure

1. Examine a slide of the liver under low magnification and locate a central vein and a portal triad. The central vein appears as a large "hole" into which a number of sinusoids empty. The portal vein may also appear as a large hole, but it can easily be distinguished from the central vein because the hepatic artery and bile duct are in close proximity to the portal vein.

2. Observe a portal triad under high magnification and distinguish between the portal vein, hepatic artery, and bile duct (fig. 10.9). The portal vein is the largest of the three. The hepatic artery can be identified by its layer of smooth muscle and lining of simple squamous endothelium; by contrast, the bile duct lacks smooth muscle and is lined with simple columnar epithelium.

D. Pancreas

The pancreas is both an exocrine and an endocrine gland. Most of the pancreas is composed of exocrine tissue, with secretory cells arranged in clusters, or **acini** (figs. 10.10 and 10.11). Each pancreatic acinus secretes **pancreatic juice** into an opening of the *pancreatic duct,* which carries these secretions to the duodenum. Pancreatic juice contains water, bicarbonate, and a variety of digestive enzymes, including *trypsin, lipase,* and *amylase* (for the digestion of protein, fat, and carbohydrates, respectively).

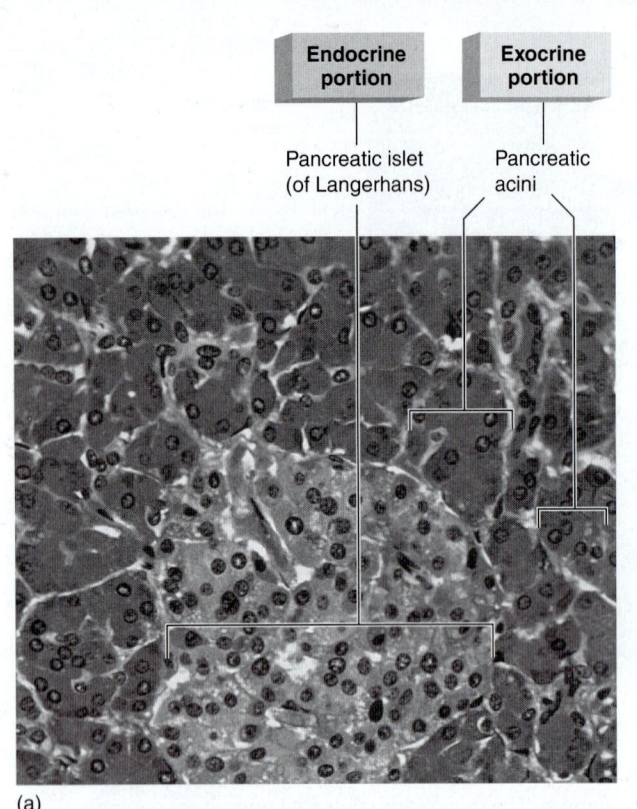

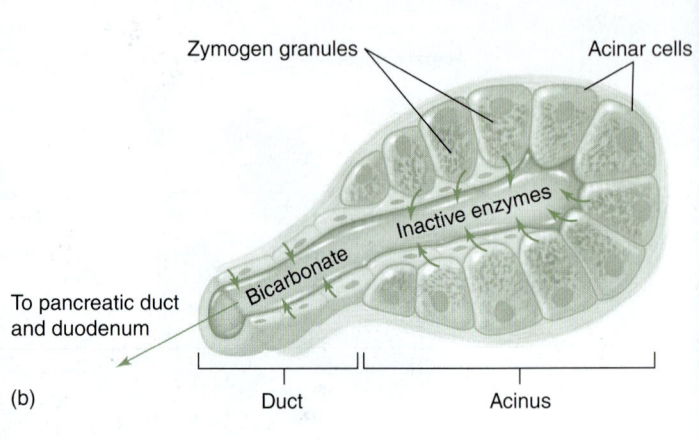

Figure 10.11 **The pancreas is both an exocrine and an endocrine gland.** (a) A photomicrograph of the endocrine and exocrine portions of the pancreas. (b) An illustration depicting the exocrine pancreatic acini, where the acinar cells produce inactive enzymes stored in zymogen granules. The inactive enzymes are secreted by way of a duct system into the duodenum.

(For a full-color version of this figure, see fig. 18.27 in *Human Physiology*, thirteenth edition, by Stuart I. Fox.)

Scattered among the exocrine acini are islands of endocrine cells. These are the **pancreatic islets,** or **islets of Langerhans** (fig. 10.11). The islets contain **alpha cells,** which secrete the hormone *glucagon,* and **beta cells,** which secrete the hormone *insulin.* These hormones are secreted into surrounding blood capillaries rather than into the pancreatic duct.

Procedure

1. Examine the pancreas under low magnification. The numerous small clusters of cells are the pancreatic acini. Occasional larger groupings of cells less intensely stained are pancreatic islets of Langerhans (fig. 10.11).
2. Scan the slide for a pancreatic duct in cross section. When one has been found, change to high magnification and observe its simple columnar epithelium.

Clinical Applications

Inflammation of the liver, or **hepatitis,** may be caused by bacterial or viral infections, alcohol abuse, allergy, or drugs. This condition is usually reversible. In **cirrhosis,** however, large areas of liver tissue are destroyed and replaced with permanent connective tissue and "regenerative nodules" of hepatocytes that lack the platelike structure of normal liver tissue. Among its many functions, the liver produces plasma *albumin* and converts ammonia to *urea;* therefore, liver disease may be accompanied by a decrease in the plasma albumin concentration and in the appearance of ammonia in the blood.

Inflammation of the pancreas, or **pancreatitis,** can result from the action of digestive enzymes on pancreatic tissue. This occurs when conditions such as alcoholism, gallstones, traumatic injury, infections, or toxicosis from various drugs provoke activation of digestive enzymes within the pancreas. These digestive enzymes are normally inactive until they enter the duodenum, but activated enzymes may reflux from the duodenum into the pancreatic duct. This produces an inflammation reaction accompanied by a "leakage" of enzymes into the blood, detected clinically by a rise in the concentration of pancreatic amylase and lipase in the plasma.

Laboratory Report 10.1

Name _____
Date _____
Section _____

REVIEW ACTIVITIES FOR EXERCISE 10.1

Test Your Knowledge

1. Identify the cells of the gastric mucosa that secrete:
   (a) HCl _____
   (b) pepsinogen _____
   (c) histamine _____

2. The three components of the mucosa layer of the digestive tract are the _____, the _____, and the _____.

3. Microscopic fingerlike projections of mucosa in the small intestine are called _____.

4. The foldings of the plasma membrane of intestinal epithelial cells that produce the "brush border" are called _____.

5. Blood is transported from the intestine to the liver in a large vessel known as the _____.

6. Once blood has reached the liver, it travels through large capillaries called _____.

7. The microscopic exocrine units of the pancreas are called _____; the endocrine structures are known as the _____.

8. The glands in the duodenum that secrete an alkaline mucus: _____.

Test Your Understanding

9. Describe the structural adaptations of the small intestine that help increase the surface area and the rate at which digestion products can be absorbed.

10. Describe the composition of pancreatic juice, and explain its functions.

11. Describe the microscopic anatomy of the liver, and explain how this structure (a) allows the liver to modify the chemical composition of the blood and (b) keeps blood separate from bile.

Test Your Analytical Ability

12. Describe the location of the digestive enzymes produced by the small intestine, and explain how these cooperate with pancreatic juice enzymes.

13. Suppose you were to grind up and homogenate a pancreas. Do you think it would be possible to isolate insulin from this homogenate? (*Hint:* Remember that the pancreas is also an exocrine gland, producing pancreatic juice.) Explain your answer.

Clinical Investigation Questions

14. What is the lamina propria and where is it located? Given that a high number of eosinophils in the lamina propria suggests an allergic reaction, how might that lead to intestinal problems?

15. Leakage of plasma proteins into the intestinal lumen is a "protein-losing enteropathy." How might this relate to the child's edema, intestinal biopsy observations, and blood test results?

EXERCISE 10.2

Digestion of Carbohydrate, Protein, and Fat

MATERIALS

1. Water-bath (set at 37°C), Bunsen burners, test tubes, test-tube clamp, graduated cylinders, droppers
2. Starch solution: dissolve 1.0 g per 100 mL water, over heat
3. Iodine solution (Lugol's reagent): dissolve 1.0 g iodine and 2.0 g potassium iodide in 300 mL of water
4. Benedict's reagent: dissolve 50.0 g sodium carbonate, 85.0 g sodium citrate, and 8.5 g copper sulfate in 5.0 L of water
5. As an alternative to saliva, α-amylase (type VI-B from porcine or human pancreas), or fungal amylase, may be used. Dilute to a concentration of about 300 units per mL, equivalent to saliva. Amylase may be purchased as a solid with 500,000 units of activity (order A-3176) from Sigma Chemical Company.
6. Freezer or ice bath; white of hard-boiled eggs
7. Pepsin (5 g per 100 mL), 2N HCl, 10 N NaOH
8. pH meter or short-range pH paper
9. Pancreatin solution (1 g per 100 mL); dairy cream or vegetable oil

Saliva contains salivary amylase, an enzyme that digests starch into sugars. Gastric juice contains hydrochloric acid (HCl) and pepsin, an enzyme that hydrolyzes peptide bonds to convert large proteins into smaller peptides. Pancreatic juice contains lipase, an enzyme that digests fats emulsified by bile salts into free fatty acids and monoglycerides. The digestive activity of these enzymes can be assessed by measuring the disappearance of substrates or the appearance of the products of the enzyme.

LEARNING OUTCOMES

You should be able to:

1. Explain the digestive action of salivary amylase, pepsin, and pancreatic lipase. Describe the procedures demonstrating the presence of these enzymes.
2. Explain how the activity of salivary amylase and pepsin is influenced by changes in pH and temperature.
3. Describe the ability of pepsin to digest large proteins.
4. Discuss why the stomach does not normally digest itself, and explain how peptic ulcers may be formed.
5. Define *emulsification,* and explain the individual roles of bile and pancreatic lipase in the digestion of fat.
6. Following fat digestion, describe how fatty acids and glycerol are absorbed and transported in the body.

Textbook/Multimedia Correlations

Before performing this exercise, you should study the introductory material presented here. Further information relating to this exercise can be found in these pages of *Human Physiology,* thirteenth edition, by Stuart I. Fox:

- *Digestion and Absorption of Food.* Chapter 18, p. 649.
- Multimedia correlations are provided in *MediaPhys 3.0:* Topics 15.9–15.55.

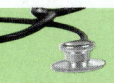

Clinical Investigation

A patient with a yellowish discoloration of the skin and sclera of the eyes had an elevated blood bilirubin concentration, and his stools demonstrated steatorrhea. He was concerned because he had been taking Zantac or Prilosec for a long time to treat an acid stomach. The physician said that the patient would probably need surgery.

- Identify what produced the yellowish discoloration and explain what could account for this symptom and the steatorrhea.
- Explain the actions of Zantac and Prilosec and whether or not these actions relate to the patient's probable need for surgery.

A. Digestion of Carbohydrate (Starch) by Salivary Amylase

The digestion of a carbohydrate such as starch begins in the mouth, where it is mixed with saliva containing the enzyme **salivary amylase**, or *ptyalin*. Starch, a long chain of repeating glucose subunits, is hydrolyzed by amylase first into shorter polysaccharide chains and eventually into the disaccharide maltose, which consists of two glucose subunits (see table 10.2 and fig. 10.12). Maltose, glucose, and other monosaccharides are known as *reducing sugars*.

In this exercise, the effects of pH and temperature on the activity of amylase will be tested by checking for the disappearance of substrate (starch) and the appearance of product (maltose) at the end of an incubation period. The appearance of a reducing sugar (maltose) in the incubation medium will be determined by the *Benedict's test*, where an alkaline solution of cupric ions (Cu^{2+}) is reduced to cuprous ions (Cu^{+}), forming a yellow-colored precipitate of cuprous oxide (Cu_2O).

Clinical Applications

Although starch digestion begins in the mouth with the action of salivary amylase, this is usually of minor importance in digestion (unless one chews excessively). Most of the digestion of polysaccharides and complex sugars to monosaccharides occurs in the small intestine, due to the actions of pancreatic amylase and brush-border enzymes. In saliva, the hydrolytic action of salivary amylase in combination with the buffering action of saliva may help prevent the accumulation of fermentable carbohydrates between the gums (gingiva) and the teeth, thus serving to protect against the growth of harmful bacteria that could cause dental cavities **(caries).**

Procedure

(See Fig. 10.13)

Step 1:
(1) Label four clean test tubes 1–4.
(2) Obtain 10 mL of saliva in a small, graduated cylinder. (Salivation can be aided by chewing a piece of paraffin.) If only 5 mL of saliva is obtained, dilute the saliva with an equal volume of distilled water.

Note: *As an alternative to saliva, the instructor may provide a solution of amylase derived from the pig or human pancreas.*

Step 2:
(1) Add 3.0 mL of distilled water to tube 1.
(2) Add 3.0 mL of saliva to tubes 2 and 3.
(3) Add 3 drops of concentrated HCl to tube 3.
(4) Boil the remaining saliva in a separate Pyrex test tube by passing the tube through the flame of a Bunsen burner. Use a test-tube clamp and keep the tube at an angle, pointed away from yourself and others. When cool, add 3.0 mL of this boiled saliva to tube 4.

Step 3: Add 5.0 mL of cooked starch (provided by the instructor) to each of the four tubes.

Step 4: Incubate all tubes for 1–1½ hours in a 37°C water-bath.

Step 5: Divide the contents of each sample by pouring half into four new test tubes.

Step 6: Test one set of four solutions for starch by adding a few drops of iodine solution *(Lugol's reagent)*. A positive test is indicated by the development of a purplish black color.

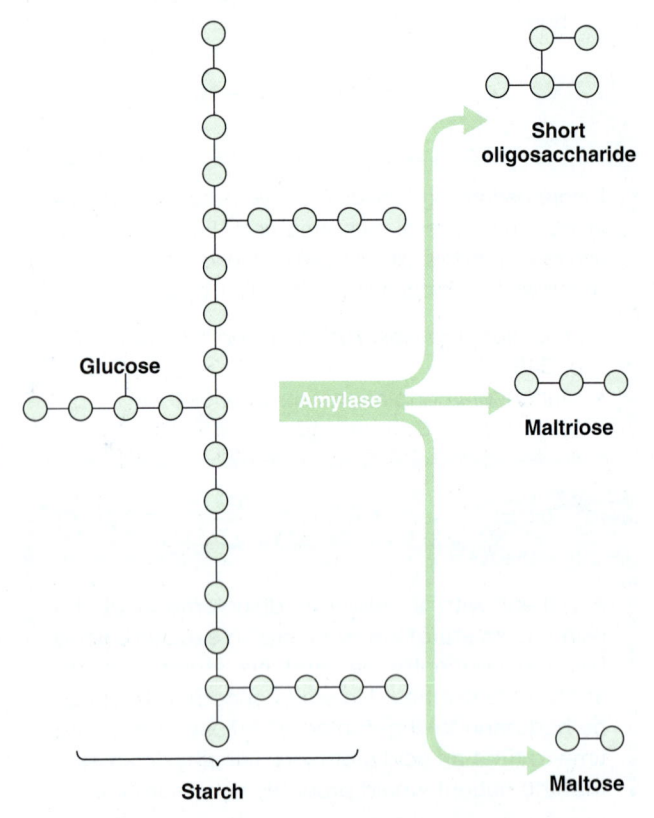

Figure 10.12 **The action of amylase.** Amylase digests starch into maltose, maltriose, and short oligosaccharides containing branch points in the chain of glucose molecules. Glucose and maltose are particularly strong reducing sugars in the Benedict's test for sugar in this exercise.

(For a full-color version of this figure, see fig. 18.32 in *Human Physiology*, thirteenth edition, by Stuart I. Fox.)

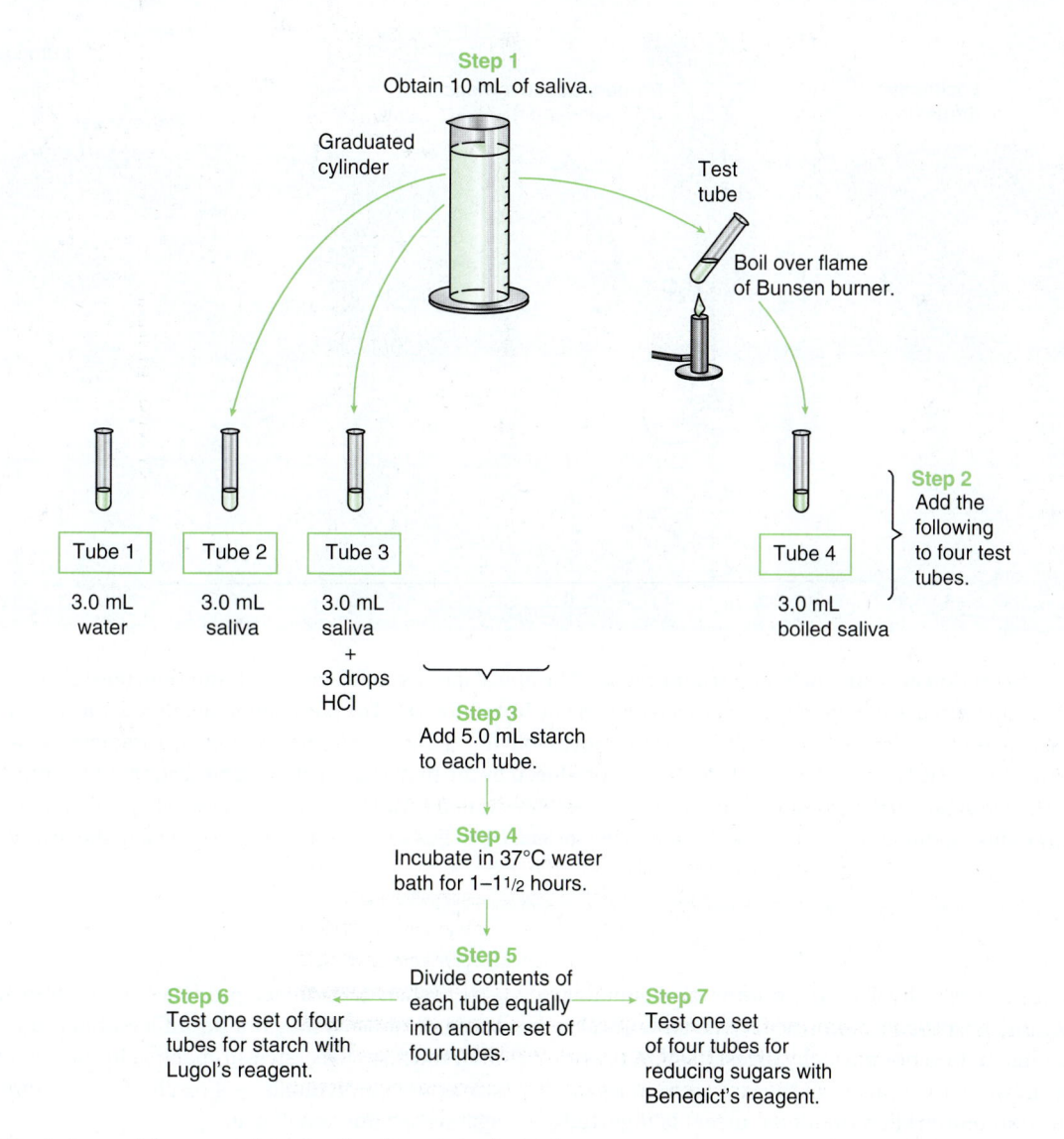

Figure 10.13 A chart of the procedure for the digestion of starch.

Step 1 Obtain 10 mL of saliva.

Step 2 Add the following to four test tubes.
- Tube 1: 3.0 mL water
- Tube 2: 3.0 mL saliva
- Tube 3: 3.0 mL saliva + 3 drops HCl
- Tube 4: 3.0 mL boiled saliva (Boil over flame of Bunsen burner.)

Step 3 Add 5.0 mL starch to each tube.

Step 4 Incubate in 37°C water bath for 1–1½ hours.

Step 5 Divide contents of each tube equally into another set of four tubes.

Step 6 Test one set of four tubes for starch with Lugol's reagent.

Step 7 Test one set of four tubes for reducing sugars with Benedict's reagent.

Step 7: Test the other set of four solutions for reducing sugars in the following way:

(1) Add 5.0 mL of Benedict's reagent to each of the four test tubes and immerse them in a rapidly boiling water-bath for 2 minutes.

(2) Remove the tubes from the boiling water with a test-tube clamp and rate the amount of reducing sugar present according to the following scale:

Blue (no maltose)	—
Green	+
Yellow	++
Orange	+++
Red (most maltose)	++++

Enter your results in the data table in your laboratory report.

B. DIGESTION OF PROTEIN (EGG ALBUMIN) BY PEPSIN

Although amylase is most active at the pH of saliva (pH 6–7), the enzyme **pepsin** has a pH optimum adapted to the normal pH of the stomach (pH less than 2—see table 10.2). The low pH of the stomach is due to the secretion of hydrochloric acid (HCl) by parietal cells in the gastric glands (fig. 10.14). The strong acidity of the stomach coagulates proteins, which facilitates their digestion by pepsin and later by other proteolytic enzymes in the small intestine.

Pepsin, secreted by the *chief cells* of the gastric glands, is responsible for the digestion of less than 15% of ingested protein. Removal of the entire stomach (complete gastrectomy) thus has little effect on protein digestion. The major site of protein digestion is the small intestine, where the enzymes *trypsin* and *chymotrypsin* (secreted by the pancreas) and the *dipeptidases* (fixed in the intestinal

375

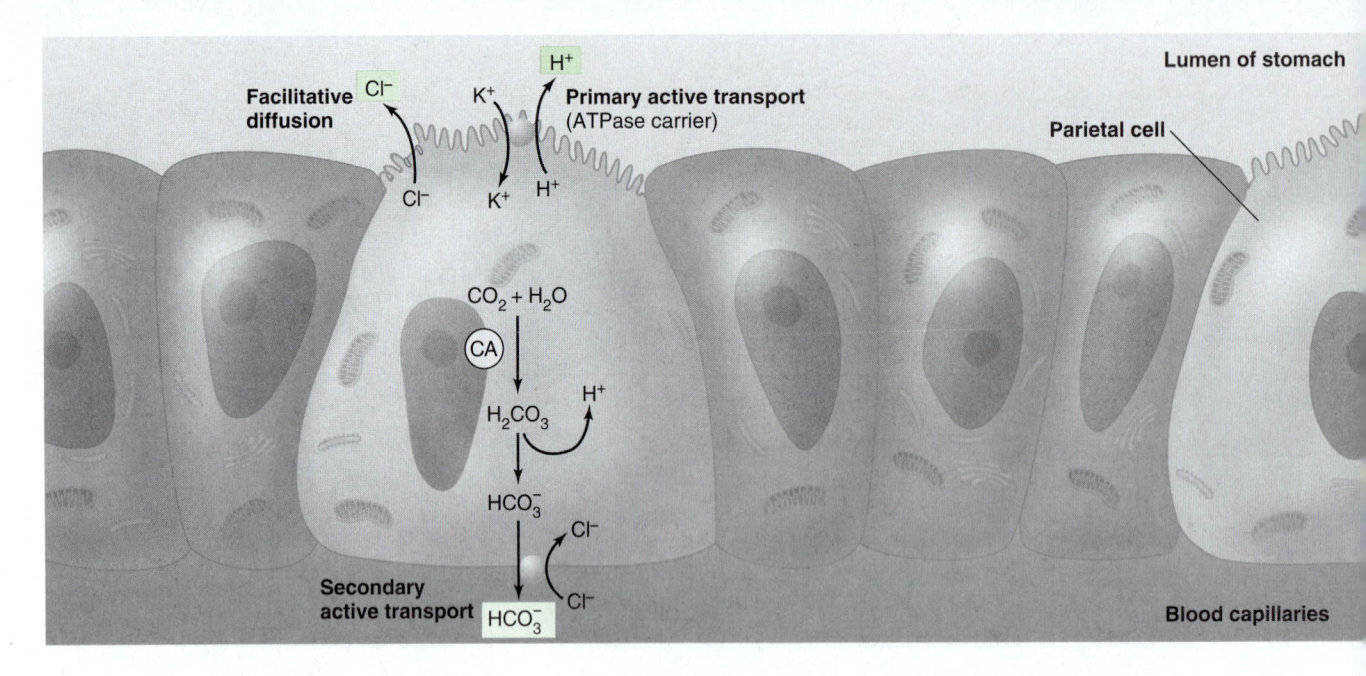

Figure 10.14 Secretion of gastric acid by parietal cells. The apical membrane (facing the lumen) secretes H$^+$ in exchange for K$^+$ using a primary active transport carrier powered by the hydrolysis of ATP. The basolateral membrane (facing the blood) secretes bicarbonate (HCO$_3^-$) in exchange for Cl$^-$. The Cl$^-$ moves into the cell against its electrochemical gradient, powered by the downhill movement of HCO$_3^-$ out of the cell. This HCO$_3^-$ is produced by the dissociation of carbonic acid (H$_2$CO$_3$), formed from CO$_2$ and H$_2$O by the action of the enzyme carbonic anhydrase (abbreviated CA). The Cl$^-$ then leaves the apical portion of the membrane by diffusion through a membrane channel. The parietal cells thus secrete HCl into the stomach lumen as they secrete HCO$_3^-$ into the blood.

(For a full-color version of this figure, see fig. 18.7 in *Human Physiology*, thirteenth edition, by Stuart I. Fox.)

brush-border mucosa) hydrolyze proteins and smaller polypeptides into absorbable amino acids (see table 10.2).

The stomach does not normally digest itself. A *peptic ulcer* may form when the mucosa of the stomach (in a gastric ulcer) or duodenum (in a duodenal ulcer) is digested by the strongly acidic gastric juice. The first line of defense against such damage is a stable gel of mucus that is adherent (stuck) to the gastric epithelial surface. This *adherent layer of mucus* contains alkaline bicarbonate (HCO$_3^-$), secreted by the apical membranes of the epithelial cells.

Also, the adherent layer of gastric mucus is the major barrier to potential damage to the stomach caused by pepsin. Little attention has historically been paid to pepsin's ability to cause damage, but there is evidence that it could pose a significant threat. Indeed, the damage to the esophagus caused by gastroesophageal reflux could be due more to pepsin than to acid. The adherent layer of mucus in the stomach protects the gastric lining from pepsin by so slowing its diffusion that it doesn't normally reach the epithelial cells.

Although the adherent layer of alkaline mucus is the first line of defense against acid and pepsin damage to the stomach, there are other important protective mechanisms. These include the presence of tight junctions between adjacent epithelial cells, which prevent acid and pepsin from leaking past the epithelial barrier and destroying underyling tissues. Also, a rapid rate of epithelial cell division replaces the entire gastric epithelium every 3 days, so that damaged cells can be rapidly replaced.

When the stomach produces an excess of acid, protective mechanisms may not be sufficient to protect the intestinal mucosa. Excessive stomach acid may be produced in susceptible individuals as a result of vagus stimulation and aggravate a duodenal ulcer.

When the acidic content of the stomach (called *chyme*) enters the duodenum, the intestine is stimulated to release the hormone *secretin*, which inhibits the gastric secretion of acid and stimulates the release of alkaline pancreatic juice (see table 10.1). The acidic chyme, therefore, is diluted and neutralized in the small intestine.

The duodenum is normally protected from gastric acid by the adherent layer of mucus on its epithelium. Duodenal cells secrete bicarbonate into this adherent mucous layer, so that the surface epithelium is normally exposed to a neutral pH. Additional protection against gastric acid is provided through bicarbonate secreted by Brunner's glands in the submucosa, which are glands unique to the duodenum. Finally, the acidic chyme is neutralized by the buffering action of bicarbonate in alkaline pancreatic juice, released into the duodenum upon the arrival of the acidic chyme from the stomach. Duodenal ulcers may result from excessive gastric acid secretion and/or inadequate secretion of bicarbonate in the duodenum. Indeed, some studies have demonstrated that people with duodenal ulcers have a reduced secretion of duodenal bicarbonate in response to acid in the lumen.

Clinical Applications

People with **gastritis** and **peptic ulcers** should avoid substances that stimulate gastric acid secretion, including coffee and alcohol, and must often take antacids (such as Tums), H_2-histamine receptor blockers (such as Zantac), or proton pump inhibitors (such as Prilosec).

It has been known for some time that most people who have peptic ulcers are infected with a bacterium known as *Helicobacter pylori,* which resides in the gastrointestinal tract of almost half the adult population worldwide. The 2005 Nobel Prize in Physiology or Medicine was awarded to two scientists who discovered that this common bacterium, rather than emotional stress or spicy food, is the most common cause of peptic ulcers of the stomach and duodenum. As a result of this discovery, many ulcers can now be effectively treated with a drug regimen that consists of a *proton pump inhibitor* (a drug such as Prilosec that inhibits the K^+/H^+ pumps) combined with two different antibiotics (such as amoxicillin and clarithromycin) to suppress the *H. pylori* infection. The most common cause of peptic ulcers that are not due to *H. pylori* infection is excessive use of *nonsteroidal anti-inflammatory drugs (NSAIDs),* such as aspirin and ibuprofen. These ulcers are usually treated with proton pump inhibitors.

Gastroesophageal reflux disease (GERD) is a common disorder in which the reflux of acidic gastric juice into the esophagus causes frequent heartburn or complications such as laryngitis, cough, and injuries to the esophagus. GERD can be reduced by not eating at least 3 hours before going to bed, and is often treated with proton pump inhibitors such as Prilosec (omeprazole) and Prevacid (lansoprazole). Drugs that block H_2 histamine receptors in the gastric mucosa, such as Tagamet and Zantac, are also used.

Procedure

(See Fig. 10.15)

Step 1: Label five clean test tubes 1–5. Using a sharp scalpel or razor blade, cut five slices of egg white about the size of a fingernail and as thin as possible. It is essential that the slices be very thin and uniform in size. Place a slice of egg white in each of the five test tubes.

Step 2:

(1) Add 1 drop of distilled water to tube 1.

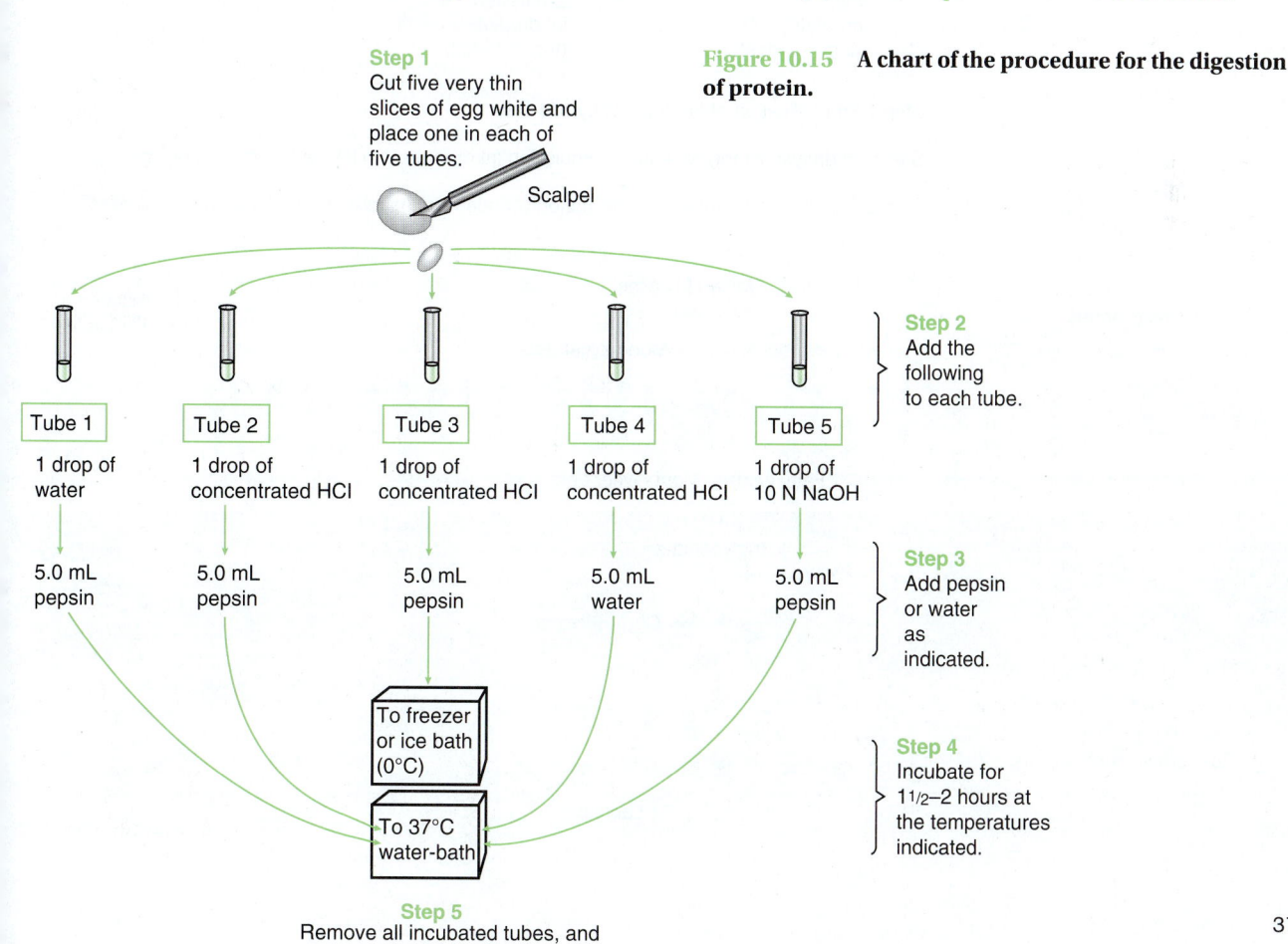

Figure 10.15 A chart of the procedure for the digestion of protein.

Step 1
Cut five very thin slices of egg white and place one in each of five tubes.
Scalpel

Tube 1	Tube 2	Tube 3	Tube 4	Tube 5
1 drop of water	1 drop of concentrated HCl	1 drop of concentrated HCl	1 drop of concentrated HCl	1 drop of 10 N NaOH
5.0 mL pepsin	5.0 mL pepsin	5.0 mL pepsin	5.0 mL water	5.0 mL pepsin

Step 2 Add the following to each tube.

Step 3 Add pepsin or water as indicated.

To freezer or ice bath (0°C)

To 37°C water-bath

Step 4 Incubate for 1½–2 hours at the temperatures indicated.

Step 5
Remove all incubated tubes, and observe appearance of egg white.

(2) Add 1-drop of concentrated hydrochloric acid (HCl) to tubes 2, 3, and 4.

(3) Add 1 drop of concentrated (10 N) NaOH to tube 5.

Step 3:

(1) Add 5.0 mL of pepsin solution to tubes 1, 2, 3, and 5.

(2) Add 5.0 mL of distilled water to tube 4.

Step 4:

(1) Place tubes 1, 2, 4, and 5 in a 37°C water-bath. Place tube 3 in a freezer or ice bath.

(2) Incubate all tubes for 1–1½ hours, and then remove the tubes (thaw the one that was frozen).

Step 5: Record the appearance of the egg white in the data table in your laboratory report.

C. Digestion of Triglycerides by Pancreatic Juice and Bile

Although the stomach produces a gastric lipase, the major digestion of triglycerides (fats and oils) occurs in the small intestine through the action of **pancreatic lipase** (see table 10.2). The digestion of fat in the small intestine is dependent upon the presence of *bile,* produced by the liver and transported to the duodenum via the bile duct (fig. 10.16). (The gallbladder serves only to store and concentrate the bile.)

Figure 10.16 Fat digestion and absorption. (a) The steps in the digestion of fat. (b) The process of fat absorption into the intestinal epithelial cells and secretion into lymphatic capillaries (lacteals).

(For full-color versions of these figures, see figs. 18.35 and 18.36 in *Human Physiology,* thirteenth edition, by Stuart I. Fox.)

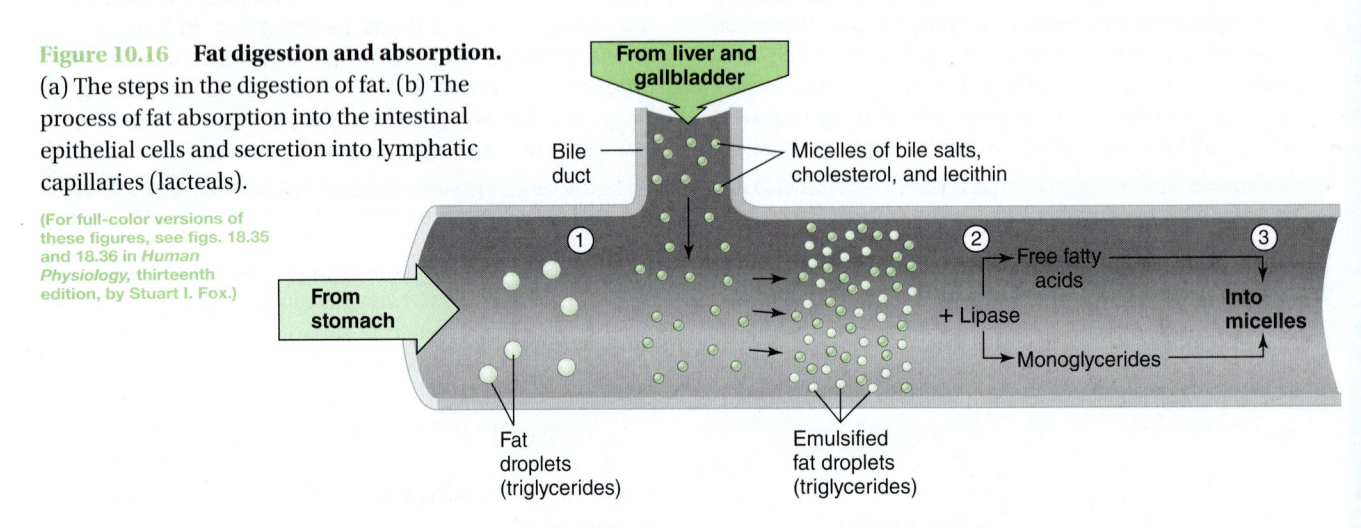

Step 1 Emulsification of fat droplets by bile salts

Step 2 Hydrolysis of triglycerides in emulsified fat droplets into fatty acid and monoglycerides

Step 3 Dissolving of fatty acids and monoglycerides into micelles to produce "mixed micelles"

(a)

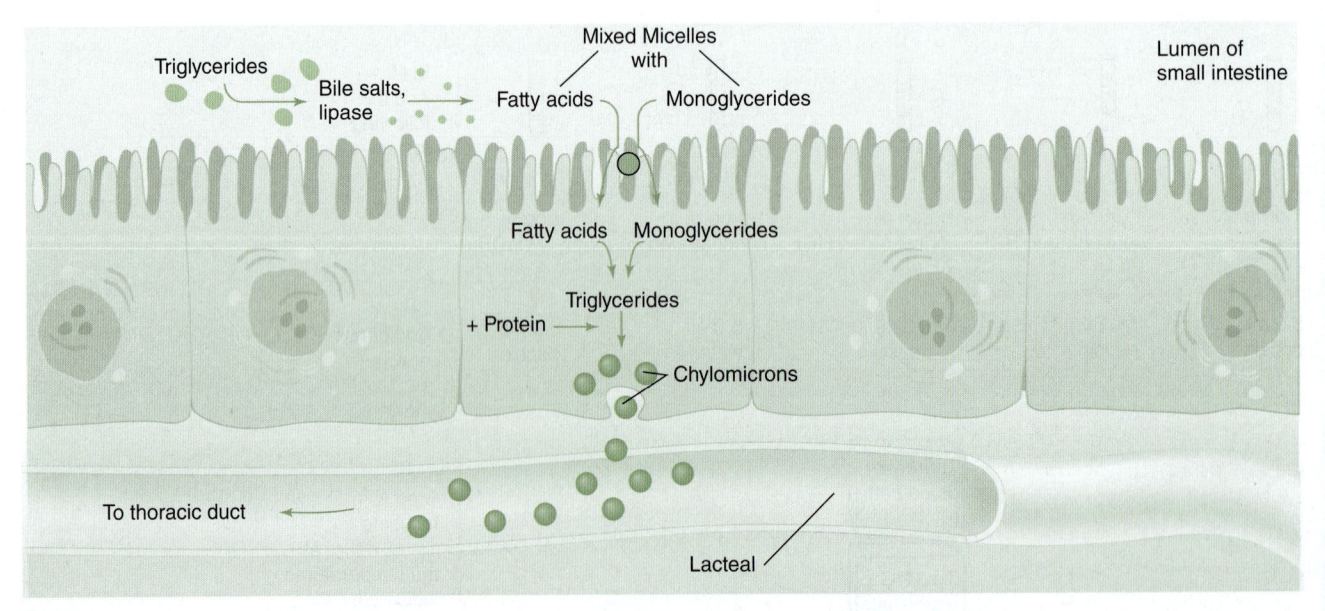

(b)

378

Fat is not soluble in water, so dietary fat enters the duodenum in the form of large fat droplets containing the fat-soluble vitamins A, D, E, and K. The detergent action of bile salts lowers the surface tension of these large droplets, breaking them up into smaller droplets in a process called **emulsification**. Following this process, more surface area is presented to the lipase enzymes, promoting the digestion of fat into monoglycerides and fatty acids (fig. 10.17), and the release of the fat-soluble vitamins.

The absorption of fat is more complicated than that of the water-soluble molecules. The monoglycerides and free fatty acids produced by lipase action aggregate to form spherical structures *(micelles),* absorbed by the intestinal epithelium. Once in the epithelial cell, the fatty acids and monoglycerides are used to resynthesize triglycerides. These newly made triglycerides are then bound to carrier proteins, forming tiny particles called *chylomicrons* that are secreted into the lymphatic capillaries (lacteals) of the intestinal villi (see fig. 10.16). From there, chylomicrons are carried via the lymph to veins. Unlike the other products of digestion, therefore, lipids enter the blood as polymers rather than monomers. It should be emphasized, however, that all foodstuffs, including fats, must be completely digested into their monomers before they can be absorbed by the digestive epithelium.

In this exercise, we will test the digestion of fat into free fatty acids and monoglycerides by measuring the decrease in pH produced by the liberation of free fatty acids as the digestion of triglycerides proceeds.

Clinical Applications

Cholesterol has an extremely low water solubility (20 μg/L), but it can be present in bile at 40 g/L because it aggregates with bile salts and lecithin in the centers of micelles. **Gallstones** commonly contain cholesterol as their major component, and the formation of gallstones is promoted by an excessive concentration of cholesterol in the bile. Gallstones are small, hard mineral deposits *(calculi)* that can produce painful symptoms by obstructing the cystic or common bile ducts. Blockage of the bile duct with a gallstone can result in the inadequate flow of bile to the intestine, producing obstructive jaundice and steatorrhea. *Obstructive jaundice* is caused by an elevation in the blood levels of the bile pigment bilirubin due to blockage of the bile duct. High bilirubin levels produce a yellowish discoloration of the skin, the sclera of the eyes, and the mucous membrane. *Steatorrhea,* the appearance of fat in the feces due to the inadequate digestion and absorption of fat, is associated with a deficiency in the uptake of fat-soluble vitamins A, D, E, and K. Vitamin K is necessary for normal blood clotting, so this condition can be serious. The most common and effective treatment for gallstones is surgical removal of the gallbladder using a procedure called *laparoscopic cholecystectomy.*

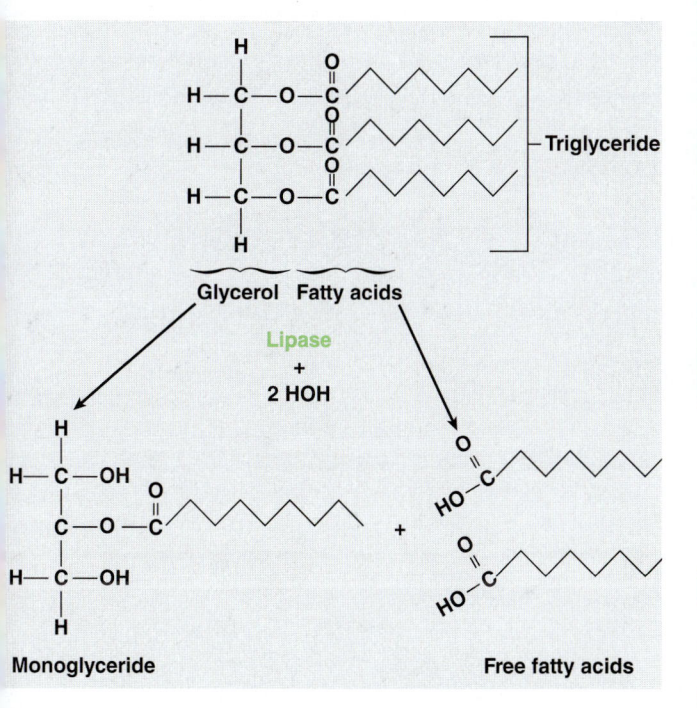

Figure 10.17 **The digestion of triglycerides.** Pancreatic lipase digests fat (triglycerides) by cleaving off the first and third fatty acids. This produces free fatty acids and monoglycerides. Sawtooth lines indicate hydrocarbon chains in the fatty acids.

(For a full-color version of this figure, see fig. 18.34 in *Human Physiology,* thirteenth edition, by Stuart I. Fox.)

Procedure

(See Fig. 10.18)

Step 1: Add 3.0 mL of dairy cream or vegetable oil to three test tubes, numbered 1–3.

Step 2: Add the following:

(1) To tube 1, add 5.0 mL of water and a few grains of bile salts

(2) To tube 2, add 5.0 mL of pancreatin solution

(3) To tube 3, add 5.0 mL of pancreatin solution and a few grains of bile salts

Step 3: Incubate the tubes at 37°C for 1 hour, checking the pH of the solutions at 20-minute intervals with a pH meter or with short-range pH paper.

Step 4: Record your data in the table in your laboratory report.

Step 1
3.0 mL cream or vegetable oil

| Tube 1 | Tube 2 | Tube 3 |

Step 2 Add the following to each tube.
- Tube 1: 5.0 mL water + bile salts (few grains)
- Tube 2: 5.0 mL pancreatin solution
- Tube 3: 5.0 mL pancreatin solution + bile salts (few grains)

Step 3
Incubate test tubes at 37°C.

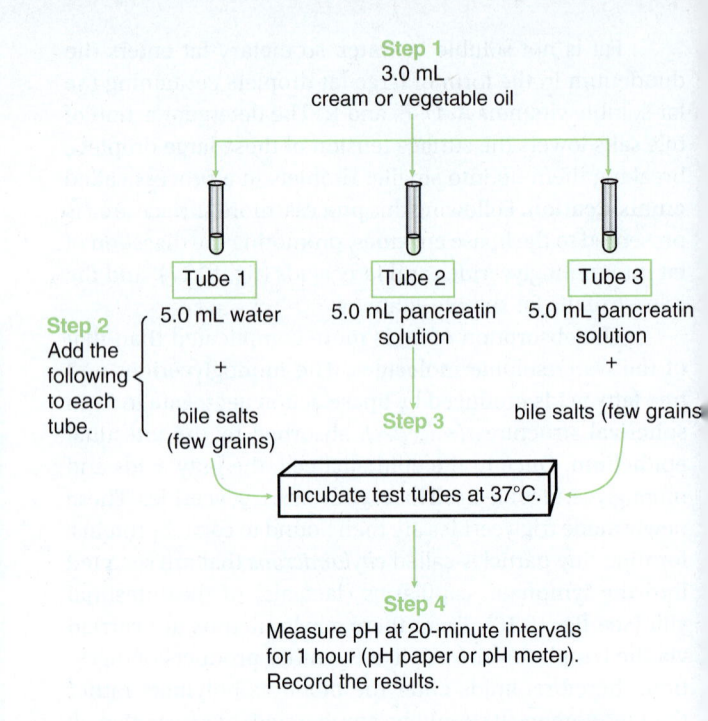

Step 4
Measure pH at 20-minute intervals for 1 hour (pH paper or pH meter). Record the results.

Figure 10.18 **A chart of the procedure for fat digestion.**

Laboratory Report 10.2

Name _____
Date _____
Section _____

DATA FROM EXERCISE 10.2

A. Digestion of Carbohydrate (Starch) by Salivary Amylase

1. Enter your data in this table using the rating method described in the procedure.

Contents Before Incubation	Starch After Incubation	Maltose After Incubation
Tube 1: Starch + distilled water		
Tube 2: Starch + saliva		
Tube 3: Starch + saliva + HCl		
Tube 4: Starch + boiled saliva		

B. Digestion of Protein (Egg Albumin) by Pepsin

1. Enter your observations in this data table.

Incubation Condition	Appearance of Egg White After Incubation
Tube 1: Protein + pepsin at 37°C	
Tube 2: Protein + pepsin + HCl at 37°C	
Tube 3: Protein + pepsin + HCl at 0°C	
Tube 4: Protein + HCl at 37°C	
Tube 5: Protein + pepsin + NaOH at 37°C	

C. Digestion of Triglycerides by Pancreatic Juice and Bile

1. Record your data in this table.

Time	pH		
	Tube 1: Fat + Bile Salts	Tube 2: Fat + Pancreatin	Tube 3: Fat + Bile Salts + Pancreatin
0 minutes			
20 minutes			
40 minutes			
60 minutes			

REVIEW ACTIVITIES FOR EXERCISE 10.2

Test Your Knowledge

1. Starch is partially digested into maltose by the action of _____
2. The enzyme in gastric juice that partially digests proteins is _____; this enzyme has a pH optimum of _____.
3. Which food group—carbohydrates, lipids, or proteins—is not digested significantly until it reaches the small intestine? _____
4. Bile is produced by the _____ and stored in the _____.
5. What is the function of bile salts? _____
6. The particles consisting of a combination of triglycerides and protein, secreted by intestinal epithelial cells into the central lacteals of the villi, are called _____.

Test Your Understanding

7. In procedure A, which tube(s) contained the most starch following incubation? Which tube(s) contained the most reducing sugars? What conclusion can you draw from these results?

8. In procedure B, which tube showed the most digestion of egg albumin? What can you conclude about the pH optimum of pepsin?

9. Compare the effects of HCl on protein digestion by pepsin and on starch digestion by salivary amylase. Explain the physiological significance of these effects.

10. In procedure C, which test tube displayed the most rapid fall in pH? Explain the reason for this, including an explanation of how the digestion of fat can cause a fall in the pH of the solution, and the function of bile salts.

11. How does the digestion and absorption of fat differ from the digestion and absorption of glucose and amino acids?

12. Why doesn't the stomach normally digest itself? Why doesn't gastric juice normally digest the duodenum?

Test Your Analytical Ability

13. Suppose, in performing procedure A, the test for starch and for reducing sugars both came out positive after a 1-hour incubation, but after the tubes incubated for 2 hours, the test for starch came out negative. How could you explain these results?

14. Reviewing your data from procedure A, what do you think happens to the digestion of a bite of bread after you swallow it?

15. Reviewing your data from procedure B, explain why frozen food keeps longer than food kept at room temperature.

16. A person with gallstones may have jaundice and an abnormally long clotting time. Explain the possible relationship between gallstones, jaundice, and blood clotting.

Clinical Investigation Questions

17. An excess of what molecule produced the yellowish discoloration, and what is a likely reason for this to occur? How does the steatorrhea relate to this reason?

18. What is the mechanism of action of Zantac and Prilosec, and what do these treat? Does this relate to the reason that the patient may need surgery? Explain.

EXERCISE 10.3

Nutrient Assessment, BMR, and Body Composition[1]

MATERIALS

1. Home scale or physicians height-weight scale
2. Tape measure, fat calipers (if available)
3. Calorie counting guide such as the U.S. Department of Agriculture Handbook, cookbooks, or popular diet books
4. Alternatively, caloric values of food, and the caloric expenditure of exercise, can be obtained from the Web. One good source is www.caloriecontrol.org.
5. A great deal of nutritional information, in addition to the caloric value of many foods, can be obtained from the Web site of the Nutrient Data Laboratory at www.ars.usda.gov/nutrientdata. Look under the heading "database products" on the home page, then click on "Nutritive Value of Foods." You must have Adobe Acrobat Reader, which can be downloaded from this site.

Energy consumed in food and expended by the metabolic activities of the body is measured in kilocalories. Weight is gained when the energy consumption is greater than the energy expenditure, and weight is lost when the reverse is true.

LEARNING OUTCOMES

You should be able to:

1. Describe the different nutrient classes, and list the calories per gram for carbohydrates, lipids, and fats.
2. Demonstrate how a dietary record is used to assess food and fluid consumption.
3. Define BMR, and demonstrate two different methods for estimating BMR.
4. Define activity factor (AF), and demonstrate how to estimate the number of calories burned for various activities.
5. Calculate and balance the number of calories consumed in the diet and calories expended in activities over three days.
6. Describe how each pound of body weight gain or loss is related to calories consumed or expended.

Textbook/Multimedia Correlations

Before performing this exercise, you should study the introductory material presented here. Further information relating to this exercise can be found in these pages of *Human Physiology*, thirteenth edition, by Stuart I. Fox:

- *Nutritional Requirement.* Chapter 19, p. 661.
- *Regulatory Functions of Adipose Tissue.* Chapter 19, p. 669.
- *Caloric Expenditures.* Chapter 19, p. 674.
- Multimedia correlations are provided in *Physiology Interactive Lab Simulations (Ph.i.L.S.):* Basal Metabolic Rate and Body Size (exercise 11).

Clinical Investigation

A man had a hip length of 40 inches and a waist measurement at its widest circumference of 44 inches. He was five feet, nine inches tall and weighed 218 pounds. He was taking medicine for hypertension, and his blood tests revealed a high fasting plasma triglyceride level. Although his plasma glucose was normal, he said his father had diabetes mellitus.

- Calculate this man's waist-to-hip ratio and body mass index (BMI) and indicate if he is normal, overweight, or obese.
- Explain the health significance of this man's measurements and family history.

Metabolism refers to the chemical transformations that occur in the body and is subdivided into **anabolism** and **catabolism**. Anabolic reactions are those that build large molecules out of smaller ones and generally require the input of energy (from ATP). Catabolic reactions are those that break down (hydrolyze) larger molecules into their subunits. These reactions occur in the body cells, which may release the subunit molecules (including glucose, amino acids, and fatty acids) into the blood.

---
1. Courtesy of Dr. Lawrence G. Thouin, Jr., Pierce College.

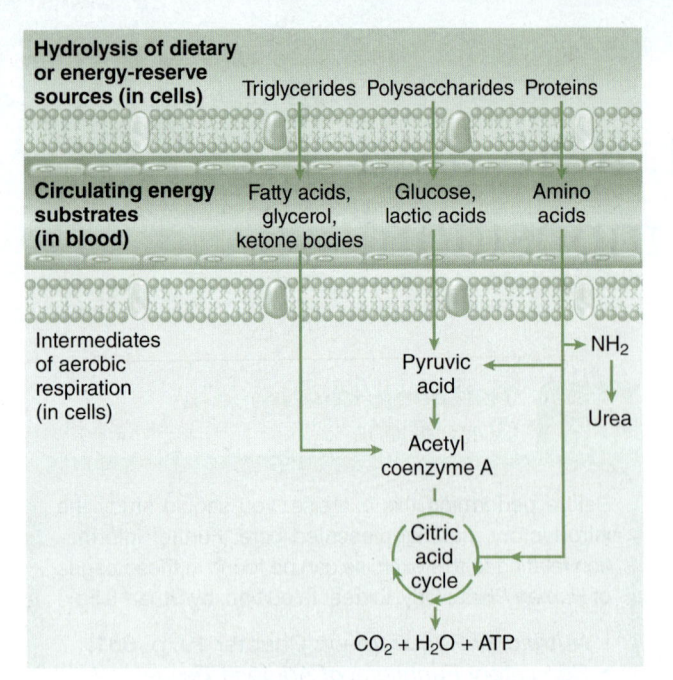

Figure 10.19 A flowchart of energy pathways in the body. The molecules indicated in the top and bottom are those found within cells, while the molecules in the middle are those that circulate in the blood.

(For a full-color version of this figure, see fig. 19.2 in *Human Physiology*, thirteenth edition, by Stuart I. Fox.)

The blood carries these *circulating energy substrates* to other cells in the body, which may use them in cellular respiration to produce ATP (fig. 10.19). In summary, anabolic reactions store energy within molecules of fat, glycogen, and protein, whereas catabolic reactions break these molecules down to derive energy from them.

The balance between anabolism and catabolism is influenced by the intake of energy from food molecules, and is regulated by a variety of hormones (fig. 10.20). When more food energy is taken into the body than is used, anabolism predominates and the body gains weight (mainly due to increased fat storage). When less food energy is consumed than is used, catabolism predominates to provide the needed energy from the molecules stored in the body. These shifts in metabolism are regulated by the levels of hormone secretion indicated in figure 10.20. When the body is neither gaining nor losing weight, anabolism and catabolism are in balance.

A. Body Composition Analysis

Body composition refers to relative proportion of both the actively metabolizing tissues, or *lean body mass*, and fat tissue in the body. Although some adipose tissue is essential, an excess of fat is detrimental to health. An estimated 25% of the U.S. population is *obese* (more than 20% above the "ideal body weight"), with greater risk of health problems.

Body composition can be estimated using a variety of techniques. One estimate of body fat is a height-to-weight ratio called the **body mass index (BMI)**. BMI equals the body weight in kilograms divided by height in meters squared. *Skinfold calipers* measure the thickness of the subcutaneous fat layer at particular body sites. These measurements are then compared to norms for total body fat. Underwater weighing, one of the most accurate methods of determining body fat, is also the most expensive and inconvenient. *Bioelectric impedance* instruments send a small current through skin electrodes to measure the resistance in the electrolyte-rich body fluids. This method estimates body water, and then uses a computer data bank to derive body fat indirectly (lean body mass has more water than fat tissue). *Infrared wands* derive the percent body fat from density data taken over the biceps brachii muscle.

Obesity has been defined by health agencies in different ways. The World Health Organization classifies people with a BMI of 30 or over as being at high risk for the diseases of obesity. According to the standards set by the National Institutes of Health, a healthy weight is indicated by a BMI between 19 and 25 (a BMI in the range of 25.0 to 29.9 is described as "overweight;" a BMI of over 30 is "obese"). According to one study, however, the lowest death rates from all causes occurred in men with a BMI in the range of 22.5 to 24.9, and in women with a BMI in the range of 22.0 to 23.4. Surveys indicate that over 60% of the adult population in the United States are either overweight (with a BMI greater than 25) or obese (with a BMI greater than 30).

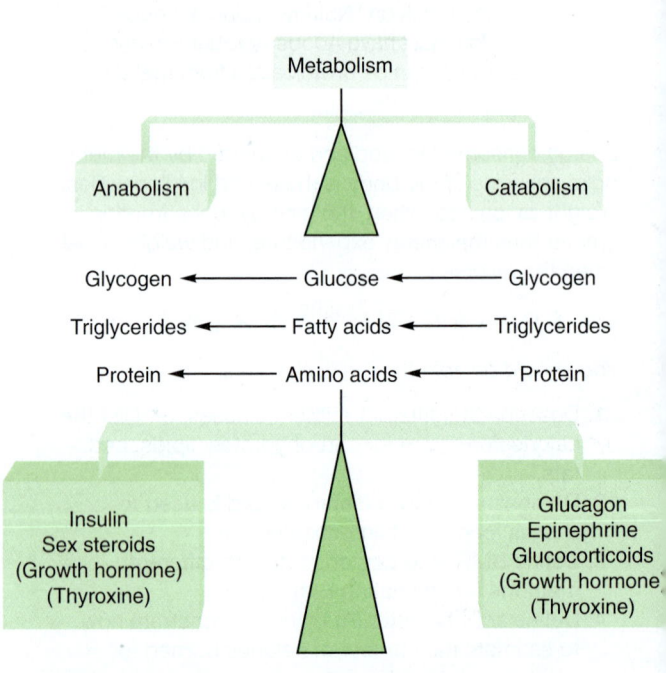

Figure 10.20 The regulation of metabolic balance. The balance of metabolism can be tilted toward anabolism (synthesis of energy reserves) or catabolism (use of energy reserves) by the combined actions of various hormones. Growth hormone and thyroxine have both anabolic and catabolic effects.

(For a full-color version of this figure, see fig. 19.6 in *Human Physiology*, thirteenth edition, by Stuart I. Fox.)

Clinical Applications

People with **apple-shaped bodies** (more fat around the abdominal area) seem to be more likely to develop cardiovascular disease, hypertension, and diabetes mellitus than those with **pear-shaped bodies** (more fat in the hips, buttocks, and thighs). Since males tend to become apple shaped, and females tend to become pear shaped, it appears that sex hormones help to direct the distribution of fat. For reasons still unknown, abdominal fat (stored deep in the body within the greater omentum) poses a greater health risk than the subcutaneous fat stored under the skin in the hips and thighs. For example, the pear-to-apple shift in fat distribution seen in postmenopausal women is accompanied by an increase in the risk of diseases, such as cardiovascular disease (usually more common in males) and breast cancer.

Procedure

(Body Mass Index)

1. The body mass index, or BMI, is obtained by the following formula:

$$BMI = \frac{w}{h^2}$$

   where:
   w = weight in kilograms (pounds divided by 2.2)
   h = height in meters (inches divided by 39.4)

2. Alternatively, you can enter your height and weight in the calculator provided at this Web site: *www.caloriecontrol.org/bmi.html*

3. Record your BMI here: _____

Normal Values A BMI of over 30 places a person at high risk for the diseases of obesity. A BMI of under 27 is considered healthy, and a BMI in the range of 23–25 appears to be optimum for health.

Procedure

(Waist-To-Hip Ratio)

1. Stand and measure your waist at the navel.
   Record this value (in cm): _____ cm.
2. Measure your hips at the greatest circumference (buttocks).
   Record this value (in cm): _____ cm.
3. Divide the waist circumference by the hip circumference to get the waist-to-hip ratio. Record this waist-to-hip ratio: _____

Normal Values According to the American Heart Association, a waist-to-hip ratio above 1.0 for men and above 0.8 for women is associated with an increased risk of cardiovascular disease.

Clinical Applications

Obesity, especially involving visceral fat, promotes insulin resistance and type 2 diabetes mellitus. Physicians have noticed that insulin resistance and type 2 diabetes mellitus are often associated with other problems, including hypertension and *dyslipidemia:* high blood triglyceride levels and low levels of HDL (high density lipoprotein—the "good cholesterol"). As might be expected from this, there is also a greater risk of atherosclerosis. This constellation of symptoms that surround type 2 diabetes, and thus also obesity, has been termed the **metabolic syndrome.** A person has the metabolic syndrome when there is *central obesity* (defined by waist circumference greater than specific values, which differ by sex and ethnicity) and two other conditions in a list that includes symptoms described above and hypertension. One study estimated that the metabolic syndrome raises the risk of coronary heart disease and stroke by a factor of three.

B. Energy Intake and the Three-Day Dietary Record

Both the chemical energy consumed in foods and the metabolic energy expended by the cell are measured in **kilocalories (kcal)**, or **Calories (C).** The major sources of food calories are carbohydrates, fats (lipids), and proteins. When allowances are made for inefficiency in the assimilation of each nutrient, one gram *(g)* of each of the three energy nutrients provides the body with approximately the following number of kilocalories:

$$1 \text{ g carbohydrate} = 4.0 \text{ kcal}$$
$$1 \text{ g fat} = 9.0 \text{ kcal}$$
$$1 \text{ g protein} = 4.0 \text{ kcal}$$

The primary **carbohydrates** in food are the *sugars* (such as glucose, fructose, and sucrose) and *complex carbohydrates* (such as starches and dietary fiber). To meet the energy requirements of children and adults, recommended dietary allowances suggest that more than half (about 55%) of the calories consumed per day come from carbohydrate sources in the diet. Emphasis should be on the increased consumption of complex carbohydrates, especially dietary fibers found in fruits, vegetables, legumes, and whole-grain cereals. In addition to providing a source of calories, dietary fiber has been associated with improving overall health by promoting normal stool elimination, enhancing satiety, and lowering plasma cholesterol levels.

Lipids are generally divided into *triglycerides, phospholipids,* and *sterols* (steroids). The digestion, emulsification, and absorption of lipids also facilitates the absorption of *fat-soluble vitamins A, D, E,* and *K* and the *essential fatty acids.*

Mammals lack the enzymes needed to insert a double bond at the n-6 or n-3 positions in fatty acids and thus must eat the **essential fatty acids**—*linoleic acid* and *alpha linolenic acid*—in the diet. Many studies demonstrate that the n-3 fatty acids—also called the **omega-3 fatty acids** (*eicosapentaeonoic acid*, or *EPA*, and *docosahaenoic acid*, or *DHA*)—may help inhibit platelet function in thrombus formation, the pro2gression of atherosclerosis, and/or ventricular arrhythmias. Because these fatty acids are plentiful in fish, the American Heart Association recommends that people eat at least two meals of fish a week.

Animal fats, solid at room temperature, are more saturated than vegetable oils because the hardness of the triglyceride is determined partly by the degree of saturation. *Trans fats*, also solid at room temperature, are produced artificially by partially hydrogenating vegetable oils (this is how margarine is made). This results in **trans fatty acids**, in which the single hydrogen bonded to each carbon is located on the opposite side of the double bond between carbons, and the carbon atoms from a straight chain. In contrast, most naturally occurring unsaturated fatty acids have their hydrogens on the same side as the double bond (are *cis fatty acids*), and their carbons form a repeating bent pattern. Trans fats are used in almost all commercially prepared fried and baked foods. Saturated fat and trans fatty acids have been shown to raise LDL cholesterol (the "bad cholesterol"), lower HDL cholesterol (the "good cholesterol"), and thereby to increase the risk of coronary heart disease. The Food and Drug Administration (FDA) now requires all manufacturers to list trans fats on their food labels.

According to U.S. Department of Agriculture (USDA) surveys, about 15% of the total food energy intake of the average American is derived from **protein.** Most of this protein, about 65%, is derived from animal sources, primarily meat and dairy products, with only about 20% from cereal grains. Despite increased protein requirements for certain populations, such as growing children, pregnant or lactating females, and the elderly, the typical American diet normally meets or exceeds the requirements. Interestingly, there is little evidence that physical exercise increases the need for protein, other than that required during the initial conditioning period. Therefore, individuals eating a typical American diet need make no adjustment to the recommended allowance for protein.

Nutrients that do not contribute energy to the body but that still are required to maintain body functions are vitamins, minerals, and water. The **fat-soluble vitamins** (A, D, E, and K) are absorbed from the small intestine with other food lipids and are concentrated to some degree in adipose tissue. The **water-soluble vitamins** are thiamine (B_1), riboflavin (B_2), niacin (B_3), pantothenic acid (B_5), pyridoxine (B_6), biotin (B_7), folate (B_9), cyanocobalamin (B_{12}), and vitamin C. Most water-soluble vitamins serve as coenzymes that assist enzymes in the regulation of metabolism. The **major minerals**, required in higher quantities, are calcium, phosphorus, magnesium, and the electrolytes (sodium, chloride, and potassium). The **minor minerals,** or **trace elements,** are required in lesser quantities. They include iron, zinc, iodine, selenium, copper, manganese, fluoride, chromium, and molybdenum. Recommended quantities of these nutrients are normally met when a variety of foods are consumed. For the average person eating a typical American diet, therefore, no supplementation is recommended. Indeed, high intakes of certain nutrients in supplement form can be toxic.

Water is also an essential dietary nutrient. Although assessments of adequate water intake involve many complex factors, the general recommendation is a minimum of 1.0 mL of water per kilocalorie of energy expended per day. Therefore, an individual expending 2,000 kilocalories per day should consume at least 2 liters of water.

Clinical Applications

Triglycerides are the major lipid components of foods and the most concentrated source of energy (9 kcal/g). The average American currently derives about 36% of the daily total dietary calories from fats. High dietary fat and cholesterol intakes have been associated with an increased risk of cardiovascular disease and cancer. As recommended by the Food and Nutrition Board's Committee on Diet and Health, the fat content of the U.S. diet should be lowered so as not to exceed 30% of the caloric intake (10% of fat calories from *saturated* fatty acids, 10% from *polyunsaturated* fatty acids, and 10% from *monounsaturated* fatty acids). Dietary cholesterol should be less than 300 mg/day.

Procedure

1. Complete the 3-day dietary record (provided in the laboratory report). The 3 days must be consecutive (attempt to include at least one weekend day). At roughly the same time each day, try to weigh yourself under consistent conditions (comparable clothing, same scale, etc.). Record your weight in pounds in the space provided for that day.
2. Record all foods and fluids consumed each day in the appropriate columns, noting the approximate time of day. Estimate food quantities by weight (such as ounces) and fluids by volume (e.g., cups, or liters), depending upon the units listed in your calorie guide. Record the total volume of fluids consumed at the bottom of the column.
3. Look up the estimated number of calories for each food or fluid noted in your diet record. The caloric values can be looked up in reference books or on the net, for example at *www.caloriescount.org/calculator.html*. Record the total calorie intake at the bottom of the column.
4. Comment on where the food or fluid was consumed. Do you always eat sitting down at a table, in a quiet, relaxing environment? Or are you sometimes in the car, between classes, in front of the TV, or in bed studying?
5. Comment on why you ate or drank. Do you always eat because you are hungry? On occasion, do you eat because you are bored, or because the food is there, or because someone else is paying for it?

C. ENERGY OUTPUT: ESTIMATES OF THE BMR AND ACTIVITY

The total energy expended each day includes the energy required at rest and that expended during physical activity.

For most people, the calories consumed at rest make up most of the total daily energy expenditure. This energy is used to pump blood, inflate the lungs, transport ions, and carry on the other functions of life. Measurement of this resting energy expenditure shortly after awakening and at least 12 hours after the last meal is known as the **basal metabolic rate (BMR)**. With all other factors equal, BMR is influenced most by the amount of actively metabolizing tissues, or *lean body mass*. The BMR is higher in younger, more muscular people, and in males (who have a higher average muscle mass than females). The BMR is also influenced by **thyroxine**. People who are *hypothyroid* have a low BMR; those who are *hyperthyroid* have a high BMR.

While most of our caloric output is spent at rest, most people are physically active and expend calories beyond the BMR. The additional number of activity calories expended will vary with the individual and with the duration, intensity, and types of activities performed. This increased caloric output can be estimated by multiplying an *activity factor* (AF) by the BMR. In general, the average sedentary person raises the total number of calories burned per day to about 130% (AF = 1.3) of the estimated BMR. Moderately active people may raise daily expenditures upwards of 150% (AF = 1.5) of the BMR estimates. Top athletes may double the BMR estimates, or more (AF ≥ 2.0). Aerobic activities, such as running, swimming, bicycling, and dancing, normally burn more calories than anaerobic activities, such as weight lifting.

Activity	Activity Factor (AF)
Lying in bed all day—equal to BMR	1.00–1.29
Mild activity—normal routine, no exercise	1.30–1.49
Moderate activity—1 hour of aerobic exercise	1.50–1.69
Heavy activity—2 to 4 hours of aerobic exercise	1.70–1.99
Rigorous athletic training	2.0 and above

Exercise 10.3 allows you to determine how many calories you gained or lost each day for 3 days and from this value how much weight you gained or lost. If you continue this exercise for weeks or months, you can accurately track your energy balance as a means of dieting. Studies suggest that those who log their food intake and energy expenditures are more successful dieters than those who do not. There are also several computer and Web-based programs that, like exercise 10.3, require dieters to enter their daily food intake and activities. For example, a free program at *www.chosemyplate.gov* is provided by the USDA (choose the SuperTracker option). The time and effort required to log daily food intake and activities is well spent, considering the increased risk that obesity poses for cardiovascular disease and type 2 diabetes.

Procedure

Step 1. Estimate your basal metabolic rate (BMR) using two different methods:

Method 1: Your weight in kilograms (2.2 lb/kg) _____ (kg):
Female BMR = 0.7 kcal/kg/hr
Male BMR = 1.0 kcal/kg/hr

Use the conversion factors to calculate your kilocalories per hour. Then, multiply this figure by 24 (hours per day), and enter your answer in the space provided here. In one day (24 hours), your BMR is approximately _____ kcal.

Method 2: Estimate your *Ideal Body Weight (IBW)* in pounds.

Female IBW = 100 lb for the first 5 feet in height + 5 lb per inch above 5 feet in height

Male IBW = 106 lb for the first 5 feet in height + 6 lb per inch above 5 feet in height

Your Ideal Body Weight is approximately _____ lb.

Next, multiply your IBW by 10 for your daily estimated BMR. In one day (24 hours), your BMR is approximately _____ kcal.

Note: *These BMRs are only estimates, so a difference between the two values is to be expected. Select the one BMR estimate you feel is most accurate, write that number in the space provided on the dietary record in the laboratory report, and use it for calculations.*

Step 2. For each day in your 3-day dietary record, select one activity factor (AF) from the chart that best reflects your total activity for that 24-hour period, with 1.30 typical for the casual college routine (without exercise); and write that decimal for each day in the AF box of the dietary record.

Note: *For variety (and more fun) attempt to vary your activities and the AF each day—another reason to include one weekend day in this report.*

Step 3. Calculate the total number of calories expended (output) each day as:
Total calories expended (kcal) =
Activity AF (from box) × BMR (from step 1, method 1 or 2)
Write this total in line 3 of the dietary record.

Step 4. Subtract total calories expended from total calories consumed to determine the caloric balance lost or gained that day (in your record subtract line 3 from line 1). Write the number of excess calories in line 4 of the dietary record; circle either calories "gained" or "lost."

Step 5. Assuming that 1 lb of body tissue (not just fat) gained or lost represents approximately 3,500 kcal, convert the excess calories from line 4 into body weight:

Body weight (lost/gained) = _____ (kcal)
÷ 3,500 (kcal/lb) = _____ (lb)

Write the pounds lost/gained that day in line 5 of the dietary record.

Step 6. Complete the evaluation section in the laboratory report, which follows the 3-day dietary record.

3-Day Dietary Record

Day # _____   Day of Week _____   Your Weight (lb) _____

Time (a.m./p.m.)	Foods-Units (oz, g, . . .)	Fluid-Units (cup, mL, . . .)	Calories (kcal)	Where?	Why?	Activities
		Total fluid:	Total kcal:			

1. Total calories consumed (intake): _____ kcal
2. Estimated BMR at rest: _____ kcal (step 1, method 1 or 2)
3. Total calories expended (output): _____ kcal (step 3)
4. Today's caloric balance: _____ kcal (step 4)   gained / lost (circle one)
5. Given that 1 lb of body weight (not just fat) is equal to approximately 3,500 kcal gained or lost, how many pounds of body weight were gained or lost today? _____ lb.
6. Enter values from today's report into the Dietary Record Evaluation form in the laboratory report.

AF

3-Day Dietary Record

Day # _____   Day of Week _____   Your Weight (lb) _____

Time (a.m./p.m.)	Foods-Units (oz, g, . . .)	Fluid-Units (cup, mL, . . .)	Calories (kcal)	Where?	Why?	Activities
		Total fluid:	Total kcal:			

1. Total calories consumed (intake): _____ kcal
2. Estimated BMR at rest: _____ kcal (step 1, method 1 or 2)
3. Total calories expended (output): _____ kcal (step 3)
4. Today's caloric balance: _____ kcal (step 4)   gained / lost (circle one)

AF ☐

5. Given that 1 lb of body weight (not just fat) is equal to approximately 3,500 kcal gained or lost, how many pounds of body weight were gained or lost today? _____ lb.
6. Enter values from today's report into the Dietary Record Evaluation form in the laboratory report.

3-Day Dietary Record

Day # _____   Day of Week _____   Your Weight (lb) _____

Time (a.m./p.m.)	Foods-Units (oz, g, . . .)	Fluid-Units (cup, mL, . . .)	Calories (kcal)	Where?	Why?	Activities
		Total fluid:	Total kcal:			

1. Total calories consumed (intake): _____ kcal
2. Estimated BMR at rest: _____ kcal (step 1, method 1 or 2)
3. Total calories expended (output): _____ kcal (step 3)
4. Today's caloric balance: _____ kcal (step 4)   gained / lost (circle one)
5. Given that 1 lb of body weight (not just fat) is equal to approximately 3,500 kcal gained or lost, how many pounds of body weight were gained or lost today? _____ lb.
6. Enter values from today's report into the Dietary Record Evaluation form in the laboratory report.

AF ☐

Laboratory Report 10.3

Name _____
Date _____
Section _____

DATA FROM EXERCISE 10.3

3-Day Dietary Record Evaluation

1. Record the total number of calories consumed over 3 days:
   Day 1 _____ Day 2 _____ Day 3 _____ = kcal

2. Record the total number of calories expended over 3 days:
   Day 1 _____ Day 2 _____ Day 3 _____ = kcal

3. Subtract the total kilocalories expended from the total kilocalories consumed (line 2 from line 1), and indicate whether the difference represents a gain (+) or a loss (−).
   _____ total kcal gained / lost (circle one) over 3 days

4. Divide your total from step 3 by 3,500 kcal/lb to convert kilocalories to pounds. Indicate whether this represents a weight gain or a weight loss.
   _____ lb gained/lost (circle one) over 3 days

5. Determine your measured weight change by subtracting your weight on day 3 from your weight on day 1. Compare your calculated weight gain/loss with your actual weight gain/loss, and discuss many of the variables that cause the numbers to differ. (*Hint:* Explain individual differences as well as variables built into the exercise.)

6. Study your 3-day dietary record. Make a detailed list of at least five specific suggestions for improving this diet. (*Hint:* Discuss nutrient classes, except water; make food exchanges for improvements, and include where and why entries. Exercise is not dietary, and water is discussed next.)

   (a)

   (b)

   (c)

   (d)

   (e)

7. Record the total volume of fluids consumed over 3 days:
Day 1 _____ Day 2 _____ Day 3 _____ = _____ Total fluid (cups, oz, etc.)

Based on the recommendation of at least 1.0 mL of water per kilocalorie of energy expended each day, did you meet this requirement each day? If not, propose a plan that would allow you to meet this requirement. Were diuretics such as caffeine or alcohol included? If so, what adjustments should be made to your daily fluid requirements?

REVIEW ACTIVITIES FOR EXERCISE 10.3

Test Your Knowledge

1. Chemical reactions in the body that break down fat, glycogen, and protein for energy comprise a subcategory of metabolism known as _____.
2. Fat provides _____ kilocalories per gram, whereas carbohydrates and protein provide _____ kilocalories per gram.
3. For optimum health, most of the calories consumed in a diet should come from _____ (carbohydrates/fat/protein).
4. Vitamins A, D, E, and K are grouped together as the _____ vitamins.
5. The lowest rate of energy expenditure of the body is called the _____.
6. A measurement of body fat that involves a height-to-weight ratio is called the _____.

Test Your Understanding

7. Distinguish between anabolism and catabolism. What can you do today to tilt your total body metabolism toward anabolism? The secretion of which hormones would be increased and which would be decreased when you did this?

8. Define *basal metabolic rate (BMR)*, and explain how physical exercise can influence body weight.

9. Explain how the BMI measurement relates to the terms "overweight" and "obese." Identify the health conditions associated with obesity.

10. What are "saturated fats"? What are "trans fats"? Explain how too much consumption of these affects health.

Test Your Analytical Ability

11. Suppose two people have the same height and weight, but person A has significantly higher bioelectrical impedance than person B. How do the BMI and percentage of body fat compare in these two people? Explain.

12. How would the two people in question 11 compare in their waist-to-hip ratios and their risks of developing type 2 diabetes mellitus and the metabolic syndrome? Explain.

Test Your Quantitative Ability

13. A college student who gained unwanted weight wishes to lose 20 pounds before the graduation ceremonies, 10 months away. By how many *kilocalories per week* would the student have to reduce intake or increase expenditures to lose the 20 pounds?

14. How many *kilocalories per day* would the student have to lose over the 10-month period? Is this a practical weight loss program? Explain fully.

15. If this student wants to lose all of this weight by only removing fat calories from foods consumed, how many *grams of fat per day* must be removed from the diet? Is this a practical weight loss program? Explain fully.

Clinical Investigation Questions

16. What is this person's waist-to-hip ratio and BMI? Are these values normal or do they indicate that the person is overweight or obese?

17. What are the health issues suggested by this person's measurements and family history? What can be done to mitigate the dangers he faces? Explain.

Section 11

Reproductive System

Unlike the physiology of other body systems, the function of the reproductive system is not to maintain homeostasis, but rather to ensure the continuity of the species. The study of the reproductive system includes the anatomy and physiology of many other body systems. Neural and endocrine regulatory mechanisms are very active and an integral part of reproductive physiology. The cardiovascular, respiratory, muscular, digestive, and other systems are likewise involved in the physiology of reproduction.

The primary sex organs are the **gonads:** *testes* in the male and *ovaries* in the female. The gonads produce **gametes** (*sperm* and *ova*) and **sex steroid hormones.** The functions of the gonads in both sexes are regulated by gonadotropic hormones (FSH and LH) secreted by the anterior pituitary gland. Secretion of these gonadotropic hormones, in turn, is regulated by hormones produced in the hypothalamus and by feedback effects from the sex steroids. The regulation of ovarian function by FSH and LH follows a cyclical pattern in females, producing the menstrual cycle, whereas the stimulation of the testes is noncyclical.

Associated with the functions of the gonads are the **accessory sex organs.** These include the uterus, uterine (fallopian) tubes, vagina, labia, and clitoris in a female (fig. 11.1) and the epididymis, vas (ductus) deferens, seminal vesicles, prostate, ejaculatory ducts, and penis in a male (fig. 11.2).

The uterine tubes in a female serve to convey the ovum toward the uterus. **Fertilization** normally occurs in the uterine tube, whereas **implantation** of the embryo occurs a week later in the inner lining, or *endometrium,* of the uterus. The endometrium undergoes monthly cycles of shedding (in *menstruation*) and growth as a result of the cyclical changes in the secretion of sex steroid hormones from the ovary.

Sperm produced in the seminiferous tubules of the testes within the scrotum are conveyed to a tubular structure, the *epididymis,* where they undergo maturation. The vas deferens transports the sperm out of the epididymis and into the body cavity, where fluid from the seminal vesicles and prostate is added to form *semen,* or *seminal fluid.* The ejaculatory duct fuses with the urethra so that semen can be ejaculated out of the urethra at the tip of the penis. Although millions of sperm are ejaculated into the female vagina, only about 100 survive to reach a uterine tube and only one sperm cell, or *spermatozoon,* is allowed to fertilize the ovum ovulated by a single follicle in the ovary. This fertilized ovum, or *zygote,* given the proper functioning of the *placenta* within the uterus, can develop into a new and genetically unique human being.

Exercise 11.1	Ovarian Cycle as Studied Using a Vaginal Smear of the Rat
Exercise 11.2	Human Chorionic Gonadotropin and the Pregnancy Test
Exercise 11.3	Patterns of Heredity

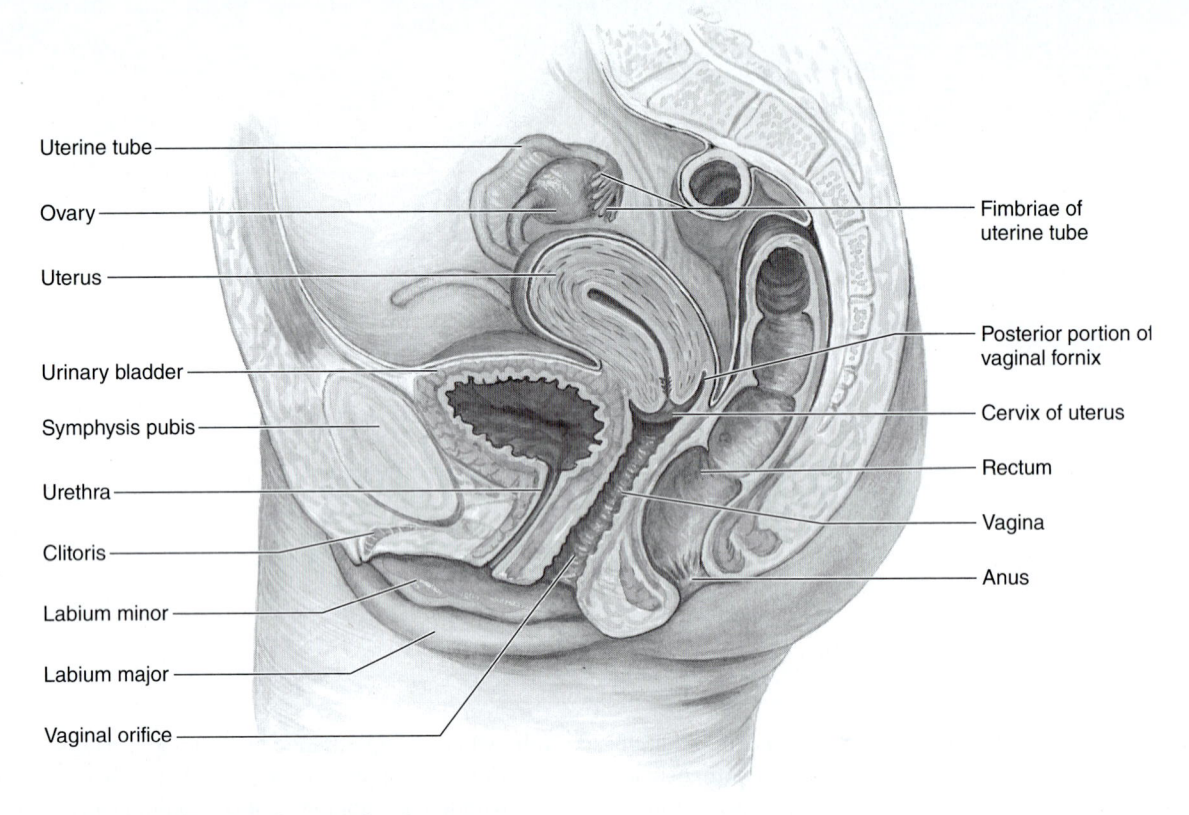

Figure 11.1 **Organs of the female reproductive system.**
(For a full-color version of this figure, see fig. 20.24 in *Human Physiology*, thirteenth edition, by Stuart I. Fox.)

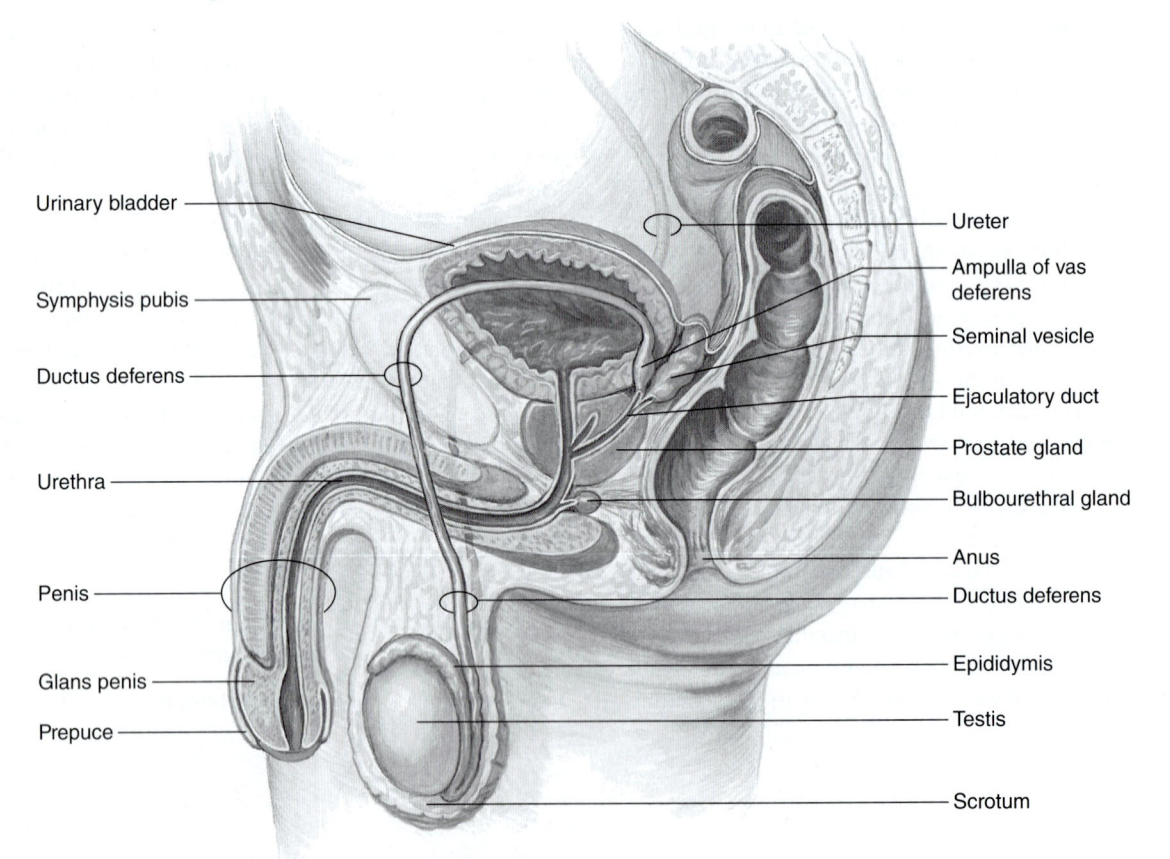

Figure 11.2 **Organs of the male reproductive system.**
(For a full-color version of this figure, see fig. 20.19 in *Human Physiology*, thirteenth edition, by Stuart I. Fox.)

EXERCISE 11.1

Ovarian Cycle as Studied Using a Vaginal Smear of the Rat

MATERIALS

1. Young female rats
2. Isotonic saline and cotton swabs (or, alternatively, eyedroppers may be used)
3. Giemsa's stain (dilute concentrate 1:50) and absolute methyl alcohol in staining jars; the staining of the slides is optional
4. Microscopes and microscope slides

The cyclic changes in ovarian hormone secretion cause cyclic changes in the epithelium of the female reproductive tract. By observing exfoliated epithelial cells, the stages of the ovarian cycle and the levels of ovarian hormone secretion can be determined.

LEARNING OUTCOMES

You should be able to:

1. Identify the phases of the ovarian cycle.
2. Describe the changes that occur in the endometrium and correlate these changes with the stages of the ovarian cycle.
3. Describe the appearance of a vaginal smear at different stages of the cycle, and explain the clinical usefulness of a vaginal smear.

Textbook/Multimedia Correlations

Before performing this exercise, you should study the introductory material presented here. Further information relating to this exercise can be found in these pages of *Human Physiology*, thirteenth edition, by Stuart I. Fox:

- *Menstrual Cycle*. Chapter 20, p. 727.
- Multimedia correlations are provided in *MediaPhys 3.0*: Topics 14.22–14.36.

Clinical Investigation

A vaginal smear was performed on a 38-year-old patient with vaginitis and yeast cells *(Candida albicans)* were observed in the smear. The patient voiced concern that she might be entering an early menopause, and the physician responded that he didn't think so because he saw primarily superficial squamous cells and some intermediate cells, but not parabasal cells, in the smear.

- Identify the layers of the human vaginal epithelium and explain what can cause different layers to appear in a vaginal smear.
- Explain why the physician did not think the patient had entered menopause.

The amount of gonadotropic hormones (FSH and LH) secreted by the anterior pituitary of females increases and decreases in a cyclical fashion. The secretion of estrogen and progesterone by the ovary will follow the same cycle. In most mammals, sexual receptivity (heat, or *estrus*) occurs during a specific part of the cycle. This pattern is called an **estrous cycle.** In humans and old-world primates, sexual receptivity occurs throughout the cycle, with monthly bleeding occurring at the beginning of each cycle. These cycles are called **menstrual cycles** (*menses* means "monthly") (fig. 11.3).

The uterus is one of the target organs of the ovarian hormones. As the secretions of estrogen and progesterone increase during the cycle, the inner lining of the uterus (the endometrium) increases in thickness (fig. 11.3). The ovarian hormones are preparing the uterus for possible implantation of the developing embryo should fertilization occur. If fertilization does not occur, the cyclical decrease in plasma levels of estrogen and progesterone causes the necrosis (cellular death) and sloughing off of the upper two-thirds of the endometrium. The cycle is ready to begin anew (fig. 11.3).

The rise and fall in the secretion of estrogen and progesterone from the ovaries during the course of a menstrual cycle causes cyclic changes in the endometrium, as illustrated in figure 11.3. At the same time, the rise and fall in estrogen

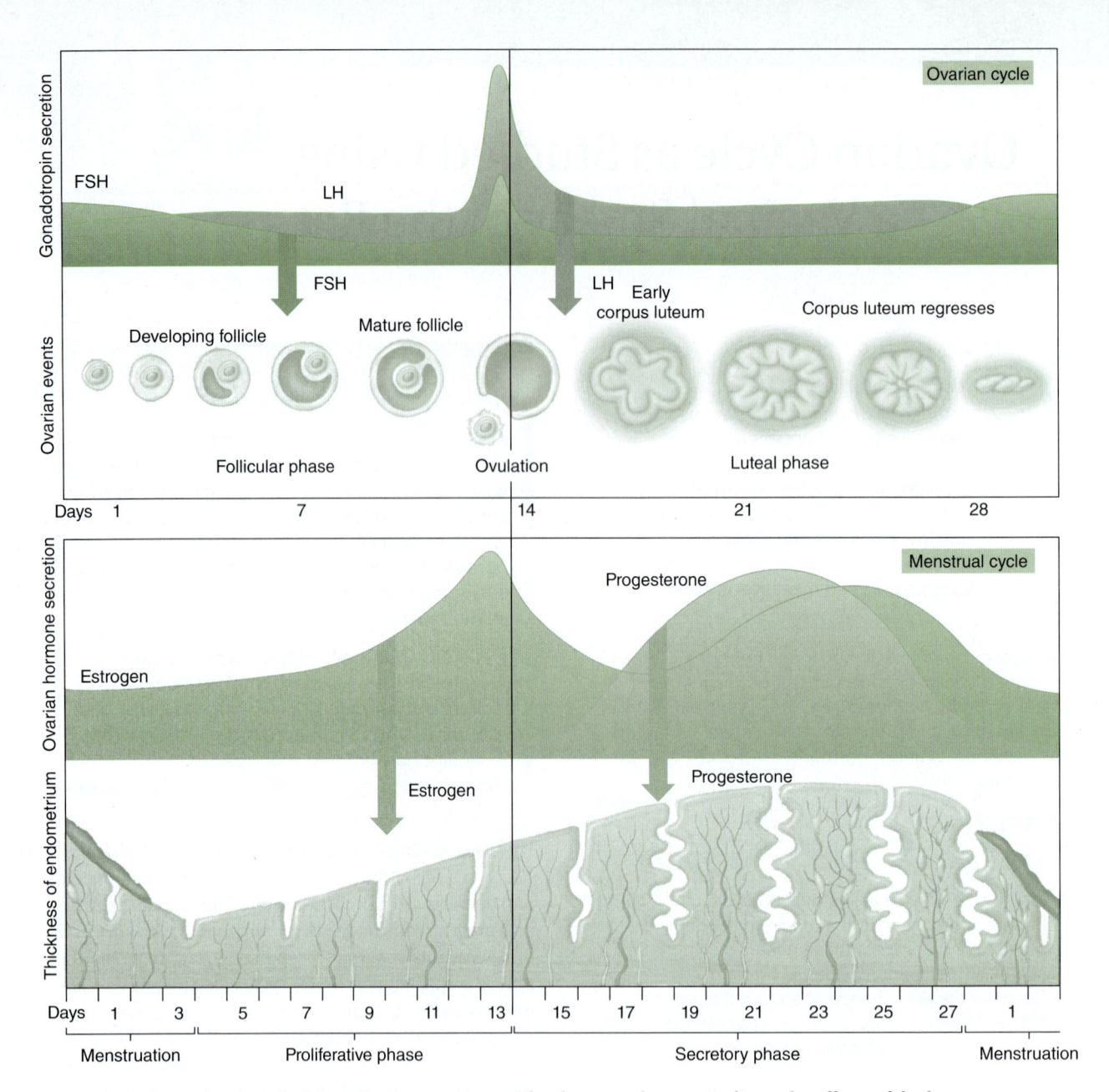

Figure 11.3 The cycle of ovulation and menstruation. The downward arrows indicate the effects of the hormones.
(For a full-color version of this figure, see fig. 20.33 in *Human Physiology*, thirteenth edition, by Stuart I. Fox.)

and progesterone during a menstrual or estrus cycle produce cyclic changes in the vaginal mucosa. This causes different cells to be shed, or *exfoliated,* at different stages of the cycle, and these can be observed by taking a **vaginal smear.**

The estrous cycle of a rat is usually completed in 4 to 5 days. The cycle is roughly divisible into four stages that can be identified by a vaginal smear (fig. 11.4).

1. **Proestrus.** Proestrus is the beginning of a new cycle. The follicles of the ovary start to mature under the influence of the gonadotropic hormones, and the ovary starts to increase its secretion of estrogen.
Vaginal smear. Nucleated epithelial cells (fig. 11.4*a,b*).
2. **Estrus.** The uterus is enlarged and distended because of the accumulation of fluid. Estrogen secretion is at its height. The rat thus becomes sexually receptive as ovulation occurs. *Vaginal smear.* Squamous cornified cells (fig. 11.4*c*).
3. **Metestrus (or *Diestrus day 1*).** Metestrus is the stage that follows ovulation. The ovary contains functioning corpora lutea secreting progesterone, preparing the uterus for implantation.
Vaginal smear. Many leukocytes, some cornified epithelial cells.
4. **Diestrus (or *Diestrus day 2*).** The corpora lutea regress, and the declining secretion of estrogen and progesterone causes regression of the uterine endometrium.
Vaginal smear. Entirely leukocytes.

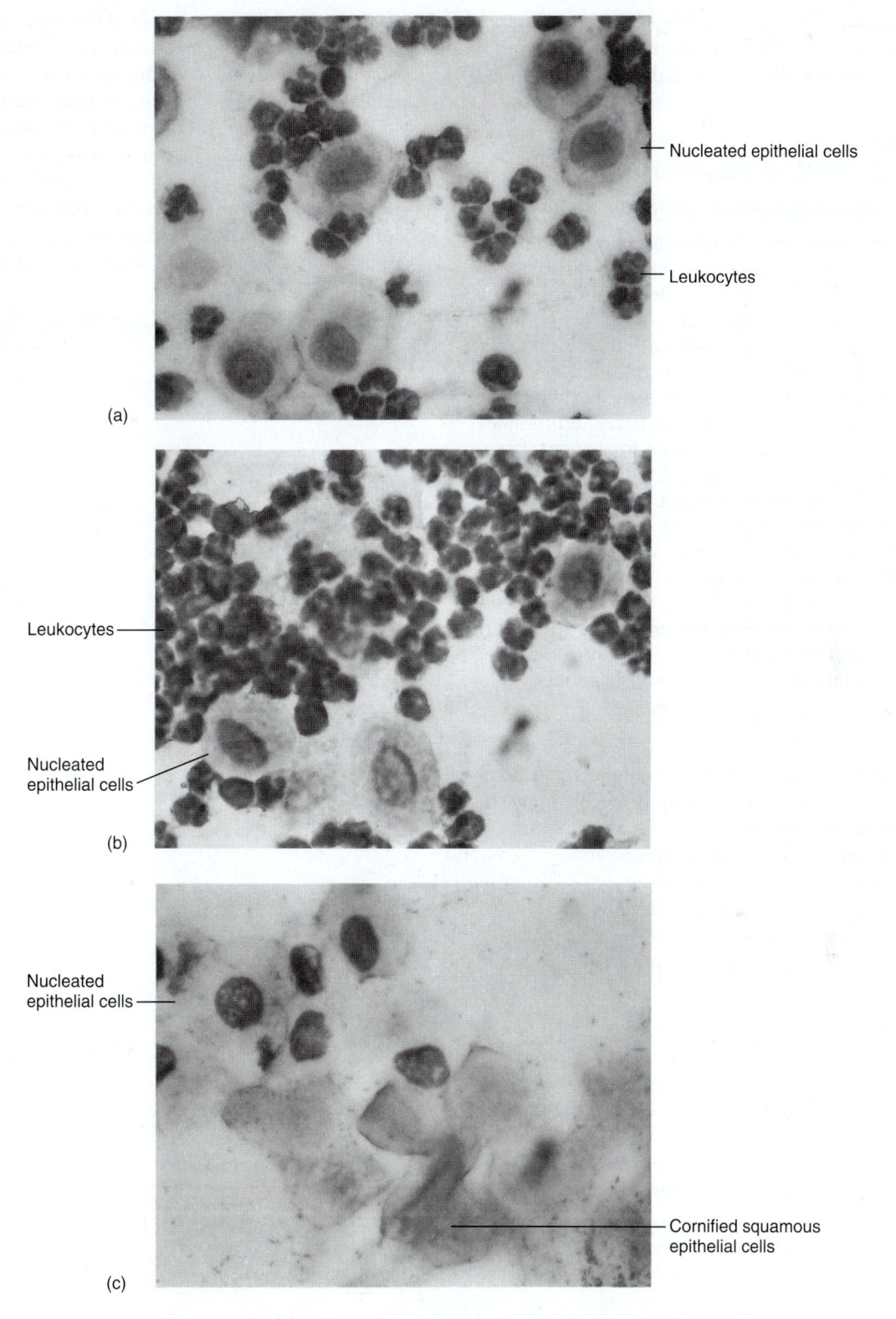

Figure 11.4 The vaginal smear of a rat. (a,b) Nucleated epithelial cells and leukocytes; (c) nucleated epithelial cells and cornified cells.

The stages of the estrous cycle correspond to the phases of the menstrual cycle as follows (fig. 11.3):

Proestrus corresponds to the *follicular phase*.
Estrus corresponds to the *ovulatory phase*.
Metestrus corresponds to the *luteal phase*.
Diestrus corresponds to the *menstrual phase*.

In humans, the vaginal epithelium undergoes cyclic changes similar to those seen in rats. However, the changes in humans follow the longer menstrual cycle, and the cell types seen in the human vaginal smear are somewhat different from those seen in the rat vaginal smear.

Clinical Applications

Vaginal smears can be used clinically to determine the stage of the menstrual cycle, the effectiveness of exogenous ("from outside") hormone treatments, and the stage and health of pregnancy. Vaginal smears are also helpful in the diagnosis of pathological states, such as *primary* or *secondary amenorrhea* (the absence of a menstrual period when one would normally be expected), inflammation, and cancer. Because malignant tissue exfoliates to a greater extent than normal tissue, this technique may discover cancers too small to detect by other means. The staining technique most often used clinically for vaginal smears was developed by an American physician, George Papanicolaou. Since the introduction of this technique in 1942, the **Pap smear** has become synonymous with the clinical vaginal smear. The Pap smear is used to screen for cervical cancer.

In humans, the vaginal epithelium contains four cell layers when it is stimulated to thicken by high estrogen secretion. From the surface inwards, these layers are: superficial squamous cells, an intermediate layer, a parabasal layer, and a basal layer. The topmost squamous cell layer *exfoliates* (sheds) and is seen in a vaginal smear. A vaginal smear during the *proliferative phase* of the uterine cycle (corresponding to the follicular phase of the ovaries, fig. 11.3), when the estrogen level is high, consists almost entirely of superficial squamous cells. Under the influence of progesterone during the *secretory phase* of the cycle (corresponding to the luteal stage of the ovaries, fig. 11.3) and during pregnancy, the top layer of cells in the vaginal epithelium is the intermediate layer. These cells are seen in a vaginal smear taken during the second half of the cycle and during pregnancy. In the absence of all sex steroids, such as prior to the onset of puberty in females, or *menarche,* the vaginal epithelium is atrophic, and parabasal cells are seen in a vaginal smear.

Procedure

1. Remove a female rat from its cage, and hold it in one hand using proper technique demonstrated by the instructor.
2. Moisten the cotton tip of a swab with isotonic saline, and insert it into the vagina. Smear this on a clean microscope slide. (One swab is sufficient to make approximately six slides.) Alternatively, insert an eyedropper containing about 1/mL of isotonic saline into the vagina. Gently squeeze the bulb two or three times, and then expel the slurry onto a microscope slide.
3. Stain the slides as listed here (optional):
   (a) Dry the slide in air.
   (b) Immerse the slide in absolute methyl alcohol (5 seconds).
   (c) Air dry the slide.
   (d) Place it in Giemsa's stain (1:50) for 30 minutes. This staining step is optional, as even unstained cells can easily be seen.
   (e) Rinse it in tap water.
   (f) Dry the slide and observe it without a coverslip under the microscope.

Laboratory Report 11.1

Name _____
Date _____
Section _____

REVIEW ACTIVITIES FOR EXERCISE 11.1

Test Your Knowledge

1. The stage of the estrous cycle characterized by the appearance of cornified epithelial cells is called the _____ phase.

2. In the stage described in question 1, which ovarian hormone is secreted in high amounts? _____

3. The phase of the human menstrual cycle that corresponds to the phase named in question 1 is the _____ phase.

4. The stage of the estrous cycle in which estrogen and progesterone secretion declines is called the _____ phase.

5. The phase of the human menstrual cycle that corresponds to the phase named in question 4 is the _____ phase.

6. The proliferative phase of the endometrium occurs during the _____ phase of the ovarian cycle.

7. The secretory phase of the endometrium occurs during the _____ phase of the ovarian cycle.

Test Your Understanding

8. Describe the events that occur in the ovaries as a rat goes through the estrous cycle. Relate these changes to the cyclic changes of the vaginal epithelium.

403

9. Describe the changes that occur in the human endometrium during the menstrual cycle. Relate these changes to the events that occur in the human ovaries during the menstrual cycle.

Test Your Analytical Ability

10. What would a vaginal smear performed on an ovariectomized rat (one with its ovaries removed) look like? Explain.

11. The contraceptive pill is composed of estrogen and progesterone. Suppose your laboratory rat, with an estrous cycle of 4 days, were to take the pill on a regimen of "3 days on, 1 day off." What would the vaginal smears taken on each of the 4 days look like? Explain.

Clinical Investigation Questions

12. What are the different layers of the human vaginal epithelium? When do particular cells of this epithelium appear in a vaginal smear, and for what reasons?

13. Why did the physician believe that the patient still had menstrual cycles and had not yet entered menopause? Explain.

EXERCISE 11.2

Human Chorionic Gonadotropin and the Pregnancy Test

MATERIALS

1. Urine collection cup
2. Pregnancy kit, agglutination type (DAP test kit—Wampole; Qupid test—Stanbio; ICON II—Hybritech; or similar kit)
3. Over-the-counter home pregnancy kit (e.p.t. or similar kit)
4. A sample of urine from a pregnant woman (if available)

Shortly after fertilization occurs, cells that are going to become part of the placenta secrete a hormone called human chorionic gonadotropin (hCG). Pregnancy is commonly tested by an assay for hCG in the plasma or urine.

LEARNING OUTCOMES

You should be able to:

1. Describe the fate of the corpus luteum at the end of a nonfertile cycle.
2. Describe the source of hCG and the physiological role of this hormone in pregnancy.
3. Demonstrate a pregnancy test, and explain how this procedure determines pregnancy.

Textbook/Multimedia Correlations

Before performing this exercise, you should study the introductory material presented here. Further information relating to this exercise can be found in these pages of *Human Physiology*, thirteenth edition, by Stuart I. Fox:

- *Implantation of the Blastocyst and Formation of the Placenta.* Chapter 20, p. 739.

Clinical Investigation

A student in a physiology laboratory performed a pregnancy test, and after this came back positive she went to her physician. The physician also performed a pregnancy test and told the student, "Congratulations, the rabbit died."

- Explain the scientific basis for the pregnancy test.
- Explain the reference to "the rabbit died" and how rabbits are used in modern pregnancy tests.

About a week after fertilization occurs, the developing embryo implants into the wall of the endometrium (a process called *implantation*). If the uterine cycle followed the nonfertile pattern, there would be a fall in ovarian hormone secretion due to inadequate stimulation by pituitary gonadotropins (because FSH and LH are suppressed by negative feedback in the luteal phase). The fall in ovarian hormone secretion would then result in the shedding of the upper two-thirds of the endometrium (menstruation) and a spontaneous abortion. After implantation, the secretion of the ovarian hormones must thus be maintained despite the lack of anterior pituitary gonadotropin stimulation.

The tiny implanted embryo saves itself from being aborted by secreting a hormone that indirectly prevents menstruation. Cells of the embryo, which will later become part of the **placenta,** secrete **chorionic gonadotropin,** or **hCG** (the *h* stands for "human"), identical to the anterior pituitary luteinizing hormone (LH) in its effects. Like LH, therefore, hCG stimulates the corpus luteum of the ovary to continue secreting large amounts of estrogen and progesterone for the first 10 weeks of pregnancy. At this time, the growing placenta becomes the primary source of both estrogen and progesterone and takes over support of the pregnancy until birth.

The secretion of hCG doubles approximately every 48 hours over the first 6 weeks of pregnancy, reaches a peak at about 10 weeks, and then rapidly declines (fig. 11.5). Tests of hCG levels in blood are the most sensitive, providing quantitative measurements of hCG that can give information

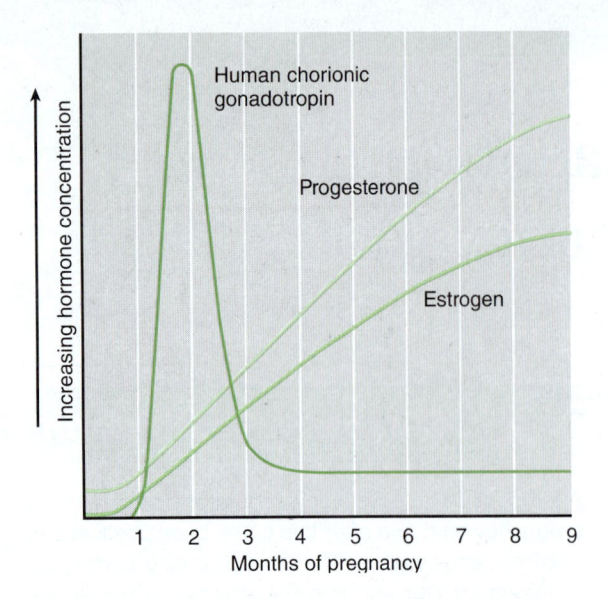

Figure 11.5 The secretion of human chorionic gonadotropin (hCG). This hormone is secreted by the embryo during the first trimester of pregnancy, and it maintains the mother's corpus luteum for the first 5 1/2 weeks. After that time, the placenta becomes the major sex-hormone-producing gland, secreting increasing amounts of estrogen and progesterone throughout pregnancy.

(For a full-color version of this figure, see fig. 20.44 in *Human Physiology*, thirteenth edition, by Stuart I. Fox.)

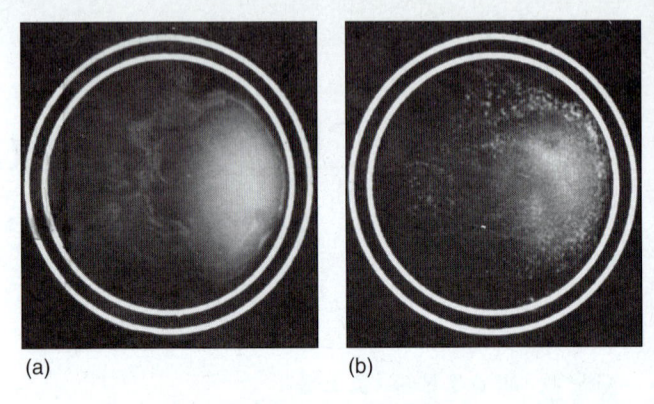

Figure 11.6 A pregnancy test. (a) Negative—no agglutination reaction occurs when urine is added to a control solution containing white latex particles with rabbit gamma globulin (from a rabbit not sensitized to hCG). (b) Positive—agglutination of latex particles occurs when urine from a pregnant woman is added to latex coated with antibodies from a rabbit sensitized to hCG.

regarding the health of the pregnancy. Most commonly, however, pregnancy tests employ urine samples. Over-the-counter home pregnancy tests can detect hCG in urine one to two weeks following conception and are generally used in the week following the first missed menstrual period.

Clinical Applications

hCG acts like LH, so injections of hCG are sometimes given clinically to trigger ovulation in women. Interestingly, this LH-like action of hCG was the basis of the old **rabbit test** for pregnancy. Beginning in the 1930s, the phrase "the rabbit died" meant that the woman was pregnant. Rabbits are reflex ovulators—if they don't have sexual intercourse, they don't ovulate. In this test, a virgin rabbit was injected with urine from a possibly pregnant woman. If the woman was pregnant and had hCG in her urine, the rabbit ovulated and produced a corpus luteum. The rabbit had to be killed to examine its ovaries; thus, the rabbit died in the test whether the woman was pregnant or not. Although it has been more than 40 years since the rabbit test was replaced by immunological detection of hCG, the phrase "the rabbit died" may still be heard to announce a positive pregnancy test.

Modern pregnancy tests still use animals (rabbits, goats, or others), but these are used in the production of the test kit, and the animals don't have to be killed in the process. Companies that produce the test kit inject the animals with hCG, a foreign antigen that stimulates the animal's immune system to produce antibodies against the hCG. The antibodies circulate in the blood plasma, and blood samples are taken to derive the antibodies for the test. All modern pregnancy tests detect the presence of hCG in a blood or urine sample by means of antigen-antibody bonding. Tests that employ antibodies to detect a specific molecule are known as **immunoassays.**

There are two polypeptide subunits, alpha and beta, in the hCG glycoprotein molecule. The alpha subunit of hCG is also present in TSH, FSH, and LH; it is the beta subunit that is unique to hCG. Thus, antibodies directed against the beta subunit of hCG provide the least amount of cross-reaction with other hormones. Accurate and sensitive immunoassays for hCG in pregnancy tests employ antibodies produced by a clone of lymphocytes—termed *monoclonal antibodies*—against the specific beta subunit of hCG.

All pregnancy tests involve the use of antibodies against hCG, but the way these are used varies with the type of test. Although not widely used anymore, the easiest tests to understand involve agglutination reactions. In these tests, antibodies are stuck onto tiny white latex particles to make them visible. If hCG is present in the urine sample, the hCG binds to the anti-hCG antibodies and causes the latex particles to agglutinate (fig. 11.6), much like the agglutination reaction used for blood typing (see exercise 6.3).

Over-the-counter home pregnancy kits are also based on this antigen-antibody reaction. They are easier to perform, but the way these tests work is more complex than with the simpler agglutination reaction. In the common one-step types of tests, the test apparatus has two windows. One window has

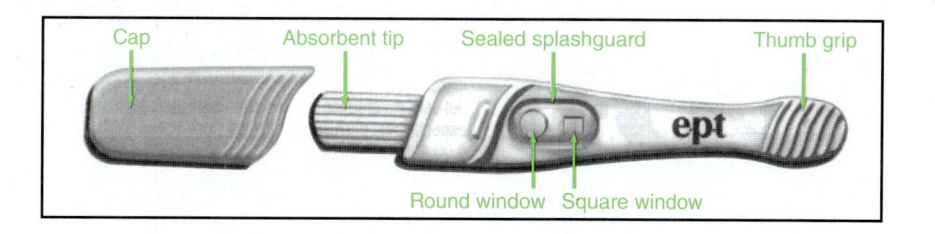

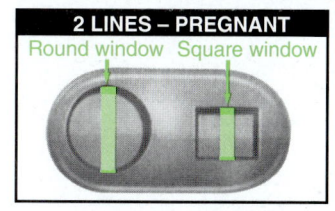

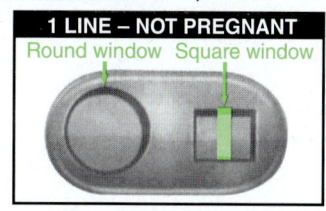

Figure 11.7 **A typical over-the-counter home pregnancy test.** This pregnancy test (e.p.t., Warner-Lambert Consumer Healthcare) is similar to the more than 20 different brands currently available in drugstores, supermarkets, and over the Internet. Follow the specific procedures that come with the specific pregnancy test to obtain accurate results. In the procedure shown, the round window is the *test window;* the square window is the *reference window.* Another simple home pregnancy kit is Clearblue Easy Digital Pregnancy Test.

a visible line; this is a "control" that demonstrates how a positive response in the other window should appear. The other window has a membrane with an invisible line. Antibodies against hCG are attached to the membrane along that line. When urine containing hCG moves along the membrane, antigen-antibody bonds form. Latex particles may be used to help visualize the line in a positive pregnancy test (fig. 11.7), but the exact way that each test works is proprietary information (not available to the public).

Procedure

Agglutination Pregnancy Test

1. Allow the urine sample to reach room temperature.
2. Fill the plastic reservoir provided in the pregnancy kit with urine, and insert the filtering attachment.
3. By gently squeezing the reservoir, expel the urine onto two circles on the disposable slide provided.
4. Shake the latex control (latex particles with gamma globulin antibody), and add 1 drop to the first circle. Mix the urine and reagent with an applicator stick by spreading the mixture over the entire circle.
5. Shake the bottle of reagent (latex particles with antibodies against hCG), and add 1 drop to the second circle. Mix as before.
6. Rock the slide gently for 1 minute; then look for agglutination. If *negative:* the solution will remain milky (fig. 11.6a). If *positive:* the solution will appear grainy (fig. 11.6b).

Home Pregnancy Test

1. Remove the test apparatus from its foil package and follow the directions as outlined for the specific pregnancy test. This generally involves putting the absorbent tip of the apparatus into a flow of urine or dipping the tip into a container of urine.
2. Follow the remaining steps of the procedure. This generally involves placing the apparatus on a flat surface for 3 minutes and then reading the results. An example of such results is illustrated in figure 11.7.

Laboratory Report 11.2

Name _____

Date _____

Section _____

REVIEW ACTIVITIES FOR EXERCISE 11.2

Test Your Knowledge

1. Menstruation is caused by a(n) _____ (increase/decrease) in the secretion of estrogen and progesterone.

2. The structure that secretes estrogen and progesterone for the first 10 weeks of pregnancy is the _____.

3. The hormone tested for in a pregnancy test is _____.

4. The hormone named in question 3 is produced by the _____.

5. The hormone named in question 3 has an action similar to which pituitary hormone? _____.

Test Your Understanding

6. Describe the formation, function, and fate of the corpus luteum during an unfertile menstrual cycle. What happens to the corpus luteum if fertilization occurs?

7. Why is this pregnancy test called an immunoassay? Explain how this test works to detect pregnancy.

8. What secretes hCG, and what is its physiological function? Describe how hCG secretion changes during pregnancy. Now, use this information to explain how a woman in her eighth month of pregnancy, and obviously showing, could have a negative pregnancy test.

Test Your Analytical Ability

9. Why are most pregnancy tests not valid if they are performed too soon after conception?

10. Suppose a man has a tumor that secretes hCG. Would he give a positive result on a pregnancy test? Would hCG have any physiological effect on him? Explain.

Clinical Investigation Questions

11. What do modern pregnancy tests contain, what do they react with in urine, and why does this reaction indicate pregnancy?

12. What is the historical reference of the phrase "the rabbit died?" Are rabbits still used for pregnancy tests? Explain.

Patterns of Heredity

EXERCISE 11.3

MATERIALS

1. Phenylthiocarbamide (PTC) paper (VWR Scientific Products, Ward's)
2. Sickle cell turbidity test (Chembio Diagnostic Systems, Inc.); prepared slides of sickle cell anemia and normal blood
3. Ishihara color-blindness charts

The ways in which many aspects of body structure and function are inherited can be understood by applying relatively simple concepts. The patterns of heredity are important in anatomy and physiology because of the numerous developmental and functional disorders that have a genetic basis. The knowledge of which disorders and diseases are inherited finds practical application in the genetic counseling of prospective parents.

LEARNING OUTCOMES

You should be able to:

1. Define the terms *dominant, recessive, homozygous,* and *heterozygous.*
2. Distinguish between autosomal and sex-linked inheritance.
3. Explain the nature of sickle-cell anemia and describe how it is inherited.
4. Describe how hemophilia and color blindness are inherited.

Textbook/Multimedia Correlations

Before performing this exercise, you should study the introductory material presented here. Further information relating to this exercise can be found in these pages of *Human Physiology*, thirteenth edition, by Stuart I. Fox:

- *DNA Synthesis and Cell Division.* Chapter 3, p. 73.
- *Cones and Color Vision.* Chapter 10, p. 301.
- *Inherited Defects in Hemoglobin Structure and Function.* Chapter 16, p. 563.

Clinical Investigation

An 11-year-old color-blind boy is treated with hydroxyurea for sickle-cell disease. The physician told him that each of his parents has the sickle-cell trait, but his color blindness was inherited only from his mother.

- Explain the cause of sickle-cell anemia and how treatment with hydroxyurea helps.
- Explain the physician's description of how the boy inherited sickle-cell disease and color blindness.

A person inherits two sets of genes controlling every trait: one from the mother and one from the father (if these genes are *autosomal*—that is, not located on the sex chromosomes). If both genes are identical, the person is said to be **homozygous** for that trait. A person homozygous for normal adult hemoglobin *A*, for example, has the **genotype** *AA*; a person homozygous for the sickled hemoglobin *S* has the genotype *SS*.

If a person inherits the gene for hemoglobin *A* from one parent and the gene for hemoglobin *S* from the other parent, this person is said to be **heterozygous** for that trait and has the genotype *AS*. This person is a carrier and has the sickle-cell *trait* but does not have sickle-cell *disease*. The **phenotype** (in this case the absence of sickle-cell disease) is the same for the heterozygote as it is for the homozygous normal person. Thus, the gene for hemoglobin *A* is **dominant** to the gene for hemoglobin *S* (or, stated another way, the gene for hemoglobin *S* is **recessive** to the gene for hemoglobin *A*).

Although the heterozygote does not display the phenotype of sickle cell disease, this person is a carrier of the sickle cell trait because one-half of the gametes will contain the gene for hemoglobin *A* and one-half will contain the gene for hemoglobin *S*. (In the process of gamete formation, known as *meiosis,* the chromosome number is halved.) If this individual mates with one who is homozygous *AA*, the probability is half the progeny will be homozygous *AA* and half will be heterozygous *AS*.

```
        AS        ×        AA         Genotype
              (mated with)            of parents
       ╱  ╲          ╱  ╲
      ½A   ½S      ½A    ½A           Genotype
                                      of gametes


  ½A × ½A = ¼AA ⎫
               ⎬ 2/4 AA
  ½A × ½A = ¼AA ⎭                     Genotype
                                      of offspring
  ½S × ½A = ¼AS ⎫                     (progeny)
               ⎬ 2/4 AS
  ½S × ½A = ¼AS ⎭
```

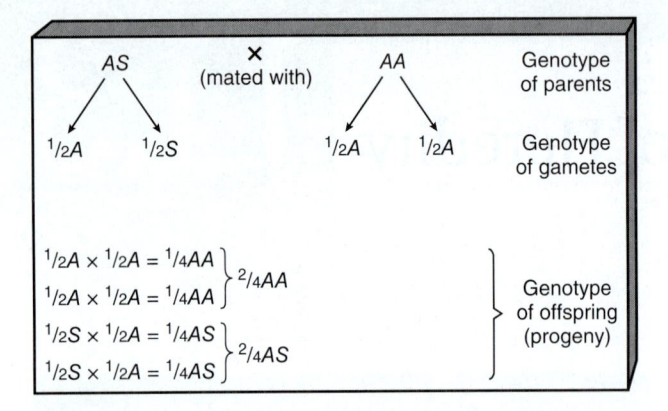

Clinical Applications

Most of the concepts of heredity discussed in this exercise were discovered in the 1860s by an Austrian monk named Gregor Mendel; consequently, these patterns of heredity are often called *simple Mendelian heredity*. A proper knowledge of these patterns is needed for genetic counseling of carriers of genetic diseases. If both parents are carriers of such diseases as sickle-cell anemia, Tay-Sachs disease, phenylketonuria (PKU), and others inherited as *autosomal recessive* traits, they should be aware that there is a 25% chance that their children will get the disease. If only one parent is a carrier, they should know that there is no chance of their children getting the disease. Further, couples should be informed that whether they have no children or a dozen, the probability that their next child will get the disease will always remain the same.

If two heterozygous *AS* individuals mate, one-fourth of the progeny will have the genotype *AA*, one-fourth will have the genotype *SS*, and one-half will have the genotype *AS*. Although individuals with the homozygous genotype *AA* and the heterozygous genotype *AS* are healthy, there is a one-in-four (25%) probability that a child from this mating will have the phenotype of sickle-cell disease (genotype *SS*).

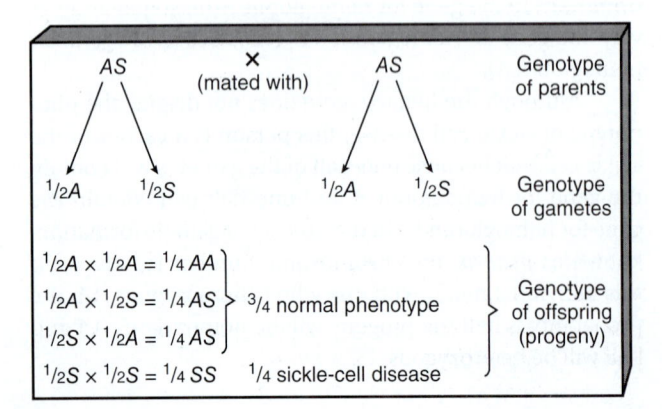

```
        AS        ×        AS         Genotype
              (mated with)            of parents
       ╱  ╲               ╱  ╲
      ½A   ½S           ½A    ½S      Genotype
                                      of gametes


  ½A × ½A = ¼ AA ⎫
  ½A × ½S = ¼ AS ⎬ ¾ normal phenotype  Genotype
  ½S × ½A = ¼ AS ⎭                     of offspring
                                       (progeny)
  ½S × ½S = ¼ SS    ¼ sickle-cell disease
```

412

A. SICKLE-CELL ANEMIA

Sickle-cell anemia is an autosomal recessive disease affecting 8–11% of the African-American population. In this disease, a single base change in the DNA, through the mechanisms of transcription and translation, results in the production of an abnormal hemoglobin (hemoglobin S). Hemoglobin S differs from the normal adult hemoglobin (hemoglobin A) by the substitution of one amino acid for another (valine for glutamic acid) in one position of the protein. There is evidence that the heterozygous condition (sometimes called *sickle-cell trait*) confers protection against malaria and so could be advantageous in regions where malaria is a significant danger.

A quick test for sickle-cell anemia is based on the fact that, under conditions of reduced oxygen tension, hemoglobin S is less soluble than hemoglobin A and tends to make a solution turbid, or cloudy (fig. 11.8a).

Clinical Applications

Under conditions of low P_{O_2}, when the hemoglobin is deoxygenated, hemoglobin S polymerizes into long fibers. This gives the red blood cells the sickle shape characteristic of sickle-cell anemia. It also reduces their flexibility, which hinders their ability to pass through narrow vessels and thereby reduces blood flow through the organs. The long fibers of hemoglobin S also damage the plasma membrane of red blood cells and promote hemolysis, which leads to a variety of complications. Further, the damaged red blood cells can injure the vascular endothelium and cause additional symptoms of sickle-cell disease. Sickle-cell anemia is treated with the drug *hydroxyurea*, which stimulates the production of hemoglobin gamma chains instead of beta chains. This results in the production of red blood cells containing fetal hemoglobin (hemoglobin F), with fewer red blood cells containing hemoglobin S.

Procedure

1. Fill a calibrated capillary tube with blood up to the line. Then, expel the blood into a test tube containing 2.0 mL of test reagent (contains sodium dithionite, which produces low oxygen tension).
2. If the solution does not become cloudy within 5 minutes the test is negative.
3. Place a drop of solution on a slide, add a coverslip and compare your cells with those in figure 11.8b. If available, compare your sample to that of a prepared slide of sickle-cell anemia.
4. Record your data in the laboratory report.

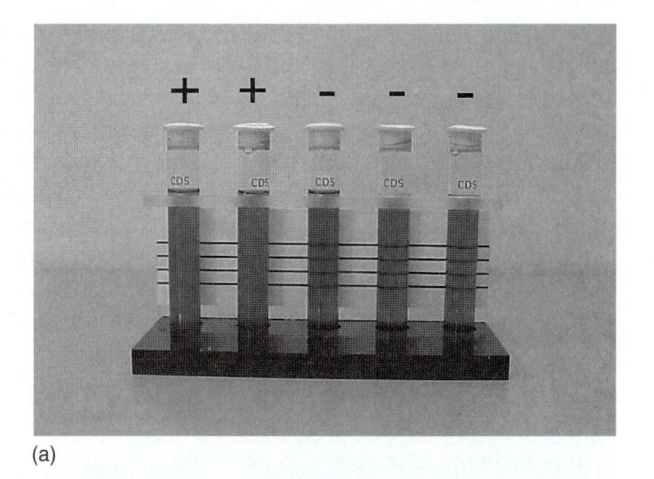

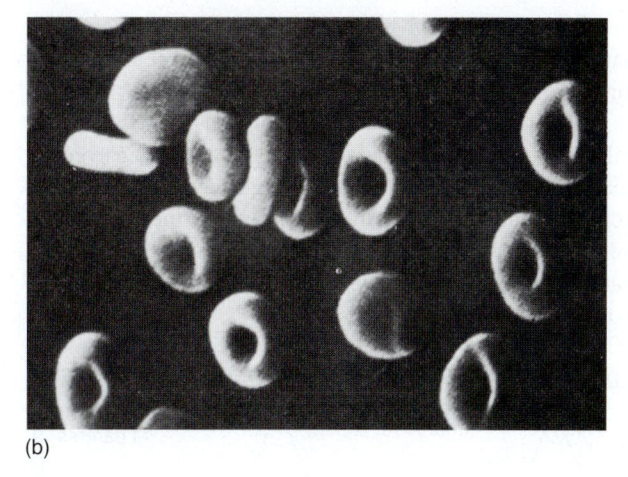

Figure 11.8 Sickle-cell anemia. (a) A turbidity test for sickle-cell anemia. Tubes indicated with a "+" are too cloudy to see through, indicating the presence of hemoglobin S. (b) Normal red blood cells under a scanning electron microscope. (c) Sickled red blood cells under a scanning electron microscope.

B. Inheritance of PTC Taste

The ability to taste PTC paper (phenylthiocarbamide) is inherited as an autosomal dominant trait. Therefore, if T is a taster and t is a nontaster, tasters have the genotype TT or Tt and nontasters have the genotype tt.

About 75% of the worldwide population perceives PTC as a bitter taste, whereas the remaining 25% does not perceive any distinctive taste. Analysis of the genetic basis of PTC tasting has revealed that three genetic changes produce the nontaster phenotype. Genetic analysis has led some scientists to conclude that all nontasters may be descended from a single individual (the "founder"), who likely lived more than 100,000 years ago.

Procedure

1. Taste the PTC paper by leaving a strip of it on the tongue for a minute or so. If the paper has an unpleasantly bitter taste, you are a taster.
2. Determine the number of tasters and nontasters, calculate the proportion of each in the class, and enter this data in your laboratory report.

C. Sex-Linked Traits: Inheritance of Color Blindness

The sex of an individual is determined by one pair of the twenty-three pairs of chromosomes inherited from the parents. These are the sex chromosomes, X and Y. The female has the genotype XX and the male has the genotype XY. Traits determined by genes located on the X sex chromosome (as opposed to the other, autosomal chromosomes) are called **sex-linked traits.**

Unlike the patterns of heredity previously considered, where the genes are carried on autosomal chromosomes, the inheritance of genes carried on the X chromosome follows a different pattern for males than for females. This is because a male inherits one X chromosome (and only one set of sex-linked traits) from his mother, whereas a female gets her two X chromosomes by inheriting one X chromosome from each parent.

The genes for color vision and for some of the blood-clotting factors are carried on the X chromosome, where the phenotypes for **color blindness** and **hemophilia** are recessive to the normal phenotypes. A normal female may have either the homozygous or the heterozygous ("carrier") genotypes, whereas a male must have either the normal or the affected phenotypes.

Let's suppose a man with the normal phenotype mates with a woman who is a carrier for color blindness (C is normal, c is color blind).

The probability that a child formed from this union will be color blind is one in four (25%). And in the event of a color-blind offspring, that offspring is certain to be male (100% probability). All female children formed from this union will have the normal phenotype, but the probability that a given female child will be a carrier for color blindness is one in two (50%).

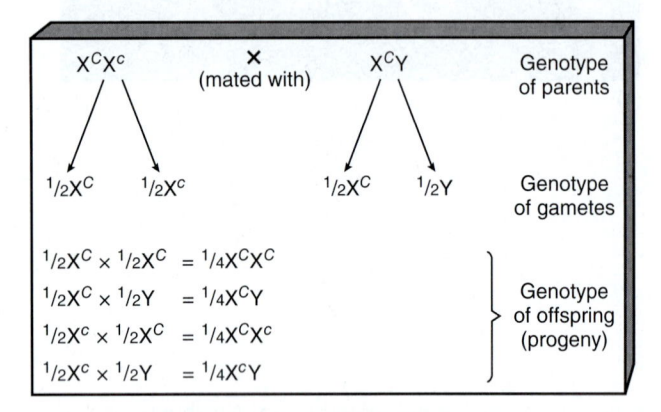

$$X^C X^c \quad \times \quad X^C Y \qquad \text{Genotype of parents}$$
$$\swarrow \searrow \qquad \swarrow \searrow$$
$$\tfrac{1}{2}X^C \quad \tfrac{1}{2}X^c \qquad \tfrac{1}{2}X^C \quad \tfrac{1}{2}Y \qquad \text{Genotype of gametes}$$

$$\tfrac{1}{2}X^C \times \tfrac{1}{2}X^C = \tfrac{1}{4}X^C X^C$$
$$\tfrac{1}{2}X^C \times \tfrac{1}{2}Y = \tfrac{1}{4}X^C Y$$
$$\tfrac{1}{2}X^c \times \tfrac{1}{2}X^C = \tfrac{1}{4}X^C X^c$$
$$\tfrac{1}{2}X^c \times \tfrac{1}{2}Y = \tfrac{1}{4}X^c Y$$

Genotype of offspring (progeny)

The perception of color is due to the action of certain photoreceptor cells, known as **cones,** in the retina of the eye. According to the *Young-Helmholtz theory* of color vision, the perception of all the colors of the visible spectrum is due to the stimulation of only three types of cones—*blue, green,* and *red*. Their names refer to the regions of the wavelength spectrum at which each type of cone is maximally stimulated. When one of these three types of cones is defective owing to the inheritance of a sex-linked recessive trait, the ability to distinguish certain colors is diminished.

Procedure

In the *Ishihara test*, colored dots are arranged in a series of circles in such a way that a person with normal vision can see a number embedded within each circle. By contrast, a color-blind person will see only an apparently random array of colored dots.

Laboratory Report 11.3

DATA FOR EXERCISE 11.3

A. Sickle-Cell Anemia
1. Was your test positive or negative? _____
2. Describe the appearance of your red blood cells in the microscope; compare to those of a prepared slide of sickle-cell disease.

B. Inheritance of PTC Taste
1. Are you a taster? _____
2. Enter the number of tasters and nontasters in your class in the following table.
3. Calculate the proportion of tasters (the number of tasters divided by the total number of students). Enter this value in this table.

Phenotype	Number in Class	Proportion of Tasters
Tasters		
Nontasters		

C. Sex-Linked Traits: Inheritance of Color Blindness
1. Are you color-blind? YES or NO (circle one). If YES, what type of color blindness do you have? _____

REVIEW ACTIVITIES FOR EXERCISE 11.3

Test Your Knowledge

1. Chromosomes other than the sex chromosomes are called _____ chromosomes.
2. If a person has two identical genes for a trait, the person is said to be _____ for that trait.
3. If a person inherits a different gene from one parent than the other for a trait, the person is said to be _____ for that trait.
4. Genes inherited on the X chromosomes code for _____ traits.
5. The physical manifestation of a genotype is called a _____.

Test Your Understanding

6. Describe the inheritance of sickle-cell disease. Include the terms genotype, phenotype, homozygous, heterozygous, dominant, and recessive in your description.

7. Describe the inheritance of color blindness, and explain why color blindness is much more common in men than women.

Test Your Analytical Ability

8. If a man with sickle-cell disease marries a woman with sickle-cell trait, what is the probability that their children will have (a) sickle-cell trait and (b) sickle-cell disease?

9. A man with normal blood clotting marries a woman who is a carrier for hemophilia, a sex-linked trait. What is the probability that their first child will have hemophilia? If this child is hemophilic, what is its sex? What is the probability that this couple's next child will have hemophilia? Explain.

Clinical Investigation Questions

10. What causes sickle-cell anemia? How does treatment with hydroxyurea help this condition?

11. What is sickle-cell trait and why did the physician state that the boy inherited his sickle-cell disease from both of his parents? Why did the physician state that the boy inherited his color blindness from his mother?

Appendix 1

Basic Chemistry

Atoms

All of the matter on earth, living as well as nonliving, is composed of about 100 different types of atoms. Each atom consists of a central positively charged *nucleus* surrounded by a region containing one or more swiftly moving, negatively charged *electrons.* A given electron can occupy any position in a certain volume of space surrounding the nucleus. The outer boundary of this volume of space is called the *orbital* of the electron. Orbitals are like *energy shells,* or barriers, beyond which the electron usually does not pass; they are often represented as a series of concentric circles around the nucleus. The nucleus makes up most of the mass of the atom but accounts for only a tiny fraction of its volume. It consists of positively charged *protons* and, with one exception (H^1), noncharged particles known as *neutrons*.

H^1	H^2 (Deuterium)	H^3 (Tritium)
1 proton	1 proton	1 proton
1 electron	1 electron	1 electron
	1 neutron	2 neutrons

A given element may exist in different forms (**isotopes**) because of the presence of different numbers of neutrons in the nucleus.

The superscript above the symbol of the element is known as the *mass number* and indicates the total number of protons and neutrons in the isotope. Although some isotopes are stable, others (e.g., H^3, or tritium) are unstable and undergo radioactive decomposition, emitting gamma rays (very high-energy light) or beta particles (high-energy electrons).

The *atomic number* of an element refers to the number of its protons. The *atomic weight* of an element refers to its weight relative to that of carbon, given as 12, and is approximately equal to the number of protons and neutrons in the element.

Element	Symbol	Atomic Number	Atomic Weight
Hydrogen	H	1	1.01
Carbon	C	6	12.01
Nitrogen	N	7	14.01
Oxygen	O	8	16.00
Sodium	Na	11	23.00
Magnesium	Mg	12	24.31
Phosphorous	P	15	30.97
Sulfur	S	16	32.06
Chlorine	Cl	17	35.45
Potassium	K	19	39.10
Calcium	Ca	20	40.08
Iron	Fe	26	55.85
Copper	Cu	29	63.54
Iodine	I	53	126.90

Chemical Bonds

Each electron shell surrounding the nucleus of an atom can accommodate only a limited number of electrons. From the inner shell outward, the maximum number of electrons is 2, 8, 18, 32, and so forth. The chemical properties of the atoms are determined by the number of electrons in the outer shell. If this number is less than the maximum, the difference can be made up by sharing electrons with another atom. Bonds formed by the mutual sharing of electrons are very strong and are called **covalent bonds.**

Hydrogen, for example, has only one electron and needs one more electron to complete its outer shell. Oxygen has eight electrons, two in its inner shell and six in its outer shell; it needs two more electrons to complete its outer shell. This requirement can be met by sharing electrons with two hydrogen atoms, forming a molecule of water.

$$H \cdot + H \cdot + \cdot \ddot{\underset{\cdot\cdot}{O}} \cdot \longrightarrow H : \ddot{\underset{\cdot\cdot}{O}} : H, \text{ or } H-O-H, \text{ or } H_2O$$

A-1

Oxygen gas is composed of oxygen molecules formed by the covalent bonding of two oxygen atoms. In this case, two pairs of electrons are shared by the two atoms, forming a double bond between them.

$$\ddot{\mathrm{O}}: + :\ddot{\mathrm{O}}: \longrightarrow :\ddot{\mathrm{O}}::\ddot{\mathrm{O}}:, \quad \text{or} \quad O=O, \quad \text{or} \quad O_2$$

An atom of nitrogen has seven electrons, two in its inner shell and five in its outer shell. It requires three electrons to complete its outer shell. This requirement may be met by sharing electrons with three hydrogen atoms, forming a molecule of ammonia, or by sharing three pairs of electrons with another atom of nitrogen, forming a molecule of nitrogen gas.

$$:\ddot{\mathrm{N}}\cdot + 3\mathrm{H}\cdot \longrightarrow \mathrm{H}:\ddot{\mathrm{N}}:\mathrm{H},\ \text{or}\ \mathrm{H}-\mathrm{N}-\mathrm{H},\ \text{or}\ NH_3$$

$$\cdot\ddot{\mathrm{N}}: + :\ddot{\mathrm{N}}\cdot \longrightarrow :\ddot{\mathrm{N}}:::\ddot{\mathrm{N}}:,\ \text{or}\ N\equiv N,\ \text{or}\ N_2$$

When the electrons are not shared equally, but instead are held by only one of the two nuclei, the atom that captures the electron has a negative charge and the atom that loses the electron has a positive charge. These charged atoms (called *ions*) may be held together by a weak electronic attraction known as an **ionic bond**.

Ions and Electrolytes

When a compound held together by weak ionic bonds is dissolved in water, it dissociates into positively charged ions *(cations)* and negatively charged ions *(anions)*. These ions can conduct electricity, and hence the original ionic compound is called an **electrolyte**. The most ubiquitous electrolyte is common table salt (NaCl).

NaCl ⟶ Na⁺ + Cl⁻
Ionic compound   Cation   Anion

Some atoms form ionic bonds as a group with other atoms and remain grouped when the ionic compound dissociates. These groups are called *radicals*. Examples of radicals include sulfate (SO_4^{2-}) phosphate (PO_4^{3-}), ammonium (NH_4^+), and hydroxyl (OH^-).

$(NH_4)_2SO_4$ ⟶ 2 NH_4^+ + SO_4^{2-}
Ammonium sulfate   Ammonium   Sulfate

The sulfate radical has two negative charges, and two ammonium radicals are needed to retain electrical neutrality.

Cation	Symbol	Anion	Symbol
Sodium	Na⁺	Chloride	Cl⁻
Potassium	K⁺	Sulfate	SO_4^{2-}
Calcium	Ca^{2+}	Bicarbonate	HCO_3^-
Magnesium	Mg^{2+}	Phosphate	PO_4^{3-}
Hydrogen	H⁺	Hydroxyl	OH⁻
Ammonium	NH_4^+	Carbonate	CO_3^{2-}

pH and Buffers

The hydrogen ion concentration of a solution can vary between 10^{-14} molar and zero (see exercise 2.6 for a discussion of molarity). Pure water, which has a hydrogen ion concentration of 10^{-7} molar, is considered neutral.

$$H-O-H \longrightarrow H^+ + OH^-$$

Any substance that increases the H^+ concentration is called an **acid**, and any substance that decreases the H^+ concentration is called a **base**. Bases decrease the H^+ concentration by adding OH^- to the solution. The OH^- can combine with free hydrogen ions to form water.

HCl ⟶ H⁺ + Cl⁻
Hydrochloric acid

NaOH ⟶ Na⁺ + OH⁻
Sodium hydroxide
(a base)

When equal amounts of hydrogen cation and hydroxyl anion are added to a solution, the acid and base neutralize each other, forming water and a **salt**.

HCl + NaOH ⟶ NaCl + H_2O
Acid   Base   Salt   Water

A convenient way of expressing the hydrogen ion concentration of a solution is by means of the symbol **pH,** the negative logarithm of the hydrogen ion concentration.

$$pH = \log \frac{1}{[H^+]}$$

Thus, pure water, with 10^{-7} moles of hydrogen ions/L, has a pH of 7.000. The pH is an inverse function of the H^+ concentration, so an increase in the hydrogen concentration above that of water (i.e., an *acidic* solution) is indicated by a pH of less than 7.000, whereas a decrease in the H^+ concentration (i.e., a *basic* solution) has a pH between 7.000 and 14. A solution that has 10^{-2} moles of hydrogen ions/L (pH 2) is acidic, whereas one that has 10^{-12} moles of hydrogen ions/L (pH 12) is basic.

Acid	Symbol	Base	Symbol
Hydrochloric acid	HCl	Sodium hydroxide	NaOH
Phosphoric acid	H_3PO_4	Potassium hydroxide	KOH
Nitric acid	HNO_3	Calcium hydroxide	$Ca(OH)_2$
Sulfuric acid	H_2SO_4	Ammonium hydroxide	NH_4OH

A **buffer** is a compound that serves to prevent drastic pH changes when acids or bases are added to a solution. It does this by replacing a strong acid or base (one that ionizes completely) with a weak acid or base (one that does not completely ionize).

$$NaHCO_3 + HCl \longrightarrow H_2CO_3 + NaCl$$
Sodium bicarbonate buffer + Hydrochloric acid → Carbonic acid + Sodium chloride

The strong hydrochloric acid was replaced by the weaker carbonic acid, thus minimizing the change in pH that would have been induced had HCl been added to the solution in the absence of buffer. The carbonic acid/bicarbonate *buffer system* also minimizes the effect of added base on the pH of the solution.

$$H_2CO_3 + NaOH \longrightarrow NaHCO_3 + H_2O$$
Carbonic acid + Sodium hydroxide → Sodium bicarbonate + Water

Organic Chemistry

The chemistry of organic compounds is based on the ability of *carbon atoms* to form chains and rings with other carbon atoms. Carbon, which has six electrons (two in the first shell and four in the second shell), requires four more electrons to complete its outer shell; hence, it is said to have four *bonding sites.*

Hydrogen (one bonding site)   H : H, or H — H, or H_2

Oxygen (two bonding sites)   H : O : H, or H — O — H, or H_2O

Nitrogen (three bonding sites)   H : N : H, or H — N — H, or NH_3

Carbon (four bonding sites)   H : C : H, or H — C — H, or CH_4

Carbon atoms can be covalently bonded to each other by sharing one pair of electrons (single bond) or by sharing two pairs of electrons (double bonds). Carbon-carbon double bonds are called sites of *unsaturation,* because they do not have the maximum number of hydrogen atoms.

—C—C—H or CH_3CH_3 C=C or CH_2CH_2

Carbon atoms can be covalently bonded together to form long chains or rings.

H—C—C—C—C—C—C—H

or

$CH_3 — CH_2 — CH_2 — CH_2 — CH_2 — CH_3$

or

C_6H_{14} (hexane)

or

C_6H_{12} (cyclohexane)

In the shorthand structural formulas for cyclic carbon compounds, the carbon atoms are represented by the corners of the figure and hydrogen atoms are not indicated.

Cyclic carbon compounds based on the structure of benzene are known as *aromatic* compounds. The common feature of their structural formula is the presence of three alternating double bonds in a six-carbon ring. This structural formula is in a sense misleading, because all the carbons in the aromatic ring are equivalent; hence, double bonds can be indicated between any two carbons in the ring.

or   or C_6H_6 (benzene)

FUNCTIONAL GROUPS

When carbon atoms are bonded together to form chains or rings, the remaining free bonding sites are available to combine with hydrogen atoms or with other compounds known as *functional groups*. These functional groups are generally more chemically reactive than the hydrocarbon backbone.

Some classes of organic compounds are named according to their functional groups. *Ketones*, for example, have a carbonyl group within the carbon chain, whereas *aldehydes* have a carbonyl group at one end of the chain. *Alcohols* have a hydroxyl group at one end of the chain, whereas *acids* have a carboxyl group at one end of the carbon chain.

Molecules identical in the type and arrangement of their atoms but which differ with respect to the spatial orientation of key functional groups are called **optical isomers.** This name derives from the fact that these isomers can rotate plane-polarized light to the right or to the left, depending on the orientation of the functional group. The two optical isomers of the simple sugars and the amino acids are named *D* (right-handed) or

L (left-handed), according to their similarity to a reference molecule.

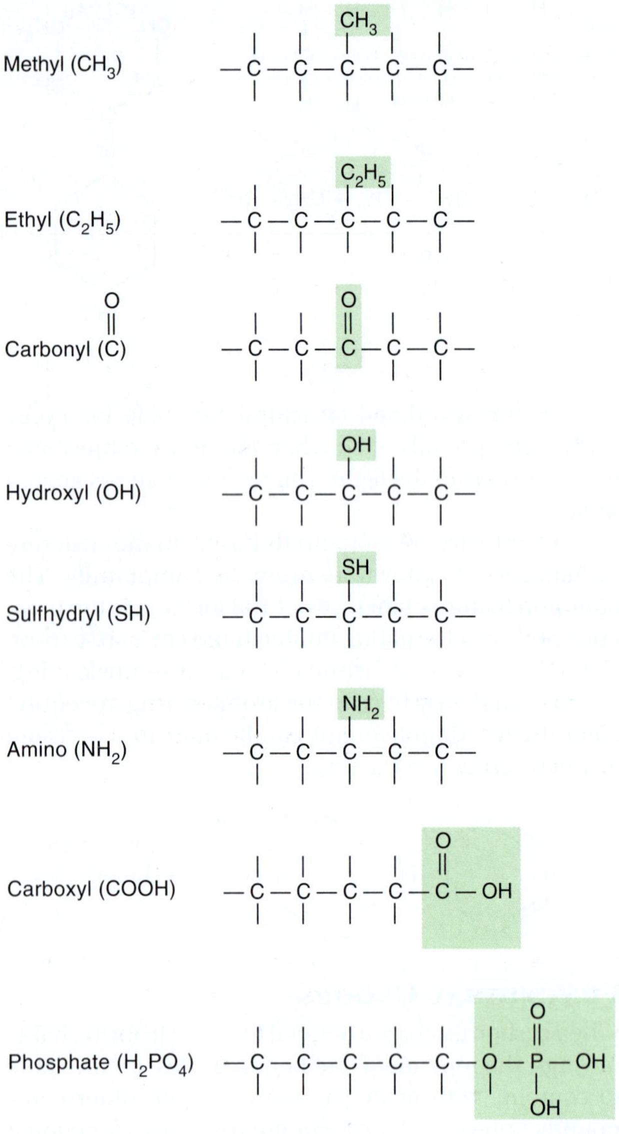

Although a synthetic mixture of simple sugars or amino acids will contain equal amounts of both forms, only one of these two optical isomers can be used by enzymes in cellular metabolism. Thus, all of the physiologically significant simple sugars are D-isomers, whereas all of the physiologically significant amino acids are L-isomers.

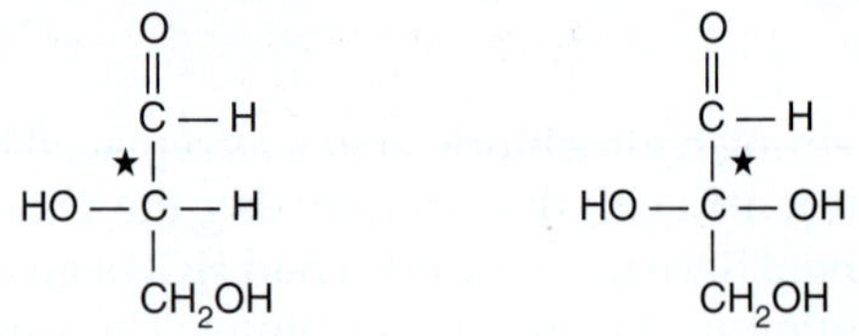

Methyl (CH$_3$)

Ethyl (C$_2$H$_5$)

Carbonyl (C)

Hydroxyl (OH)

Sulfhydryl (SH)

Amino (NH$_2$)

Carboxyl (COOH)

Phosphate (H$_2$PO$_4$)

L-glyceraldehyde

D-glyceraldehyde

Ketone

Aldehyde

Alcohol

Organic acid

Credits

Photo Credits

Design Element

"Clinical Investigation" (stethoscope): © Nathan Blaney/Getty Images RF.

Chapter 1

Figure 1.1: Reichert Scientific Instruments; 1.4(top): © Photo Researchers, Inc.; (middle & bottom), 1.5(all): © Ed Reschke; 1.6b: Reprinted from CELL, Vol. 112, 2003, pp. 535-548, M. Perez-Moreno et al, Sticky business..., Copyright 2003, with permission from Elsevier; 1.7(Loose): © Biology Media/Photo Researchers, Inc.; (Dense, Reticular): © Ed Reschke; (Adipose): © Ed Reschke/Peter Arnold/Getty Images; 1.8(top): © Ed Reschke; (middle, bottom): © Ed Reschke/Peter Arnold/Getty Images; 1.9a, 1.10: © Ed Reschke; 1.11(both): © John Cunningham/Visuals Unlimited.

Chapter 2

Chapter 2.1a,b: © Dr. Stuart Fox; 2.2a: Courtesy of Bausch & Lomb, Analytical Systems Division; 2.2b-2.6, : © Dr. Stuart Fox; 2.7, 2.9a-h: Courtesy Pall Gelman Sciences; 2.14, 2.15: © Dr. Stuart Fox.

Chapter 3

Figure 3.11: © Dr. Stuart Fox; 3.14a-e: © Dr. Sheril Burton; 3.18: © Dr. Stuart Fox; 3.24b: © Steve Allen/Brand X Pictures/Getty Images; 3.31, 3.32: © Dr. Stuart Fox.

Chapter 4

Figure 4.2a-4.6: © Ed Reschke; 4.10a: © R. Kessel/Visuals Unlimited; 4.10b: © Dr. Stuart Fox.

Chapter 5

Figure 5.2, 5.3: Courtesy of Narco Bio-Systems, John F. Pritchett, photographer; 5.4: © BIOPAC Systems, Inc.; 5.5: © Dr. Stuart Fox; 5.6: The IWorx system components, Courtesy of iWorx; 5.7a-5.9c: © Dr. Stuart Fox; 5.10: BIOPAC Systems, Inc.; 5.15, 5.16: © Dr. Stuart Fox; 5.17b: BIOPAC Systems, Inc.; 5.19a,b: © The McGraw-Hill Companies, Inc./Kent Van De Graaff; 5.20, 5.21a,b: © Dr. Stuart Fox.

Chapter 6

Figure 6.2: © Dr. Stuart Fox.

Chapter 7

Figure 7.2a-7.4b & 7.6a: © Dr. Stuart Fox; 7.6b: BIOPAC Systems, Inc.; 7.22a: © Dr. Stuart Fox; 7.23a: © Ruth Jenkinson/Photo Researchers, Inc.

Chapter 8

Figure 8.1: Courtesy of Warren E. Collins, Inc.; 8.2: Courtesy of Phipps & Bird, Inc./Intelitool; 8.14: © BIOPAC Systems, Inc.

Chapter 9

Figure 9.1b: © Biology Media/Photo Researchers, Inc.; 9.4a,b: © Dr. Stuart Fox; 9.7(all): Courtesy of Abbott Laboratories; 9.9(all): Courtesy of Hycor Biomedical Inc.

Chapter 10

Figure 10.3 & 10.4: © Ed Reschke; 10.6: From Mariano S.H. DiFore, Atlas of Human Histology, 1981. Copyright © Williams and Wilkens, a Waverly Company; 10.7a: From P.R. McCurdy, Sickle Cell Disease, © 1973 Medcom, Inc.; 10.7b: © Visuals Unlimited; 10.8b: © Victor B. Eichler; 10.11a: © Ed Reschke.

Chapter 11

Figure 11.4a-11.6b: © Dr. Stuart Fox; 11.8a: Courtesy of Chembio Diagnostic Systems, Medford, New York; 11.8b,c: From P.R. McCurdy, Sickle Cell Disease, © 1973 Medcom, Inc.

Plate

Plate 2(all): © Dr. Stuart Fox.

Index

Note: Page references followed by the letters *f* and *t* indicate figures and tables, respectively.

A

A bands, 174*f*
Abdominal muscles, 298, 298*f*
Abducens nerve, damage to, 120
A blood type, 225, 225*t*
ABO antigen system, 203, 225
Absorbance, 35-36, 37*f*
Absorption, 14
   in digestive tract, 35, 361
Absorption maximum, of cones, 123
Absorption spectrum, 322
   for hemoglobin, 207, 322, 322*f*
Accessory sex organs, 397
Accommodation, in eye, 118-119, 119*f*
ACE. *See* Angiotensin converting enzyme
Acetic acid, and taste perception, 141
Acetoacetic acid, in ketonuria, 353
Acetone, 168, 353
Acetylcholine (ACh), 93, 180, 182
Acetylcholine (ACh) receptors
   muscarinic, 241
   nicotinic, 180
ACh. *See* Acetylcholine
Achilles reflex, 101, 102*f*
Acid(s), 327-328, 328*t*, A-2-A-3
   and taste perception, 139
Acid-base balance
   disorders of, 329, 329*t*, 330
   maintenance of, 328*f*
   organic, A-4
   renal regulation of, 335
   respiration and, 327-330
Acidic solutions, 327, A-2
Acidophils, 154
Acidosis, 329
   respiratory, 329, 329*t*, 330
Acid phosphatase, 63
Acinar cells, 370*f*
Acini, of pancreas, 150, 150*f*, 369-370, 369*f*, 370*f*
Acne, exogenous androgens and, 159
Acromegaly, 153
ACTH. *See* Adrenocorticotropic hormone
Actin, 173, 180-181, 182*f*
Action potentials
   all-or-none law of, 88, 88*f*
   conduction of, 86-88, 87*f*, 89*f*
   definition of, 87
   in heart, 238
   local anesthetics and, 89
   in muscles, 173, 180-182, 181*f*, 182*f*, 193-194
   recording, 86-90, 88*f*, 90*f*
   stimulus frequency and, 88
   stimulus strength and, 88
   in synaptic transmission, 93, 94*f*
   in unmyelinated axon, 89*f*
Activated partial thromboplastin time (APTT)
   normal value for, 232
   test for, 231, 232
Active sites, 61

Active transport, 33, 76, 76*f*, 86
Activity. *See also* Exercise
   estimation of, 388-392
   units of, 62
Activity factor (ΛF), 389
Adaptation
   dark, 120
   sensory, 108
Adenine, 69-70
Adenohypophysis. *See* Pituitary gland, anterior
Adenomas, 153
Adenosine diphosphate (ADP)
   in ATP production, 293
   in blood clotting, 229, 230*f*
   in muscle fatigue, 187
Adenosine triphosphate (ATP)
   in muscle fatigue, 186
   production of, 293, 386
   as universal energy carrier, 293
ADH. *See* Antidiuretic hormone
Adherens junctions, 14, 16*f*
Adherent layer of mucus, 376
Adipocytes, 16, 17*f*
Adipose tissue, 16, 17*f*
   in body composition, 386-387
   endocrine function of, 147*t*, 148
   in skin, 20, 21*f*
ADP. *See* Adenosine diphosphate
Adrenal gland, 151-152
   cortex of, 151-152
      endocrine function of, 147*t*, 151-152, 157-160
      histology of, 151-152, 151*f*
   medulla of, 151-152
      disorders of, 153
      endocrine function of, 147*t*, 151-152
      histology of, 151-152, 151*f*
Adrenal hyperplasia, 158, 160
Adrenocorticotropic hormone (ACTH), 151-153, 160
Adrenocorticotropin. *See* Adrenocorticotropic hormone
Aerobic capacity, 288
Aerobic respiration, 293, 313-314, 327
AF. *See* Activity factor
Afferent arteriole, 336*f*
Afferent (sensory) pathways, 85
Afterimage, 123
   negative, 123
   positive, 123
Agglutination reaction
   in pregnancy test, 406-407, 406*f*
   in red blood cells, 223, 224*f*
Agonist muscles, 196
Agranular endoplasmic reticulum, 6*f*, 7*t*
Agranular leukocytes, 213
Air trapping, 305
Alanine, 40
Albinism, 70
Albumin(s), 40, 54
   digestion of, 375-378, 377*f*

   functions of, 54
   production of, 370
Albuminuria, 40
Albuterol, 295, 306
Alcohols, A-3, A-4
Aldehydes, A-3, A-4
Aldosterone, 159-160
   effects of, 147*t*, 338-339
   in fluid/electrolyte balance, 337-339, 339*f*, 340
   secretion of, 151, 159-160, 337-338
   structure of, 159
Aldosteronism, 159-160
Alkaline phosphatase, in serum
   measurement of, 63
   normal range for, 62, 63
Alkalosis, 187, 329
   respiratory, 329, 329*t*, 330
Alkaptonuria, 50, 70, 71
All-or-none law, of action potentials, 88, 88*f*
Alpha cells, 150*f*, 151, 370
Alpha chains, of hemoglobin, 204*f*, 206
Alpha globulins, 40
   alpha-1, 54
   alpha-2, 54
Alpha waves, 95-96, 95*f*
Alveolar minute volume, 313
Alveoli, of lungs, 313
Amenorrhea
   primary, 402
   secondary, 402
American Heart Association, 388
Amino acids, 35, 35*t*, 39-40
   in blood, 50
   charge of, 53-54
   functional groups of, 40, 49, 54, A-3, A-4
   identification of, 49-50
   isomers of, A-3, A-4
   pH of, 53-54
   separation of, 49-50
   thin-layer chromatography of, 49-50
Amino group, 39-40, 53-54, A-4
Aminopeptidase, 367*t*
Ammonia molecule
   formation of, A-2
Ammonium radical, A-2
Ammonium-magnesium phosphate crystals, 355*f*
Amphoteric molecules, 53
Amplifier, in physiograph, 176, 176*f*
Ampullae
   of inner ear, 135*f*
   of vas deferens, 398*f*
Amylase
   pancreatic, 367*t*, 369
   salivary, 367*t*, 373-375, 374*f*, 375*f*
Anabolic steroids, 159
Anabolism (anabolic reaction), 293, 385-386, 386*f*
Anaerobic metabolism, 293
Anaerobic threshold, 288

I-1

Anal canal, 362f
Anaphase, 7t, 8f
Anaphase I, 7t, 9f
Anaphase II, 7t, 9f
Androgenic hormones, 157–158
Androgens, 150, 151–152, 157–158
　exogenous, 159
Androstenedione, 158
Anemia, 208
　macrocytic, 208
　microcytic hypochromic, 208
　normocytic normochromic, 208
　pernicious, 208
　sickle-cell, 411–412, 413f
Anesthetics, local, 89
Angina pectoris, 110, 288
Angiotensin converting enzyme (ACE), 338
Angiotensin converting enzyme (ACE) inhibitors, 337
Angiotensin I, 338
Angiotensin II, 159, 338, 339f
Angiotensinogen, 338
Anhydrase, carbonic, 327–328
Animal experiments, in pregnancy test development, 406
Anions, A-2
Ankle-jerk reflex, 101, 102f
Antacids, 377
Antagonist effects, 26
Antagonistic effectors, 26
Antagonistic muscles, 195, 196
Antibodies, 54, 214f, 215
　monoclonal, 406
Anticoagulants, 206, 230–231
Antidiuretic hormone (ADH), 337–338
　effects of, 147t
　in fluid/electrolyte balance, 337–338, 338f, 339f, 340
　inadequate, in diabetes insipidus, 338
　secretion of, 152, 337–338
Antigens, 214
　in blood typing, 223–225
Antrum, ovarian, 149, 149f
Anus, 361, 362f, 398f
Anvil (incus), 129, 130f
Aorta, 236f
Aortic bodies, 315, 315f
Aortic semilunar valve, 236f, 273
Apex, of heart, 248f
Appendix, 362f, 366
Apple-shaped bodies, 387
APTT. *See* Activated partial thromboplastin time
Aqueous humor, 115
Arcuate artery, 336f
Arcuate vein, 336f
Areolar connective tissue, 16, 17f
Arginine, 50
Arm extension, EMG during, 196
Arm flexion, EMG during, 196
Aromatic compounds, A-3
Arrector pili muscle, 20, 21f
Arrhythmias, 251–253
Arterial pressure, mean, 282–284
Arterial tree, 281
Arterioles, 235
　afferent, 336f
　efferent, 336f
　in skin, 21f
Artery(ies), 235
　arcuate, 336f
　brachial, 281
　central, 116f
　hepatic, 368–369, 368f

　interlobar, 336f
　interlobular, 336f
　pulmonary, 236f
　renal, 336f
Artifact, stimulus, 87
Asthma, 305–306
Astigmatism, 115–118, 117f
Astigmatism chart, 118
Astrocytes, 20f
Asynchronous activation, 186
Asystole, 253
Atheromas, 39
Atherosclerosis, 39, 169, 252, 253, 260, 388
Athlete's bradycardia, 288
Atom(s), A-1
Atomic number, A-1
Atomic weight, A-1
ATP. *See* Adenosine triphosphate
Atrial fibrillation, 251–252, 252f
Atrial flutter, 251–252, 252f
Atrial natriuretic hormone, 147t
Atrial systole, 248
Atrial tachycardia, paroxysmal, 253
Atrioventricular (AV) block, 251–252
　first-degree, 252, 253f
　second-degree, 252, 253f
　third-degree (complete), 252, 253f
Atrioventricular (AV) bundle, 248, 248f, 265, 266
Atrioventricular (AV) node, 248, 248f
Atrioventricular valves, 260f, 273, 274f, 275f
Atrium (atria), 235
　contractions of, 238, 274f
　left, 235, 236f
　right, 235, 236f
Atropa belladonna, 241
Atrophy, 153
Atropine, cardiac effects of, 241
Auditory meatus, external, 129, 130f
Auditory nerve, 135f
Auditory tube, 130f
Auerbach's plexus, 363, 364f, 366f
Auricle, 130f
Auscultation, of heart sounds, 273–275, 274f
Automatic devices, for dispensing fluids, 35f
Automaticity, 237–238, 247
Autonomic motor nerves, 85
Autosomal trait, 411
　recessive, 412
AV. *See under* Atrioventricular
Average value, 26
Aversion conditioning, 196
Avogadro's number, 79
Axon(s), 19–20, 20f, 86, 87f
　collateral, 87f
　function of, 19–20, 86
　myelinated, 20
　regeneration of, 20
　structure of, 87f
　unmyelinated, 20, 89f
Axon hillock, 86, 87f
Axon initial segment, 86, 87f

B

Babinski's sign, 101–102, 102f
Bacteria
　in intestine, 367
　in urine, 353, 354f
Balance, 135–136
Basal metabolic rate (BMR)
　definition of, 314, 361
　estimation of, 388–392
Base(s), chemical, 327, 328t
Base(s), nucleotide, 69–70
Base(s), pH, A-2–A-3

Basement membrane, 13, 14f, 15f
Base triplet, 70
Basic solutions, 327, A-2
Basilar membrane, 129, 131f
Basophil(s), 154, 213
Basophil leukocytosis, 216
Bath, negative feedback control of temperature in, 25–27
B blood type, 225, 225t
B cells, 214–215, 214f
Beats, ectopic, 240, 252–253
Beats per minute (pulse rate), 27
Beer's law, 35–36, 322, 348
Belladonna, 241
Benedict's test, 374, 374f
Benign hypertension, 283
Benzene, A-3
Beta cells, 150f, 151, 153, 167, 370
Beta chains, of hemoglobin, 204f, 206
Beta globulins, 40, 54
Beta particles, A-1
Beta waves, 95f, 96
Bicarbonate, 327–329
Biceps brachii muscle, 194, 195–196
Biceps-jerk reflex, 101, 102f
Bicuspid valve, 236f
Bifocals, 118
Bigeminy, 252, 253f
Bile
　fat digestion by, 378–380
　gallstones and, 379
　production of, 368–369
　secretion of, 368–369
　triglyceride digestion by, 378–380, 378f, 379f
Bile canaliculi, 368f, 369
Bile duct, 368, 378
　common, 362f, 368, 369f
Bile ductule, 368f
Bilirubin, 206
　conjugated, 206, 353
　free, 206, 353
　in urine, 351, 353
Bilirubinuria, 353
Binaural localization, 132
Biochemical measurements, 33
Bioelectric impedance, 386
Biofeedback
　electroencephalogram and, 95, 197
　electromyogram and, 196–197
Biopac system
　for blood pressure measurement, 284, 284f
　for ECG, 261, 261f
　for EEG, 93, 96, 96f
　for EMG, 196
　hand dynamometer in, 188, 188f
　pulse transducer in, 261, 261f
　for recording heart contractions, 239–240, 239f, 240f
　for recording muscle contractions, 176–177, 177f, 178f, 180, 180f, 187, 188, 188f
　for spirometry, 295, 297f, 298–299, 299f, 307, 307f
Bipolar leads, for ECG/EKG, 248, 249f
Bipolar neurons, 122
Bird spirometer, 296
Bitter taste, 139, 140f
Bladder. *See* Gallbladder; Urinary bladder
Blank tube, 37
Blindness
　color, 124, 413–414
　glaucoma and, 119
Blind spot, 122–123, 122f
Blood, 203–234

I-2

amino acids in, 50
buffers in, 328–330
centrifuged, 18f, 204f
circulating energy substrates in, 386, 386f
composition of, 17, 18f, 203, 204f
as connective tissue, 16–17, 18f
functions of, 235
hydrostatic pressure of, 40, 41f
oxygen transport in, 205–209
partial pressure of gases in, 313, 314f
pH of, 294, 327–330, 329t
precautions in handling, 215
samples of, obtaining, 206–207, 206f
Blood clot, 54, 203
Blood clotting system, 229–232
clotting factors in, 230, 231f
disorders of, 230, 231t
extrinsic pathway of, 230, 231f
intrinsic pathway of, 230, 231, 231f
Blood flow
laminar, 281–282
peripheral resistance and, 235, 281
turbulent, 282
Blood plasma. See Plasma
Blood pressure
blood volume and, 340
classifications of, 284t
depressed, 340
diastolic, 282–284, 282f, 284t
elevated, 169, 283
measurements of, 281–284, 282f, 284f
negative feedback control of, 27–28, 27f
normal values for, 284t
systolic, 282–284, 282f, 284t
Blood smear, 18f, 215, 216–217, 216f
Blood types, 203, 223–225
ABO antigen system of, 203, 225
incidence of, 225t
Rh factor in, 203, 223–225
Blood urea nitrogen (BUN), 346
Blood volume, renal regulation of, 335, 340
Blue cones, 123–124, 123f, 414
B lymphocytes. See B cells
BMI. See Body mass index
BMR. See Basal metabolic rate
Body composition
analysis of, 386–387
definition of, 386
Body mass, lean, 386, 388
Body mass index (BMI), 386–387
Body shape, 387
Body weight, ideal, 389
Bonds, chemical A-1–A-2
Bonds, peptide, 39–40
Bone, 16–17, 18f
Botox, 182
Botulinum toxin, 182
Bowman's capsule, 335
Boyle's law, 297
Brachial artery, blood flow in, 281
Bradycardia
athlete's, 288
definition of, 251
Brain
electroencephalogram of, 93–96, 175
biofeedback and, 95, 197
definition of, 93
electrode placement in, 95–96, 96f
waveforms of, 95–96, 95f
endocrine function of, 148
sensations and, 85
Brain mapping, 107f, 108
Breathing. See Ventilation
Bronchiolitis, chronic obstructive, 306

Bronchitis, 305–306
Bronchodilator, 306
Brunner's glands, 366, 366f, 376
"Brush border," 366
Brush-border enzymes, 366, 374
BTPS factor, 300–305, 307, 317
Buffer(s), 328–330, A-2–A-3
Buffer system, A-3
Bulbourethral gland, 398f
BUN. See Blood urea nitrogen
Bundle branch, 248, 248f
Bundle-branch block, 266
Bundle of His, 248, 248f, 265

C

Caffeine, cardiac effects of, 240
Calcitonin, 147t, 152
Calcium ions
cardiac effects of, 242
in muscle contractions, 173, 180–181, 181f, 182f
in muscle fatigue, 186
Calcium oxalate crystals, 355f, 356f
Calcium phosphate crystals, 355f
Calipers, skinfold, 386
Calories, 387, 388–392
cAMP. See Cyclic AMP
Canaliculi
bile, 368f, 369
bone, 17, 18f
Canal of Schlemm, 119
Capillaries, 235, 236f
peritubular, 336f
Capillary tube, 206–207, 207f
Capsule
of adrenal gland, 151, 151f
glomerular, 335, 336f
Carbohydrates
complex, 387
dietary intake of, 387
digestion of, 373–375, 374f, 375f
metabolism of, 33
as nutrients, 361
in plasma, 37–38
Carbon
bonding sites of, A-3
chemistry of, A-3–A-4
Carbon atoms, A-3
Carbon dioxide
partial pressure of, 313, 314f, 315–316
production of, exercise and, 330
respiratory exchange/elimination of, 293–294, 313–316, 327–328, 330
in ventilation regulation, 330
Carbonic acid, 327–329
Carbonic anhydrase, 327–328
Carbon monoxide, 207
Carbon monoxide poisoning, 321–322
Carbonyl group, A-3–A-4
Carboxyhemoglobin, 207, 321–322
Carboxyl group, 39–40, 54, A-3–A-4
Carboxypeptidase, 367t
Cardiac cycle, 247–248
Cardiac muscle, 19, 19f, 173
Cardiac output, 235, 281
Cardiac (pulse) rate
calculation of, 251
exercise and, 259, 287–290
maximum, 287, 287t
normal values for, 28
position changes and, 288–290
resting
negative feedback control of, 27–28, 27f
normal values for, 28

training, 287–288, 287t
Cardiocomp, 251, 266
Cardiovascular system, 235–292
blood pressure measurements in, 281–284
drug effects on heart, 237–242
electrocardiogram of heart, 247–253
exercise (physical fitness) and, 259–261, 287–290
function of, 235
heart sounds in, 273–276
structure of, 235, 236f
Caries, 374
Carotid bodies, 315, 315f
Carrier, 412
Cartilage, 16, 18f
Casts, in urine, 351–356, 354f–356f
Catabolism, 293, 385–386, 386f
Catalase, 63
Catalysts, 61
Cataracts, 117
Cavities, of eye, 115
CCK. See Cholecystokinin
Cecum, 362f
Cell(s), 1–10
division of, 6–10, 7t, 8f–9f
microscopic examination of, 2–10
proper functioning of, 33
structure of, 1, 6–10, 6f, 7t
Cell body, 19–20, 20f, 86, 87f
Cell-mediated immunity, 215
Cell membrane. See Plasma (cell) membrane
Cellular respiration, 293
Cellulose acetate electrophoresis, 56, 56f, 57f
Celsius, 4
Central artery, 116f
Central canal, 16–17, 18f
Central chemoreceptors, 315, 316f
Central lacteal, 368
Central nervous system (CNS), 20, 86
Central obesity, 169, 387
Central vein, 116f, 368–369, 368f
Centriole, 6f, 8f
Centrosome, 7t, 8f
Cerebral cortex
motor areas of, 107, 107f
sensory areas of, 107, 107f
Cerebrovascular accident, 39
Cerumen, 132
Cervix, 398f
Chambers
in eye, 115, 116f
in heart, 235
Cheek cells, microscopic examination of, 5–6
Cheeks, taste buds on, 139
Chemical bonds, A-1–A-2
Chemical senses, 141
Chemistry
basic, A-1–A-4
organic, A-3–A-4
Chemoreceptors, 314–315, 316f
carbon dioxide and, 315, 316f
central, 315, 316f
peripheral, 315, 315f, 316f
Chewing, 361
Chief cells, 365, 365f, 375
Chloride, in urine, 339–340
Cholecystectomy, laparoscopic, 379
Cholecystokinin (CCK), 147t, 362t
Cholesterol
abnormal levels of, 169, 387
dietary, 388
exogenous androgens and, 159
and gallstones, 379
HDL (good) vs. LDL (bad), 39

I-3

Cholesterol *(continued)*
  obesity and, 169
  in plasma
    high levels of, 39
    homeostasis of, 33
    measurement of, 34, 38–39
    normal values for, 39
  in steroid biosynthesis, 158*f*
  structure of, 39
  in urine, 355*f*
Chondrocytes, 16, 18*f*
Chorionic gonadotropin, human (hCG), 405–407, 406*f*
Choroid, 115, 116*f*
Christmas disease, 230
Chromaffin cells, 152
Chromatid pairs, 8*f*
Chromatin, 6*f*, 7*t*, 8*f*
Chromatogram, 50, 160
Chromatography, thin-layer
  of amino acids, 49–50
  of steroids, 157, 160–161
Chromophils, 154, 154*f*
Chromophobes, 154, 154*f*
Chromosomes, 1, 6–10, 8*f*–9*f*
  diploid number of, 150
  haploid number of, 150
  homologous, 6
  pairs of, 6
  sex, 413–414
Chronic obstructive bronchiolitis, 306
Chronic obstructive pulmonary disease (COPD), 306
Chylomicrons, 378*f*, 379
Chyme, 376
Chymotrypsin, 367*t*, 375–376
Cilia, 7*t*, 14, 15*f*
Ciliary body, 115, 116*f*
Ciliary muscle, 118, 119*f*
Ciliated columnar epithelium
  pseudostratified, 14, 15*f*
  simple, 14, 15*f*
Circular muscles, 120, 364*f*
Circulating energy substrates, 386, 386*f*
Circulatory system, 235, 236*f*
  in kidneys, 336*f*
Circus rhythm, 253
Cirrhosis, 54, 64, 370
Cis fatty acids, 388
Citric acid, as anticoagulant, 231
Clinical death, in pithing frog, 177
Clitoris, 397, 398*f*
Clonal selection theory, 214
Clonus, 196
Closed-angle glaucoma, 119
Clostridium botulinum, 182
Clotting. *See* Blood clot; Blood clotting system
Clotting factors, 230
CNS. *See* Central nervous system
Cocaine, 89
Cochlea, 129, 130*f*, 135*f*
Cochlear duct, 131*f*
Cochlear fluid, 129–131. *See also* Endolymph
Cochlear implants, 132
Cochlear nerve, 130*f*
Coenzymes, 62
Cofactors, 62
Cold sensation, 105–107
Colitis, ulcerative, 367
Collagen, 16, 17*f*, 18*f*, 20
  in blood clotting, 229–230, 230*f*
Collateral axon, 87*f*
Collecting duct, renal, 336*f*

Collins respirometer, 295–296, 296*f*, 299–307, 300*f*, 301*f*, 317
Colloid, 152, 152*f*
Colloid osmotic pressure, 40, 41*f*
Colon, 366
  ascending, 362*f*
  descending, 362*f*
  endocrine function of, 362*t*
  histology of, 366, 367*f*
  sigmoid, 362*f*
  transverse, 362*f*
Color, wavelengths of, 36
Color blindness, 124, 413–414
Colorimeter, 35–37, 37*f*
  for hemoglobin measurement, 207, 322
  procedure for standardizing, 36–37
Color vision, 122, 123–124, 413–414
  trichromatic, 123
  Young-Helmholtz theory of, 123, 414
Columnar epithelium, 13–14
  pseudostratified ciliated, 14, 15*f*
  simple, 13–14, 15*f*
  simple ciliated, 14, 15*f*
Common bile duct, 362*f*, 368, 369*f*
Common hepatic duct, 368
Complement, 214*f*, 215
Complete tetanus, 187
Complex carbohydrates, 387
Compound microscope, 2, 3*f*
Computerized data acquisition and analysis, 176, 239
Concentration, tonicity and, 79–80
Concentration gradients, 75–76
Concentric (shortening) contraction, 194
Condensation (dehydration synthesis), 34–35, 37, 293
Conditioning
  aversion, 196
  operant, 196
Conduction, of sound waves, 129–131
Conduction deafness, 129, 131, 132
Conduction system, of heart, 247–248, 248*f*
Cones, 85, 122, 122*f*
  absorption maximum of, 123
  in color vision, 122, 123–124, 414
  in photopic (day) vision, 122
  types of, 123–124, 123*f*, 414
Conjugated bilirubin, 206, 353
Conjunctiva, 116*f*
Connective tissue, 13, 16–17
  dense, 16, 17*f*
  loose (areolar), 16, 17*f*
  proper, 16, 17*f*
  reticular, 16, 17*f*
  in skin, 20, 21*f*
Consensual reaction, 121
Constancy, dynamic, 25, 26*f*
Contralateral muscles, 99
Control center, respiratory, 314–315, 316*f*
Convergence, in eye, 120, 122*f*
Convoluted tubule, 336*f*
  distal, 336*f*
  proximal, 336*f*
COPD. *See* Chronic obstructive pulmonary disease
Cornea, 115, 116*f*
Cornified layer, 13
Cornified squamous epithelial cells, in vaginal smear, 400, 401*f*
Corona radiata, 149, 149*f*
Coronary thrombosis, 260
Cor pulmonale, 306

Corpuscular hemoglobin concentration, mean. *See* Mean corpuscular hemoglobin concentration
Corpuscular volume, mean. *See* Mean corpuscular volume
Corpus luteum, 148, 159, 400*f*, 405–406
Corti, organ of, 129, 131*f*
Corticospinal tract, 101
Corticosteroid hormones, 151–152, 157, 159–160
Corticosterone, 152, 160, 161
Cortisol. *See* Hydrocortisone
Cortisone, 160, 161
Coumarin, 230
Coupler, in physiograph, 176, 176*f*
Covalent bonds, A-1–A-2
Cranial nerves
  in ear, 130*f*, 135, 135*f*
  in tongue, 141, 141*f*
Creatinine, 335, 346, 347
Crenation, 80, 80*f*
Crohn's disease, 367
Cross-bridges, in muscle contraction, 181–182, 182*f*
Crossed-extensor reflex, 99, 100*f*
Crypt of Lieberkühn, 364*f*, 366, 366*f*
Crystallin, 117
Crystals, in urine, 353, 355*f*, 356*f*
Cuboidal epithelium, 13
  simple, 13, 14*f*
Cumulus oophorus, 149, 149*f*
Cupula, 135–136, 136*f*
Cushing's syndrome, 158, 160, 216
Cutaneous receptors, 105–110, 106*f*, 106*t*
Cutaneous reflex, 101–102, 102*f*
Cutaneous sensation
  modalities of, 105
  receptors for, 85, 105–110, 106*f*, 106*t*
Cuvette, 36–37
Cyclic AMP (cAMP), 240
Cyclic carbon compounds, A-3
Cystic duct, 368, 369*f*
Cystine, in urine, 71, 355*f*
Cystinuria, 71, 355*f*
Cytoplasm, 1, 6, 6*f*, 7*t*
Cytosine, 69–70
Cytotoxic T cells, 215

D

Dark adaptation, 120
Data acquisition and analysis, computerized, 176, 239
Daughter cells, 6–10, 9*f*
Day vision, 122
D cells, 365
Deafness
  conduction, 129, 131, 132
  sensory, 129, 131
Death
  clinical, in pithing frog, 177
  sudden cardiac, 253
Defibrillation, electrical, 253
Dehydration synthesis (condensation), 34–35, 37, 293
Dehydroepiandrosterone (DHEA), 158
Delta waves, 95–96, 95*f*
Dendrites, 19–20, 20*f*, 86, 87*f*
Dense connective tissue, 16, 17*f*
Densitometer, 54
Deoxycorticosterone (DOC), 151, 159, 161
Deoxyhemoglobin, 207, 321–322
Deoxyribonucleic acid. *See* DNA
Deoxyribonucleotides, 35*t*

I-4

Depolarization
    in heart, 247–248
        mean axis of, 265–267, 266*f*–268*f*
    in membrane potentials, 87–88, 88*f*
    in muscles, 173
Dermis, 20, 21*f*
Desmosomes, 14, 16*f*
Dextrose 5% in water solution, 80
DHA. *See* Docosahaenoic acid
DHEA. *See* Dehydroepiandrosterone
DHT. *See* Dihydrotestosterone
Diabetes insipidus, 338
Diabetes mellitus, 38, 153, 167–169, 352
    beta cells in, 153
    diagnosis of, 28
    glucose concentration in, 28, 28*f*, 35, 38
    obesity and, 169, 387
    test for, 352
    type 1, 38, 168–169, 169*t*
    type 2, 38, 169, 169*t*, 352, 387
Dialysis, 78
Dialysis tubing, 78, 78*f*, 79*f*
Diapedesis, 213, 214*f*
Diaphragm, 297–298, 298*f*, 362*f*
Diastole, 247, 273
Diastolic pressure, 282–284, 282*f*, 284*t*
Dicumarol, 230
Diestrus, 400–401
Diet, 361
Dietary record, three-day, 388, 389–392
Diffraction grating, 36
Diffusion, 75–76. *See also* Osmosis
Digestion, 35, 361–396
    of carbohydrates, 373–375, 374*f*, 375*f*
    definition of, 361
    of proteins, 373, 375–378, 377*f*
    of triglycerides, 378–380, 378*f*, 379*f*
Digestive system
    absorption in, 35
    development of, 361
    as "disassembly" line, 361
    enzymes in, 366, 367*t*, 369–370, 373, 374
    histology of, 363–370
    hormones in, 361, 362*t*
    layers (tunics) of, 363, 364*f*
    structure of, 361, 362*f*
Digitalis, cardiac effects of, 242, 252
Digitalis glycosides, 242
Digoxin, 241
Dihydrotestosterone (DHT), 161
1,25-Dihydroxyvitamin D$_1$, 147*t*
Dimensional analysis, 4–5
Dimers, 35
Diopters, 116, 118, 121
Dipeptidases, 375–376
Diploid cell, 6, 150
Disaccharides, 37
D-isomers, A-3–A-4
Dissociation, 80
Distal convoluted tubule, 336*f*
Diverticula, 361
Dizziness, 136
DNA, 1
    base sequence on, 69–70
    in enzyme production, 33
DOC. *See* Deoxycorticosterone
Docosahaenoic acid (DHA), 388
Dominant genes, 411
Dorsal root, 99
Double reciprocal innervation, 100*f*
Dramamine, 136
Drug(s)
    cardiac effects of, 237–242

    definition of, 237
    parasympathomimetic, 241
    pharmacological effects of, 237
    physiological effects of, 237
    sympathomimetic, 241
"Drug" crystal, in urine, 356*f*
Dub sound, 273
Duchenne's muscular dystrophy, 187
Duct(s). *See also specific ducts*
    formation of, 145
Ductus (vas) deferens, 150, 397, 398*f*
Duodenal glands, 366, 366*f*, 376
Duodenal papilla, 369*f*
Duodenal ulcers, 376
Duodenum, 362*f*, 365–368, 366*f*, 369*f*
D5W (5% dextrose in water), 80
Dynamic constancy, 25, 26*f*
Dynamometer, 188, 188*f*
Dyslipidemia, 169, 387
Dystrophin, 187

E

Ear(s), 129–138
    in balance and equilibrium, 135–136
    binaural localization of sound in, 132
    conduction of sound waves in, 129–131
    hearing function of, 129–132
    structure of, 129, 130*f*, 131*f*
Eardrum, 129, 130*f*
Earwax (cerumen), 132
Eccentric (lengthening) contraction, 194
ECG. *See* Electrocardiography
ECL. *See* Enterochromaffin-like cells
Ectoderm, 151
Ectopic beats, 240, 252–253
Edema, 40, 79, 80, 352
EEG. *See* Electroencephalogram
Effector(s), 25, 26*f*, 27*f*
    antagonistic, 26
Effector organs, 85, 86
Efferent arteriole, 336*f*
Efferent ductules, 150*f*
Efferent (motor) pathways, 85
Egg albumin, digestion of, 375–378, 377*f*
Egg cell. *See* Ovum
Eicosapentaenoic acid (EPA), 388
Ejaculation, 397
Ejaculatory duct, 150, 397, 398*f*
Ejection, cardiac, 274*f*
Ejection fraction, 288
EKG. *See* Electrocardiography
Elastic cartilage, 16, 18*f*
Elastic fibers (elastin), 16, 17*f*, 18*f*
Electrical axis, mean. *See* Mean electrical axis, of ventricles
Electrical defibrillation, 253
Electrocardiography (ECG/EKG), 175, 247–253, 249*f*
    abnormal patterns in, 251–253
    definition of, 248
    exercise and, 259–261, 288
    impulse conduction and, 248, 250*f*
    intraventricular pressure and, 259, 260*f*
    lead placement for, 248, 249*f*
    and mean electrical axis of heart, 265–267, 266*f*–268*f*
    myocardial ischemia and, 260, 260*f*
    normal patterns in, 248, 249*f*
    phonocardiogram and, 275–276, 275*f*
    procedure for, 248–251
Electrodes
    in action potential recordings, 89–90
    in EEG, 95–96, 96*f*
    in EMG, 195–196, 195*f*

    exploring, 187
    in finger twitches, 187–188, 187*f*
    recording, 89
    stimulating, 89
Electroencephalogram (EEG), 93–96, 175
    biofeedback and, 95, 197
    definition of, 93
    electrode placement in, 95–96, 96*f*
    waveforms of, 95–96, 95*f*
Electrolyte(s), A-2
    concentrations of, 335
    renal regulation of, 335, 337–340
    in urine, 339–340
Electrolyte gel, 195
Electromyogram (EMG), 175, 193–197
    during arm flexion and extension, 196
    biofeedback and, 196–197
    definition of, 195
    electrode placement in, 195–196, 195*f*
    recording procedures for, 195–196
Electrons, A-1
Electrophoresis
    cellulose acetate, 56, 56*f*, 57*f*
    diagnostic uses of, 54
    materials for, 53, 55*f*, 56*f*, 57*f*
    normal pattern of serum, 54*f*
    SDS-polyacrylamide gel, 54, 55*f*
    of serum proteins, 53–56
Embryo, implantation of, 397, 405
EMG. *See* Electromyogram
Emmetropia, 117, 117*f*
Emphysema, 305–306
Emulsification, 378*f*, 379
Encephalitis, 95
Endergonic reactions, 293
Endocrine glands, 85, 145
    disorders associated with, 153
    formation of, 145, 146*f*
    histology of, 148–154
    major hormones of, 147*t*
    partial list of, 147*t*
    primary effects of, 147*t*
    target organs of, 147*t*, 148
Endocrine system, 145–172
    histology of, 148–156
    in homeostasis, 85
    structure of, 145
Endogenous substances, 237
Endolymph, 129–131, 135–136, 136*f*
Endometrium, 159, 397, 405
    implantation in, 397, 405
    menstrual cycle and, 397, 399–402, 400*f*
Endoplasmic reticulum, 7*t*
    agranular (smooth), 6*f*, 7*t*
    granular (rough), 6*f*, 7*t*
Endothelium, 13
Energy
    ATP as universal carrier of, 293
    intake of, 387–388
    output of, 388–392
    pathways in body, 385–386, 386*f*
Energy shells, A-1
Energy substrates, circulating, 386, 386*f*
Enterochromaffin-like cells (ECL), 365
Enterocytes, 366
Enzyme(s), 33
    activity of
        induced-fit model of, 61
        lock-and-key model of, 61, 62*f*
        measurement of, 61–64
        pH and, 61–62
        temperature and, 61–62
    brush-border, 366, 374

I-5

Enzyme(s) *(continued)*
  in digestive system, 366, 367*t*, 369–370, 373, 374
    relative specificity of, 61
    structure of, 61
    synthesis of, 33
Enzyme-substrate complex, 61
Eosinopenia, 216
Eosinophil(s), 213–214
Eosinophil leukocytosis, 216
EPA. *See* Eicosapentaeonic acid
Epidermis, 20, 21*f*
Epididymis, 150, 150*f*, 397, 398*f*
Epiglottis, 140*f*
    taste buds on, 139
Epilepsy, 95
Epileptic seizure, petit mal, 95
Epinephrine
    cardiac effects of, 147*t*, 241
    function of, 38
    secretion of, 151
Epithelial cells
    in urine, 353, 354*f*
    in vaginal smear, 401*f*
Epithelial membranes, 5–6
Epithelial tissues, 13–15
    characteristics of, 13
    germinal, 150, 150*f*
    glandular, 13, 145, 146*f*
    microscopic examination of, 5–6, 13–15
    in skin, 20, 21*f*
Epithelium. *See* Epithelial tissues
Epitympanic recess, 130*f*
EPSP. *See* Excitatory postsynaptic potential
Equilibrium, 135–136
Equivalent weight, 339
ERV. *See* Expiratory reserve volume
Erythroblastosis fetalis, 224
Erythrocytes. *See* Red blood cell(s) (RBC)
Erythropoietin, 147*t*, 206
Esophagus, 362*f*
    histology of, 363–365, 364*f*
    layers of, 363
Essential hypertension, 283
Estetrol, 161
Estradiol, 147*t*, 158*f*, 159
Estriol, 161
Estrogens, 157–159
    in menstrual cycle, 399–400, 400*f*
    in positive feedback, 153
    in pregnancy, 405
    secretion by placenta, 161
    structure of, 158–159
    thin-layer chromatography of, 161
Estrous cycle, 399–402
Estrus, 399–402
Ethyl group, A-4
Eustachian tube, 130*f*
Excitation-contraction coupling, 173, 181–182, 181*f*, 182*f*
Excitatory postsynaptic potential (EPSP), 93, 94*f*
Excitotoxicity, 252
Exercise
    and carbon dioxide production, 330
    and cardiac rate, 287–290, 287*t*
    and electrocardiogram results, 259–261, 288
    and energy expenditure, 388–392
    and respiratory system, 313–317
Exercise testing, 288
Exergonic reactions, 293
Exfoliation, in ovarian cycle, 400, 402
Exhalation, 295–298
Exocrine glands, 145

formation of, 145, 146*f*
Exocytosis, in synaptic transmission, 93, 94*f*
Exogenous substances, 237
Expiration, muscles of, 298, 298*f*
Expiratory reserve volume (ERV), 296, 298
    calculation of, 301–302
    in spirometry recording, 297*f*, 300*f*, 302*f*
Expiratory volume, forced (FEV), 305–307, 306*f*, 307*f*
Exploring electrodes, 187
Extension, EMG during, 196
Extensors, 99, 100*f*
External auditory meatus, 129, 130*f*
Extracellular fluid, ion concentrations in, 76, 76*f*
Extracellular material (matrix), 16
Extrafusal fibers, 99–101
Extrasystoles, 252
Extravasation, 213, 214*f*
Extrinsic muscles, of eye, 120, 120*f*, 120*t*
Extrinsic pathway, of blood clotting, 230, 231*f*
Eye(s), 115–124
    accommodation in, 118–119, 119*f*
    examination with ophthalmoscope, 121–122, 121*f*
    extrinsic muscles of, 120, 120*f*, 120*t*
    refraction in, 115–118, 116*f*
    structure of, 115, 116*f*
Eye chart, Snellen, 117–118

F

Facial nerve, 141, 141*f*
Factor VIII, 230
Factor IX, 230
Factor X, 231*f*
Factor XII, 230, 231*f*
Fallopian (uterine) tube, 397, 398*f*
Farsightedness, 117, 117*f*
Fat(s)
    absorption of, 378*f*, 379
    in body composition, 386–387
    definition of, 38
    dietary intake of, 387–388, 389
    digestion of, 378–380, 378*f*, 379*f*, 380*f*
    distribution of, and body shape, 387
    neutral, 38
    as nutrients, 361
    in plasma, 38–39
    saturated, 388, 389
    trans, 388
Fat bodies, in urine, 356*f*
Fatigue, muscle, 185–188
    in frog, 187
    in human, 187–188, 187*f*, 188*f*
Fat-soluble vitamins, 379, 388
Fatty acids, 35, 35*t*
    cis, 388
    essential, 388
    free, 38, 378*f*, 379–380, 379*f*
    monounsaturated, 389
    nonesterified, 38
    omega-3, 388
    polyunsaturated, 389
    saturated, 38, 388
    trans, 388
    unsaturated, 38, 388, 389
Feedback
    negative
        in cardiac (pulse) rate, 27–28, 27*f*
        in fluid/electrolyte balance, 338*f*, 339*f*
        in glucose regulation, 28, 28*f*
        in homeostasis, 25–28, 26*f*
        in hormone regulation, 153
        mechanism of, 26

in respiration, 315
sensitivity of, 26
in water-bath regulation, 25–27
positive, 153
Femoral nerve, test for, 101, 102*f*
Fenestrae, 368
Fermentation, 293
Fertilization, 397
FEV. *See* Forced expiratory volume
FEV$_1$ test, 306–307, 307*f*, 307*t*
FFA. *See* Free fatty acids
Fiber, dietary, 387
Fibril(s), 173
Fibrillation
    atrial, 251–252, 252*f*
    ventricular, 251, 253, 253*f*
Fibrin, 54, 229–230, 231*f*
Fibrinogen, 54, 229
Fibrocartilage, 16, 18*f*
Fibrosis, pulmonary, 305
Filaments
    microfilaments, 7*t*
    myofilaments, 174*f*
    sliding, in muscle contraction, 173, 174*f*
    thick, 173
    thin, 173
Filiform papillae, 140*f*
Filtrate, 335, 338–339
Filtration, by nephrons, 335, 345–347, 346*f*
Fimbriae, of uterine tube, 398*f*
Finger twitches, 187–188, 187*f*, 188*f*
Fitness, physical. *See also* Exercise
    and cardiovascular system, 287–290
Flagella, 7*t*
Flapping tremor, 196
Flexicomp, 101
Flexion, EMG during, 196
Flexors, 99, 100*f*
Fluid balance, renal regulation of, 335, 337–340
Fluid circulation, in plasma, 40, 41*f*
Fluid dispensing, automatic devices for, 35*f*
Flutter, atrial, 251–252, 252*f*
Focal length, 116, 118
Folic acid deficiency, 208
Follicle(s)
    hair, 20
    ovarian, 148–149, 149*f*, 400*f*
    thyroid, 152, 152*f*
Follicle-stimulating hormone (FSH), 152, 159, 397, 399
Follicular phase, of menstrual cycle, 400*f*, 401
Food and Nutrition Board, 389
Foramen ovale, 274
Forced expiratory volume (FEV), 305–307, 306*f*, 307*f*
Fovea centralis, 116*f*, 120, 121, 121*f*, 122, 122*f*
FRC. *See* Functional residual capacity
Free bilirubin, 206, 353
Free fatty acids (FFA), 38, 378*f*, 379–380, 379*f*
Free nerve endings, 106*f*, 106*t*
Frog
    gastrocnemius muscle of
        contraction of, 177–180, 182
        preparation of, 177–179, 179*f*
        stimulation of, 179–180
        twitch, summation, and tetanus in, 187
    heart of
        drug effects on, 237–242
        exposure of, 238, 238*f*
        recording contractions of, 238–240, 238*f*, 239*f*, 240*f*
    muscular contractions in, 177–180
        preparation for, 177–179, 178*f*, 179*f*
        stimulation of, 179–180, 182

I-6

sciatic nerve of, 86–90, 88f, 90f, 179–180, 179f, 182
Fructose, 37
FSH. See Follicle-stimulating hormone
Functional (R) groups, 40, 49, 54, A-3–A-4
Functional residual capacity (FRC), 297, 297f
Fundus, of eye, 121–122, 121f
Fungiform papillae, 140f

G

Gallbladder, 362f, 369f, 378
Gallstones, 379
Galvanometer, 176, 176f
Gametes, 6, 397. See also Ovum; Sperm
Gamma-carboxyglutamate, 230
Gamma globulins, 40, 54
Ganglion
  parasympathetic, 240
  trigeminal, 141f
Ganglion cells, 119, 122
Gap junctions, 248
Gases, partial pressure of, in blood, 313, 314f
Gas exchange, 313, 321
Gastric acid (juice), 373, 375–376, 376f
Gastric glands, 365, 365f, 375–376
Gastric inhibitory peptide (GIP), 362t
Gastric lipase, 378
Gastric pits, 364–365, 365f
Gastric ulcers, 376
Gastrin, 147t, 362t, 365
Gastritis, 377
Gastrocnemius muscle, of frog
  contraction of, 177–180, 182
  preparation of, 177–179, 179f
  stimulation of, 179–180
  twitch, summation, and tetanus in, 187
Gastroesophageal reflux, 376
Gastroesophageal reflux disease (GERD), 377
Gastrointestinal hormones, 361, 362t
Gastrointestinal tract
  absorption in, 35, 361
  enzymes in, 366, 367t, 369–370, 373, 374
  histology of, 363–370
  layers of, 363, 364f
  structure of, 361, 362f
G cells, 365
Gel electrophoresis, SDS-polyacrylamide, 54, 55f
Gelman Sepra Tek electrophoresis system, 56f
General Conference on Weights and Measures, 3
Genes
  dominant, 411
  in enzyme production, 33
  patterns of heredity, 411–414
  recessive, 411
Genetic code, 69–70
Genetics
  of metabolism, 69–71
    disorders in, 50, 70–71
  patterns of, 411–414
Geniculate bodies, lateral, 123
Genotypes, 69, 223–224, 411–412
GERD. See Gastroesophageal reflux disease
Germinal epithelium, 150, 150f
GFR. See Glomerular filtration rate
GH. See Growth hormone
Ghrelin, 365
Gigantism, 153
Gingiva, 374
GIP. See Gastric inhibitory peptide
Glands. See also specific glands
  definition of, 145
  exocrine vs. endocrine, 145

formation of, 13, 145, 146f
  neural activation of, 85
Glandular epithelium, 13, 145, 146f
Glans penis, 398f
Glaucoma, 119
  closed-angle, 119
  open-angle, 119
Globulins
  alpha, 40, 54, 54f
  beta, 40, 54, 54f
  gamma, 40, 54, 54f
Glomerular capsule, 335, 336f
Glomerular filtrate, 335, 338–339
Glomerular filtration rate (GFR), 345–347
Glomerulonephritis, 353
Glomerulus, 335, 336f
Glossopharyngeal nerve, 141, 141f
GLP-1. See Glucagon-like peptide-1
Glucagon, 38, 147t, 150–151, 370
  negative feedback control of, 28, 28f
Glucagon-like peptide-1 (GLP-1), 362t
Glucocorticoids, 147t, 151–152, 157, 159–160
Gluconeogenesis, 345
Glucose
  abnormal levels of, 167–169
  in diabetes mellitus, 28, 28f, 35, 38, 352
  as energy source, 293
  formula for, 37
  insulin regulation of, 167–169, 168f
  negative feedback control of, 28, 28f
  in plasma
    homeostasis of, 33
    measurement of, 34, 37–38
    normal fasting range of, 38
  as reducing sugar, 374
  in urine, 167, 351, 352
Glucose tolerance, impaired, 352
Glucosuria, 167
Glutamate
  in excitotoxicity, 252
  taste of, 139
GLUT4 carrier proteins, 167, 168f
D-glyceraldehyde, A-4
L-glyceraldehyde, A-4
Glycerol, 35, 35t, 38
Glycine, 40
Glycogen, 37–38
  in muscle fatigue, 186–187
Glycogen storage disease, 71
Glycosuria, 352
GnRH. See Gonadotropin-releasing hormone
Goblet cells, 14, 15f, 366f
Golgi complex, 6f, 7t
Gonad(s), 397. See also Ovary; Testis
Gonadotropic hormones, 148, 152–153, 397, 399
Gonadotropin-releasing hormone (GnRH), 153, 154f
Gout, 353
G-proteins, and taste perception, 139–141, 140f
Graafian follicle, 148, 149f
Granular endoplasmic reticulum, 6f, 7t
Granular leukocytes, 213
Granulosa cells, 149, 149f
Granulosa cell tumors, 153
Green cones, 123–124, 123f, 414
Gristle, 16
Ground substance, 16
Growth hormone (GH), 152, 153
Guanine, 69–70
Guanylin, 362t
Gustation, 139–141, 413
Gustatory cell, 139, 140f
Gustducin, 141

Guthrie test, 71
Gynecomastia, 159
Gyrus
  postcentral, 107, 108
  precentral, 107

H

Hair, 21f
Hair bulb, 21f
Hair cells, 129, 131f, 135, 136f
Hair follicles, 20
Hammer (malleus), 129, 130f
Hamstrings, 99
Haploid cell, 10, 150
H bands, 174f
hCG. See Human chorionic gonadotropin
HDL cholesterol, 39, 169, 387, 388
Healthy, definition of, 28
Hearing, 129–132
Hearing aids, 132
Heart, 235, 236f
  abnormal rhythms of, 251–253
  conduction system of, 247–248, 248f
  contraction of, 173, 237–238, 274f
  drug effects on, 237–242
  electrocardiogram of, 247–253
  endocrine function of, 147t
  exercise (physical fitness) and, 259–261, 287–290
  innervation of, 259–260
  mean electrical axis of, 265–267, 266f–268f
  sounds of. See Heart sounds
  structure of, 235, 236f
  valves of, 235, 236f, 273. See also Valves, heart
Heart attack. See Myocardial infarction
Heart disease, ischemic, 110. See also Myocardial ischemia
Heart murmurs, 274
Heart (pulse) rate
  calculation of, 251
  exercise and, 259–260, 287–290
  maximum, 287, 287t
  normal values for, 28
  position changes and, 288–290
  resting
    negative feedback control of, 27–28, 27f
    normal values for, 28
  training, 287–288, 287t
Heart sounds, 273–276
  auscultation with stethoscope, 273–275, 274f
  first, 273
  intraventricular pressure and, 273, 274f
  recording of, 275–276, 275f
  second, 273
  splitting of, 273
Height, and vital capacity, 302–305, 302f, 303t, 304t
Height-to-weight ratio, 386–387
Helicobacter pylori, 377
Helper T cells, 215
Hematocrit
  definition of, 206
  normal values for, 206
  procedure for, 206–207, 207f
Hematuria, 353
Heme groups, 204f, 206, 353
Hemocytometer, 205, 207–208, 215, 215f
Hemoglobin
  absorption spectra for, 207, 322, 322f
  chemical forms of, 207, 321
  concentration of

I-7

Hemoglobin *(continued)*
    mean corpuscular, 207–208
    measurement of, 207
    normal values for, 207
  function of, 205–209, 293–294, 321–322
  oxygen loading onto, 294
  oxygen unloading from, 294
  percent saturation of, 321–322
  in sickle-cell anemia, 411–412, 413f
  structure of, 204f, 206
  in urine, 353
Hemoglobin A, 411–412
Hemoglobin S, 411–412
Hemoglobinuria, 353
Hemolysis, 80, 80f
Hemolytic disease of the newborn, 224
Hemophilia, 230, 231, 413
Hemostasis, 229
Heparin, 206
Hepatic artery, 368–369, 368f
Hepatic duct, 361, 369f
  common, 368
Hepatic plate, 368, 368f
Hepatic portal vein, 368–369, 368f
Hepatic triad, 368f, 369
Hepatic vein, 368
Hepatitis, 54, 64, 370
Hepatocytes, 368
Heredity, patterns of, 411–414
Heterozygous genotype, 223, 411–412
Hexosaminidase A, 71
Hippuric acid crystals, 355f
His, bundle of, 248, 248f, 265, 266
Histamine, 213, 365
Histamine H$_2$-receptor blockers, 377
Homeostasis
  definition of, 1, 25
  in diagnostic procedures, 28
  as dynamic constancy, 25, 26f
  endocrine system in, 85
  in internal chemical environment, 33
  negative feedback and, 25–28, 26f
  nervous system in, 85
  renal function and, 335–359
Home pregnancy tests, 406–407, 407f
Homocystinuria, 50
Homologous chromosomes, 6
Homozygous genotype, 223, 411–412
Homozygous recessive genotype, 223
Hormones, 145. *See also specific hormones*
  functions of, 148
  gastrointestinal, 361, 362t
  major, 147t
  in metabolic regulation, 386, 386f
  secretion of, 145
  target cells of, 145
  target organs of, 147t, 148
  transport of, 235
Human chorionic gonadotropin (hCG), 405–407, 406f
Humoral immunity, 215
Hyaline cartilage, 16, 18f
Hyaline casts, 354f, 356f
Hydrochloric acid (HCl), 329–330, 373, 376, A-3
Hydrocortisone, 38, 152, 158f, 160, 161
Hydrogen, chemistry of, A-1–A-2
Hydrogen ions
  and pH, 327, A-2–A-3
  and taste perception, 139
Hydrolysis reaction, 35
Hydrostatic pressure, of blood, 40, 41f
β-Hydroxybutyric acid, 353
Hydroxyl group, A-3–A-4
Hydroxyl radical, A-2

Hydroxyurea, for sickle-cell anemia, 412
Hyperbilirubinemia, 353
Hyperglycemia, 38, 167, 352
Hyperkalemia, 242, 340
Hypermetropia, 117, 117f
Hyperopia, 117, 117f
Hyperpigmentation, 153
Hyperpnea, 315, 330
Hyperpolarization, 93
Hyperproteinemia, 40
Hypertension, 169, 283, 284t, 340
  ACE inhibitors for, 337
  benign, 283
  blood volume and, 340
  essential, 283
  malignant, 283
  primary, 283
  secondary, 283
Hyperthyroidism, 389
Hypertonic solutions, 79–80, 80f
Hypertrophy, cardiac, 251, 266
Hyperventilation, 187, 329, 329f, 330
Hypocalcemia, 187
Hypodermis, 20, 21f
Hypoglycemia, 38, 167–169
  reactive, 169
Hypophysis. *See* Pituitary gland
Hypoproteinemia, 40
Hypotension, 340
Hypothalamohypophyseal portal system, 152–153, 153f
Hypothalamus, 85
  endocrine function of, 147t, 152–153, 153f
  as "master gland," 153
  osmoreceptors in, 338
Hypothalamus-pituitary-gonad system, 153, 154f
Hypothyroidism, 389
Hypotonic solutions, 79–80
Hypoventilation, 314, 315, 329, 329f, 330

I

I bands, 174f
IBW. *See* Ideal body weight
IC. *See* Inspiratory capacity
ICSH. *See* Interstitial cell-stimulating hormone
Ideal body weight (IBW), 389
Ileum, 362t, 365–368, 367f
Immune responses, 213–215
Immunity
  cell-mediated, 215
  humoral, 215
Immunoassays, 406
Immunological competence, 224
Impaired glucose tolerance, 352
Implantation, 397, 405
Inborn errors of metabolism, 50, 70–71
Incident light, 36
Incomplete tetanus, 187
Incus, 129, 130f
Induced-fit model, of enzyme activity, 61
Infarction, myocardial, 64
Infections, intracranial, 95
Inflammatory bowel disease, 367
Inflammatory response, 213–215, 214f
Infrared wands, 386
Infundibulum, 153f
Inhalation, 295–298
Inhibiting hormones, 147t
Inhibitory postsynaptic potential (IPSP), 93, 94f
Initial segment of axon, 86, 87f
Injury current, 181

Inner ear, 129, 130f, 135–136, 135f
Inspiration, muscles of, 297–298, 298f
Inspiratory capacity (IC), 296
  calculation of, 301
  in spirometry recording, 297f, 300f, 301f
Inspiratory reserve volume (IRV), 296, 297f, 298
Insulin, 150–151, 167
  effects of, 38, 147t, 167
  glucose regulation by, 167–169, 168f
  negative feedback control of, 28, 28f
  secretion of, 370
Insulinomas, 169
Insulin resistance, 169, 352, 387
Insulin shock, 167–169
Integrating center, 25, 26f, 27f
Intelitool physiogrip, 188, 188f
Interarterial septum, 248f
Intercalated discs, 19, 19f
Intercostal muscles
  external, 297–298, 298f
  internal, 298, 298f
Intercostal space, fifth left, 275
Interlobar artery, 336f
Interlobar vein, 336f
Interlobular artery, 336f
Interlobular ducts, 150–151
Interlobular vein, 336f
Intermediates, metabolic, 69
International System of Units (SI), 3–4, 4t
International unit (IU), 62
Interphase, 8f
Interstitial cells, 150, 150f, 158–159
Interstitial cell-stimulating hormone (ICSH), 158
Interstitial tissue, 150
Interventricular septum, 248f
Intestinal crypts, 364f, 366, 366f
Intestinal microbiota, 367
Intestinal stem cells, 366
Intestines. *See* Large intestine; Small intestine
Intracellular fluid, ion concentrations in, 76, 76f
Intracranial infections, 95
Intrafusal fibers, 99–101
Intraocular pressure, 119
Intravenous fluids, 80
Intraventricular pressure, and electrocardiogram, 260, 260f
Intrinsic factor, 208, 365
Intrinsic pathway, of blood clotting, 230, 231, 231f
Inulin, renal clearance of, 346, 347f
Invitrogen electrophoresis system, 55f
Ion(s), A-2
  concentrations in intracellular/extracellular fluid, 76, 76f
Ion channels
  and membrane potentials, 86–88, 88f
  and synaptic transmission, 94f
Ipsilateral muscles, 99
IPSP. *See* Inhibitory postsynaptic potential
Iris, 115, 116f, 120
Iron, in hemoglobin, 205–206, 207
Iron deficiency, 208
IRV. *See* Inspiratory reserve volume
Ischemia
  cerebral, 252
  myocardial, 251, 253, 260, 260f, 288
Ischemic heart disease, 110
Ishihara test, 124, 414
Islets of Langerhans, 38, 150–151, 369f, 370, 370f
  alpha cells of, 150f, 151, 370

I-8

beta cells of, 150f, 151, 153, 167, 370
  endocrine function of, 147t
  histology of, 150-151, 150f
Isomers, optical, A-3-A-4
Isometric contraction, 194, 194f
Isotonic contraction, 194, 194f
Isotonic solutions, 79-80, 80f
Isotopes, A-1
Isovolumetric contraction, 274f
Isovolumetric relaxation, 274f
Itch, sensation of, 105
IU. *See* International Unit
iWorx system
  for recording heart contractions, 239
  for recording muscle contractions, 176-177, 178f

J

Jaundice, 206, 353
  obstructive, 379
  physiological, of newborn, 206
Jejunum, 365-368
Junctional complexes, 14-15, 16f

K

Keratinized layer, 13
Ketoacidosis, 168
Ketone bodies, 168
  renal elimination of, 335
  in urine, 351, 352-353
Ketonemia, 353
Ketones, A-3-A-4
Ketonuria, 352-353
Kidney(s)
  in electrolyte balance, 335, 337-340
  in fluid balance, 335, 337-340
  functions of, 147t, 335
  and homeostasis, 335-359
  hormones in, 147t
  plasma clearance in, 345-348
  reabsorption in, 14, 335, 337-339, 346f
  structure of, 335, 336f
Kidney stones, 71, 346
Killer T cells, 215
Kilocalories, 361, 387
Klinefelter's syndrome, 153
Klüver-Bucy syndrome, 123
Knee-jerk reflex, 99, 100f, 101, 102f
Korotkoff, Nikolai S., 282
Korotkoff, sounds of, 282, 282f
Kupffer cells, 368
Kymograph, 238, 238f, 295, 307

L

Labium major, 397, 398f
Labium minor, 397, 398f
Lactase, 367t
Lactate dehydrogenase (LDH)
  action of, 61-62, 63-64
  in serum
    diagnostic uses of, 64
    measurement of, 63-64
    normal values of, 64
Lactate threshold, 288
Lactation, persistent, 153
Lacteal, 364f, 368, 378f, 379
Lactic acid, in muscle fatigue, 186
Lactic acid fermentation, 293
Lacuna, 18f
Lamellae, 18f
Lamina propria, 363, 364f, 365f, 366f
Laminar flow, 281-282
Langerhans, islets of. *See* Islets of Langerhans

Lansoprazole, 377
Laparoscopic cholecystectomy, 379
Large intestine
  histology of, 366, 367f
  layers of, 363
Laryngeal nerve, 141f
LASIK (laser-assisted in situ keratomileusis), 118
Lateral geniculate bodies, 123
Law of specific nerve energies, 85
L cones, 123-124, 123f
LDH. *See* Lactate dehydrogenase
LDL cholesterol, 39, 388
Lean body mass, 386, 388
Left atrium, 235, 236f
Left ventricle, 235, 236f
Length, metric units of, 3
Lengthening (eccentric) contraction, 194
Lens, 115, 116f
  accommodation of, 118-119, 119f
  focal length of, 116, 118
  strength of, 116
  transparency of, 117
Leptin, 147t
Lethal injections, 242
Leucine crystals, 356f
Leukocytes. *See* White blood cell(s) (WBCs)
Leukocytosis, 216
  basophil, 216
  eosinophil, 216
  lymphocyte, 216
  monocyte, 216
  neutrophil, 216
Leukopenia, 216
Levator palpebrae superioris, 120f
Leydig cells, 150, 150f, 158
LH. *See* Luteinizing hormone
Lidocaine, 89
Lieberkühn, crypt of, 364f, 366, 366f
Light
  refraction, in eye, 115-118, 116f
  wavelengths of, 36
Light absorbance, 35-36, 37f
Limb leads, for ECG/EKG, 248, 249f
Lingual nerve, 141f
Lingual tonsil, 140f
Linoleic acid, 388
α-Linolenic acid, 388
Lipase
  gastric, 378
  pancreatic, 367t, 369, 373, 378-379, 378f, 379f
Lipids. *See also* Fat(s)
  dietary intake of, 387-388, 389
  metabolism of, 33
  in plasma, 38-39
  solubility of, 38
L-isomers, A-3-A-4
Liter, 4
  milliequivalents per, 339
Liver, 362f
  catalase in, 63
  development of, 361
  disorders of, 370
  functions of, 147t, 148, 368-369
  histology of, 368-369, 368f
  hormones in, 147t
  lobules of, 368, 368f
  structure of, 368, 368f
Lobules, liver, 368, 368f
Local anesthetics, 89
Localization, of sound, 132
Lock-and-key model of enzyme activity, 61, 62f
Longitudinal muscle, 364f
Loose connective tissue, 16, 17f

Loudness, 129
Lub sound, 273
Lumen, 13, 14f, 15f
  of seminiferous tubules, 150f
Lumina, 13
Lung(s)
  capacity of
    measurement of, 295-307, 297f
    total, 296, 297f, 305t
  in circulatory system, 236f
  function of, measurement of, 295-307
  obstructive disorders of, 305-306
  restrictive disorders of, 305, 306
  volume of, measurement of, 295-307, 297f
Luteal phase, of menstrual cycle, 400f, 401
Luteinizing hormone (LH), 152, 153, 159, 397, 399
Lymph nodes, 214
Lymphocyte(s), 213-215
Lymphocyte leukocytosis, 216
Lymphokines, 215
Lysosome, 6f, 7t
Lysozyme, 366

M

Macrocytic anemia, 208
Macrophages, tissue, 214
Macula lutea, 121, 121f
Macular degeneration, 122
Malignant hypertension, 283
Malleus, 129, 130f
Maltase, 367t
Maltose, 37, 374-375, 374f
Mammary glands, 145
Mannitol, 80
Manometer, 281
Maple syrup disease, 50
Mass, metric units of, 3
Mass number, A-1
Master glands, 152, 153
Mastication, 361
Matrix, 16
Maximal response, 88f
Maximum cardiac rate, 287, 287t
MCHC. *See* Mean corpuscular hemoglobin concentration
M cones, 123-124, 123f
MCV. *See* Mean corpuscular volume
Mean arterial pressure, 282-284
Mean corpuscular hemoglobin concentration (MCHC)
  calculation of, 207-208
  normal values for, 208
Mean corpuscular volume (MCV)
  calculation of, 207-208
  normal values for, 208
Mean electrical axis, of ventricles
  measurement of, 265-267, 266f-268f
  normal values for, 265
Measurements, 3-5, 33. *See also specific measurements*
Medial popliteal nerve, test for, 101, 102f
Meditation, 95
Medulla oblongata
  chemoreceptors of, 314-315, 316f
  respiratory control center of, 314-315, 316f
Megakaryoctye, 203
Meiosis, 6-10, 7t, 9f, 150, 411
Meissner's corpuscle, 20, 106f, 106t
Meissner's plexus, 363, 364f
Melanin, 70
Melatonin, 147t
Membrane potentials, 86-88
  resting, 86, 242

I-9

Menarche, 402
Mendel, Gregor, 412
Mendelian heredity, simple, 412
Ménière's disease, 136
Menses, 399
Menstrual cycle, 397, 399–402, 400*f*
Menstrual phase, 401
Menstruation, 397, 400*f*
Merkel's discs, 106*f*, 106*t*
Mesenchyme, 16
Mesoderm, 151
Messenger RNA (mRNA), 70
Metabolic pathway, 69
Metabolic rate, 314
    basal, 314, 361, 388–392
    exercise and, 314
Metabolic syndrome, 169, 387
Metabolism
    balance in, regulation of, 385–386, 386*f*
    definition of, 293, 385
    genetic control of, 33, 69–71
    inborn errors of, 50, 70–71
    respiration and, 293–333
Metaphase, 7*t*, 8*f*
Metaphase I, 7*t*, 9*f*
Metaphase II, 7*t*, 9*f*
Meter, 3
Metestrus, 400–401
Methemoglobin, 207, 322
Methylene blue, 5
Methyl group, 40, A-4
Metric system, 3–5, 4*t*
    conversions in, 4, 4*t*
    development of, 3
MI. *See* Myocardial infarction
Micelles, 378*f*, 379
Microbiota, intestinal, 367
Microcytic hypochromic anemia, 208
Microfilaments, 7*t*
Micrometers, 5
Microns, 5
Microscope, 2–3
    care and cleaning of, 2
    components of, 2, 3*f*
    examination of cells, 2–10
    examination of tissues and organs, 13–20
    inverted image of, 2–3
    size of objects under, estimation of, 5
    visual field of, 5
Microtubule, 6*f*, 7*t*
Microvilli, 361, 366
    in taste perception, 139, 140*f*
Middle ear, 129, 130*f*
Middle ear infection, 131, 132
Milliequivalents, 339–340
Milliequivalents per liter, 339
Milliosmolality, 80
Mineral(s), 361, 388
    major, 388
    minor, 388
Mineralocorticoids, 151–152, 157, 159–160
Minute volume
    alveolar, 313
    total, 313–317
        average, 317
        exercise and, 314, 315–316
        measurement of, 316–317
Mitochondrion, 6*f*, 7*t*
Mitosis, 6–10, 7*t*, 8*f*
Mitral regurgitation, 274
Mitral stenosis, 252
Mitral valve, 273
Mitral valve prolapse, 274
Modality, 85

Molality, 79–80
Molarity, 79
Mole, 79
Monoclonal antibodies, 406
Monocyte(s), 213–214
Monocyte leukocytosis, 216
Monoglycerides, 378*f*, 379–380, 379*f*
Monomers, 34–35, 35*t*
Monosaccharides, 35, 35*t*, 37
Monosodium glutamate, 139
Monosynaptic reflexes, 99, 100*f*, 101
Monounsaturated fatty acids, 389
Monovision, 118
Motion sickness, 136
Motor cortex, 107
Motor nerves
    autonomic, 85
    in muscle contraction, 194, 194*f*
    in skin, 21*f*
    somatic, 85, 180–182
    stimulation of, 180–182
    structure of, 87*f*
Motor nuclei, phrenic, 314
Motor (efferent) pathways, 85
Motor stimulation
    in vitro, 185–186
    in vivo, 186
Motor units, 186, 194, 194*f*
    asynchronous activation of, 186
    recruitment of, 195–196
mRNA. *See* Messenger RNA
Mucosa, 363, 364*f*
Mucous neck cells, 365, 365*f*
Mucus
    adherent layer of, 376
    secretion of, 365
"Mulberry cell," 356*f*
Mumps, 153
Murmurs, heart, 274
Muscarinic receptors, 241
Muscle(s), 13, 19. *See also* Cardiac muscle; Skeletal muscle; Smooth muscle
    cells of, 19, 19*f*
    neural activation of, 85
    of respiration, 297–298, 298*f*
    in skin, 20, 21*f*
    striated, 19, 19*f*
Muscle fatigue, 185–188
    in frog, 187
    in human, 187–188, 187*f*, 188*f*
Muscle fibers, 19, 173, 174*f*
Muscle spindles, 99–101
Muscle stretch reflexes, 99, 195
Muscular dystrophy, 187
Muscularis externa, 363, 364*f*, 366*f*
Muscularis mucosa, 363, 364*f*, 366*f*
Musculocutaneous nerve, test for, 101, 102*f*
Myelin, 87*f*
Myelinated axons, 20
Myelin sheath, 20
Myenteric plexus, 363, 364*f*, 366*f*
Myocardial cells, 19, 19*f*, 247–248
Myocardial infarction (MI), 64
Myocardial ischemia, 253, 260, 260*f*, 288
Myocardium, 247–248
Myofibrils, 174*f*, 182*f*
Myofilaments, 174*f*
Myopia, 117–118
Myosin, 173
Myosin cross-bridges, 181, 182*f*

N

Narco Mark III Physiograph, 176*f*

National Conference on Cardiopulmonary Resuscitation and Emergency Cardiac Care, 251
Near point of vision, 118–119
Nearsightedness, 117–118, 117*f*
Necrosis, 251
NEFA. *See* Nonesterified fatty acids
Negative afterimage, 123
Negative feedback
    in cardiac (pulse) rate, 27–28, 27*f*
    in fluid/electrolyte balance, 338*f*, 339*f*
    in glucose regulation, 28, 28*f*
    and homeostasis, 25–28, 26*f*
    in hormone regulation, 153
    mechanism of, 26
    in respiration, 315
    sensitivity of, 26
    in water-bath regulation, 25–27
Negative feedback mechanism, 26
Neovascularization, and macular degeneration, 122
Nephron(s)
    functions of, 335, 346*f*
    secretion by, 346
    structure of, 335, 336*f*
Nephron loop, 336*f*
Nephrosis, 353
Nephrotic syndrome, 54
Nerve(s). *See also specific nerves*
    action potential of. *See* Action potentials
    cranial
        in ear, 130*f*, 135, 135*f*
        in tongue, 141, 141*f*
    motor
        autonomic, 85
        in muscle contraction, 194, 194*f*
        in skin, 21*f*
        somatic, 85, 180–182
        stimulation of, 180–182
        structure of, 87*f*
    sensory
        in skin, 21*f*
        structure of, 87*f*
Nerve cells. *See* Neurons
Nerve endings, free, 106*f*, 106*t*
Nerve energies, law of specific, 85
Nerve fibers. *See* Axon(s)
Nerve gas, 241
Nerve regeneration, 20
Nervous system, 85–144
    auditory pathways in, 129–132
    balance and equilibrium in, 135–136
    cutaneous receptors and referred pain, 105–110
    electroencephalogram in, 93–96
    in homeostasis, 85
    recording action potential in, 86–90
    reflex arc in, 99–102
    vision and visual pathways in, 115–124
Nervous tissue, 13, 19–20, 20*f*
    in skin, 20, 21*f*
NETs (neutrophil extracellular traps), 214
Neural crest ectoderm, 151
Neuroglial cell, 20, 20*f*
Neurohypophysis. *See* Pituitary gland, posterior
Neurons, 19–20, 20*f*
    action potentials of, 86–90
    postsynaptic, 93
    presynaptic, 93
    structure of, 86, 87*f*
Neurotransmitters, 93, 94*f*
Neutral fats, 38. *See also* Triglycerides
Neutral solutions, 327

Neutrons, A-1
Neutrophil(s), 213–214
Neutrophil extracellular traps, 214
Neutrophil leukocytosis, 216
Newborn
   hemolytic disease of the, 224
   physiological jaundice of, 206
Nicotine, cardiac effects of, 240–241
Nicotinic receptors, 180
Night vision, 122
Ninhydrin, 50
Nitrogen
   blood urea, 346
   chemistry of, A-2
Node of Ranvier, 87*f*
Nonesterified fatty acids (NEFA), 38
Nonpolar molecules, 76
Nonpolar solvents, solubility of compounds in, 76
Nonsteroidal anti-inflammatory drugs (NSAIDs), and peptic ulcers, 377
Norepinephrine, 151
Normal range, 28
Normal saline solution, 80
Normal values, 28
Normocytic normochromic anemia, 208
NSAIDs, and peptic ulcers, 377
Nuclear envelope, 6*f*, 7*t*
Nucleated epithelial cells, in vaginal smear, 401*f*
Nucleolus, 6*f*, 7*t*, 8*f*
Nucleotide bases, 69–70
Nucleus (atom), A-1
Nucleus (cell), 1, 6, 6*f*
Nucleus (nervous system)
   paraventricular, 153*f*
   supraoptic, 153*f*
Nutrient(s)
   assessment of, 385–392
   classes of, 361
   dietary intake of, 387–388
Nutrition, 361–396
Nystagmus, 120
   vestibular, 135–136

O

Obesity, 169, 386–387
   central, 169, 387
   WHO definition of, 386
Oblique muscles
   abdominal, 298*f*
   inferior, 120*f*, 120*t*
   superior, 120, 120*f*, 120*t*
O blood type, 225, 225*t*
Obstructive jaundice, 379
Obstructive pulmonary disorders, 305–306
Ocular muscles, 120, 120*f*, 120*t*
Oculomotor nerve, damage to, 120
Oils, 38
Olfaction, taste and, 141
Oligodendrocytes, 20
Omega-3 fatty acids, 388
Omeprazole, 377
Oncotic pressure, 40, 41*f*
Oocyte, secondary, 10, 149, 149*f*
Oogenesis, 10
Open-angle glaucoma, 119
Operant conditioning, 196
Ophthalmoscope, 121–122, 121*f*
Ophthalmoscopy, 121–122
Opsin, 123
Optic chiasma, 153*f*
Optic disc, 121, 121*f*
Optic nerve, 116*f*, 122
Optical isomers, A-3–A-4

Oral cavity, 362*f*
Orbital, electron, A-1
Organ(s)
   definition of, 1
   endocrine functions of, 147*t*, 148
   microscopic examination of, 13–20
   skin as example of, 20
   target, 147*t*, 148
   tissues in, 1
Organ of Corti, 129, 131*f*
Organelles, 6, 6*f*, 7*t*
Organic acid, A-4
Organic chemistry, A-3–A-4
Organic molecules, 34
Organic solvents, 76
Organ systems, 1
Origins, in chromatography, 49, 160
Oscilloscope
   in action potential recordings, 89–90, 90*f*
   in EEG, 93, 95–96
Osmolality, 80
Osmometers, erythrocytes as, 80, 80*f*
Osmoreceptors, 338
Osmosis, 75–78
   definition of, 77
   model of, 78*f*
   across semipermeable membrane, 77–78, 77*f*
Osmotically active solutes, 79
Osmotic pressure, 77–78
   colloid, 40, 41*f*
Ossicles, of middle ear, 129, 130*f*
Osteocytes, 16–17, 18*f*
Otitis media, 131, 132
Otolith organs, 135, 135*f*
Otosclerosis, 131
Outer ear, 129, 130*f*
Ova. *See* Ovum
Oval fat bodies, in urine, 356*f*
Oval window, 129, 130*f*
Ovarian cycle, 397, 399–402, 400*f*
Ovarian follicles, 148–149, 149*f*, 400*f*
   graafian, 148, 149*f*
   primary, 149*f*
   secondary, 149, 149*f*
Ovary, 397, 398*f*
   cell division (meiosis) in, 6–10
   disorders associated with, 153
   endocrine function of, 147*t*, 148–149
   histology of, 148–149, 149*f*
   regulation of, 397
Overweight, definition of, 386
Ovulation, 148, 153, 159
Ovulatory phase, of menstrual cycle, 400*f*, 401
Ovum (ova), 6, 148–149
   as exocrine secretion, 148
   fertilization of, 397
   production of, 6–10, 397
Oxalic acid, as anticoagulant, 231
Oxygen
   chemistry of, A-1–A-2
   consumption of, 316
   maximal uptake of, 288
   partial pressure of, 313, 314*f*
   respiratory exchange of, 293–294, 313–316
   transport in blood, 205–209, 293–294, 321–322
Oxygen-carrying capacity, 205–206
Oxygen debt, 316
Oxyhemoglobin, 207, 321–322
   absorption spectra for, 322, 322*f*
   methods for obtaining, 322
   saturation of, 321–322
Oxytocin, 147*t*, 152

P

Pacemaker, 238, 248, 248*f*, 260, 265
   artificial, 252, 253*f*
   ectopic, 240, 252–253
Pacemaker region, 238
Pacinian corpuscles, 106*f*, 106*t*
PAH. *See* Para-aminohippuric acid
Pain
   referred, 108–110, 109*f*, 110*f*
   sensation of, 105
Palatine tonsil, 140*f*
Palpitations, heart, 252
Pancreas, 362*f*, 369*f*
   alpha cells of, 150*f*, 151, 370
   beta cells of, 150*f*, 151, 153, 167, 370
   development of, 361
   endocrine function of, 147*t*, 148, 150, 369–370, 369*f*, 370*f*
   exocrine function of, 145, 150, 369–370, 369*f*, 370*f*
   histology of, 150–151, 150*f*, 369–370
   structure of, 369–370, 369*f*
Pancreatic amylase, 367*t*, 369
Pancreatic duct, 361, 368, 369–370, 369*f*, 370*f*
Pancreatic islets. *See* Islets of Langerhans
Pancreatic juice, 150, 366, 373
   composition of, 369
   production of, 361
   release of, stimulation of, 376
   secretion of, 369–370, 369*f*
   triglyceride digestion by, 378–380, 378*f*, 379*f*
Pancreatic lipase, 367*t*, 369, 373, 378–379, 378*f*, 379*f*
Pancreatitis, 370
Paneth cells, 366
Papanicolaou, George, 402
Papillae, of tongue, 140*f*
Papilledema, 121
Pap smear, 402
Para-aminohippuric acid (PAH), 346
Paracellular transport, 14–15
Parafollicular cells, 152, 152*f*
Parasympathetic ganglia, 240
Parasympathetic nerves, 27
   in heart, 259–260
Parasympathomimetic drugs, 241
Parathyroid glands, 147*t*
Parathyroid hormone, 147*t*
Paraventricular nucleus, 153*f*
Parent cells, 6–10
Parietal cells, 365, 365*f*, 375, 376*f*
Parotid gland, 362*f*
Paroxysmal atrial tachycardia, 253
Partial pressure, of gases, in blood, 313, 314*f*
Passive transport, 76
Patella, 100*f*
Patellar reflex, 99, 100*f*, 101, 102*f*
Pear-shaped bodies, 387
Pectoralis major muscles, 298, 298*f*
Penicillin, renal clearance of, 346
Penis, 397, 398*f*
Pepsin
   characteristics of, 367*t*
   name of, 62
   pH optimum of, 367*t*, 375
   protein digestion by, 373, 375–378
Pepsinogen, 365
Peptic ulcers, 376, 377
Peptide bonds, 39–40
Percent saturation, 321–322
Percent transmittance, 35–36
Perfusion, 235, 281
Peripheral chemoreceptors, 315, 315*f*, 316*f*

I-11

Peripheral nerve, 86
   dorsal root of, 99
Peripheral nervous system, 86
Peripheral nervous system (PNS), 20
Peripheral resistance, 235, 281
Peristalsis, 361
Peritubular capillaries, 336*f*
Permeability, 76, 76*f*
Pernicious anemia, 208
Peroxisome, 7*t*
Petit mal epileptic seizure, 95
pH, A-2–A-3
   of amino acids, 53–54
   of blood, 294, 327–330, 329*t*
   buffers and, 328–330, A-2–A-3
   definition of, 327
   and enzyme activity, 61–62
   optimum, 62, 62*f*
   scale, 328*t*
   of stomach, 375
   of urine, 340
Phagocytes, 214*f*
Phagocytosis, 213–215
Phantom limb phenomenon, 108
Pharmacological effects, 237
Pharynx, 362*f*
Phenotype, 69, 411
Phenylalanine, 50, 70–71
Phenylketonuria (PKU), 50, 70–71
Phenylthiocarbamide (PTC) taste, inheritance of, 413
Pheochromocytoma, 153
Phipps spirometer, 296
Phonocardiogram, electrocardiogram and, 275–276, 275*f*
Phosphatase
   acid, 63
   action of, 62
   alkaline, 63
      in serum, measurement of, 63
   name of, 62
Phosphate
   in ATP production, 293
   in urine, 356*f*
Phosphate group, A-4
Phosphate radical, A-2
Phosphodiesterase, 240
Phospholipids, 38, 387–388
Photopic vision, 122
Photopsins, 124
Photoreceptors, 85, 122, 122*f*. *See also* Cones; Rods
Phrenic motor nuclei, 314
Physiograph
   in cardiac contractions, 238–239, 239*f*
   components of, 176, 176*f*
   in EEG, 93, 95–96
   in muscle contractions, 175–177, 176*f*
Physiograph Mark III recorder, 176*f*
Physiogrip, 188, 188*f*
Physiological effects, of drugs, 237
Physiological jaundice of newborn, 206
Physiology, definition of, 1
Pigment, in blood, 206, 207
Pilocarpine, cardiac effects of, 241
Pineal gland, 147*t*
Pinna, 129
Pipettors, 35*f*
Pitch, 129
Pithing, 177–179, 178*f*, 179*f*
Pituicytes, 154
Pituitary gland
   anterior, 147*t*, 152–154
      disorders of, 153

      hypothalamic control of, 152–153, 153*f*
      histology of, 152–154, 154*f*
      as "master gland," 152, 153
   posterior, 147*t*, 152–154
      hypothalamic control of, 152, 153*f*
Pituitary stalk, 153*f*
PKU. *See* Phenylketonuria
Placenta, hormone secretion by, 159, 161, 405
Plantar reflex, 101–102, 102*f*
Plasma, 17, 203, 204*f*
   cholesterol in, 33–34, 38–39
   collection of samples, 347–348
   concentration of, renal regulation of, 337–340, 338*f*
   definition of, 54
   fluid circulation in, 40, 41*f*
   function of, 203
   glucose in, 33–34, 37–38
   proteins in, 33, 39–41, 54
      abnormal levels of, 33, 40
      functions of, 40
      measurement of, 34, 39–41
      normal fasting level of, 41
      types of, 40, 54
   regulation of, 33, 34–35
Plasma cells, 215
Plasma clearance, renal, 345–348
   of inulin, 346, 347*f*
   of para-aminohippuric acid, 346
   of penicillin, 346
   of urea, 345–348
Plasma (cell) membrane
   in cellular structure, 1, 6, 6*f*
   function of, 7*t*
   permeability of, 76, 76*f*
   structure of, 7*t*, 33, 38, 76, 77*f*
   transport across, 33, 76, 76*f*, 77
Platelet(s), 17, 18*f*, 203, 204*f*
   aggregation of, 229, 230*f*
Platelet plug, 229, 230*f*
Plicae circulares, 364*f*, 366
PNS. *See* Peripheral nervous system
Polar body, 10
Polar solvents, solubility of compounds in, 76
Polydipsia, 338
Polymers, 34–35, 35*t*
Polymorphonuclear leukocytes, 213
Polysaccharides, 37–38
Polyunsaturated fatty acids, 389
Popliteal nerve, medial, test for, 101, 102*f*
Portal system, hypothalamohypophyseal, 152–153, 153*f*
Portal triad, 368*f*, 369
Portal vein, hepatic, 368–369, 368*f*
Portal venules, 153*f*
Positive afterimage, 123
Positive feedback, 153
Postcentral gyrus, 107, 108
Postsynaptic neurons, 93
Potassium, diffusion of, depolarization and, 87, 88*f*
Potassium balance, renal regulation of, 337–338
Potassium ions
   and aldosterone secretion, 338
   cardiac effects of, 242
Potential difference, 86–90, 93, 173
Precentral gyrus, 107
Prednisone, 160
Pregnancy tests, 405–407, 406*f*, 407*f*
Pregnenolone, 158*f*
Prehormones, 161
Premature ventricular contractions (PVCs), 251–252, 253*f*

Prepuce, 398*f*
Presbyopia, 118
Pressure, sensation of, 105
Presynaptic neurons, 93
Prevacid, 377
Prilosec, 377
Primary amenorrhea, 402
Primary hypertension, 283
Primary ovarian follicles, 149*f*
Primary sex organs, 397
Primary tissues, 13
P-R interval, 248
Procaine, 89
Products, of catalytic process, 61
Proestrus, 400–401
Progesterone, 147*t*, 157, 159
   in menstrual cycle, 399–400, 400*f*
   in pregnancy, 405
   synthesis of, 158*f*
Prolactin, 152, 153
Proliferative phase, in menstrual cycle, 400*f*, 402
Prophase, 7*t*, 8*f*
Prophase I, 7*t*, 9*f*
Prophase II, 7*t*, 9*f*
Prostate cancer, 63
Prostate gland, 150, 397, 398*f*
Protein(s)
   dietary intake of, 388
   digestion of, 373, 375–378, 377*f*
   metabolism of, 33
   as nutrients, 361
   plasma, 39–41, 54
      abnormal levels of, 33, 40
      functions of, 40
      homeostasis of, 33
      measurement of, 34, 39–41
      normal fasting level of, 41
      types of, 40, 54
   receptor, 214
   serum, electrophoresis of, 53–56, 54*f*
   structure of, 39–40
   synthesis of, 70
   in urine, 351, 352
Proteinuria, 352
Prothrombin, 229, 231*f*
Prothrombin time
   normal value for, 232
   test for, 231–232
Proton(s), A-1
Proton pump inhibitors, 377
Proximal convoluted tubule, 336*f*
Pseudostratified ciliated columnar epithelium, 14, 15*f*
PTC (phenylthiocarbamide) taste, inheritance of, 413
Ptyalin. *See* Salivary amylase
Pulmonary artery, 236*f*
Pulmonary disorders, 305–306
   obstructive, 305–306
   restrictive, 298, 305, 306
Pulmonary fibrosis, 305
Pulmonary function, measurements of, 295–307
Pulmonary semilunar valve, 266, 273
Pulmonary vein, 236*f*
Pulse oximeter, 322
Pulse pressure, 282–284
Pulse rate. *See* Cardiac rate
Pulse transducer, 261, 261*f*
Punctate distribution, 105
Pupil, 115, 116*f*
   constriction of, 120–121
   dilation of, 120–121
Pupillary reflex, 120–121

Purkinje fibers, 248, 248f
Pus, 214
PVCs. *See* Premature ventricular contractions
P wave, 248, 249f, 265
Pyramid, of middle ear, 130f
Pyramidal motor tracts, 101

Q

QRS complex, 248, 249f, 265
Quinine sulfate, 141

R

R (functional) groups, A-3–A-4
Rabbit test, 406
Radial muscles, 120
Radial nerve, test for, 101, 102f
Range, 26
Ranvier, node of, 87f
Rapid filling, of ventricles, 274f
Rat, estrous cycle of, 399–402
Rathke's pouch, 152
RBCs. *See* Red blood cell(s)
Reabsorption, renal, 14, 335, 337–339, 346f
Reactants, 61
Reactive hypoglycemia, 169
Receptors
    acetylcholine
        muscarinic, 241
        nicotinic, 180
    chemoreceptors, 314–315, 316f
    cutaneous, 85, 105–110, 106f, 106t
    protein, 214
    sensory, 85
    stretch, 99
Recessive genes, 411
Recessive traits, autosomal, 412
Reciprocal innervation, double, 100f
Rectum, 362f, 398f
Rectus abdominis, 298f
Rectus muscles
    inferior, 120f, 120t
    lateral, 120, 120f, 120t
    medial, 120f, 120t
    superior, 120f, 120t
Red blood cell(s) (RBC), 17, 18f, 203, 204f
    agglutination reaction of, 223, 224f
    crenation of, 80, 80f
    destruction of, 206
    function of, 203, 205
    hemolysis of, 80, 80f
    metabolism of, 206
    as osmometers, 80, 80f
    oxygen transport in, 205–209, 293–294, 321–322
    production of, 206
    in urine, 353, 354f
Red blood cell count, 205–209, 321
    normal values for, 207
    procedure for, 206–209
Red cones, 123–124, 123f, 414
Reducing sugars, 374, 374f
5α-Reductase, 161
Referred pain, 108–110, 109f, 110f
Reflex(es)
    ankle-jerk, 101, 102f
    Babinski's, 101–102, 102f
    biceps-jerk, 101, 102f
    crossed-extensor, 99, 100f
    cutaneous, 101–102, 102f
    definition of, 99
    knee-jerk, 99, 100f, 101, 102f
    monosynaptic, 99, 100f, 101
    muscle stretch, 99, 195
    plantar, 101–102, 102f
    pupillary, 120–121
    spinal nerve stretch, tests for, 99–101
    triceps-jerk, 101, 102f
Reflex arc, 99–102
Refraction, in eye, 115–118, 116f
Refraction errors, 117–118, 117f
Refractory period, 87
Regeneration of axon, 20
Regeneration tube, 20
Regulatory T cells, 215
Releasing hormones, 147t, 152–153
Renal artery, 336f
Renal collecting duct, 336f
Renal cortex, 336f
Renal function, 335–359. *See also* Kidney(s); *specific functions*
Renal medulla, 336f
Renal pelvis, 336f
Renal plasma clearance, 345–348
    of inulin, 346, 347f
    of para-aminohippuric acid, 346
    of penicillin, 346
    of urea, 345–348
Renal plasma threshold, for glucose, 352
Renal reabsorption, 14, 335, 337–339, 346f
Renal stones, 71
Renal tubule, 335, 336f
Renal vein, 336f
Renin, 338, 339f
Repolarization
    in heart, 247–248
    in membrane potentials, 87
    in muscles, 173
Reproductive system, 397–416
    female, 397, 398f
    functions of, 397
    male, 397, 398f
Reserve volume
    expiratory, 296, 297f, 298, 300f, 301–302, 301f
    inspiratory, 296, 297f, 298
Residual capacity, functional, 297, 297f
Residual volume (RV), 297, 297f
    calculation of, 305, 305t
Resistance, peripheral, 235, 281
Respiration, 293–333
    and acid-base balance, 327–330
    aerobic, 293, 313–314, 327
    anaerobic, 293
    and blood pH, 327, 329t
    cellular, 293
    chemoreceptor control of, 314–315, 315f
    exercise and, 313–317
    measurements of pulmonary function, 295–307
    muscles of, 297–298, 298f
    and oxyhemoglobin saturation, 321–322
Respiratory acidosis, 329, 329t, 330
Respiratory alkalosis, 329, 329t, 330
Respiratory control center, 314–315, 316f
Respirometer, Collins, 295–296, 296f, 299–307, 300f, 301f, 317
Resting membrane potential, 86, 242
Resting pulse rate
    negative feedback control of, 27–28, 27f
    normal values for, 28
Restrictive pulmonary disorders, 298, 305, 306
Rete testis, 150f
Reticular connective tissue, 16, 17f
Reticular fibers, 16, 17f
Reticuloendothelial system, 206, 353
Retina, 115, 116f
    examination with ophthalmoscope, 121–122, 121f
    inverted image on, 115–116, 116f
    organization of cells in, 122, 122f
    photoreceptors of, 122
Retinal disparity, 120
Retinene, 123–124
R (functional) groups, 40, 49, 54
Rh factor, 203, 223–225
Rh negative, 223
Rhodopsin, 123
RhoGAM, 224
Rho(D) immune globulin, 224
Rh positive, 223
Rhythmicity, 248
Ribonucleic acid. *See* RNA
Ribonucleotides, 35t
Ribosome, 6f, 7t, 70
Right atrium, 235, 236f
Right ventricle, 235, 236f
Ringer's lactate, 80
Rinne's test, 131, 132f
RNA, 1, 70
Rods, 85, 122, 122f
Root hair plexus, 106f
Rough endoplasmic reticulum, 6f, 7t
Round window, 130f
Ruffini corpuscle, 106f, 106t
Rugae, 364
RV. *See* Residual volume
R wave, 265

S

Saccule, 135, 135f
Saline solution
    normal, 80
    tonicity of, 80
Salivary amylase, 367t
    carbohydrate digestion by, 373–375, 374f, 375f
    characteristics of, 367t
Salivary glands, 145
Salt
    formation of, A-2
    table, as electrolyte, A-2
Salty taste, 139, 140f
SA node. *See* Sinoatrial node
Sarcolemma, 174f, 181f, 182f
Sarcomeres, 173, 174f
Sarcoplasm, 174f
Sarcoplasmic reticulum, 173, 180–182, 181f
Saturated fatty acids/fats, 38, 388, 389
Saturation, percent, 321–322
Scala tympani, 131f
Scala vestibuli, 131f
Scalenus muscle, 298, 298f
Schlemm, canal of, 119
Schwann cells, 20, 87f
Sciatic nerve, of frog, 86–90, 88f, 90f, 179–180, 179f, 182
Sclera, 115, 116f
S cones, 123–124, 123f
Scopolamine, 136
Scotopic vision, 122
Scrotum, 397, 398f
SDS-polyacrylamide gel electrophoresis, 54, 55f
Sebaceous glands, 20, 21f, 145
Secondary amenorrhea, 402
Secondary hypertension, 283
Secondary oocyte, 10, 149, 149f
Secondary ovarian follicles, 149, 149f
Second messengers, in taste perception, 140f, 141
Secretin, 147t, 362t, 376
Secretion, by nephrons, 346
Secretory phase, in menstrual cycle, 400f, 402

I-13

Secretory vesicle, 6f
Sediment, in urine, 353-356, 354f-356f
Seizure, petit mal epileptic, 95
Selectively permeable membranes, 76
Semen, 150, 397
Semicircular canals, 130f, 135-136, 135f, 136f
Semilunar valves, 236f, 260f, 266, 273, 274f, 275f
Seminal fluid, 397
Seminal vesicles, 150, 397, 398f
Seminiferous tubules, 150, 150f
Semipermeable membranes, 76
    osmosis across, 77-78, 77f
Sensation
    cutaneous, 85, 105-110
    gustatory, 139-141
    olfactory, 141
Senses, chemical, 141
Sensitivity, of negative feedback, 26
Sensors, 25, 26f, 27f
Sensory adaptation, 108
Sensory areas, cerebral, 107, 107f
Sensory deafness, 129, 131
Sensory nerves
    in skin, 21f
    structure of, 87f
Sensory (afferent) pathways, 85
Sensory physiology, 85
    cutaneous receptors in, 105-110, 106f, 106t
    modalities of sensation, 105
    referred pain in, 108-110, 109f, 110f
Sensory receptors, 85, 105-110
Septal defects, 266
Septum
    interarterial, 248f
    interventricular, 248f
Serosa, 363, 364f
Serum
    alkaline phosphatase in, 63
    definition of, 54
    lactate dehydrogenase in, 63-64
    proteins, electrophoresis of, 53-56, 54f
Set point, 25-26, 26f
Sex chromosomes, 413-414
Sex-linked traits, 413-414
Sex organs
    accessory, 397
    primary, 397
Sex steroid hormones, 153, 154f, 397. *See also* Steroids
Sexual receptivity, 399-400
Shells, electron, A-1
Shock, insulin, 167-169
Shortening (concentric) contraction, 194
SI. *See* International System of Units
Sickle-cell anemia, 411-412, 413f
Sickle-cell trait, 411, 412
Simple epithelium, 13-14
    ciliated columnar, 14, 15f
    columnar, 13-14, 15f
    cuboidal, 13, 14f
    squamous, 13, 14f
Simple Mendelian heredity, 412
Simple squamous epithelium, 363
Sinoatrial (SA) node, 238, 248, 248f, 259-260, 265
Sinusoids
    of liver, 368, 368f
    of pituitary gland, 154
Sinus rhythm, 252
Skeletal muscle, 19, 173-202
    activation of, 85
    characteristics of, 19
    contraction of
        concentric, 194

        eccentric, 194
    electromyogram recordings of, 193-197
        isometric, 194, 194f
        isotonic, 194, 194f
        mechanism of, 173
        neural control of, 85, 175-182
        physiograph recording of, 175-177, 176f, 177f
        sliding filament model of, 173, 174f
        stimulation of, 179-180
    excitation-contraction coupling in, 173, 181-182, 181f, 182f
    fatigue of, 185-188
    structure of, 19, 19f
    summation of, 185-188
    tetanus of, 185-188
Skin
    endocrine function of, 147t, 148
    as organ, tissues of, 19, 21f
    sensory receptors in, 85, 105-110, 106f, 106t
Skinfold calipers, 386
Sleep, delta waves in, 95
Sliding filament model of muscle contraction, 173, 174f
Small intestine, 362f
    development of, 361
    endocrine function of, 147t, 362t
    histology of, 365-368, 366f, 367f
    layers of, 363
    regions of, 365
    villi and microvilli of, 361, 366, 366f
Smell, taste and, 141
Smooth endoplasmic reticulum, 6f, 7t
Smooth muscle, 19, 19f
SNARE complex, 94f
Snellen eye chart, 117-118
Sodium balance, renal regulation of, 337-339, 339f
Sodium dithionite, 322
Sodium hydrosulfite, 322
Sodium ions
    and muscle contractions, 173
    and taste perception, 139
Sodium-potassium pump
    digitalis and, 242
    and membrane potentials, 86-87
Soft palate, taste buds on, 139
Solute, 75
    osmotically active, 79
Solutions, 75
    acidic, 327
    basic, 327
    concentration and tonicity of, 79-80
    hypertonic, 79-80, 80f
    hypotonic, 79-80, 80f
    isotonic, 79-80, 80f
    neutral, 327
Solvent(s), 75
    organic, 76
    polar *vs.* nonpolar, 76
Solvent front, 50
Somatic motor nerves, 85, 180-182
Somatomedins, 147t
Somatosensory cortex, 107
Somatostatin, 365
Somatotropic (growth) hormone, 152, 153
Sound(s)
    binaural localization of, 132
    conduction of, 129-131
    heart, 273-276
    of Korotkoff, 282, 282f
    loudness of, 129
    pitch of, 129
Sour taste, 139, 140f

Specific gravity, of urine, 340, 340f
Specific nerve energies, law of, 85
Spectronic 20 colorimeter, 36-37
Spectrophotometer (colorimeter), 35-37, 37f
    for hemoglobin measurement, 207, 322
    procedure for standardizing, 36-37
Sperm
    fertilization by, 397
    production of, 6-10, 150, 397
Spermatic cord, 150f
Spermatogenesis, 10, 150
Spermatogenic cells, 150
Spermatozoa, 150, 150f, 397
Sphygmomanometer, 281-284, 282f, 284f
Spinal cord, tests for damage to, 101-102, 102f
Spinal nerve, ventral root of, 99
Spinal nerve stretch reflexes, tests for, 99-101
Spindle fibers, 8f
Spirocomp program, 296, 297f, 299, 306
Spirogram, 297f
Spirometry, 295-307
Spleen, 214
Spotting, 49, 160
Squamous epithelium, 13
    simple, 13, 14f, 363
    stratified, 13, 14f
    in vaginal smear, 400, 401f
Stain
    methylene blue, 5
    Wright's, 217
Standard, 36
Standard curve, 36
Stapedius muscle, 130f
    tendon of, 130f
Stapes, 129, 130f
Starch, digestion of, 373-375, 374f, 375f
Steatorrhea, 379
Stem cells, intestinal, 366
Stenosis, mitral, 252
Sternocleidomastoid muscle, 298, 298f
Steroids, 38-39
    anabolic, 159
    biosynthetic pathways for, 158f
    functional categories of, 157
    sex, 397
    structure of, 39, 157
    thin-layer chromatography of, 157, 160-161
    types of, 157-160
Sterols, 387-388
Stethoscope
    auscultation of heart sounds with, 273-275, 274f
    for blood pressure measurement, 281-284, 282f, 284f
Stimulus, threshold, 179
Stimulus artifact, 87
Stirrup (stapes), 129, 130f
Stomach, 362f
    endocrine function of, 147t, 148, 362t
    histology of, 363-365, 365f
    layers of, 363
    pH of, 375
    ulcers in, 376
Strabismus, 123, 182
Stratified epithelium, 13
    squamous, 13, 14f
Stratum basale, 21f
Stratum corneum, 21f
Stratum granulosum, 21f
Stratum spinosum, 21f
Stretch receptors, 99
Stretch reflexes, spinal nerve, tests for, 99-101
Striated muscle, 19, 19f, 173, 174f. *See also* Cardiac muscle; Skeletal muscle

Stroke, 39, 252
Stroke volume, 287-288
Strong acids and bases, A-3
S-T segment, 260, 260f
Sublingual gland, 362f
Submandibular gland, 362f
Submaximal response, 88f
Submucosa, 363, 364f
Submucosal plexus, 363, 364f
Substrates, 61-62
Sucrase, 367t
Sucrose, 37, 141
Sudden death, cardiac, 253
Sugars, 387
    isomers of, A-3–A-4
    reducing, 374, 374f
    simple (monosaccharides), 35, 35t, 37
Sulfate radical, A-2
Sulfhydryl group, A-4
Summation, 185-188
    in frog gastrocnemius muscle, 187
    in human muscle, 187-188, 187f, 188f
Suppressor T cells, 215
Supraoptic nucleus, 153f
Supraventricular tachycardia, 253
Suspensory ligaments, of eye, 116f, 119f
Sweat glands, 20, 21f, 145
Sweat pores, 21f
Sweet taste, 139, 140f
Sympathetic nerves, 27
    in heart, 259-260
Sympathomimetic drugs, 241
Symphysis pubis, 398f
Synapse, 93
Synaptic cleft, 93
Synaptic integration, 93, 94f
Synaptic transmission, 93, 94f
Synaptic vesicle, 93, 94f
Systole, 247-248, 273
Systolic pressure, 282-284, 282f, 284t

T

T_3. See Triiodothyronine
T_4. See Tetraiodothyronine
Table salt, A-2
Tachycardia
    paroxysmal atrial, 253
    supraventricular, 253
    ventricular, 251-253, 253f
Tagamet, 377
Target cells, 145
Target organs, 147t, 148
Taste buds, 139-141, 140f
Taste cells, 139, 140f
Taste perception, 139-141
    modalities of, 139, 140f
    olfaction and, 141
    of phenylthiocarbamide, inheritance of, 413
Taste pore, 139, 140f
Tay-Sachs disease, 71
T cells, 214-215
Tectorial membrane, 129, 131f
Teeth, 362f
Telophase, 7t, 8f
Telophase I, 7t, 9f
Telophase II, 7t, 9f
Temperature
    and enzyme activity, 61-62
    metric units of, 3-4
Temperature homeostasis, 25-27
Temperature optimum, 62, 62f
Temperature receptors
    adaptation of, 108
    distribution of, 105

    mapping of, 105-107
Temporal bone, 130f
Temporal lobe, inferior, 123
Tensor tympani muscle, 130f
    tendon of, 130f
Testis (testes), 397, 398f
    atrophy of, exogenous androgens and, 159
    cell division (meiosis) in, 6-10, 150
    disorders associated with, 153
    endocrine function of, 147t, 150, 158
    histology of, 150, 150f
    regulation of, 397
    structure of, 150, 150f
Testosterone, 147t, 150, 158f, 161
Tetanus, 185-188
    complete, 187
    in frog gastrocnemius muscle, 187
    in human muscle, 187-188, 187f, 188f
    incomplete, 187
Tetany, 186, 187
Tetracaine, 89
Tetraiodothyronine (T_4), 147t, 152, 389
Tetralogy of Fallot, 266
Theca interna, 149f
Theta waves, 95-96, 95f
Thick filaments, 173
Thin filaments, 173
Thin-layer chromatography
    of amino acids, 49-50
    of steroids, 157, 160-161
Thin-layer chromatography plate, 160
Thin-layer plate, 49-50
Thistle tube setup, 78, 78f, 79f
Thoracic volume, 297-298
Threshold potential, 87-88
Threshold stimulus, 179
Thrombin, 229, 231f
Thrombocytes. See Platelet(s)
Thromboplastin, tissue, 230, 231f
Thromboplastin time, activated partial, 231, 232
Thrombosis, coronary, 260
Thromboxane A_2, 230f
Thrombus, 252
Thymine, 69-70
Thymopoietin, 147t
Thymus
    endocrine function of, 147t
    in lymphocyte production, 214
Thyrocalcitonin (calcitonin), 147t, 152
Thyroid follicles, 152, 152f
Thyroid gland
    dysfunction of, 389
    endocrine function of, 147t
    histology of, 152, 152f
Thyroid-stimulating hormone (TSH), 152
Thyroxine. See Tetraiodothyronine
Tickle, sensation of, 105
Tidal volume (TV), 296
    calculation of, 300-301
    in spirometry recording, 297f, 300f, 301f, 315f
Tight junctions, 14, 16f, 376
Tinnitus, 136
Tissue(s), 13-20. See also specific types
    definition of, 1, 13
    microscopic examination of, 13-20
    primary, 13
    in skin, 20, 21f
    types of, 13
Tissue factor, 230, 231f
Tissue macrophages, 214
Tissue thromboplastin, 230, 231f
TLC. See Total lung capacity
T lymphocytes. See T cells
Tongue, 140f, 362f

    innervation of, 141, 141f
    taste buds on, 139, 140f
Tonicity, 79-80
Tonsils, 140f
Total lung capacity (TLC), 296, 297f, 305t
Total minute volume, 313-317
    average, 317
    exercise and, 314, 315-316
    measurement of, 316-317
Touch sensation
    receptors for
        distribution of, 105
        mapping of, 105-107
    two-point threshold in, 107-108
Trace elements, 388
Training cardiac rate, 287-288, 287t
Transcellular transport, 15
Transcription, 70
Transducer, in physiograph, 176, 176f
Transducin, 141
Trans fats, 388
Trans fatty acids, 388
Transfer RNA (tRNA), 70
Transitional epithelium, 14, 15f
Translation, 70
Transmittance, percent, 35-36
Transport
    active, 33, 76, 76f, 86
    of oxygen, 205-206, 293-294, 321-322
    paracellular, 14-15
    passive, 76
    transcellular, 15
Transport maximum, for glucose, 352
Transversus abdominis, 298f
Tremor, flapping, 196
Triceps brachii muscle, 195-196
Triceps-jerk reflex, 101, 102f
Trichromatic color vision, 123
Tricuspid valve, 236f, 273
Trigeminal ganglion, 141f
Trigeminal nerve, 141f
Triglycerides, 38
    dietary intake of, 387-388, 389
    digestion of, 378-380, 378f, 379f, 380f
Triiodothyronine (T_3), 147t, 152
Trimers, 35
Triple phosphate crystals, 356f
tRNA. See Transfer RNA
Trochlea, 120f
Trochlear nerve, damage to, 120
Trophic hormones, 147t
Tropomyosin, 173, 180-181, 182f
Troponin, 173, 180-181, 182f
Trypsin, 62, 367t, 369, 375-376
TSH. See Thyroid-stimulating hormone
Tums, 377
Tunics
    of eye, 115, 116f
    of gastrointestinal system, 363, 364f
Tuning fork, 131, 132, 132f
Turbulent flow, 282
TV. See Tidal volume
T wave, 248, 249f
Twitch, 185-188
    in frog gastrocnemius muscle, 187
    in human muscle, 187-188, 187f, 188f
Two-point threshold test, 107-108
Tympanic cavity, 130f
Tympanic membrane, 129, 130f
Tympani nerve, 141f
Tympanitis, 132
Tyrosine, 50, 70
    in urine, 355f, 356f

I-15

U

Ulcer(s)
  duodenal, 376
  gastric, 376
  peptic, 376, 377
Ulcerative colitis, 367
Ulnar nerve, referred pain from, 110, 110f
Ultrafiltrate, 335
Umami, 139, 140f
Unipolar leads, for ECG/EKG, 248, 249f
Units of activity, 62
Unmyelinated axons, 20
  conduction of action potentials in, 89f
Unopette system, 205, 207
Unsaturated fatty acids, 38, 388, 389
Unsaturation, A-3
Uracil, 70
Urea, 335
  ammonia conversion to, 370
  measurement of, 348
  normal range of, 347
  renal plasma clearance of, 345–348
Urea nitrogen, blood, 346
Ureter, 335, 336f, 398f
Urethra, 398f
Uric acid crystals, 353, 355f, 356f
Urinary bladder, 398f
Urinary tract infection (UTI), 351
Urine
  abnormal appearance of, 351t
  casts in, 351, 352, 354f–356f
  chloride concentration of, 339–340
  clinical examination of, 351–356
  collection of samples, 347–348
  formation of, 335
  glucose in, 167, 351, 352
  hemoglobin in, 353
  ketone bodies in, 352–353
  pH of, 340
  pregnancy testing for hCG in, 405–407, 406f
  protein in, 351, 352
  sediment in, 353–356, 354f–356f
  specific gravity of, 340, 340f
  volume of, 340
Urinometer float, 340, 340f
Uterine tube, 397, 398f
Uterus, 397, 398f
UTI. See Urinary tract infection
Utricle, 135
Uvula, 141f
U wave, 248

V

Vacuole, 7t
Vagina, 397, 398f
Vaginal fornix, 398f
Vaginal orifice, 398f
Vaginal smear
  in human, 401–402
  in rat, 399–402, 401f
Vagus nerve, 141f
Vallate papillae, 140f
Valves, heart, 235, 236f
  abnormalities of, 266, 274
  in heart sounds, 273, 274f, 275f
  intraventricular pressure and, 260f
Vasa recta, 336f
Vascular tree, 235
Vas (ductus) deferens, 150, 397, 398f
Vasoconstriction, 229
Vasopressin. See Antidiuretic hormone
VC. See Vital capacity

Veins, 235
  arcuate, 336f
  central, 368–369, 368f
  hepatic, 368
  hepatic portal, 368–369, 368f
  interlobar, 336f
  interlobular, 336f
  pulmonary, 236f
  renal, 336f
Vena cava
  inferior, 236f
  superior, 236f
Ventilation, 293–294
  average frequency of, 317
  carbon dioxide elimination in, 293–294, 327
  carbon dioxide in regulation of, 330
  functions of, 327
  muscles of, 297–298, 298f
  normal (unforced), 297–298
  oxygen delivery in, 293–294, 321–322, 327
Ventilation/perfusion ratio, 314
Ventilometer pen, 296f, 300
Ventral root, 99
Ventricles, heart, 235
  contractions of, 238
  hypertrophy of, 266
  left, 235
  mean electrical axis of, 265–267, 266f–268f
  pressure in
    and electrocardiogram, 260, 260f
    and heart sounds, 273, 274f
  right, 235, 236f
Ventricular contractions, premature, 251–252, 253f
Ventricular fibrillation, 251, 253, 253f
Ventricular systole, 248
Ventricular tachycardia, 251–253, 253f
Venules, 235
  portal, 153f
  in skin, 21f
Vertigo, 136
Vesicles
  secretory, 6f
  seminal, 150, 397, 398f
  synaptic, 93, 94f
Vestibular apparatus, 135–136, 135f, 136f
Vestibular membrane, 131f
Vestibular nerve, 130f, 135f
Vestibular nystagmus, 135–136
Vestibulocochlear nerve, 129, 130f, 131f, 135
Villi, 361, 366, 366f
Vision, 115–124
  color, 122, 123–124
  near point of, 118–119
  photopic, 122
  scotopic, 122
Visual acuity, 115–118
Visual cortex, 123
Visual field, of microscope, 5
Vital capacity (VC), 296
  measurement of, 297f, 300f, 302–305
  predicted exhaled in first second, 307t
  predicted values for
    for females, 303t
    height and, 302–305, 302f, 303t, 304t
    for males, 304t
    measurements below 80 percent of, 298
  timed, 305–307, 306f, 307f, 307t
Vitamin(s), 361
  fat-soluble, 379, 388
  water-soluble, 388
Vitamin A$_1$, 123
Vitamin B$_{12}$ deficiency, 208

Vitamin K deficiency, 230
Vitreous humor, 115, 116f
Voltage. See Potential difference
Volume, metric units of, 3–4
Von Willebrand's factor, 230f

W

Waist-to-hip ratio, 387
Warfarin, 230
Warmth, 105–107
Water
  as essential nutrient, 361, 388
  molecule, formation of, A-1
  pH of, A-2
Water-bath temperature, negative feedback control of, 25–27
Water-soluble vitamins, 388
Wavelengths, of light, 36
Waxy cast, 354f
WBCs. See White blood cell(s)
Weak acids and bases, A-3
Weber's test, 131, 132f
Weight
  equivalent, 339
  ideal body, 389
Wenckebach's phenomenon, 252
White blood cell(s) (WBCs), 17, 18f, 213–217
  abnormal levels of, 216
  agranular, 213
  functions of, 213–215
  granular, 213
  types of, 213–215
  in urine, 351, 353, 354f
  in vaginal smear, 401f
White blood cell count, 213–217
  clinical applications of, 216
  differential, 216–217, 216f
  total, 215
White light, 36
Wright's stain, 217

X

X chromosome, 413
  and color blindness, 124
Xenobiotics, 237

Y

Y chromosome, 413
Young-Helmholtz theory of color vision, 123, 414

Z

Zantac, 377
Z lines, 174f
Zona fasciculata, 151f, 152, 160
Zona glomerulosa, 151, 151f, 159
Zona pellucida, 149, 149f
Zona reticularis, 151f, 152, 160
Zonular fibers, 116f
Zwitterion formula, 53
Zygote, 10
Zymogen granules, 370f

I-16